XXIII Fluid Mechanics Conference (KKMP 2018)

Journal of Physics: Conference Series
Volume 1101

Zawieracie, Poland
9 – 12 September 2018

ISBN: 978-1-5108-7373-5
ISSN: 1742-6588

Printed from e-media with permission by:

Curran Associates, Inc.
57 Morehouse Lane
Red Hook, NY 12571

Some format issues inherent in the e-media version may also appear in this print version.

Printed by Curran Associates, Inc. (2018)

For permission requests, please contact the Institute of Physics
at the address below.

Institute of Physics
Dirac House, Temple Back
Bristol BS1 6BE UK

Phone: 44 1 17 929 7481
Fax: 44 1 17 920 0979

techtracking@iop.org

Additional copies of this publication are available from:

Curran Associates, Inc.
57 Morehouse Lane
Red Hook, NY 12571 USA
Phone: 845-758-0400
Fax: 845-758-2633
Email: curran@proceedings.com
Web: www.proceedings.com

Table of contents

Volume 1101

XXIII Fluid Mechanics Conference (KKMP 2018)

9–12 September 2018, Zawiercie, Poland

Accepted papers received: 7 September 2018
Published online: 23 October 2018

Preface

XXIII Fluid Mechanics Conference (KKMP 2018)

Peer review statement

Papers

Fractal reconstruction of sub-grid scales for large eddy simulation of atmospheric turbulence

Emmanuel O. Akinlabi, Marta Wacławczyk and Szymon P. Malinowski.....1

Experimental and comparative study on the two-phase pressure drop of air-water mixture in U-bend and straight pipe annuli

R Andrzejczyk and T Muszynski.....9

Numerical investigation of the gyroplane longitudinal static stability for the selected stabilizer angles

Z Czyż, T Łusiak, P Karpiński and J Czarnigowski.....17

Passive Skin Friction Control Near Turbulent Separation – Preliminary Results

A Dróżdż, W Elsner and D Sikorski.....25

Influence of wind on the movement of the load

Dawid Cekus, Renata Gnatowska and Paweł Kwiatoń.....32

Aerodynamic assessment of the cycloidal rotor fan

S Dykas, M Majkut, K Smołka, M Strozik and T Staśko.....40

Thermal dilatation of solid grains: the LBM approach

A. Grucelski.....48

Experimental and numerical investigation of cavitation on Clark Y foil

D Homa, W Wróblewski, M Majkut and M Strozik.....54

Fluid flow and heat transfer analysis of a photovoltaic module under varying environmental conditions

Marek Jaszczur, Qusay Hassan, Mateusz Szubel and Ewelina Majewska.....62

Influence of deflocculants on shear stress in hydromixture flow

Beata Jaworska and Artur Bartosik.....70

An investigation of secondary flow features in a low pressure turbine

P. Jonak, T. Borzecki and S. Kubacki.....78

Leakage flow reduction in different configuration of labyrinth seal on a turbine blade tip

Piotr Kaczyński, Ryszard Szwaba, Filip Wasilczuk, Paweł Flaszyński and Piotr Doerffer.....86

Diffuse interface models for two-phase flows in artificial compressibility approach

Adam Kajzer and Jacek Pozorski.....94

Validation of numerical models for flow simulation and wave propagation along human aorta

N Kizilova and J Mizerski.....100

Pre-swirl energy saving device in marine application

P Król and K Tesch.....109

Prediction of endwall losses in a low pressure turbine cascade with an algebraic intermittency model

S Kubacki, P Jonak and E Dick.....119

Numerical analysis of centrifugal compressor operating in near-surge conditions

M Kulak, F Grapow and G Liśkiewicz.....127

Local Nusselt number evaluation in the case of jet impingement

Tomasz Kura, Elżbieta Fornalik-Wajs, Jan Wajs and Sasa Kenjeres.....135

A system for measuring environmental data in full-scale

Tomasz Lipecki, Paulina Jamińska-Gadomska, Andrzej Sumorek, Ewa Błazik-Borowa and Jarosław Bęc.....143

Analysis of high-speed photography of cavitating flow in convergent-divergent nozzle

Seweryn Lipiński and Agnieszka Niedźwiedzka.....151

Comparison of direct flow simulation in a random packed bed of Raschig rings with a macroscale momentum-sink approach

Maciej Marek.....157

Experimental Investigations On The Momentum Pressure Drop During Flow Boiling Of R134a

T Muszynski, R Andrzejczyk and C A Dorao.....164

Application of acoustic oscillations in flame extinction in a presence of obstacle

Paweł Niegodajew, Konrad Gruszka, Renata Gnatowska and Michal Šofer.....172

Wetting phenomena and droplet-wall interaction modelling with Smoothed Particle Hydrodynamics approach

M Olejnik and J Pozorski.....180

Effect of Hot Spot Location on Flow Structure in Nozzle Guide Vane

Michal Piotrowicz, Pawel Flaszynski and Piotr Doerffer.....186

On the 3D structure of the flow-field in the vicinity of inclined plate

P Prochazka, V Uruba and V Skala.....194

Developing micro-scale heterogeneous numerical simulation of a solid oxide fuel cell anode

Tomasz A Prokop, Janusz S Szmyd and Grzegorz Brus.....202

The actuator cell model for the Darrieus wind turbine

K Rogowski.....209

Experimental and numerical investigations of Tesla turbine

K Rusin, W Wróblewski and M Strozik.....217

Application of Background Oriented Schlieren for quantitative measurement of transonic flows

L Santos and W Strycznewicz.....225

Hybrid MPI/Open-MP acceleration approach for high-order schemes for CFD

Michal Saczek, Karol Wawrzak, Artur Tyliszczak and Andrzej Boguslawski.....233

A Computational Study of the Fluid Particles Distribution in an Helicopter Wake

T P Seredyn.....241

Aeroelastic analysis of a helicopter rotor in hover

A Sieradzki.....249

Numerical investigations of an influence of the aspect ratio on the Savonius rotor performance

Krzysztof Sobczak.....256

Experimental study of the aerodynamic perturbations induced by the single normal hot-wire probe positioned normally in the vicinity of a wall

Jacek Sobczyk.....262

Numerical modelling of flow field within a packed bed of granular material

Marcin Sosnowski, Renata Gnatowska, Jacek Sobczyk and Waldemar Wodziak.....268

The effect of heat exchanger geometry on adsorption chiller performance

Marcin Sosnowski, Karolina Grabowska, Jarosław Krzywański, Wojciech Nowak, Karol Sztekler and Wojciech Kalawa.....277

Catalytic utilization of unconventional fuels in a gas turbine

M Stajnke and J Badur.....283

Numerical analysis of an impact of spray characteristics and co-flow temperature on a flame lift-off height

Jakub Stempka and Artur Tyliszczak.....291

Cascaded Lattice Boltzmann Method application in forced and natural convection

R Straka, K V Sharma, D Svyetlichnyy and F W Tavares.....300

Prediction of helicopter rotor noise in hover using FW-H analogy

T Suresh, O Szulc, P Flaszynski and P Doerffer.....309

Numerical study of potential application of active suction and blowing through blade tip perforation to reduction of helicopter rotor thickness noise

O Szulc and P Doerffer.....317

Experimental and numerical research on heat and air flow through a granular material

E Szymanek and A Tyliszczak.....323

Discrete-continuous optimisation of an axial flow blood pump

K Tesch and K Kaczorowska-Ditrich.....331

Analysis of the fuel spray diversity in the opposed-piston engine

T Tulwin and P Karpiński.....339

Extended Bubble Cavitation Model to predict water hammer in viscoelastic pipelines

Kamil Urbanowicz and Mateusz Firkowski.....348

On the structure of the boundary layer under adverse pressure gradient on an inclined plate

V Uruba, P Procházka and V Skála.....357

Turbulence Behind 3D Multi-Scale Sparse Grids

Syed M. Usama, Jackson Tellez-Alvarez, Jacek Kopec, Kamil Kwiatkowski, Jose-Manuel Redondo and Nadeem A. Malik.....365

Analysis of species diffusion and methanol decomposition source in thermocatalytic reactor based on the intermetallic phase of Ni_3Al for low Reynolds numbers

P Ziółkowski, M Stajnke, P Jóźwik, Z Bojar, PJ Ziółkowski and J Badur.....372

Applications of wave equations with logarithmic nonlinearity in fluid mechanics

Konstantin G. Zloshchastiev.....380

IOP Conf. Series: Journal of Physics: Conf. Series **1101** (2018) 011001 doi:10.1088/1742-6596/1101/1/011001

XXIII Fluid Mechanics Conference (KKMP2018)

9-12 September 2018, Zawiercie, Poland

Preface

This Journal of Physics: Conference Series contains the papers presented at XXIII Fluid Mechanics Conference (XXII FMC) held in Zawiercie in Poland during 09th-12th September of 2018. The Conference is organised by Czestochowa University of Technology (CzUT) under the auspices of the Ministry of Science and Higher Education and the Polish Academy of Sciences, Committee of Mechanics.

The Fluid Mechanics Conferences have been taking place every two years since 1974. The first Conference was inaugurated by Prof. Włodzimierz Prosnak from Warsaw University of Technology and since then the scientific centers from Lodz University of Technology, Poznan University of Technology, Silesian University of Technology, Bialystok University of Technology, AGH University of Science and Technology, Gdansk University of Technology, Military Technical Academy, Rzeszow University of Technology, Wroclaw University of Technology, Poznan University of Technology, Institute of Fluid-Flow Machinery Polish Academy of Sciences and Institute of Aviation took responsibility for this scientific event. Częstochowa Universiy of Technology is proud to be the organiser of the conference for the third time. Previous editions of Fluid Mechanics Conference organized by CzUT were held in 1978 and 1998.

The aim of the Conference is to bring together academics, scientists and experts in fluid mechanics and to provide a forum for the exposure and exchange of ideas, methods and most recent results in the field of new experimental techniques and computational methods associated with fluid mechanics and related disciplines.

Conference topics include, but are not limited to, Experimental Fluid Mechanics, Aerodynamics, General Fluid Dynamics, Flow Machinery, Hydromechanics, Atmospheric Science, Interdisciplinary Areas in Heat and Fluid Flow, Computational Fluid Dynamics, Measurement Techniques, Micro-, Nano- and Bio-flows, Multi-phase Flows, Combustion and Turbulence.

Seven keynote lectures were delivered during the conference by world leading experts. The conference attracts a lot of attention, as evidenced by the large number of submitted papers. All contributed papers were peer reviewed by the Review Committee. Accordingly, 110 papers were selected for presentation at the XXIII Fluid Mechanics Conference and the best of them are included in this edition of Journal of Physics: Conference Series. We hope that this publication will be used not only as a document of this event but also to demonstrate the new achievements in fluid mechanic research.

During the conference, Prof. Janusz W. Elsner Competition is organised for the best scientific paper in fluid mechanics addressed to young researchers. The competitors, either MSc or PhD students up to 35 years of age, should present individual or joint papers with supervisors, containing original experimental, numerical or analytical results. Competition is organised under the patronage of Polish Society for Theoretical and Applied Mechanics and European Research Community on Flow, Turbulence and Combustion. The report on this event and list of laureates are published in PTMTS and ERCOFTAC bulletins. It should be noticed that during the present edition, 8 contributions qualified by the Competition Jury for the final presentation at the Conference had to be selected out of 35 proposals submitted and since the prevailing part of them presented a high calibre research the task of the Jury was extremely difficult. The scientific level of qualified papers is as high as in previous editions and it is believed that brilliant scientific careers of Competition laureates will confirm the need for this event and the right verdict of Competition Jury.

Conference Organizers would like to express grateful appreciation to the Polish Ministry of Science and Higher Education, Committee on Mechanics of Polish Academy of Science, European Research Community on Flow, Turbulence and Combustion, Polish Society for Theoretical and Applied Mechanics for their support and patronage. Special thanks goes to the reviewers for their work in encouraging to improve and correct the submitted papers. Their contribution cannot be overestimated.

Witold Elsner

Andrzej Bogusławski

Stanisław Drobniak

Artur Tyliszczak

IOP Conf. Series: Journal of Physics: Conf. Series **1101** (2018) 011002 doi:10.1088/1742-6596/1101/1/011002

Peer review statement

All papers published in this volume of *Journal of Physics: Conference Series* have been peer reviewed through processes administered by the proceedings Editors. Reviews were conducted by expert referees to the professional and scientific standards expected of a proceedings journal published by IOP Publishing.

IOP Conf. Series: Journal of Physics: Conf. Series **1101** (2018) 012001 doi:10.1088/1742-6596/1101/1/012001

Fractal reconstruction of sub-grid scales for large eddy simulation of atmospheric turbulence

Emmanuel O. Akinlabi, Marta Wacławczyk, Szymon P. Malinowski
Institute of Geophysics, Faculty of Physics, University of Warsaw, Warsaw, Poland
E-mail: emmanuel.akinlabi@fuw.edu.pl

Abstract. We present a fractal sub-grid scale model for large eddy simulation (LES) of atmospheric flows. The fractal model is based on the fractality assumption of turbulent velocity field with a dynamical hypothesis based on energy dissipation. The fractal model reconstruct sub-grid velocity field from the knowledge of its filtered values on LES grid, by means of fractal interpolation, proposed by Scotti and Meneveau (1999). The characteristics of the reconstructed signal depends on the (free) stretching parameters, which is related to the fractal dimension of the signal. In previous studies, the stretching parameters was assumed to be constant in space and time and are obtained from experimental velocity signals of homogeneous and isotropic turbulence. To improve this method and account for the stretching parameter variability, we calculate the probability distribution function of the stretching parameter from direct numerical simulation (DNS) data of stratocumulus-top boundary layer (STBL) (courtesy of Prof. J.-P. Mellado from the Max Planck Institute of Meteorology) using the geometric method proposed by Mazel and Hayes. We perform 1D *a priori* test and compare statistics of the constructed velocity increment with DNS velocity increments.

1. Introduction

Atmospheric flows in nature display complex spatial and temporal structures over a wide range of scales from the large synoptic scales $\mathcal{O}(1000km)$ to the smallest dissipative scales $\mathcal{O}(1cm - 1mm)$ with Reynold number $L/\eta \approx \mathcal{O}(10^9)$. All scales play important role in weather prediction including the small scales, which influence droplet collision rate in clouds and affect average settling velocity of droplets with Stokes number much smaller than 1. Direct numerical simulation is an ideal approach to resolve all these scales but it imposes an unrealistic computational cost. Alternatively, large-eddy simulation (LES) allows for significantly improved accuracy in simulating atmospheric (turbulent) flows, by calculating the large scale features of the flow while the interactions between large (resolved) scales and small (unresolved) scales are accounted for by a subgrid-scale model. This type of subgrid models are called functional models. Several functional model has been proposed since the development of the first by Smagorinsky [7] but none has achieve the necessary accuracy for modeling turbulence.

However, structural sub-grid models, an alternative to the functional models, aim at mimicking (some of) the subgrid scales. These models allows for the approximate reconstruction of two-point particle statistics at the subgrid scale, such as relative small scale velocity and particle segregation patterns [12]. Example of structural models are the fractal interpolation [1] and the kinematic simulations based on Fourier modes [8, 11]. Pozorski and Rosa (2018) [11] studied the effect of filtering of the carrier fluid velocity on particle dynamics in forced isotropic turbulence. They also use kinematic simulations (KS) based on Fourier modes, as a structural subgrid model, to construct the (unresolved) small scales for particle dispersion. They concluded that the kinematic simulation could not mimic the sweeping effect of small scales by the resolved eddies.

Fractal interpolation technique (FIT) was introduced to construct synthetic, fractal subgrid-scale fields applied to large eddy simulation of both steady and freely isotropic decaying turbulence [4]. A (free) parameter used to determine the characteristics of the reconstructed (FIT) signal is called the stretching parameter. In [4], it was assumed that the stretching parameters are constant in space and time in homogeneous and isotropic turbulence. These parameters were set to $d = \pm 2^{-1/3}$. Salvetti et al (2006) [6] use the fractal interpolation technique to reconstruct the velocity field from the filtered values on a coarse (resolved) grid to track Lagrangian particles in turbulent channel flows. They concluded that the constant values of the stretching parameters are not well suited for Lagrangian tracking of particles in near-wall turbulence. Basu et al. (2004) [2] proposed an extension of this work by developing a multiaffine fractal interpolation scheme with stretching parameters $d = -0.887$ and $d = -0.676$ and showed that it preserves the higher-order structure functions and the non-Gaussian probability density function of the velocity increments. They performed an extensive *a priori* analyses of atmospheric boundary layer measurements and argued that the multiaffine closure model should give satisfactory performance in large eddy simulations.

Atmospheric turbulence is known to be inhomogeneous and possesses both internal and external intermittency. Internal intermittency means that large velocity gradients are present at small scales and the probability density function (pdf) of velocity differences at small scales are highly non-gaussian. External intermittency refers to the co-existence of both laminar and turbulent regions in the flow. These attributes make it difficult to synthesize turbulent velocity field using fractal interpolation. The approach of assuming a constant stretching parameter to construct subgrid velocity signals are unrealistic (as reported by [6]) since the local stretching parameters are likely to change randomly in space and time.

The aim of this work is to develop an improvement to the fractal interpolation technique (FIT), which can be used as a closure model for Lagrangian tracking of particles in atmospheric turbulence. To account for the variability of the stretching parameter, we compute the local stretching parameter with a method proposed by Mazel and Hayes (1992) [5] and use its pdf for the fractal interpolation. Also, we compare the statistics of the velocity increments using both the previous version of FIT, the new one and multiaffine fractal interpolation scheme with the statistics of the DNS velocity increments.

2. Fractal interpolation techniques

2.1. Basics

The fractal interpolation technique is an iterative affine mapping procedure to construct the synthetic (unknown) small-scales eddies of the velocity field $\mathbf{u}(\mathbf{x}, t)$ from the knowledge of its filtered or coarse-grained field $\tilde{\mathbf{u}}(\mathbf{x}, t)$. For the case of three interpolating points $\{(x_i, \tilde{u}_i), i = 0, 1, 2\}$, the fractal interpolation iterative function system (IFS) is of the form $\{R^2; w_j, j = 1, 2\}$, where:

$$w_j \begin{pmatrix} x \\ u \end{pmatrix} = \begin{bmatrix} a_j & 0 \\ c_j & d_j \end{bmatrix} \begin{pmatrix} x \\ u \end{pmatrix} + \begin{pmatrix} e_j \\ f_j \end{pmatrix}, \quad j = 1, 2 \tag{1}$$

with constraints

$$w_j \begin{pmatrix} x_0 \\ \tilde{u}_0 \end{pmatrix} = \begin{pmatrix} x_{j-1} \\ \tilde{u}_{j-1} \end{pmatrix} \quad \text{and} \quad w_j \begin{pmatrix} x_2 \\ \tilde{u}_2 \end{pmatrix} = \begin{pmatrix} x_j \\ \tilde{u}_j \end{pmatrix}, \quad j = 1, 2 \tag{2}$$

The parameters a_j, c_j, e_j and f_j can be determined in terms of d_j (called the stretching parameter). The two stretching parameters determine the characteristics of the reconstructed signal and are constrained to be real and lie in the interval (-1,1). Its value is independent of the interpolation points. Thus, once the stretching parameter is chosen, the remaining parameters a_j, c_j, e_j and f_j are given as

$$a_j = \frac{x_j - x_{j-1}}{x_2 - x_0} \tag{3}$$

$$e_j = \frac{x_2 x_j - x_0 x_j}{x_2 - x_0} \tag{4}$$

$$c_j = \frac{\tilde{u_j} - \tilde{u}_{j-1}}{x_2 - x_0} - d_j \frac{\tilde{u_2} - \tilde{u_0}}{x_2 - x_0} \tag{5}$$

$$f_j = \frac{x_2 \tilde{u}_{j-1} - x_0 \tilde{u_j}}{x_2 - x_0} - d_j \frac{x_2 \tilde{u_0} - x_0 \tilde{u_2}}{x_2 - x_0} \tag{6}$$

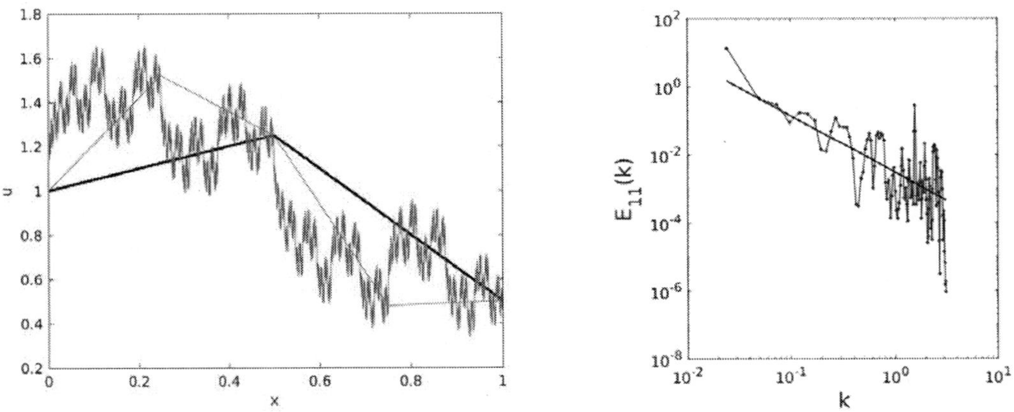

Figure 1. a) Different stages during the construction of a fractal function which interpolates between $(x_i, \bar{u}_i) = (0,1)$ to $(x_i, \bar{u}_i) = (1, 0.5)$ after 0,1 and 10 iterations with stretching parameter $d = \pm 2^{-1/3}$ b) Energy spectrum of the constructed signal showing $-5/3$ slope.

More details can be found in [1]. To generate a synthetic small scale field, this mapping is simply iterated many more times as shown in figure 1. It was shown in [4] that the stretching parameter depends on the fractal dimension of the signal (scaling exponent spectrum) D as:

$$D = 1 + log_N \sum_{n=1}^{N} |d_n| \tag{7}$$

where $N =$ the number of anchor points $- 1$. The attractor of the above IFS, G, is the graph of a continuous function $u : [x_0, x_2] \to R$, which interpolates the data points $(x_i, \tilde{u_i})$, if the stretching parameter d_j obey $0 \leq |d_j| < 1$. Also, if

$$\sum_{j=1}^{N} |d_j| > 1 \tag{8}$$

and $(x_i, \tilde{u_i})$, are not collinear, then the fractal dimension of G is the unique real solution D of

$$\sum_{j=1}^{N} |d_j| a_j^{D-1} = 1 \tag{9}$$

2.2. Stretching parameter estimation

As discussed in [5], the local stretching parameter d can be calculated using an analytic and geometric approach. Both method gives approximate results as shown in figure 3-6 of [5]. For this study, we use the geometric approach, which can be understood as a reverse process of fractal interpolation. For example, for 5 interpolation points $\{(x_i, u_i), i = 0, 1, 2, 3, 4\}$, let μ be the vertical distance between the middle interpolation point (x_2, u_2) and a straight line between the end points (x_0, u_0) and (x_4, u_4). The value of μ is positive if the interpolation points are above the straight line and negative otherwise. Let ν_1 be the vertical distance between the first three interpolation points $\{(x_i, u_i), i = 0, 1, 2\}$ and a straight line between (x_0, u_0) and (x_2, u_2) while ν_2 be the vertical distance between the last three interpolation points $\{(x_i, u_i), i = 2, 3, 4\}$ and a straight line between (x_2, u_2) and (x_4, u_4). Both ν_1 and ν_2 are positive if their respective interpolation points are above their respective straight lines and negative otherwise. Then the stretching parameters d_1 and d_2 are ν_1/μ and ν_2/μ respectively. An illustration of this calculation is presented in figure 2.

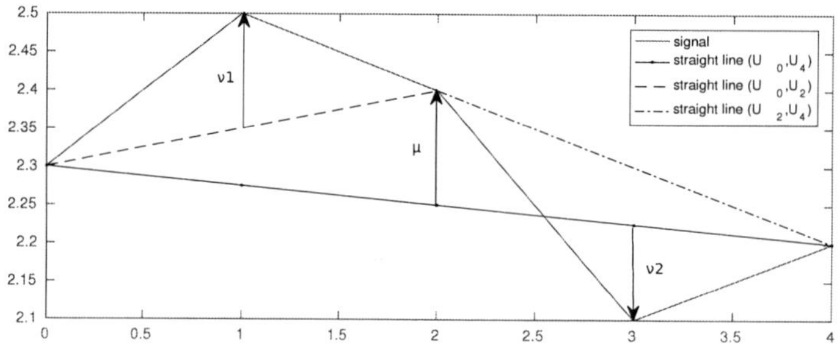

Figure 2. Geometric approach for stretching parameter calculation.

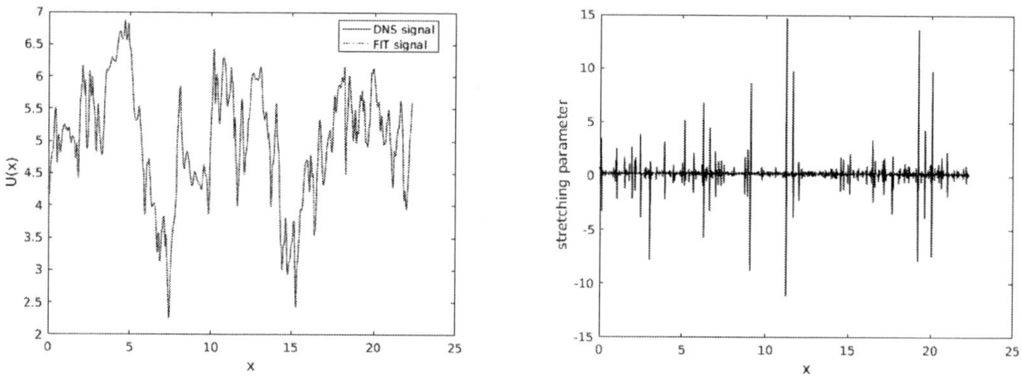

Figure 3. a) 1D DNS velocity signals showing the original and the FIT reconstructed signal b) The stretching parameters used to construct FIT signal in figure (a).

Figure 3a shows the 1D original and reconstructed FIT signal with calculated d values. If the procedure for d retrieval is correct both the original and the FIT signal should be identical, which is the case as observed in figure 3a. The stretching parameter as shown in figure 3b varies significantly in space and has values outside the interval $(-1, 1)$ because the original signal is highly intermittent. Regions of high velocity gradients in intermittent signals can not be reproduced with FIT since d should only take values within the interval $(-1, 1)$. If we place a constraint, such that $|d| \leq 1$, only some of the sub-grid scales will be reproduced. We compare this geometric approach of calculating the stretching parameter with constant values proposed by [4] and [2] in section 3.2.

3. Results

Direct numerical simulation (DNS) data of stratocumulus Cloud-Top for DYCOMS-II RF01 [3] (courtesy of Prof. J.-P. Mellado from the Max Planck Institute of Meteorology) is used to calculate the stretching parameters spatial probability distribution, using the geometric approach. Details of the simulation is outside the scope of this study and readers are referred to [3] for more details.

3.1. Stretching parameter spatial probability distribution function

Horizontal profile at a height corresponding to in-cloud region with highest turbulent intensity is used for the calculation of d. First, we investigate the variability of d as the DNS velocity signal is filtered, starting with the fully resolved DNS signal and filtering successively down to wavenumber from the inertial range. In each iteration step, the signal is filtered with a finite impulse response (FIR) filter of order 30 (a form of a low pass filter) designed using the Hamming window method and we calculate the local estimate of d with the geometric approach explained in section 2.2. The decimation factor is set to equals twice the grid size of the previous iteration step. This was done with decimation function in MATLAB® software. Stretching parameter values outside the interval $(-1, 1)$ were neglected and absolute value of d was used to calculate the spatial probability distribution function.

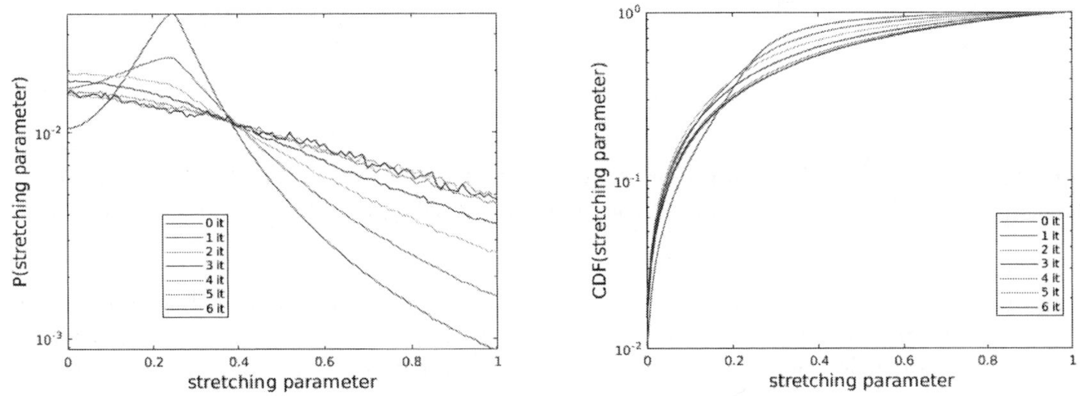

Figure 4. a) Spatial probability distribution of stretching parameter for 1D DNS velocity signals at different iteration step b) Its cumulative probability distribution at different iteration step

Figure 4 shows the spatial probability distribution of stretching parameter for 1D DNS velocity signals for different iteration step and their cumulative probability distribution function. In figure 4a, the pdfs change significantly in the first four iterations but seems to be self-similar when filtered with wavenumbers in the inertial range. Figure 5 shows the average stretching parameter and fractal dimension of 1D DNS velocity signal for different iteration steps. In figure 5a, the average stretching parameter are seen to be smaller than 0.5 as compared to figure 2 of [6] where an average value of 0.6 and 0.7 were reported for x and y directions, respectively. Our result with somewhat smaller d might be due to the presence of external intermittency (laminar

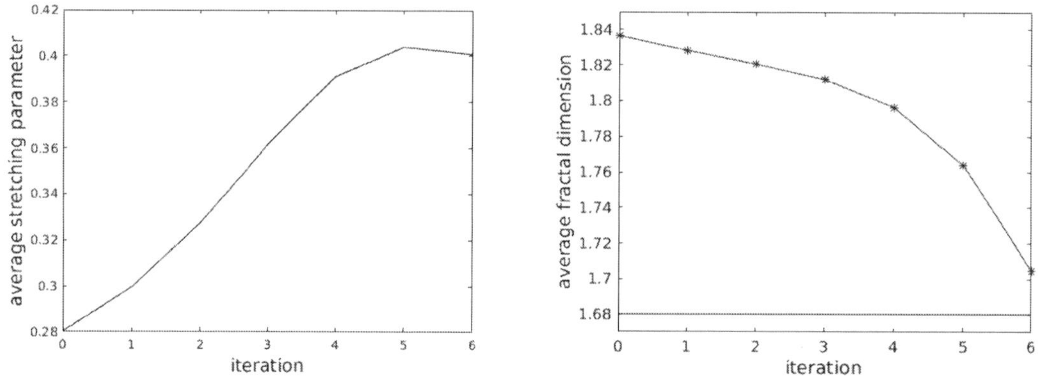

Figure 5. a) Average stretching parameter of $1D$ DNS velocity signals for different iteration step b) Its average fractal dimension showing the 5/3 scaling.

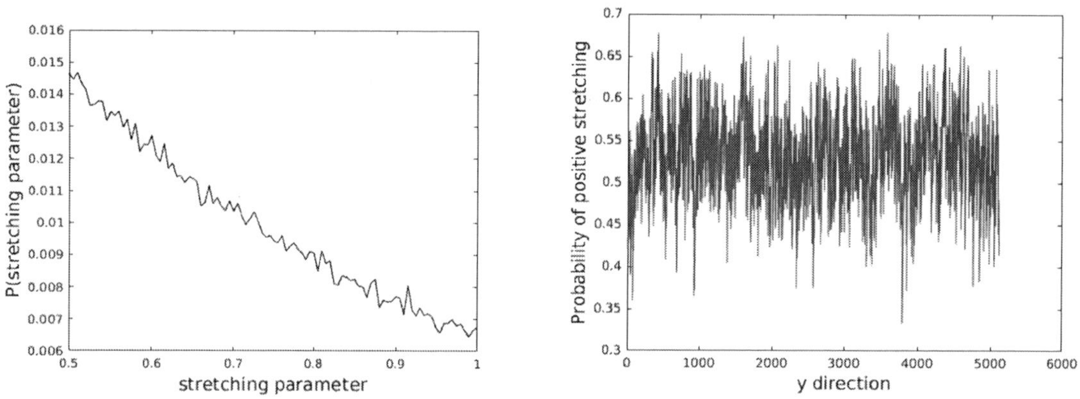

Figure 6. a) Pdf of stretching parameter d within the interval [0.5, 1] estimated from 1D DNS velocity signal (b) Probability of positive stretching parameter

regions will give zero or close to zero stretching parameter). If we ignore $|d| < 0.5$, we will obtain similar average stretching parameter as in [6] (see fig. 6a). Still, with our current results the average fractal dimension decreases to 5/3 scaling, as expected in the inertial range, see figure 5b.

Since this work focuses on constructing inertial range subgrid scales of atmospheric turbulence, we focus on the pdf of d when filtered with wavenumbers from the inertial range. The pdf of d, as shown in figure 6a, will be used to construct sub-grid scale for the filtered DNS velocity signal in section 3.2. We also investigate the probability of having positive (or negative) stretching parameter and figure 6b shows that, on the average, there exist an equal probability of having positive and negative stretching parameters.

3.2. 1D fractal interpolation of DNS of stratocumulus cloud-top for DYCOMS-II RF01

We filter each 1D DNS velocity signal of stratocumulus cloud-top in the y direction with inertial range wavenumber and use the pdf of d in figure 6a to construct subfilter scale using fractal interpolation technique. We present the spectra and pdfs of the velocity increments of DNS, filtered and FIT reconstructed signal. We also present the spectra and pdfs of the velocity

IOP Conf. Series: Journal of Physics: Conf. Series **1101** (2018) 012001 doi:10.1088/1742-6596/1101/1/012001

increments of FIT reconstructed signal using a constant values of d as previously proposed.

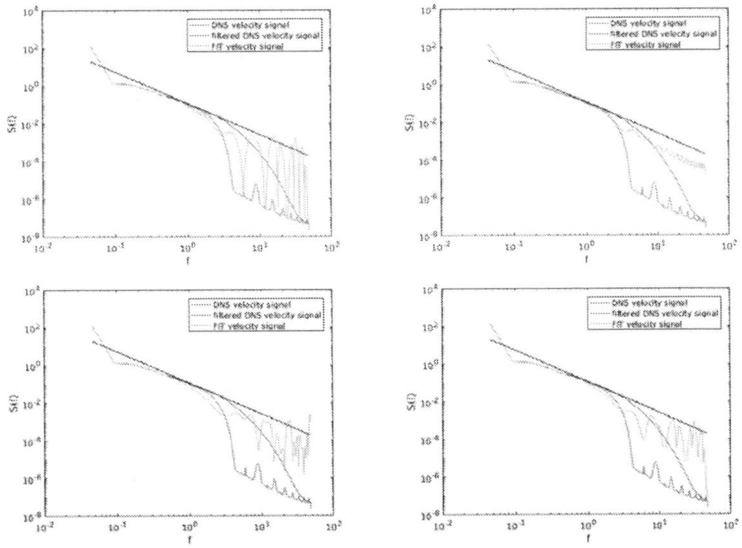

Figure 7. Energy spectra showing $-5/3$ scope (a) with constant stretching parameter $d = \pm 2^{-1/3}$ (b) with constant stretching parameters $d = 0$ and $d = 0.93$ (c) with constant stretching parameters $d = -0.887$ and $d = -0.676$ (d) with stretching parameters from section 3.1

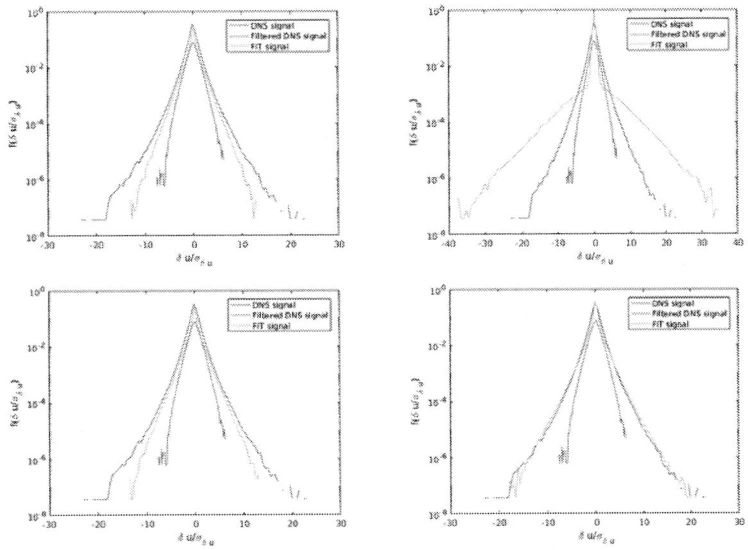

Figure 8. Normalized Pdfs of DNS, filtered and FIT velocity increments (δu) signal (a) with constant stretching parameter $d = \pm 2^{-1/3}$ (b) with constant stretching parameters $d = 0$ and $d = 0.93$ (c) with constant stretching parameters $d = -0.887$ and $d = -0.676$ (d) with stretching parameters from section 3.1

Here, we focus on extending the inertial range scales and not explicit recovery of the dissipative range of the spectra. Figure 7 shows the energy spectra of DNS, filtered and FIT velocity signals. All the energy spectra exhibit periodic modulation and this is due to the dyadic nature of the fractal interpolation technique (see figure 7 and 8 of [4]). Basu et al. (2004) [2] avoid this periodic modulation by applying the discrete Haar wavelet transform. Since the purpose of this work is to obtain statistics of velocity component of atmospheric turbulence, we do not follow this approach and account for the Fourier spectra only. Figure 8 shows the normalized pdfs of velocity increment (δu) for DNS, filtered and FIT velocity signals. The best fit with DNS data is observed for the reconstruction with the new pdf of stretching parameter, shown in figure 8d.

4. Conclusion

In this work, we present a fractal subgrid scale model for large eddy simulation of atmospheric flows. With fractal interpolation technique, we construct subgrid velocity field from the knowledge of its filtered or LES grid proposed by [4]. The (free) stretching parameter determines the characteristics of the reconstructed signal which can be derived from the fractal dimension. In previous literature, the stretching parameter is chosen to be constant in time and space. To account for its spatial variability, we estimate the spatial pdf of the stretching parameter from DNS data of stratocumulus-top boundary layer, using the geometric method proposed by [5]. We compare constant estimate of the stretching parameter with spatial pdf of the stretching parameter to construct sub-grid velocity. We present the energy spectra and pdfs of the normalized velocity increments.

References

[1] M.F Barnley: Fractal Everywhere, Academy Press, Boston, MA, 1993.

[2] S. Basu, E. Foufoula-Georgiou and F. Porte-Agel: Synthetic Turbulence, Fractal interpolation and Large Eddy Simulation, Physical Review E70, 026310, 2004.

[3] Mellado J. P.: Cloud-top entrainment in stratocumulus clouds, Annual Review of Fluid Mechanics, 49, 145-169, 2017

[4] A. Scotti , C. Meneveau: A fractal model for large eddy simulation of turbulent flow, Physica D, 127 198232, 1999.

[5] D. S. Mazel and M. H. Hayes: Using Iterated Function systems to model discrete sequences, IEEE Transactions on signal processing, 40(7), 1992

[6] M. V. Salvetti, C. Marchioli and A. Soldati: Lagrangian tracking of particles in large eddy simulation with fractal interpolation, Conference on Turbulence and Interactions TI2006, 2006.

[7] J. Smagorinsky: General circulation experiments with the primitive equations: I. The basic equations, Mon. Weather Rev. 91(99), 1963.

[8] M. Łuniewski and J. Pozorski: Spatial distribution and settling velocity of heavy particles in synthetic turbulent fields, 5 Trans. Inst. Fluid-Flow Mach. 135 87100, 2017.

[9] C. Beghein, C. Allery, M. Wacławczyk and J. Pozorski: Application of POD-based dynamical systems to dispersion and deposition of particles in turbulent channel flow, International Journal of Multiphase Flow, 58, 97-113, 2014 `https://doi.org/10.1016/j.ijmultiphaseflow.2013.09.001`.

[10] G. C. Burton and W. J. A. Dahm, Multifractal subgrid-scale modeling for large-eddy simulation. i. Model development and a priori testing, Physics of Fluids, 17(7):075111, 2005.

[11] J. Pozorski and B. Rosa: The motion of settling particles in isotropic turbulence: filtering impact and kinematic simulations as subfilter model, Direct and Large Eddy Simulation XI, 6 pp. Springer, 2018 (in press)

[12] J-P Minier and J. Pozorski:Particles in Wall-Bounded Turbulent Flows: Deposition, Re-Suspension and Agglomeration, Springer, 571, 0254-1971, 2017. doi: 10.1007/978-3-319-41567-3

IOP Conf. Series: Journal of Physics: Conf. Series **1101** (2018) 012002 doi:10.1088/1742-6596/1101/1/012002

Experimental and comparative study on the two-phase pressure drop of air-water mixture in U-bend and straight pipe annuli

R Andrzejczyk, T Muszynski

Gdansk University of Technology, Faculty of Mechanical Engineering, Department of Energy and Industrial Apparatus, Narutowicza 11/12, 80-233 Gdansk, Poland
email: rafal.andrzejczyk@pg.edu.pl

Abstract. In this paper, the experimental and theoretical analysis of pressure drop in single-phase and two phase-flow were presented for straight and U-bend smooth tube annulus and tube annulus with wire coil insert. Experiments for various boundary conditions were performed. In case of U-tube and straight tube with and without turbulator, tests were made for the water-water and air-water systems. The study covered a wide measuring range, i.e. V_w = 9*10-5-8.87*10-6 m3/s - for water, and V_a=5.55*10-5 m3/s.-for air. The test elements were made from a copper pipe with an external diameter of 10 mm and 18 mm and wall thickness 1 mm. The helicoidal vortex generator was made from brass wire with a diameter of 2.4 mm, coil diameter 13 mm and pitch 11 mm. For these geometries, the values of pressure drop and heat flux were determined. Obtained experimental results were compared with correlations from literature. The best coherence with database were obtain for Lockhart-Martinelli and Sugawar et al. models for two-phase flow regime.

1. Introduction

Nowadays, the rapid development of practical engineering applications for mini and micro-devices, micro-systems, advanced material designs, manufacturing electronic microchips increases the demand for better understanding of fluid mechanics [1]. The prediction of single phase and the two-phase pressure gradient is an essential step in the design of a variety of equipment in the power and process engineering, renewable energy systems and heating, ventilation, air conditioning or refrigerating systems. However it should be emphasized that understanding of fluid mechanics are especially very important in case of using different heat transfer enhancement techniques in heat exchangers [2,3]. In case of the most heat transfer enhancement techniques, the heat transfer coefficient enhancement achieved is accompanied by a significant increase in the pressure drop. Therefore properly prediction of heat transfer coefficient as well as pressure drops are mandatory to determine the conditions under which the use of this methods are favorable.

There are many studies in open literature on pressure drop calculation in single phase flow. The most important parameter to determine pressure loses is friction factor *(f)*. Most of the works concerned experimental or numerical investigations of this parameter. The literature review of selected works is presented in table 1.

For last few years also many works were made in the field of pressure drops determination in conventional and mini-microchannels with swirl flow inserts such as twisted tapes, coil wires, static mixers et all [8]. The authors in this study focused on experimental investigation in single and two-phase flow in the pipe annulus with wire coil insert.

Table 1. Single-phase friction factor correlations

Study	Re × 10³	ε	f
Fang et al. [4]	3 < Re< 10⁵	0.0 <ε <0.05	$f = 1.613\left[ln\left(0.234\varepsilon^{1.1007} - \frac{60.525}{Re^{1.1105}} + \frac{56.291}{Re^{1.0712}}\right)\right]^{-2}$
Blausius	Re≤20	Smooth tubes	$f = \frac{0.3164}{Re^{0.25}}$
	20≤Re≤2000		$f = \frac{0.184}{Re^{0.2}}$
Moody [5]	4<Re<5x10⁵	0 <ε <0.01	$f = 0.0055\left[1 + \left(2 \cdot 10^4 \cdot \frac{\varepsilon}{d} + \frac{10^6}{Re}\right)^{\frac{1}{3}}\right]$
Wood [6]	4<Re<5x10⁴	10⁻⁵ <ε <0.04	$f = 0.53 \cdot \varepsilon + 0.094 \cdot \varepsilon^{0.225} + 88 \cdot \varepsilon^{0.44} \cdot Re^{-1.62 \cdot \varepsilon^{0.134}}$
Swamee and Jain [7]			$f = \frac{0.25}{\left[log\left(\frac{\varepsilon/d}{3.7} + \frac{5.74}{Re^{0.9}}\right)\right]^2}$

In the table below an overview of selected works is presented. It should be noted that most of existing studies were made for extensive range of wire coil parameter but for limited range of tube diameter and Prandtl number. Generally, the majority of investigations were made only for air flow. This creates a problem for correct selection correlation for an engineering calculations.

Table 2. Single-phase friction factor correlations for tube with coiled wire inserts

Study	Re × 10³	d [mm]	Test fluid (Pr)	f
Garcia et. al.[9]	0.8 < Re< 90	18	2.8 < Pr < 50	$f = 5.76 \cdot \left(\frac{e}{d}\right)^{0.95} \cdot \left(\frac{p}{d}\right)^{-1.21} \cdot Re^{-0.217}$
Sharafeldeen [10]	1.4 ≤ Re ≤ 42	45	Air (0.7)	$f = 0.3251 \cdot Re^{-0.101} \cdot \left(\frac{e}{d}\right)^{0.196} \cdot \left(\frac{p}{d}\right)^{-0.211}$
Gunes [11]	3.5≤ Re≤27	56	Air (0.7)	$f = 3.970492 \cdot \left(\frac{p}{d}\right)^{-0.31182} \cdot Re^{-0.367485} \cdot \left(\frac{s}{d}\right)^{-0.157719}$
Slaiman [12]	5-40	11 and 14	Water (2.55-2.98)	$f = 3.6346 \cdot \left(\frac{e}{d}\right)^{0.8912} \cdot Re^{-0.0964} \cdot \left(\frac{p}{d}\right)^{-0.7856}$
Keklikcioglu, and Ozceyhan [13]	3.4<Re<27	56	Air (0.7)	$f = 6.423 \cdot Re^{-0.301} \cdot \left(\frac{p}{d}\right)^{-0.587} \cdot \left(\frac{s}{d}\right)^{-0.106}$
Akhavan-Behabadi et al. [14]	0.02<Re<0.5	26.04	Oil (120<Pr<300)	$f = 16.8/Re^{0.96}$
Yakut and Sahin [15]	5<Re<17	50	Air (0.7)	$f = 4.44 \cdot Re^{-0.218} \cdot \left(\frac{p}{d}\right)^{-0.223}$
Keklikcioglu, and Ozceyhan [16]	2.8<Re<27.8	56	Air (0.7)	$f = 72.599 \cdot Re^{-0.514} \cdot \left(\frac{e}{d}\right)^{0.486} \cdot \left(\frac{p}{d}\right)^{-0.367}$

This problem is even more difficult in case of two phase flow conditions. A number of studies has been done concerned at two-phase pressure drop (TFPD) in straight pipe [18] as well as in curved pipe [19]. However, most of the studies were focused on two-phase pressure drop during the flow in circular mini and microchannels [20]. The two-phase flow multiplier is defined as a ratio of friction pressure drop in the two-phase flow, $\left(\frac{dP}{dz}\right)_{TP}$ to the friction pressure drop in the flow of either liquid of as $\left(\frac{dP}{dz}\right)_0$

$$\Phi^2 = \frac{\left(\frac{dP}{dz}\right)_{TP}}{\left(\frac{dP}{dz}\right)_0} \tag{1}$$

Generally published predictive methods are limited in validity to specific working fluids and ranges of operating parameters for the data upon which these methods are based, see table 3. It should be emphasize, that still there is a need of pressure drop databases covering broad ranges of experimental conditions and a reasonable understanding of the phenomena accompanying are necessary. It is crucial important especially in case of more complicated geometries, e.g. U-bend tubes, coiled tubes, tube with swirl flow devices, displaced enhancement devices [21,22]. In this paper, the comprehensive study of two-phase pressure drop for air-water mixture in tube annulus has been shown. Experimental works were performed for smooth straight and U-bend pipes as well as straight and U-bend pipes with wire-coil inserted for wide measuring range.

Table 3. Two-phase frictional pressure gradient correlations

Study	d_h	Test fluid	Φ^2
Lockhart and Martinelli [23]	1.49-25.83 mm	Air-water, oils, hydrocarbon	$\Phi^2{}_{ML\_LO} = 1 + \frac{C}{X} + \frac{1}{X^2}$ for $Re_l > 4000$ $\Phi_{ML\_LO} = 1 + CX + X^2$ for $Re_l < 4000$, $X = \left((dp/dz)_l/(dp/dz)_g\right)^{0.5}$ $C_{lt}=5, C_{tl}=10, C_{ll}=12, C_{tt}=20$
Mishima and Hibiki [24]	1.05–4.08 mm	Air-water	$\Phi^2{}_M = 1 + \frac{C}{X} + \frac{1}{X^2}$ $C = 21 \cdot (1 - \exp \cdot (-0.319/D_e))$
Sugawara et al.[25]	0.7-9.1 mm	Air-water	$\Phi^2{}_M = 1 + \frac{C}{X} + \frac{1}{X^2}$ $C = 21 \cdot (1 - \exp \cdot (-0.333/D_e))$

The primary objectives of the present study are:

I. Providing an experimental database including multiple data points for two-phase total pressure drop during air-water flow at annuli in smooth straight pipe, U-bend pipe and in straight and U-bend pipe with wire-coil inserted.

II. Comparing experimental data with well-known correlations for TFPDs from open literature.

III. Conducting a systematic assessment of predictive techniques for a two-phase pressure drop.

IV. Comprehensive validation influences of mass flux, coiled wire effects for TFPDs.

2. Experimental setup

In order to obtain experimental values of two-phase pressure drop the special experimental facility was done (see fig.1). Construction of the test stand allows to change parameters of water flow as well as air flow parameters. Pressure resistances (ΔP) were measured using a differential pressure transducer in range 0-3 bar and 0.25 accuracy class made by PELTRON. The volumetric flow rate of water (V_w) was measured with a ROL 16 Rotameter in class 2.5 volumetric flow rate of air (V_a) was measured with a ROL o6 Rotameter in class 2.5. Authors decided to supply all elements from district water system due to the stable temperatures of that source (+/- 0.5 K). Because of high and stable pressure levels in the system (P>4bar),it was possible to achieve turbulent as well as laminar flow conditions. All experiments were performed with steady-state conditions and in each time taken twice for the same thermal flow parameters. The paper presented preliminary study for this reason the measurement accuracy were sufficient. The main purpose of presented study was to find future experimental investigations direction. What is more also the cfd calculation will be done into a future to better understand physic of all phenomes.

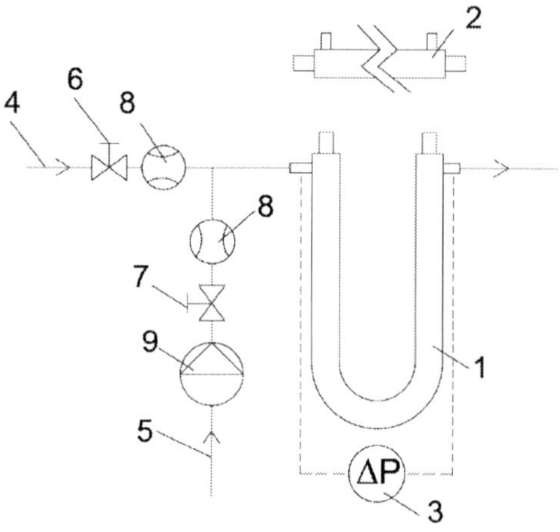

Figure 1. Test facility: 1 – U-tube HX, 2 – Tube in tube HX, 3 – differential pressure transducer , 4 –tap water, 5 –air,6 – cold water control valve, 7 – air control valve, 8 – rotameter, 9 – air compressor

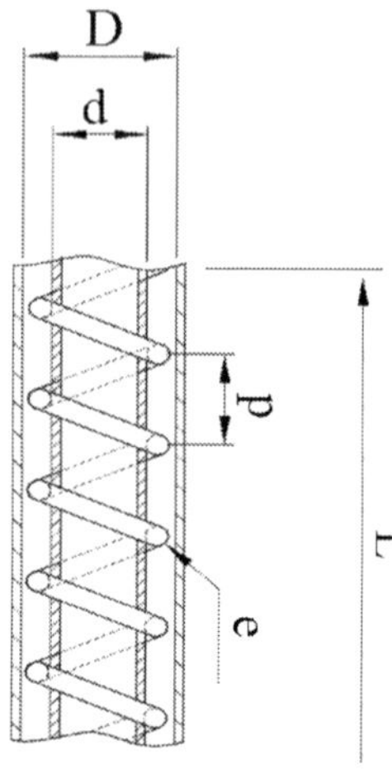

Figure 2. Turbulator dimensions and cross-section of the apparatus with a view of the helical turbulent element

Table 4. Uncertainties of selected parameters

Symbol	Operating range	Uncertainty
V_w	$9*10^{-5}$-$8.87*10^{-6}$ m^3/s	Maximum error = 2.5%
V_a	$5.55*10^{-5}$ m^3/s	Maximum error = 2.5%
ΔP	0-200 kPa	Maximum error = 0.5 kPa
Re	827-18423	Maximum error =5.7%

3. Experimental procedure

Reynolds number at single phase regime was calculated as the mass flow rate through equivalent diameter:

$$Re = \frac{G \cdot D_e}{\mu} \tag{2}$$

Equivalent diameter depends on the volume of the annulus divided by tube length and circumference.

$$De = \frac{4 \cdot V_{sh}}{\pi \cdot D \cdot L} \tag{3}$$

The volume available for the flow of fluid in the annulus, V_{sh} can be calculated knowing geometrical dimensions of the tested element.

$$V_{sh} = \frac{\pi}{4} \cdot D^2 \cdot L - \left(\frac{\pi}{4} \cdot d^2 L + \frac{\pi}{4} \cdot e^2 L\right) \tag{4}$$

The measured pressure drop is the sum of friction pressure drop (ΔP_{frict}), expansion (ΔP_{exp}) and contraction (ΔP_{contr}) losses due to the headers at both ends of the test section [26]:

$$\Delta P = \Delta P_{frict} + \Delta P_{exp} + \Delta P_{contr} \tag{5}$$

The pressure drop due to contraction was estimated using a flow model recommended by Hewitt et al. [26] for single-phase flow.

$$\Delta P_{contr} = \frac{G^2}{2 \cdot \rho} \cdot \left[\left(\frac{1}{C_{con}} - 1\right) + 1 - \gamma^2\right] \tag{6}$$

where γ is the area ratio ($A_{intake\ manifold}/A_{shell}$) and C_{con} is the coefficient of contraction, which, in turn, is a function of this area ratio:

$$C_{con} = \frac{1}{0.639 \cdot (1-\gamma)^{0.5} + 1} \tag{7}$$

For the expansion into the header from the test section, the following flow model recommended by Hewitt et al. [26] was also used:

$$\Delta P_{exp} = \frac{G^2 \cdot \gamma \cdot (1-\gamma) \cdot \Psi_s}{\rho} \tag{8}$$

where Ψ_s, the separated flow multiplier, is also a function of the phase densities and the quality. In single flow case, those multiplier and quality are equal to unity. The friction factor was calculated as below [27]:

$$f = \frac{\Delta P}{\left(\frac{\rho \cdot w^2}{2}\right) \cdot \left(\frac{L}{D_e}\right)} \tag{9}$$

The Lockhart-Martinelli parameter was calculated directly from definition as a relation between single phase liquid pressure drop and single phase gas pressure drop [28]:

$$X = \sqrt{\frac{(dp/dz)_l}{(dp/dz)_g}} \tag{10}$$

The pressure drop for a single-phase flow, can be obtained from the following expression:

$$\left(\frac{dP}{dz}\right)_0 = f \cdot \frac{G^2}{2 \cdot \rho} \cdot \frac{L}{D_e} \tag{11}$$

Finally the pressure drop for two-phase flow was calculated as follow:

$$\left(\frac{dP}{dz}\right)_{TP} = \Phi^2 \cdot \left(\frac{dP}{dz}\right)_l \tag{12}$$

4. Experimental results

It is worth to note that in case of single phase flow (without turbulator and with wire coil insert) experimental correlations have significant difference values of friction factors. In case of smooth tubes (straight and U-bend) only the Swamee and Jain correlation good fits to experimental data. Other experimental correlations underestimating friction factor. In case of single-phase flow with wire coil insert none of the experimental models have a satisfactorily fit to experimental results for friction factor. Nevertheless the best coherency have Keklikcioglu, and Ozceyhan model.

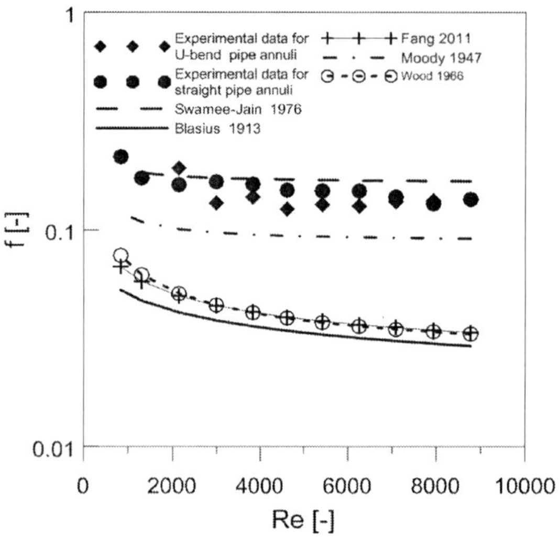

Figure 3. Hydraulic characteristic for annuli without coil wire insert: friction factors vs Reynolds numbers

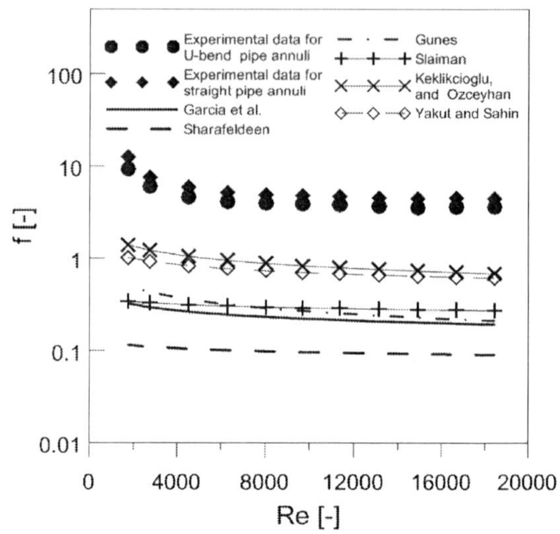

Figure 4. Hydraulic characteristic for annuli with coil wire insert: friction factors vs Reynolds numbers

Based on the above analysis, it was decided that in the calculations of two-phase flow resistance were used only correlations for friction factors that had the best coherency with experimental data. As can be seen from calculation results Lockhart-Martinelli (LM) correlation are best fit to experimental data for two-phase air-water flow in smooth pipe annuli. However, for small values of X parameter obtained results are underestimate compare to experimental data. It could be explained by the higher percentage of measurement errors in the obtained results. In case of two-phase flow for annuli with coil wire inserted the best coherency with experimental data was obtained for Sugawara et al correlation. It should be emphasize that this correlation is a modification of LM correlation for small diameter channels and equivalent diameter for annuli with coil wire insert was close to 8 mm.

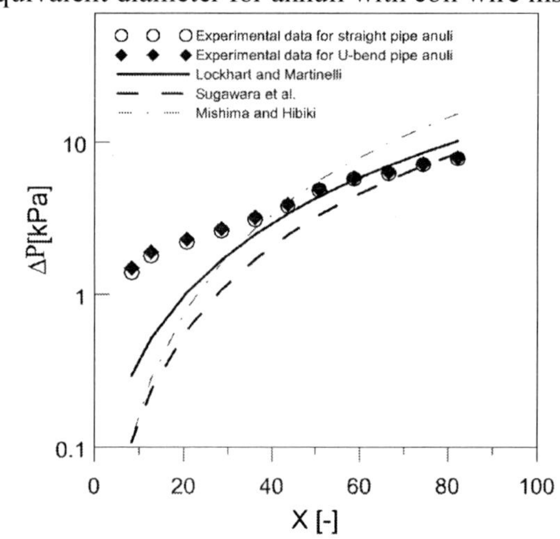

Figure 5. Two phase pressure drop versus Lockhart Martinelli parameter for annuli without coil wire insert

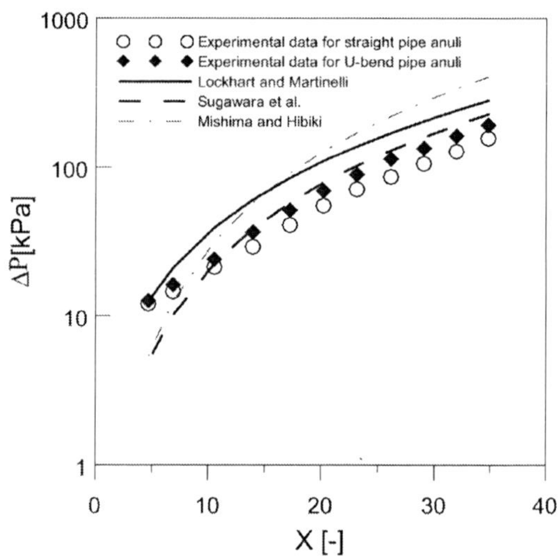

Figure 6. Two phase pressure drop versus Lockhart Martinelli parameter for annuli with coil wire insert

5. Conclusions

Based on collected experimental data, frictional pressure drop for two-phase flow, single phase flow and single friction factors were calculated. Examination of correlation for single phase friction factor has shown significant difference of calculation results. As can be seen from calculation results LM correlation are best fit to experimental data for two-phase air-water flow in smooth pipe annuli. In case of two-phase flow for annuli with coil wire inserted the best coherency with experimental data was obtained for Sugawara et al correlation. Undoubtedly, still important is an issue to examine similarities and differences between well-known friction factor correlations to avoid misusing of them. Generally published predictive methods are limited in validity to specific working fluids and ranges of operating and geometrical parameters. Incorrect use of correlation for single phase friction factor could dramatically increase calculation errors especially in case of two-phase flow. To avoid this problems during selecting a given correlation, very important is taking into account its scope of applicability as well as flow and geometrical parameters.

Nomenclature

A - heat transfer area [m^2]
C – coefficient in LM parameter [-]
C_{con} - coefficient of contraction [-]
d – diameter of the inner tube [m]
D_e – equivalent diameter [m]
$(dP/dz)_{TP}$ – pressure drop for two-phase [Pa/m]
$(dP/dz)_0$ – pressure drop for single phase [Pa/m]
e – wire diameter [m]
f – friction factor [-]
G – mass flux [kg/m^2s]
L – length [m]
p – wire pitch, [m]
ΔP – pressure drop [Pa]
Pr – Prandtl number [-]
Re – Reynolds number [-]
w – welocity [m/s]
V – volume [m^3]
V_{sh} – volume available in the annulus [m^3]
X Lockhart-Martinelli parameter [-]

Greek symbols

γ area ratio [-]
Φ^2 two-phase flow multiplier [-]
Ψ_s the separated flow multiplier [-]
ε roughness factor [-]
μ viscosity [Pas]
ρ density [kg/m^3]

Superscripts

a air
frict friction
exp expansion
g gas
contr contraction
l liquid
lt laminar/turbulent
ll- laminar/laminar
tt- turbulent/turbulent
w water

6. References

[1] Andrzejczyk R, Muszynski T and Dorao C A 2017 Experimental investigations on adiabatic frictional pressure drops of R134a during flow in 5 mm diameter channel *Exp. Therm. Fluid Sci.* **83** 78–87
[2] Andrzejczyk R and Muszynski T 2017 Thermodynamic and geometrical characteristics of mixed convection heat transfer in the shell and coil tube heat exchanger with baffles *Appl. Therm. Eng.* **121** 115–25
[3] Andrzejczyk R and Muszynski T 2018 An Experimental investigation on the effect of NEW continuous core-baffle geometry on the mixed convection heat transfer in shell and coil heat exchanger *Appl. Therm. Eng.*
[4] Fang X, Xu Y and Zhou Z 2011 New correlations of single-phase friction factor for turbulent pipe flow and evaluation of existing single-phase friction factor correlations *Nucl. Eng. Des.* **241** 897–902
[5] Moody L F 1947 An approximate formula for pipe friction factors *Trans. ASME* **69** 1005–11
[6] Wood D J 1966 An explicit friction factor relationship *Civ. Eng* **36** 60–1
[7] Oke I A, Ojo S O and Adeosun O O 2015 Performance evaluation for darcy friction factor formulae using colebrook-white as reference *Ife J. Sci.* **17** 75–86
[8] Liu S and Sakr M 2013 A comprehensive review on passive heat transfer enhancements in pipe exchangers *Renew. Sustain. Energy Rev.* **19** 64–81
[9] Garcia A, Vicente P G and Viedma A 2005 Experimental study of heat transfer enhancement with wire coil inserts in laminar-transition-turbulent regimes at different Prandtl numbers *Int. J.*

Heat Mass Transf. **48** 4640–51

[10] Sharafeldeen M A, Berbish N S, Moawed M A and Ali R K 2017 Experimental investigation of heat transfer and pressure drop of turbulent flow inside tube with inserted helical coils *Heat Mass Transf.* **53** 1265–76

[11] Gunes S, Ozceyhan V and Buyukalaca O 2010 Heat transfer enhancement in a tube with equilateral triangle cross sectioned coiled wire inserts *Exp. Therm. Fluid Sci.* **34** 684–91

[12] Slaiman Q J M and Znad A N 2017 Enhancement of Heat Transfer in The Tube-Sid of A Double Pipe Heat Exchanger by Wire Coils *Al-Nahrain J. Eng. Sci.* **16** 51–7

[13] Keklikcioglu O and Ozceyhan V 2016 Experimental investigation on heat transfer enhancement of a tube with coiled-wire inserts installed with a separation from the tube wall *Int. Commun. Heat Mass Transf.* **78** 88–94

[14] Akhavan-Behabadi M A, Kumar R, Salimpour M R and Azimi R 2010 Pressure drop and heat transfer augmentation due to coiled wire inserts during laminar flow of oil inside a horizontal tube *Int. J. Therm. Sci.* **49** 373–9

[15] Yakut K and Sahin B 2004 The effects of vortex characteristics on performance of coiled wire turbulators used for heat transfer augmentation *Appl. Therm. Eng.* **24** 2427–38

[16] Keklikcioglu O and Ozceyhan V 2018 Experimental investigation on heat transfer enhancement in a circular tube with equilateral triangle cross sectioned coiled-wire inserts *Appl. Therm. Eng.* **131** 686–95

[17] Muszynski T, Andrzejczyk R and Dorao C A 2017 Detailed experimental investigations on frictional pressure drop of R134a during flow boiling in 5 mm diameter channel: The influence of acceleration pressure drop component *Int. J. Refrig.* **82**

[18] Andrzejczyk R and Muszynski T 2017 The performance of H2O, R134a, SES36, ethanol, and HFE7100 two-phase closed thermosyphons for varying operating parameters and geometry *Arch. Thermodyn.* **38** 3–21

[19] Autee A T and Giri S V 2016 Experimental study on two-phase flow pressure drop in small diameter bends *Perspect. Sci.* **8** 621–5

[20] Mikielewicz D, Wajs J, Andrzejczyk R and Klugmann M 2016 Pressure drop of HFE7000 and HFE7100 during flow condensation in minichannels *Int. J. Refrig.* **68**

[21] Andrzejczyk R and Muszyński T 2016 Performance analyses of helical coil heat exchangers. The effect of external coil surface modification on heat exchanger effectiveness *Arch. Thermodyn.* **37** 137–59

[22] Muszynski T and Andrzejczyk R 2016 Heat transfer characteristics of hybrid microjet - Microchannel cooling module *Appl. Therm. Eng.* **93** 1360–6

[23] Lockhart R W and Martinelli R C 1949 Proposed correlation of data for isothermal two-phase, two-component flow in pipes *Chem. Eng. Prog* **45** 39–48

[24] Mishima K and Hibiki T 1996 Some characteristics of air-water two-phase flow in small diameter vertical tubes *Int. J. Multiph. Flow* **22** 703–12

[25] Kaji M, Sawai T, Kagi Y and Ueda T 2010 Heat transfer and fluid dynamics of air–water two-phase flow in micro-channels *Exp. Therm. Fluid Sci.* **34** 446–53

[26] Hewitt G F, Shires G L and Bott T R 1994 *Process heat transfer* vol 113 (CRC press Boca Raton, FL)

[27] Sheikholeslami M, Gorji-Bandpy M and Ganji D D 2015 Review of heat transfer enhancement methods: focus on passive methods using swirl flow devices *Renew. Sustain. Energy Rev.* **49** 444–69

[28] Chen J J J and Spedding P L 1981 An extension of the Lockhart-Martinelli theory of two phase pressure drop and holdup *Int. J. Multiph. Flow* **7** 659–75

Acknowledgments

The work presented in the paper was partially funded by statuary activity of faculty of Mechanical Engineering of Gdansk University of Technology.

IOP Conf. Series: Journal of Physics: Conf. Series **1101** (2018) 012003 doi:10.1088/1742-6596/1101/1/012003

Numerical investigation of the gyroplane longitudinal static stability for the selected stabilizer angles

Z Czyż[1], T Łusiak[2], P Karpiński[1] and J Czarnigowski[1]

[1] Lublin University of Technology, Faculty of Mechanical Engineering, Department of Thermodynamics, Fluid Mechanics and Aviation Propulsion Systems, Nadbystrzycka 36 Str., 20-618 Lublin, Poland

[2] Polish Air Force Academy, Department of Airframe and Engine, 35 Dywizjonu 303 Str., 08-521 Dęblin, Poland

z.czyz@pollub.pl

Abstract. Stability is a key aspect for aircraft as it directly affects its flight balance and manoeuvre capabilities. This numerical research with Ansys Fluent investigates an airflow around a model of the designed gyrocopter – Aduster to examine its static longitudinal stability for the given ranges of horizontal stabilizer angles and angles of attack. Our calculations give certain aerodynamic forces and moments. The paper investigates drag, lift, a pitching moment at the given angles as well as gives the calculated coefficients of those forces and the pitching moment. The research results and the stability criterion enable us to analyse our gyrocopter's static longitudinal stability.

1. Introduction

Gyrocopters have recently been gaining importance in the general aviation, especially in the air cargo sector and sport. Compared to light fixed-wing aircraft, they are distinguished by their short take-off and landing; and compared to light helicopters, their purchasing and operating costs are lower and operation is simpler. Currently, numerical calculations are very advanced and enable a precise modeling of gyrocopter flight mechanics. However, flight phenomena can be precisely described if certain additional data is applied. To validate a model, data about physical phenomena and processes that determine a behavior of aircraft, including geometry, mass, inertia and at least aerodynamic characteristics of rotors and airfoils are indispensable. The work in [1] presents the aspects of a mathematical modeling of gyrocopter flight dynamics including trim, stability and control with a flight test data validation. The same author in his previous work [2] applied an advanced mathematical model of a gyrocopter and aerodynamic forces and moments to investigate gyrocopter longitudinal stability. According to this research, a rotorspeed degree of freedom is an indispensable parameter in this type of modeling. The work [3] discusses the research into gyrocopter stability and flight control based on flight test data. The paper [4] presents a mathematical model of a vehicular towed autogyro system. To control the pitch angle so maintaining longitudinal stability, a PID controller is applied. The pitch angle was controlled with an inner altitude control loop. Another study using a mathematical model of gyrocopter flight dynamics is in [5] where a non-linear mathematical model of a small unmanned gyrocopter was created to investigate how a location of the center of gravity affects stability. This research was initiated in the research in [6] which had studied the drag acting on the gyrocopter components. A similar type of research into the drag and lift acting on the gyrocopter is discussed in [7]. Here, the research focuses on the static longitudinal stability of the research aircraft to investigate the impact of the angle of the horizontal stabilizer on stability and the generated total lift within a given range of angles of attack.

CFD (Computational Fluid Dynamics) calculations are a widely-used method to research aircraft. They are cheap, fast and easy to examine even large aircraft structures. Parameters like aerodynamic forces and moments for given airflow conditions are calculated here. The work in [8] applied this method to optimize a rotor design in a light gyrocopter. The author developed a set of dedicated gyrocopter airfoils as a competitive alternative to classical aerodynamic airfoils used in designing gyrocopter blades. The same applies to main rotor blades created with newly designed airfoils.

A CFD method was also applied to evaluate how a main rotor, its configuration and flight conditions influence gyrocopter aerodynamic characteristics and stability [9]. It was found that the airflow generated by a main rotor changes the most important aerodynamic coefficients and stability derivatives of the gyroplane body. The work in [10] presents the methodology of designing and optimizing gyroplane rotors with a CFD method. The authors in [11] used a CFD technique to create a dynamic model of the designed unmanned gyrocopter. The researchers in [12] made calculations of a cooling system in the designed gyrocopter drive. However the results of CFD engine cooling system presented in [13] show that the cooling capabilities are strongly dependent on surrounding geometries.

Numerical research can be validated with wind tunnel experiments, e.g. the wind tunnel research in [14] enabled the aerodynamic characteristics of the researched model of a gyrocopter. The pitching moment coefficient versus an angle of attack for different gyrocopter configurations, i.e. with or without stabilizers. The research shows that the fuselage reduces nearly by half the degree of the pitching moment generated by the tail beam. The fuselage without tail beams was unstable within the entire range of angles of attack. Wind tunnel research can be extended by a PIV visualization of an airflow around a research object. A PIV method was discussed in [15] to specify aerodynamic characteristics of a quad-rotor autogyro model. Moreover, the research in [16] uses a PIV method to describe the velocity field of the airflow around this quad-rotor autogyro model. The calculations enabled them to determine the component velocities as a function of a vertical coordinate. Accordingly, a comparative analysis of the research aircraft configuration was possible. Wind tunnel research for larger objects is much more expensive so alternative methods to generate an airflow around a research object, e.g. using a motor vehicle emerge. Such an approach is discussed in the work [17] where the drag of a gyrocopter model was examined on a platform mounted on a car roof. A moving vehicle caused an external airflow around the model. A difference between this experiment and a computer simulation consisted chiefly in airflow disturbances around the vehicle. The authors suggested an additional plate to be placed to separate the airflow around the research area and the vehicle. Such a plate can keep a research object in a non-disturbed flow as it is isolated from a vehicle.

2. Gyrocopter model and research plan

A solid model of the research gyrocopter was created in CATIA V5. The aircraft was segmented as shown in Figure 1. This segmentation enabled us to identify aerodynamic forces and moments acting on its individual parts. Figure 2 shows the dimensions of the research gyrocopter and its position relative to the global coordinate system (center of gravity). It is a Cartesian right-handed longitudinal system with an x-axis along the axis of symmetry of the gyrocopter, the transverse y-axis and the vertical z-axis. The aerodynamic moment generated relative to the y-axis is a pitching moment. Angle α is clockwise positively calculated. Sideslip angle β for this test series was 0°. Horizontal stabilizer angle γ varied in our calculations.

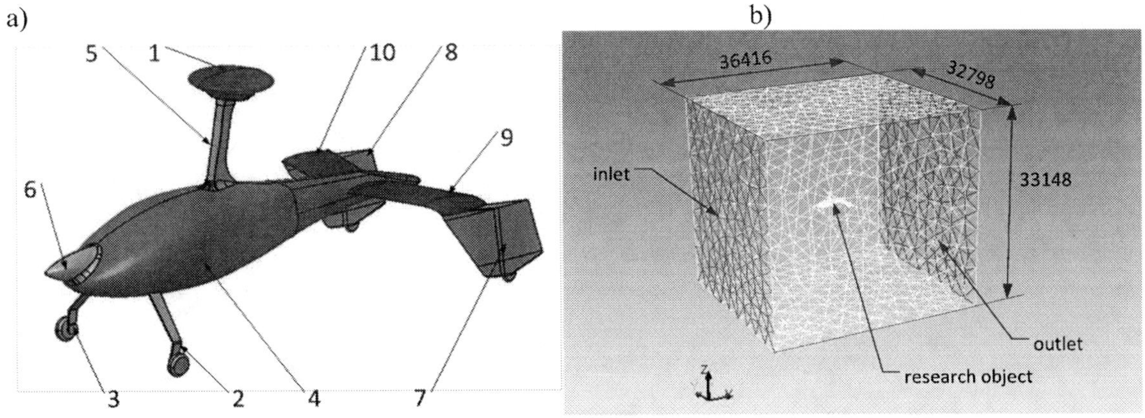

Figure 1. Solid model of the research gyrocopter (a): 1 - head, 2 - left leg, 3 – right leg, 5 – mast, 6 – nose, 7 – vertical stabilizer (left), 8 – vertical stabilizer (right), 9 – horizontal stabilizer (left), 10 – horizontal stabilizer (right) [1] and the computational domain defined in Ansys Fluent with the selected boundary conditions and dimensions (b).

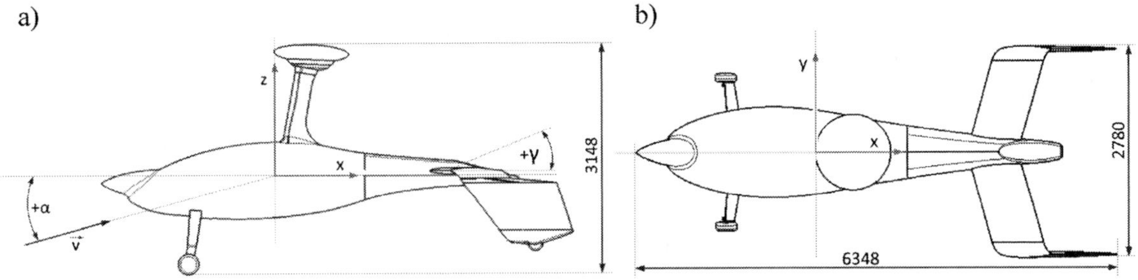

Figure 2. Global coordinate system and dimensions of the research gyrocopter model.

A solid model of the research gyrocopter was prepared for CFD calculations by being imported into Ansys Fluent in a Design Modeler module and removing all types of discontinuities, holes, sharp and fragmented edges and surfaces. Such a solid gyrocopter model was centrally placed in a rectangular processing domain of 36416 x 33148 x 32798 mm (length x height x width) that was created with a function of *Enclosure* (Figure 1). On the walls of the domain, an velocity inlet and an pressure outlet were defined. The components of the gyrocopter were set as wall. The type of calculations was pressure-based in steady conditions. The airflow was as ideal gas. A turbulence model was *k-ω* SST which uses two equations, i.e. one of them describes turbulence kinetic energy *k* and the other one describes a specific dissipation of kinetic energy *ω*. This model is also capable of restricting shear stress transport (SST).

Turbulence intensity was assumed as 1%, a turbulence viscosity coefficient as 5, air temperature as 288 K (15 °C), dynamic viscosity as $1.7894 \cdot 10^{-5}$ kg/ms, and air velocity as 28 m/s (approx. 100 km/h). A mesh of the model was created of tetrahedrons with an advanced function of Curvature (Figure 3). On the surface of the entire gyrocopter, a wall layer was created by means of Inflation and Smooth Transition. There was defined 7 layers of a 1.15 growth rate. The mesh effect was investigated. The domain was calculated with different number of elements. The number of elements and elements size which does not affect the results of calculations has been found. The mesh consisted of approx. 6.7 million elements.

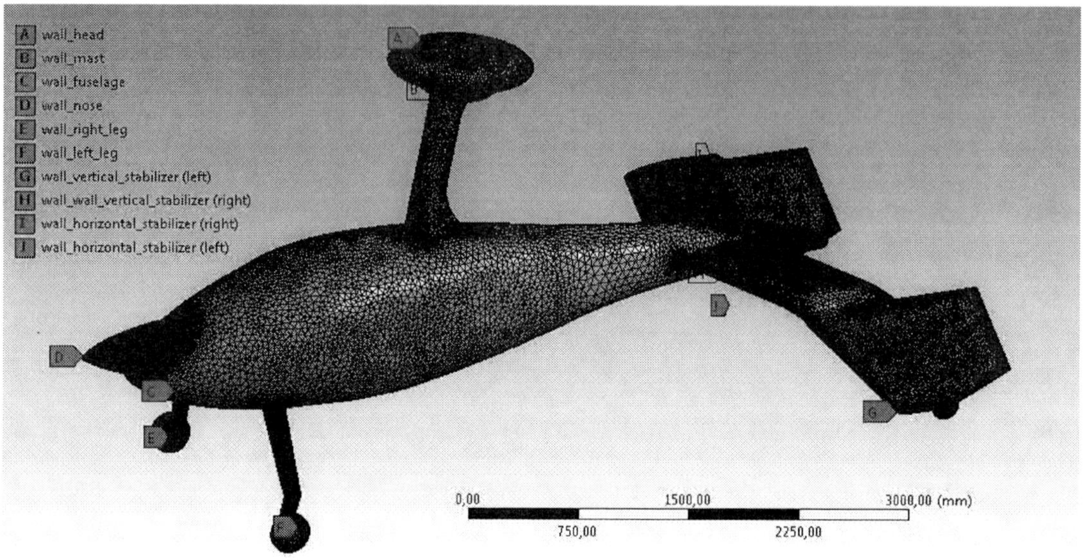

Figure 3. Generated mesh – a general view.

For this model of the gyrocopter, 55 measuring points were defined by an angle of attack α and horizontal stabilizer angle γ. The angle of attack ranged from -25° to + 25°, every 5° and the horizontal stabilizer angle ranged from -10° to + 10°, every 5°. For each measuring point, the drag, lift and pitching moment that acted on the gyrocopter components were recorded to calculate corresponding aerodynamic coefficients from formulas (1) and (2):

$$C_{x,z} = \frac{P_{x,z}}{0.5\rho v^2 \pi R^2} \tag{1}$$

$$C_{my} = \frac{M_y}{0.5\rho v^2 \pi R^3} \tag{2}$$

where: $\rho = 1.226$ kg/m^3 – air density, $v = 28$ m/s – airflow velocity, $R = 5$ m – rotor radius.

The stability criterion is the case when the total pitching moment acting on the aircraft is negative when disturbance occurs (3):

$$\frac{\delta M_y}{\delta \alpha} < 0 \tag{3}$$

Static longitudinal stability was examined according to this stability criterion.

3. Results and discussion

The numerical calculations gave the values of drag, lift and pitching moment as a function of angle of attack at different horizontal stabilizer angles applied to individual gyrocopter components. Then, the total values of the aerodynamic forces, pitching moment and the corresponding aerodynamic coefficients were calculated. Additionally, the pressure contours and velocity contours in the plane of symmetry of the research gyrocopter for the given angles and settings of the horizontal stabilizer were plotted.

Figure 4 shows the total drag acting on the research gyroplane (a) and the drag coefficient (b). If a horizontal stabilizer is added, the drag at high angles of attack significantly increases. Its lowest value, i.e. approx. 200 N, regardless of a horizontal stabilizer angle, occurs near an angle of attack equal to 0. Increased and decreased angles of attack result in higher drag. The highest drag, i.e. more than 800 N is generated at extremal angles of attack, i.e. -25° and + 25°.

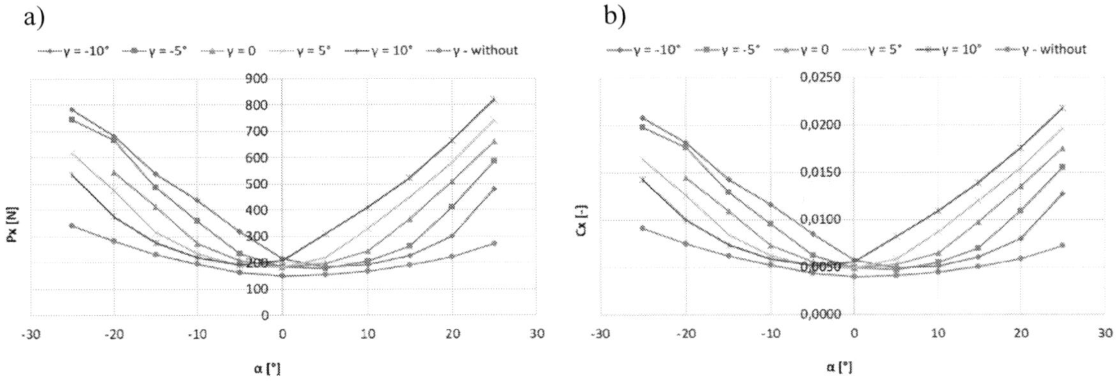

Figure 4. Total drag (a) and drag coefficient (b) versus the angle of attack at the given horizontal stabilizer angles.

Figure 5 shows the total lift acting on the research gyrocopter and the lift coefficient. If there is no horizontal stabilizer, lift increases approximately linearly as the angle of attack increases. If a horizontal stabilizer mounted, there are three areas where the lift increases as the angle of attack increases. The nature of this growth changes according to the value of the angles of attack and horizontal stabilizer angles. For each of the horizontal stabilizer angles, an area of a linear correlation between the lift and an angle of attack can be distinguished. As horizontal stabilizer angles increase, this area shifts towards lower angles of attack. The angle of lift equals zero behaves in a similar way. The largest positive lift, i.e. approx. 1200 N occurs at the largest angle of attack, whereas the highest negative lift, i.e. approx. - 1000 N occurs at the smallest angle of attack.

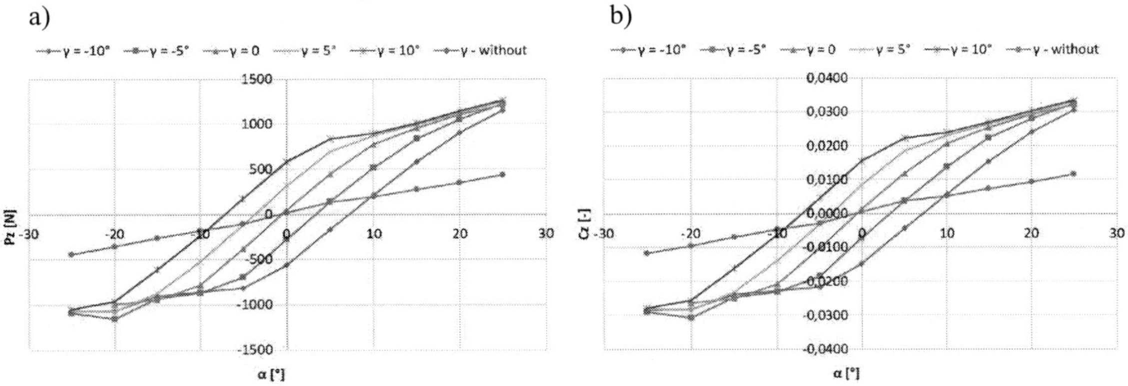

Figure 5. Total lift (a) and the lift coefficient (b) versus the angle of attack at the given horizontal stabilizer angles.

Figure 6 shows the pitching moment acting on the research gyrocopter and the pitching moment coefficient. The research in [1] shows that at high absolute angles of attack the horizontal stabilizer has the largest share in generating the pitching moment. If there is no horizontal stabilizer, the pitching moment increases roughly linearly as the angle of attack increases and is 0 at an angle of attack of approx. 4°. The stabilizer reverses this correlation and three areas with a decreasing pitching moment as an angle of attack increases occur. The nature of this decrease changes according to the angle of attack and the horizontal stabilizer angle. For each of the horizontal stabilizer angles, just as it is for lift, an area of the linear correlation between the pitching moment and an angle of attack can be distinguished. As the horizontal stabilizer angle increases, this area shifts towards smaller angles of attack. The angle corresponding to the maximum pitching moment (positive) shifts towards smaller angles as the horizontal stabilizer angle decreases. The minimal pitching moment (negative) occurs at the largest angle of attack regardless of the horizontal stabilizer angle. The contour of the pitching moment is similar to the stabilizer angle ranging between -5° and -10° at the angle of attack less than -10° because at an horizontal stabilizer angle of -10° and the given range of the angle of attack, there is probably the phenomenon of air stream detachment, which decreases lift acting on the stabilizer and reduces the pitching moment so stability is lost.

a) b)

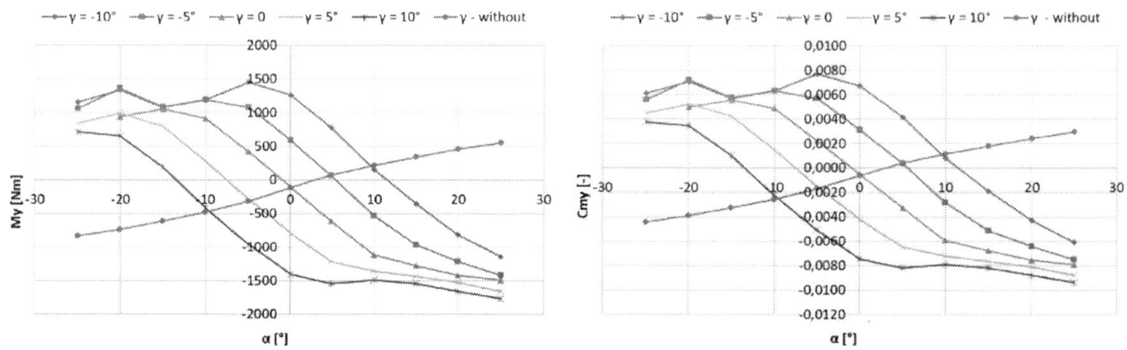

Figure 6. Pitching moment (a) and the pitching moment coefficient (b) versus the angle of attack for the given horizontal stabilizer angles.

The method of least squares enabled the approximating functions of the pitching moment as a function of the angle of attack for each horizontal stabilizer angle. Later, the derivatives of these functions after the angle of attack were calculated and their values at points corresponding to the defined measuring points were calculated. The approximating functions of the pitching moment according to the angle of attack for the given horizontal stabilizer angles γ are given in equations (4)-(8):

$$\gamma = \textbf{-10°}: \quad C_{my}(\alpha) = -9.23921568627466\cdot 10^{-11}\cdot\alpha^6 + 1.38461538461564\cdot 10^{-10}\cdot\alpha^5 + \\ + 1.06396681749624\cdot 10^{-7}\cdot\alpha^4 - 4.20163170163386\cdot 10^{-8}\cdot\alpha^3 - \\ - 4.11777375565614\cdot 10^{-5}\cdot\alpha^2 - 0.000276308857808854\cdot\alpha^1 + 0.00677889757301523 \tag{4}$$

$$\gamma = \textbf{-5°}: \quad C_{my}(\alpha) = -6.31111111111114\cdot 10^{-11}\cdot\alpha^6 - 2.30769230769235\cdot 10^{-10}\cdot\alpha^5 + \\ + 6.87094017094019\cdot 10^{-8}\cdot\alpha^4 + 5.08566433566437\cdot 10^{-7}\cdot\alpha^3 - \\ - 0.0000253538150738151\cdot\alpha^2 - 0.00049215034965035\cdot\alpha^1 + 0.00348181818181818 \tag{5}$$

$$\gamma = \textbf{0°}: \quad C_{my}(\alpha) = -1.77777777777745\cdot 10^{-12}\cdot\alpha^6 - 4.20512820512804\cdot 10^{-10}\cdot\alpha^5 - \\ - 5.82905982906027\cdot 10^{-9}\cdot\alpha^4 + 8.88461538461527\cdot 10^{-7}\cdot\alpha^3 + \\ + 8.51732711732826\cdot\alpha^2 - 0.000599151515151513\cdot\alpha^1 - 0.000524475524475529 \tag{6}$$

$$\gamma = \textbf{5°}: \quad C_{my}(\alpha) = 1.82483660130736\cdot 10^{-11}\cdot\alpha^6 - 1.58974358974371\cdot 10^{-10}\cdot\alpha^5 - \\ - 3.56264454499766\cdot 10^{-8}\cdot\alpha^4 + 4.52972027972038\cdot 10^{-7}\cdot\alpha^3 + \\ + 0.0000186614031719918\cdot\alpha^2 - 0.000487238927738929\cdot\alpha^1 - 0.0043563554092966 \tag{7}$$

$$\gamma = \textbf{10°}: \quad C_{my}(\alpha) = 3.05359477124189\cdot 10^{-11}\cdot\alpha^6 + 3.89743589743589\cdot 10^{-10}\cdot\alpha^5 - \\ - 5.00130718954254\cdot 10^{-8}\cdot\alpha^4 - 2.24475524475524\cdot 10^{-7}\cdot\alpha^3 + \\ + 0.0000265526733397323\cdot\alpha^2 - 0.000276240093240093\cdot\alpha^1 - 0.00730843274372687 \tag{8}$$

Figure 7-8 depict the examples of the pressure contours and velocity contours on the plane of symmetry of the research gyrocopter at the extremal angle of attack and horizontal stabilizer angle.

a) b)

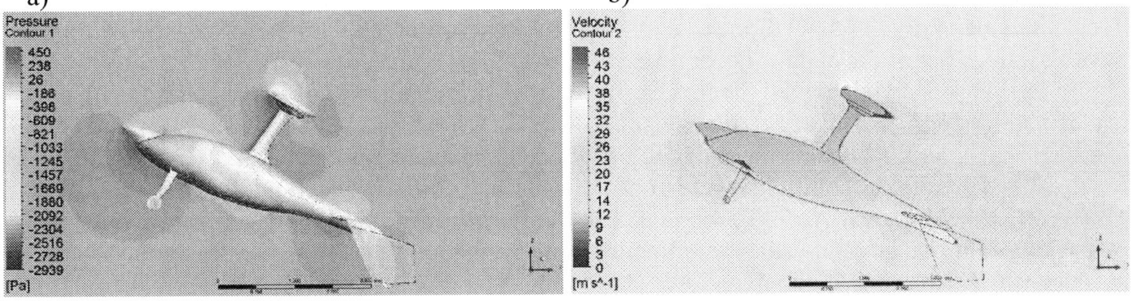

Figure 7. Pressure contour (a) and velocity contour (b) on the plane of symmetry of the gyrocopter at $\alpha = +25°$, $\gamma = -10°$.

a) b)

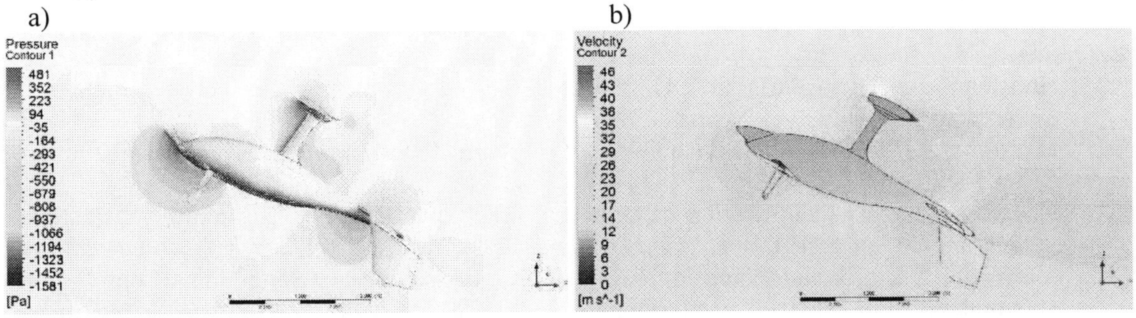

Figure 8. Contour of pressure (a) and contour of velocity (b) on the plane of symmetry of the gyrocopter at $\alpha = +25°$, $\gamma = +10°$.

Figure 9 shows derivatives δMy/δα according to angles of attack α and horizontal stabilizer angles γ. The positive derivative was marked with a red color.

a) b)

α(°)	γ (°)				
	-10	-5	0	5	10
-25	7.4E-04	6.8E-04	6.7E-04	2.8E-04	7.3E-05
-20	-2.0E-04	-3.9E-05	3.2E-04	-2.8E-05	-2.8E-04
-15	-5.0E-05	-8.7E-05	-4.5E-05	-3.8E-04	-5.9E-04
-10	1.7E-04	-8.1E-05	-3.5E-04	-6.0E-04	-6.7E-04
-5	8.1E-05	-2.3E-04	-5.4E-04	-6.2E-04	-5.3E-04
0	-2.8E-04	-4.9E-04	-6.0E-04	-4.9E-04	-2.8E-04
5	-6.4E-04	-6.8E-04	-5.3E-04	-2.8E-04	-5.1E-05
10	-7.4E-04	-6.2E-04	-3.6E-04	-1.2E-04	2.5E-05
15	-4.9E-04	-3.3E-04	-1.7E-04	-6.0E-05	-6.9E-05
20	-2.3E-04	-9.4E-05	-5.6E-05	-1.1E-04	-1.9E-04
25	-9.1E-04	-6.6E-04	-1.8E-04	-1.7E-04	5.5E-05

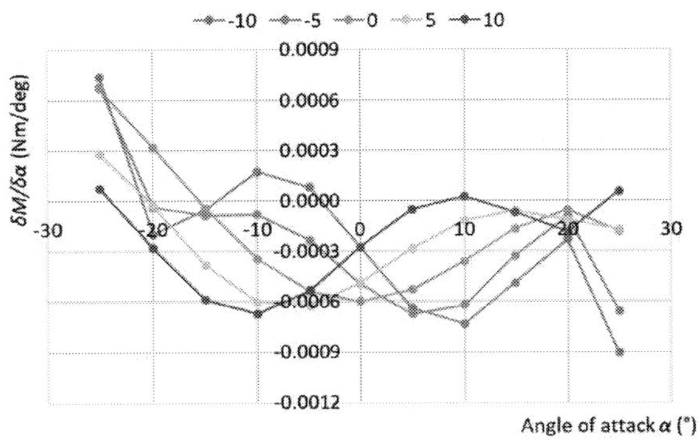

Figure 9. Derivatives δMy/δα versus angles of attack α and horizontal stabilizer angles γ: (a) and (b).

4. Conclusions

This paper discusses the static longitudinal stability of the gyrocopter at different horizontal stabilizer angles and angles of attack. The drag, lift and pitching moment at the given stabilizer angles and angles of attack were achieved from the numerical calculations of the created model with Ansys Fluent. The investigation focused on how the aerodynamic characteristics of the research gyrocopter are impacted by the horizontal stabilizer angles. The aircraft stability criterion based on the derivative of the pitching moment enabled us to examine the stability of the research gyrocopter. The calculations enabled us to identify angles of attack at which the research object lost its stability. For all of the examined horizontal stabilizer angles, derivative δMy/δα was positive, i.e. stability was lost at angle $\alpha = -25°$. In addition, at $\gamma = 0°$, this derivative remained positive at the next angle of attack $\alpha = -20°$. There were four more positive values out of the angles, i.e. at $\alpha = -10°$ and -5° if $\gamma = -10°$, at $\alpha = 10°$ and 25° if $\gamma = 10°$ (see Figure 12). This is mainly due to the loss of lift force on the stabilizer. For the angle of attack $\alpha = -10°$, $\alpha = -5°$ and $\gamma = -10°$ the real angle of attack of the stabilizer relative to airflow equals respectively -15° and -20°. Then the lift force on the stabilizer is reduced. An analogous situation occurs for the $\gamma = 10°$. It should be noted that stability is lost not only at the given points but within a range of angles. Figure 12 (right) depicts derivatives δMy/δα versus the angle of attack and this correlation enables us to estimate stability ranges and values at which the object remains neutrally stable. It is important to have as much stability margin as possible; however, too high stability is adverse to aircraft maneuverability.

Acknowledgements
This work has been financed under Grant Agreement No. POIR.01.01.01-00-0056/16.

References
[1] Houston S and Thomson D 2017 On the modelling of gyroplane flight dynamics. *Progress in Aerospace Sciences* vol 88 pp 43-58
[2] Houston S S 1996 Longitudinal stability of gyroplanes *The Aeronautical Journal* vol 100 no 991 pp 1-6
[3] Houston S S 1998 Identification of autogyro longitudinal stability and control characteristics *Journal of Guidance, Control, and Dynamics* vol 21 no 3 pp 391-399
[4] Ma Y Cai Z Liu N and Wang Y 2016 System composition and longitudinal motion control simulation of vehicular towed autogyro *Guidance, Navigation and Control Conference* pp 1018-1023
[5] Xiang C Wang X Ma Y and Wang Y 2014 Modeling and modal responses analysis of an unmanned small-scaled gyroplane *Guidance, Navigation and Control Conference* pp 262-268
[6] Czyż Z Karpiński P Łusiak T and Szczepanik T 2017 Numerical analysis of the influence of particular autogyro parts on the aerodynamic forces *ITM Web of Conferences* vol 15 p 07008
[7] Czyż Z Ilhan I Akcay M and Czarnigowski J 2017 Air flow analysis around the autogyro fuselage *Journal of Technology and Exploitation in Mechanical Engineering* vol 3 no 1 pp 13-20
[8] Stalewski W 2017 Improvement and optimisation of gyroplane performance *Prace Instytutu Lotnictwa*
[9] Figat M 2017 Aerodynamics analysis of the main rotor influence on the static stability of the gyroplane *Aircraft Engineering and Aerospace Technology* vol 89 no 5 pp 663-670
[10] Stalewski W 2016 Aerodynamic design of modern gyroplane main rotors *Prace Instytutu Lotnictwa.*
[11] Lin Q Cai Z and Wang Y 2014 Design, model and attitude control of a model-scaled gyroplane *Guidance, Navigation and Control Conference* pp. 1282-1287
[12] Grabowski L Czyz Z and Kruszczynski K 2014 Numerical Analysis of Cooling Effects of a Cylinders in Aircraft SI Engine *SAE Technical Paper* 2014-01-2883
[13] Pietrykowski K Tulwin T 2015 Aircraft Radial Engine CFD Cooling Model SAE INTERNATIONAL JOURNAL OF ENGINES vol 8 no 1 pp 82-88
[14] Coton F N and Smrcek L 1998 Aerodynamic characteristics of a gyroplane configuration *Journal of Aircraft* vol 35 no 2 pp 274-279
[15] Krzysiak A 2017 Wind tunnel tests of quad-rotor autogyro model *Journal of KONES* vol 24 no 1 pp 231-238
[16] Czyż Z and Stryczniewicz W 2018 Investigation of Aerodynamic Interference in a Multirotor by PIV Method *Advances in Science and Technology. Research Journal* vol 12 no 1 pp 106–114
[17] Czyż Z Magryta P and Szlachetka M 2015 Experimental investigation of the impact of flight speed on drag force in the autogyro model. *Advances in Science and Technology. Research Journal* vol 9 no 26 pp 89-95

Passive Skin Friction Control Near Turbulent Separation – Preliminary Results

A Dróżdż[1] , W Elsner[1] and D Sikorski

[1] Faculty of Mechanical Engineering and Computer Science, Częstochowa University of Technology, Institute of Thermal Machinery, Częstochowa, Poland

E-mail: arturdr@imc.pcz.pl

Abstract. The paper deals with experimental analysis of passive skin friction control method applied for a high Reynolds number turbulent boundary layer close to separation. A proposed new method is based on the mechanism of scale interaction observed in the wall bounded flows known as amplitude modulation of small scales by large-scale motion. The applied control technique introduces artificially amplification of the modulation mechanism using a corrugated surface in the direction of flow at specific ratio of surface waving parameters. Preliminary results show that the method can be use in an efficient way to postpone the turbulent boundary layer separation for high Reynolds number, where methods based on classical turbulizers fail. This method introduces reorganization of turbulence production, which introduces asymmetry between sweep and ejection events. This results in positive skewness of streamwise fluctuations and additionally increases the momentum near the wall. The momentum is increased only in the inner part of boundary layer. The persistent of the developed flow control method is permanent farther downstream the corrugated wall.

1. Introduction

A broad variety of control methods aiming at the reduction of the global drag has been investigated in the past. Most of them do not offer real opportunities for practical implementation or are not efficient enough, what applies, in particular, to active control methods [1][2]. From practical point of view more valuable are passive control methods because of zero energy expenditure in comparison with energy consuming active methods.

The selection of appropriate methods for flow control concerns in particular high Reynolds number turbulent boundary layers (TBLs) that are under the influence of adverse pressure gradient conditions leading to separation. The majority of flow control studies concerning skin-friction drag reduction have attempted to manipulate structures within the near-wall region [2]. In case of turbulent separation it should be emphasized that, unlike the methods aimed at skin friction drag reduction, the method that increases skin friction should be used. This may be achieved by some form of roughness as it is well established that vortex generation caused e.g. by surface roughness can by-pass laminar-turbulent transition and delay flow separation. However, for high Reynolds numbers (i.e. above chord Reynolds number, $Re_c = 10^5$) using any kind of turbulizators are ineffective [1]. That is why there is a need to search for new methods. It is well known that turbulent coherent structures play a crucial role in the momentum transport of wall-bounded turbulent flow and generation of skin-friction drag. So, the alternative method can be based on the newly discovered phenomenon of amplitude modulation imposed by large-scale structures. The detailed description of this phenomenon has been presented in the paper

of Mathis et al. 2009 [3].The large-scale motions are most energetic in the outer region of a high-Reynolds-number or of an APG TBLs, but they also have an influence in the inner-region. For relatively high Reynolds number TBL, the large-scale motion causes the sequential increase and decrease of flow velocity near the wall and in areas of high-speed fluid (increase of mean shear) the higher small scales amplitude and the higher convection velocity are observed [4][5].
Studies presented in the paper concern a new passive method of controlling TBL with detachment being the result of strong adverse pressure gradient (APG). The motivation of the study originates from the work of Dróżdż and Elsner 2017 [4], who observed the increase of convection velocity of small scale structures being the effect of amplitude modulation phenomenon. The physical mechanism standing behind the rise of total convection velocity is the specific forcing of the near-wall small-scale turbulence by large scales in the regions of locally elevated velocity. This specific forcing enhances not only the energy but also convection velocity[4], which induces increase in momentum near the wall. This phenomenon is amplified as a function of the Reynolds number which to some degree was confirmed in Ref. [6].

The idea of the proposed concept is to artificial stimulate turbulent structures near the wall by appropriately designed wall topology e.g. stationary wavy surface (corrugated wall), aimed at enhancement the amplitude modulation mechanism. The suitable corrugated surface can affect the small-scales in the same manner as the large-scale structures i.e. increasing their activity in high speed zones (crests) forcing increased convection and thus increasing the wall-shear stresses. The random in nature turbulence production will be in this way ordered resulting in the assumed increase of wall-shear stresses. Effect of wavy wall on turbulent flow was already studied experimentally among the other by Segunda et al. [7], where the impact of wavy wall periods number on the recirculation zone in consecutive periods were experimentally analysed and turbulence modelling were evaluated. Recent papers by Hamed et al. [8][9] shows the influence of ratio of amplitude of waviness to incoming boundary layer thickness for the 2D and 3D wavy wall and conclude that there is lower modulation of boundary layer parameters for 3D wavy wall. However, there is no systematic optimisation analysis of this problem in the literature, in particular no attempt to assess the ratio of amplitude to period for high Reynolds number flow. The more so, there is also no data available on the analysis of the effectiveness of surface modification for a strong pressure gradient flows and close to separation.

2. Experimental setup and apparatus

The experiment was performed in an open-circuit wind tunnel, where the TBL was developed along the flat plate, which was 5035 mm long, what allows to reach boundary layer thickness, $\delta \approx$ 80mm at the inlet to the test section. The inlet rectangular channel has two pairs of suction gaps aimed, at maintaining overpressure, to control the two-dimensionality of the flow by minimizing the boundary layers on the side walls. Triangular corner inserts were used in the whole inlet channel to reduce the effect of secondary vortices developing along rectangular channel. A slight inclination of the upper wall helped maintain zero pressure gradient ($\frac{dP_\infty}{dx} \approx 0$, where P_∞ is external static pressure and x is the streamwise direction) conditions at the inlet. The wavy wall was introduced on the distance 650 mm just before the inlet plane. The specially design diffuser shape test section with a length of 1.835 m (see Figure 1) equipped with a perforated movable upper wall allows to generate on the bottom wall the turbulent boundary layer, which is at the verge of separation. Wall perforation of 10.1% was adopted, characterized by 0.5mm circular holes. Modification of the shape and position of the upper wall and the suction flux allows for generation of a wide range of pressure gradient conditions, while the zero pressure gradient conditions were maintained at the inlet channel. With specific pressure conditions, it is possible to generate, on the bottom wall, a turbulent boundary layer which is at the verge of separation. Full separation on the lower flat plate does not occur even for the suction case. After that point, due to the cessation of suction, the flow returns to the attached state.

The velocity measurements were performed with hot-wire anemometry Dantec Dynamics Streamline Pro. A single hot-wire probe of a diameter d = 3 μm and length l = 0.4 mm was used. The hot-wire bridge was connected to a 16 bit A/D converter. Acquisition was maintained at the frequency of 25 kHz, with minimum 50 s sampling records. The ambient conditions were carefully controlled during the

IOP Conf. Series: Journal of Physics: Conf. Series **1101** (2018) 012004 doi:10.1088/1742-6596/1101/1/012004

measurements. To have the verified reference friction velocity u_τ along the flow the fringe skin friction (FSF) technique was applied introduced by Tanner and Blows in 1976 [10], who related the evolution of oil droplet thickness to skin friction τ_W. For the purpose of skin friction measurement, the optical equipment was installed in the wind tunnel under the glass plate (see Figure 1a). It consisted of remotely controlled commercial camera equipped with Macro lens and SOX Whitecroft Lighting sodium lamp, emitting the monochromatic light of the wavelength $\lambda = 0.5893$ μm used to illuminate the oil droplet. For the current measurements the OM50 silicone oil with the viscosity of about 0.048 Pas was used. The accuracy of wall shear data was in the range of 1%. In the course of a single profile measurement the scatter of ambient temperature at the end of the test section did not exceed ±0.2° and the temperature difference between the flowing air and the wall was also below 0.1°C. Further details of the methodology and post processing was presented in Ref. [11]. The level of uncertainty was equal 1% of the initial value 0.2 Pa. High accuracy was obtained using at least 40 pictures taken each time of the measurement and at least 30 of them were processed. In order to obtain maximum of skin friction downstream the corrugation the measurement of skin friction using OFI method was performed in location 800 mm (150 mm downstream the corrugation) by changing sing wave amplitude.

a)

b)

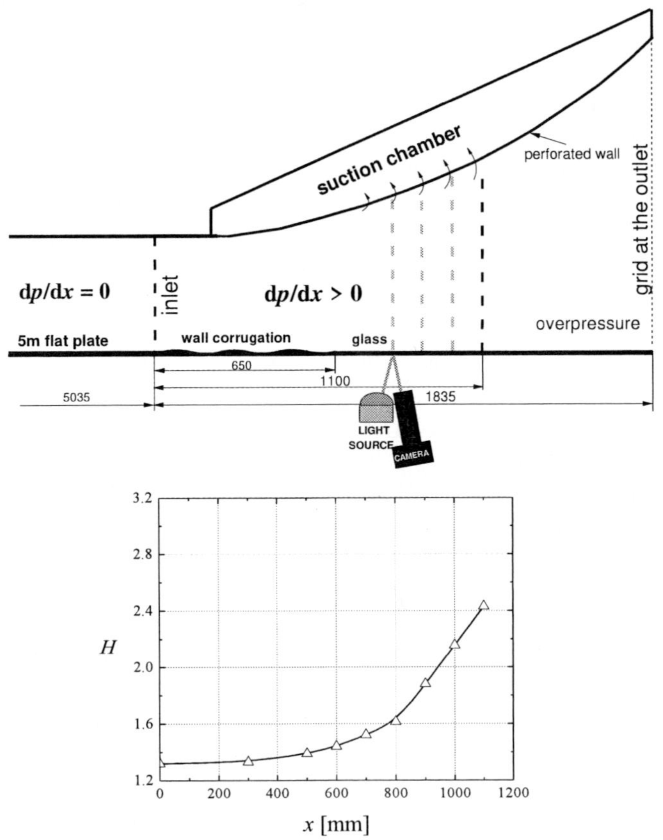

Figure 1: Test section and oil film experimental setup a); shape factor H distribution b).

Basic inflow parameters have been summarized in Table 1, where Tu is turbulence intensity, U_{in} mean velocity outside of TBL, u_τ friction velocity $\left(u_\tau = \sqrt{\frac{\tau_w}{\rho}}\right)$, θ momentum loss thickness and $Re_\theta = \frac{U_{in}\theta}{\nu}$ is Reynolds number, where ν is the kinematic viscosity. The pressure distribution imposed by the upper wall and the active suction is presented in figure 1b. Further details of the experiment can be found in Ref. [6].

Table 1. Inlet conditions of ZPG turbulent boundary layer

Tu [%]	U_{in} [m/s]	u_τ [m/s]	θ [mm]	Re_θ [-]
0.7%	20	0.72	10.3	10150

As the corrugation, the sine wave was used. The corrugation was introduced at the beginning of adverse pressure gradient on the distance 650 mm (see Figure 1a). The 100 mm wide 2D wavy wall was introduced in the center of 250 mm wide flat plate then the side edges of the flat plate and wavy surface ware rounded by the elastic tape. To characterize the local flow conditions, the shape factor $H = \delta^*/\theta$, where δ^* is displacement thickness as a function of the streamwise position for the reference case (flat wall) is shown in Fig. 2b. The parameter has a constant value slightly above 1.4 up to $x = 500$ mm, when it starts to increase its value. However, a sharp almost linear increase of H can be observed from $x \approx 800$ mm, indicating the strong process of the flow destabilization. For streamwise distance of 1100 mm, the so called intermittent transitory detachment (ITD) is reached.

3. Results

The measurements were performed for the inlet Reynolds number $Re_\theta = 10150$ which is equivalent to Reynolds number based on friction velocity equal 3300. Four different wavy surfaces was used during step by step optimisation characterised by different A/λ ratio, where A is the amplitude and λ is the wave period, in the following order of A/λ = 0.05 then 0.08, 0.16, and the last was 0.12. On each step the skin friction was estimated using oil-film interferometry technique and the velocity profile using hot wire was measured. The results presented in the Figure 2 show the wall shear stress distribution as a function of A/λ measured 150 mm downstream the corrugated surface. The initial increase of wall shear stress with the increase of A/λ ratio is observed (Figure 2). A significant rise, by 42.5%, of the wall shear stress occurs in the range up to A/λ equal to 0.135. It is clearly visible that this effect disappears for higher values of the reduced amplitude. It should be noted that the wall stress zero point at $A/\lambda = 0.24$ has been entered into the graph to ensure approximation of the distribution using 3rd order polynomial. The above results show that the applied method, by appropriate selection of the geometrical parameter, allows to obtain a significant increase of wall shear stresses and thus affect the delay of turbulent detachment.

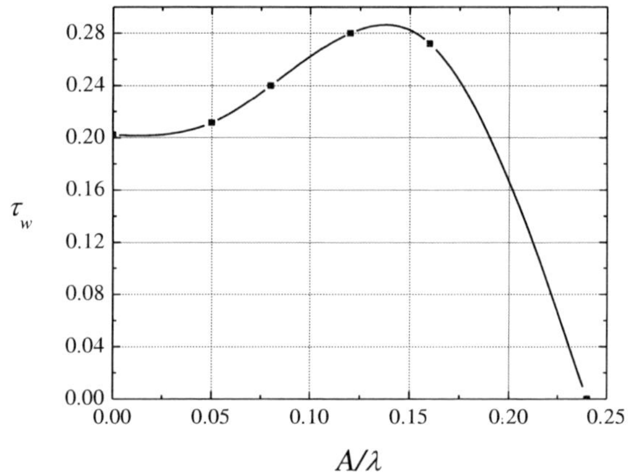

Figure 2. Wall shear stress downstream of corrugation wall as a function of A/λ

In order to assess how the corrugated wall affect the boundary layer, it was decided to analyse the mean and fluctuation velocity profiles as well as skewness factor at a few distances downstream the modification of the plate surface for the parameter $A/\lambda = 0.12$, for which the response of the boundary layer was the strongest. Figure 3 presents the comparison of mean velocity profiles measured 800, 900, 1000 and 1100 mm from the inlet in each case compared with the reference data taken on flat plate (Fig. 3). In the figure, for both cases, the wall distance was scaled by δ, although for corrugated wall the boundary thickness difference is up to 10% than for the flat plate. The mean profiles where reduced by U_∞, which is the mean velocity outside the boundary layer thickness and it is the same for both cases.

For the first travers (Figure 3a), the significant increase of velocity is observed in the inner region of boundary layer (<0.3δ), which coincides well with the area of the boundary layer where the effect of amplitude modulation is usually observed. This suggests that the amplification of this mechanism for the specific parameters of corrugated wall has been achieved. The analysis of downstream development of boundary layer reveals that this impact is permanent, because the effect of the increase in velocity due to wavy wall is present at each consecutive traverse (Figure 3b). The more so, the area of flow modification increasing also in the direction normal to the wall, what indicate an enhancement of the momentum transport to the wall by strengthening of small scale sweep events.

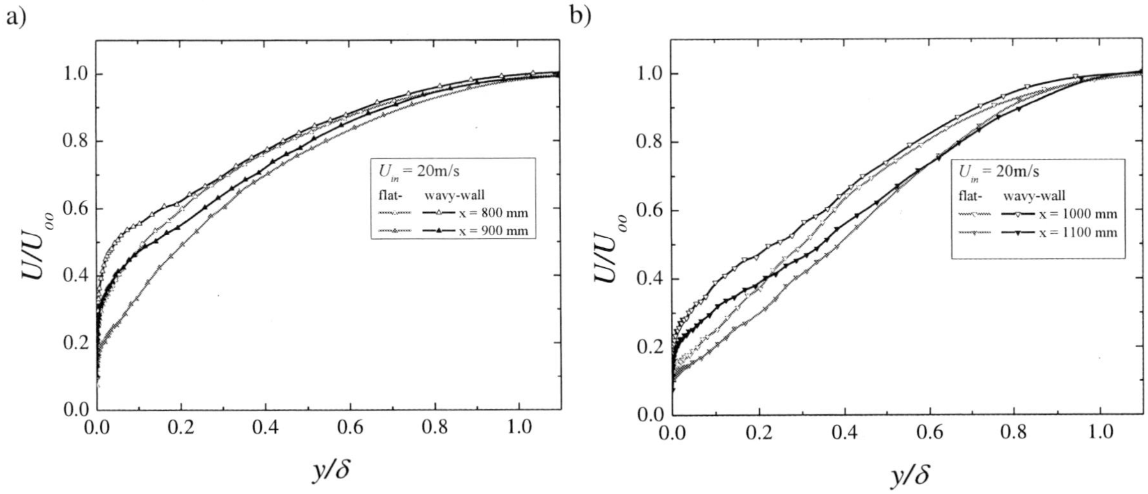

Figure 3. Assessment of the persistence of a disturbance: mean flow velocity downstream corrugation.

In order to confirm this statement, the following figures show the streamwise Reynolds stress scaled on $U_o = 2(U_\infty - U_{0.5\delta})$ and the skewness factor for two chosen location i.e. 800 and 1100 mm from the inlet (Figure 4). As above, the new data were compared with those obtained for a flat plate. The outer scaling by U_o is the best for comparison between flat and wavy wall as it is based on the outer part of mean velocity profile, which is unaffected by the wavy wall (see Fig. 3). When analysing the distributions of Reynolds stress *u'u'* (Fig. 4a) it can be noticed that the outer part of the curve is modified in terms of location of the outer maximum, as for wavy wall (dark profiles) it is located slightly farther from the wall. This is very likely to be the result of the impact of the crest of the wavy surface. However, already above 0.5δ all the profiles perfectly collapse using U_o. It means that the most affected area is the near-wall region, where the increase of *u'u'* occurs due to increased strain rate. Downstream the flow the shift of streamwise Reynolds stress maximum towards the centre of the boundary layer is observed. For corrugated wall the increased *u'u'* stress value is observed below the outer maximum of *u'u'*. The skewness factor analysis (Figure 4b) shows that for location 800 mm the changes are not so substantial as for two lower order moments, however further downstream (x = 1100 mm) the skewness factor is again significantly affected in virtually the entire boundary layer thickness. It means that the proposed method provides long lasting effect on the near wall flow. The confirmation of the impact to be induced is the strong drop of the skewness value, which lead to the detachment delays. The turbulent structures, generated by the increased strain rate in modified momentum zone at location just after corrugation, induces the lower sweep event (in comparison to those for the flat-plate), which is manifested by the lower skewness.

The results of this experiment supports the findings of Segunda et al. [7] in channel turbulent flow, where the recirculation zone in the wavy wall valleys decreases in increasing the number of wave periods. However, the difference between experiments in the literature is that that the observed reduction of recirculation zone is assign to the increased of turbulent kinetic energy in the preceding recirculation zone, where the separation bubble occurs. For the optimum A/λ ratio (which is also function of Reynolds number) the flow is maintained at the edge of occurring of recirculation zone. The effect of increased momentum could be assign either to modulation of convection velocity or/and to production of TKE in preceding recirculation zone.

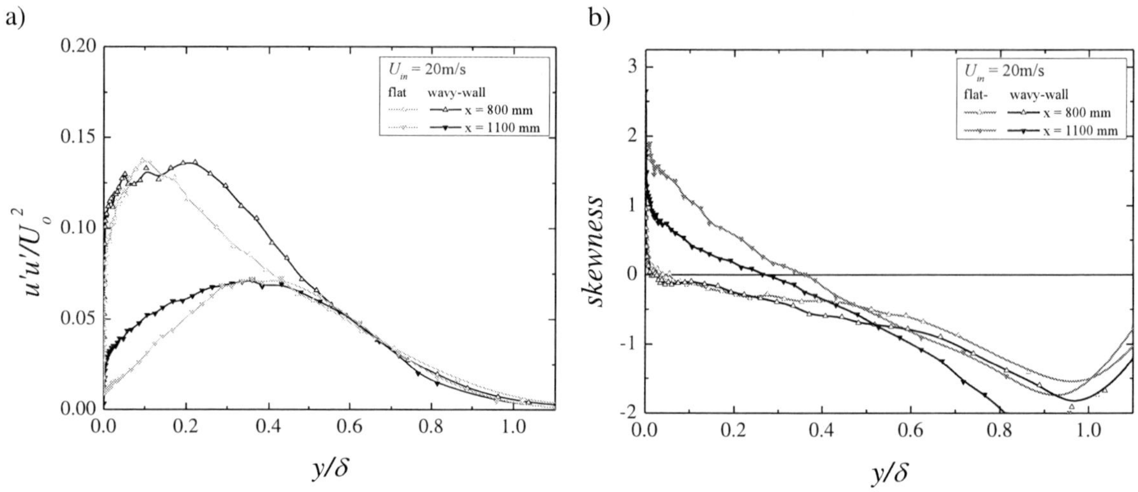

Figure 4. Effect of flow control method persistence: streamwise Reynolds stress a) and skewness factor b) downstream corrugation.

4. Conclusions

It was demonstrated that the for the specific amplitude and the period of wavy wall the maximum of the skin friction value can be achieved when dealing with high Reynolds number turbulent boundary layer in adverse pressure gradient leading to separation. Preliminary results shows that the method can be use in an efficient way to postpone the turbulent boundary layer separation, where classical methods based on turbulizators fail. The optimized wavy wall shape substantially increases the momentum near the wall for high Reynolds number turbulent boundary layer and acts in similar way as the amplitude modulation mechanism leading to increase in convection velocity of turbulence. The significant increase of velocity is observed in the inner region of boundary layer ($<0.3\delta$), which coincides well with the area of the boundary layer where the effect of amplitude modulation is usually observed. Such reorganization of turbulence production introduces dominance of the sweep over the ejection, which is manifested in positive skewness of streamwise fluctuations. Increased sweep enhanced momentum transfer to the wall and cause the higher turbulence production downstream the flow. The proposed method provides long lasting effect on the near wall flow.

5. References

[1] Gad-el-Hak M and Bushnell D M 1991 Separation Control : Review *J. Fluids Eng.* **113** 5–30
[2] Ghebali S, Chernyshenko S I and Leschziner M A 2017 Turbulent skin-friction reduction by wavy surfaces *Phys. Fluids* **29** 105102
[3] Mathis R, Hutchins N and Marusic I 2009 Large-scale amplitude modulation of the small-scale structures in turbulent boundary layers *J. Fluid Mech.* **628** 311
[4] Dróżdż A and Elsner W 2017 Amplitude modulation and its relation to streamwise convection velocity *Int. J. Heat Fluid Flow* **63** 67–74
[5] Dróżdż A and Elsner W 2017 Convection velocity variation as a result of amplitude modulation phenomena *Progress in Turbulence VII* ed P J Örlü R., Talamelli A., Oberlack M. (Springer Proceedings in Physics, vol 196. Springer, Cham) pp 33–8
[6] Dróżdż A and Elsner W 2017 An experimental study of turbulent boundary layers approaching separation *Int. J. Heat Fluid Flow* **68C** 337–47
[7] Segunda V M, Ormiston S J and Tachie M F 2018 European Journal of Mechanics / B Fluids Experimental and numerical investigation of developing turbulent flow over a wavy wall in a horizontal channel *Eur. J. Mech. / B Fluids* **68** 128–43
[8] Hamed A M, Kamdar A, Castillo L and Chamorro L P 2015 Turbulent boundary layer over 2D and 3D large-scale wavy walls *Phys. Fluids* **27** 106601

[9] Hamed A M, Castillo L and Chamorro L P 2017 Turbulent boundary layer response to large-scale wavy topographies *Phys. Fluids* **29**
[10] Tanner L H and Blows L G 1976 A study of the motion of oil films on surfaces in air flow, with application to the measurement of skin friction *J. Phys. E.* **9** 194–202
[11] Dróżdż A, Elsner W and Sikorski D 2018 Skin friction estimation in a strong decelerating flow *J. Theor. Appl. Mech.* **56**

Acknowledgments

The investigations presented in this paper have been obtained with funding from the statutory fund BS/MN 1-103-304/2017

IOP Conf. Series: Journal of Physics: Conf. Series **1101** (2018) 012005 doi:10.1088/1742-6596/1101/1/012005

Influence of wind on the movement of the load

Dawid Cekus[1], Renata Gnatowska[2] and Paweł Kwiatoń[1]

[1] Institute of Mechanics and Machine Design Fundamentals, Czestochowa University of Technology, Czestochowa, Poland
[2] Institute of Thermal Machinery, Czestochowa University of Technology, Czestochowa, Poland

E-mail: kwiaton@imipkm.pcz.pl

Abstract. This paper presents theoretical considerations and numerical calculations concerning wind influence on the movement of the load during working cycle. During the test, the load is treated as a rigid body. The first part of this work includes the analysis of the aluminium cubical model of the load, which could move only in one plane, striving to increase the stability of motion and prevent vibrations. Numerical analysis was performed in Matlab/Simulink program and compared with experimental results, which were carried out in a low speed wind tunnel equipped with PIV system. The study was carried out for two cases: with and without seeding. The second part of this work includes the application of this research method for analysing the laboratory mobile crane's duty cycle. Numerical simulations were carried out on a modified simulation model. The sample numerical results present the trajectory and positions of the load carried on the non-deformable rope taking into account the wind influence.

1. Introduction

The problem of load movement analysis is a complex issue in the field of multi-body dynamics of machines and transport devices. Movement of the carried load has a significant impact on the duty cycle of these devices and the cargo fluctuations can lead, in extreme cases, to loss of stability. This work concerns the analysis of the movement of the load carried by a mobile crane taking into account wind influence. The phenomenon of wind effect is often neglected during the analysis of the cargo motion [1]. This interaction can cause failures of cranes due to wind effect [2, 3]. In the literature and norms there are specific critical wind values at which the transporting device can work. The influence of stability and external factors on its dynamic behavior were discussed in works [4-6].

The movement of the load in the form of a rigid body, taking into account the influence of wind pressure, has been analyzed in this work. In the first part of the work, the load movement was considered as the pendulum motion on a rigid rope [7] with the impact of external forces. The simulation research have been compared with the results of an experiment conducted in a wind tunnel and the PIV method. This method makes it possible to measure and visualize the velocity of the flow field using laser light [8]. In the second part, the equations determined in the first part were used to analyze the crane's duty cycle. The solution of the initial problem of the load movement was numerically calculated using the ode45 procedure in the Matlab/Simulink program. This function implements the Runge-Kutta method with a variable time step for efficient numerical calculation. The necessary condition of using this procedure is to provide system into the form of first order differential equations [9].

IOP Conf. Series: Journal of Physics: Conf. Series **1101** (2018) 012005 doi:10.1088/1742-6596/1101/1/012005

2. Analytical description of the problem

The phenomenon of aerodynamic drag has often been overlooked during research on the transport of loads. The wind influence on the movement of the load can be taken into account by adding additional external force $\mathbf{F}_w$ [10]:

$$\mathbf{F}_w = \frac{1}{2}\rho_{air} A V^2 C_F \tag{1}$$

where: ρ_{air} is the density of air, A is a reference area, V is the wind velocity relative to object and C_F is a dimensionless force coefficient. Simulation model that takes into account the impact of wind during the working cycle is shown in Figure 1a.

The cargo model with the movable local rectangular coordinate system (C) and forces acting on it during duty cycle is shown in Figure 1b. The load is suspended on a non-deformable line ρ_l. In order to define the position of the load during working cycle, it is necessary to determine the orientation of the C-coordinate system with respect to the global system. For this purpose, it was necessary to use Bryant angles, which allow determining the angular velocities and accelerations associated with the mobile coordinate system [4, 5].

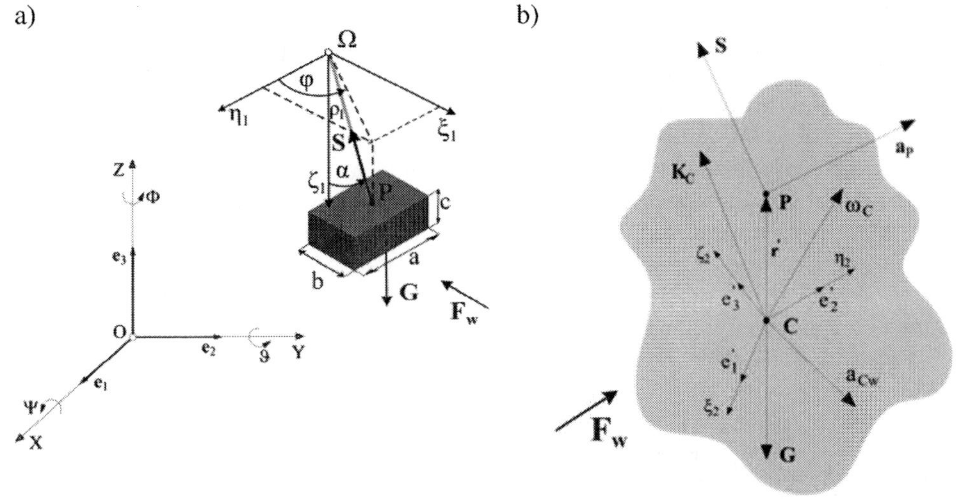

Figure 1. a) Computational model of the load, b) Forces acting on the load

Components of vector $\boldsymbol{\omega}$ in the moving coordinate system can be calculated from the formula:

$$\begin{aligned} \omega_x &\equiv \omega_1 = \dot{\Psi}\cos\vartheta\cos\Phi + \dot{\vartheta}\sin\Phi \\ \omega_y &\equiv \omega_2 = -\dot{\Psi}\cos\vartheta\sin\Phi + \dot{\vartheta}\cos\Phi \\ \omega_z &\equiv \omega_3 = \dot{\Psi}\sin\vartheta + \dot{\Phi} \end{aligned} \tag{2a-c}$$

Rotation about a fixed point can be described by the vector equation [3]:

$$\mathbf{M}_C = \frac{d}{dt}\mathbf{K}_C, \tag{3}$$

where: $\mathbf{K}_C$ - angular momentum vector which can be described as:

$$\mathbf{K}_C = J_C \boldsymbol{\omega}_C \tag{4}$$

where: J_C – moments of inertia matrix, $\boldsymbol{\omega}_C$ – absolute angular velocity of body rotation around its center of mass.

Using the second law of Newton's dynamics, the vector of moments acting on a solid body can be written as [4]:

$$\mathbf{M}_C = \mathbf{r} \times \mathbf{F}, \tag{5}$$

where: **F** – vector sum of the tension in the rope, force of inertia of the load and the wind force.

Dynamic equations of motion for the cargo, treated as a rigid body, can be presented in the form using Einstein's summation convention [4, 5]:

$$M_\alpha = J_{\alpha\gamma}\varepsilon_\gamma + \varepsilon_{\alpha\beta\gamma}\omega_\beta J_{\gamma\kappa}\omega_\kappa, \quad \alpha = 1,2,3 \tag{6}$$

where: **J** is the moments of inertia matrix, and ε_{jkm} is Ricci symbol.

Taking into account the angular velocity and accelerations of the load (2), moment components and their transformations (3), as well as forces acting on the load (4), a system of three second order differential equations is obtained:

$$\mathbf{D}\ddot{\mathbf{X}} = \mathbf{E}. \tag{7}$$

3. Applied research methods

Using the above relationships and introducing them to the simulation model [4], the initial-value problem of the cargo movement has been solved numerically in the Matlab/Simulink program. In the first phase the load motion was treated as the pendulum motion, that could move only in one plane. The load was suspended using a metal rod, which properties correspond to a non-deformable line [7]. The simulation results were compared with experimental results, which were conducted in a low speed wind tunnel equipped with a Particle Image Velocimetry (PIV) system. With use Particle Image Velocimetry can be measured instantaneous velocity distribution, vortex properties and calculate the forces acting on the object. Simulation parameters and material properties are shown in Table 1.

Table 1. Material properties and simulation parameters[a].

Mass of the load	Density of material load's	Length of the line	Load dimensions	Dimensionless wind force coefficient	Wind velocity	Deflection of angle
0.5kg	2800kg/m^3	0.18m	0.06x0.06x0.05m	2.05	0m/s 6m/s 12m/s	7° 21°

[a] Constant and controlled temperature and humidity (for experimental results).

3.1. Simulation results

The numerical simulations were performed using Runge-Kutta fourth-order method. In order to solve the initial problem with this method of the movement, the previous formulas should be reduced to the system of first order differential equations by substitutions [4, 9]:

$$X_1 = \Psi_1, X_2 = \Theta_2, X_3 = \Phi_3,\ X_4 = \dot{\Psi}_1, X_5 = \dot{\Theta}_2, X_6 = \dot{\Phi}_3. \tag{8}$$

The movement of the pendulum with the length of l with the suspended mass m can be represented by the equation [7]:

$$ml\ddot{\theta} = -mg\sin\theta, \tag{9}$$

The simulations were made for two values of deflection of an angle: 7° (Fig. 2) and 21° (Fig. 3). The motion of the lifted load was considered for the time $t = 5$s. Based on the performed test (frequency of vibrations), it can be concluded that the consideration of wind force has been correctly interpreted and implied. Due to the motion analysis in only one direction, only values of one Bryant angle variable were obtained (θ).

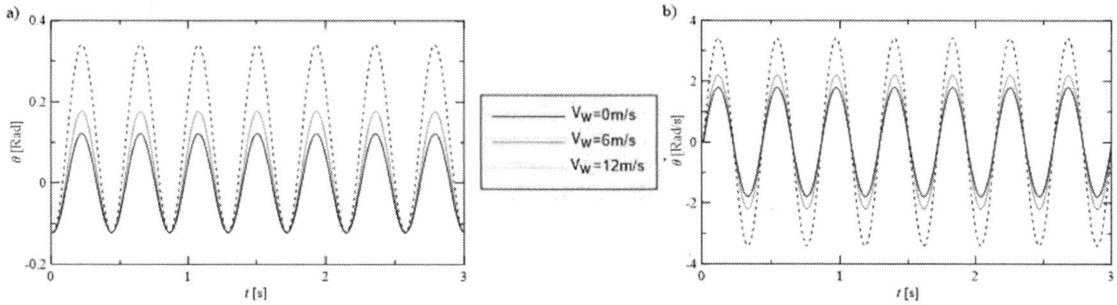

Figure 2. Changes of: a) Bryant angles, b) generalized velocity, depending on the strength of the wind (deflection of angle: 7°)

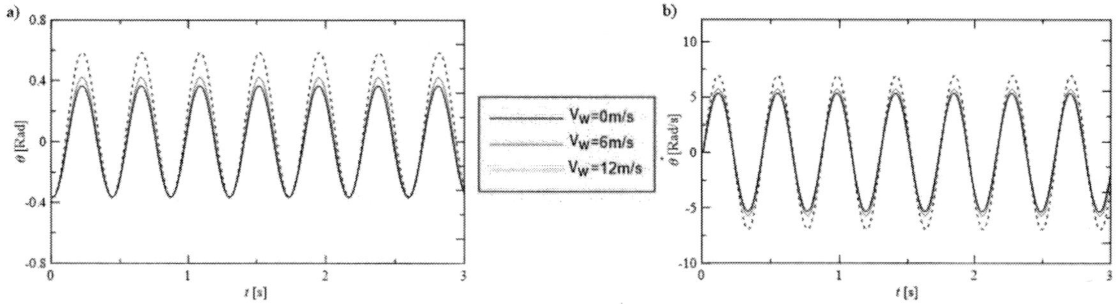

Figure 3. Changes of: a) Bryant angles, b) generalized velocity, depending on the strength of the wind (deflection of angle: 21°)

3.2. Experimental results

The experimental tests, confirms the dynamic of load motion. Differences in results are caused by a different type of analyzed pendulum: mathematical and physical. During the experiments, the aluminum cuboidal load model could move only in one plane, like the pendulum. The PIV measurements were carried out for two inflow velocities 6 m/s and 12 m/s, with acquisition frequency 15 Hz and constant and controlled temperature and humidity. The equipment PIV parameters was described in [11]. Recorded data were processed with the use of the adaptive correlation algorithm and as a result, instantaneous vortices fields were achieved (Fig. 4).

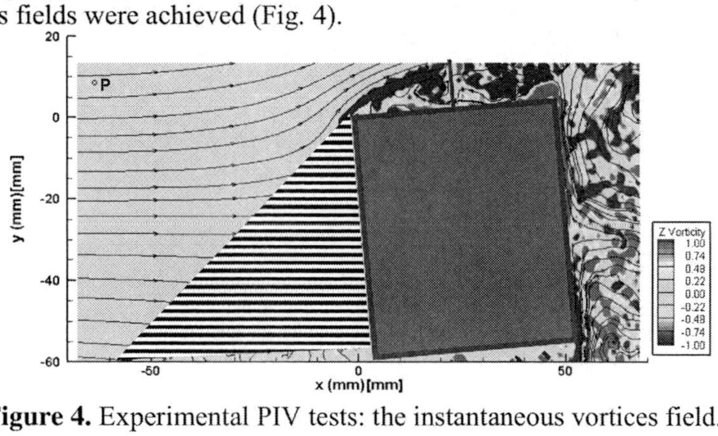

Figure 4. Experimental PIV tests: the instantaneous vortices field.

The load motion is subject to friction and air drag (the cross-sectional area of the string and load), so the amplitude of it swings declines. Figure 5 presents the oscillations of longitudinal and perpendicular velocity components in point P (see Fig. 4) with respect to the time. The position of the point P is in front of the rotating load, which is rotating in plane parallel to the incoming velocity. As the load motion does the back and forth, the velocity is continuously changing. There will be times at which the velocity is a negative value (for moving leftward) and other times at which it will be a positive value (for moving rightward).

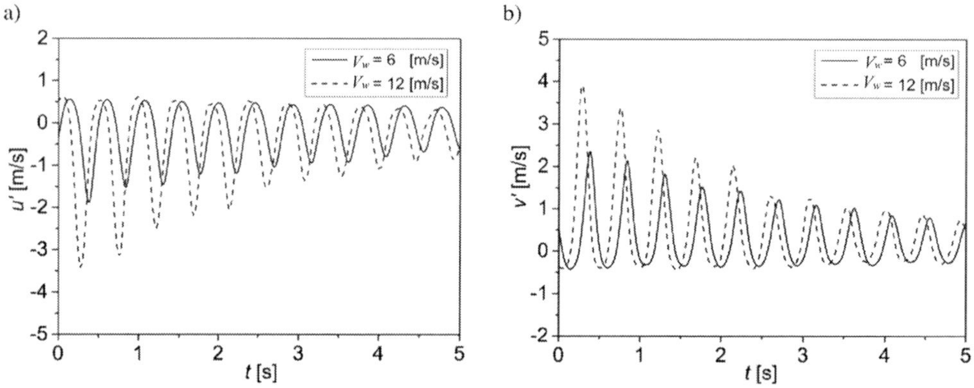

Figure 5. Experimental PIV tests:the oscillations of a) longitudinal and b) transverse velocity components in point P.

4. Application of the research method

The simulation model of a mobile crane was created in SolidWorks environment and transferred to Matlab/Simulink program using CAD process of implementation [4, 12]. SimMechanics models allow for tracking many parameters (trajectory, velocity or acceleration) of any elements of a complex system [12]. Thanks to this approach, it is possible to avoid a number of complex analytical calculations. Bryant angles and their related velocity to the OXYZ coordinate system were used to analyze the working cycle of mobile crane (Figure 6).

Table 2. Material properties and simulation parameters.

Mass of the load	Density of material load's	Diameter of the rope	Length of the rope	Load dimensions	Dimensionless wind force coefficient	Wind velocity
50kg	2800kg/m^3	0.013m	1.5m	0.3x0.3x0.2m	2.05	0m/s, 6m/s 12m/s

IOP Conf. Series: Journal of Physics: Conf. Series **1101** (2018) 012005 doi:10.1088/1742-6596/1101/1/012005

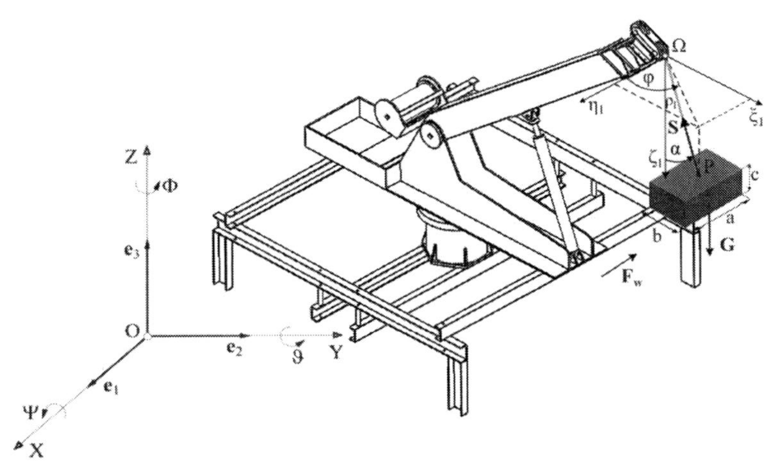

Figure 6. Simulation model of mobile crane

Taking into account the equation of motion determined during the analysis of the pendulum, the system had to be supplemented with additional equations related to the mobile crane motion. System of equations of load motion carried by laboratory mobile crane [4]:

$$\ddot{\xi} = a_{\Omega X} + \frac{\xi}{\rho_l^2}\left[\rho_l\ddot{\rho}_l + \dot{\rho}_l + \xi a_{\Omega X} + \eta a_{\Omega Z} + \zeta\left(a_{\Omega Y} + g\right) - \left(\dot{\xi} + \dot{\eta} + \dot{\zeta}\right)\right]$$

$$\ddot{\eta} = a_{\Omega Z} + \frac{\eta}{\rho_l^2}\left[\rho_l\ddot{\rho}_l + \dot{\rho}_l + \xi a_{\Omega X} + \eta a_{\Omega Z} + \zeta\left(a_{\Omega Y} + g\right) - \left(\dot{\xi} + \dot{\eta} + \dot{\zeta}\right)\right] \qquad (10a\text{-}c)$$

$$\ddot{\zeta} = a_{\Omega Y} - g - \frac{\zeta}{\rho_l^2}\left[\rho_l\ddot{\rho}_l + \dot{\rho}_l + \xi a_{\Omega X} + \eta a_{\Omega Z} + \zeta\left(a_{\Omega Y} + g\right) - \left(\dot{\xi} + \dot{\eta} + \dot{\zeta}\right)\right]$$

The analyzed motion began from zero initial conditions for the time equal 15s. Control function were adopted as a trapezoid pulses and start-up, steady and braking states were included. The working cycle included operation: boom inclination angle (t_{start}=0s, t_{stop}=5s, v_{max}=0.08rad/s), boom length changes (t_{start}=5s, t_{stop}=10s, v_{max}=0.2m/s) and 5s free oscillations of the lifted cargo. The initial problem was solved using ode 45 procedure in Matlab/Simulink enviroment.

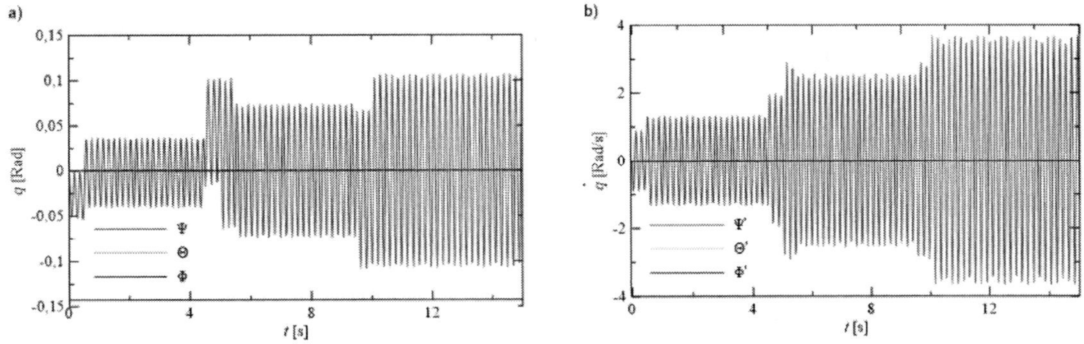

Figure 7. Changes of: a) Bryant angles, b) generalized velocity ($\mathbf{V}_w$=0m/s)

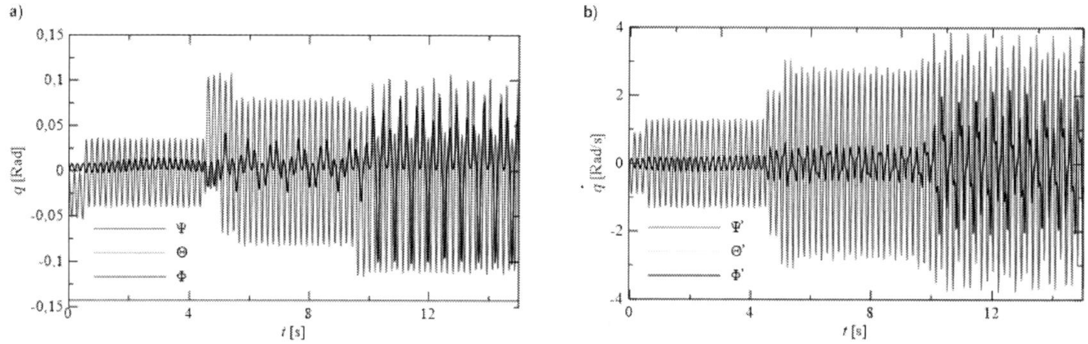

Figure 8. Changes of: a) Bryant angles, b) generalized velocity ($\mathbf{V}_w$=6m/s)

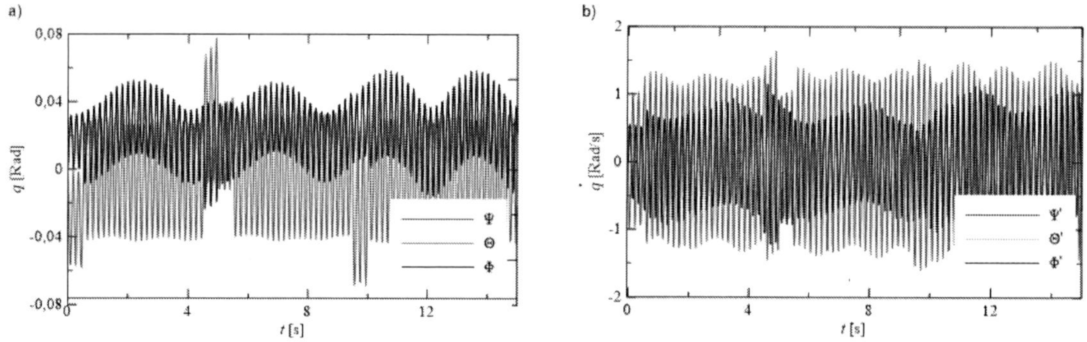

Figure 9. Changes of: a) Bryant angles, b) generalized velocity ($\mathbf{V}_w$=12m/s)

Based on the performed tests, it can be concluded that the wind force has a significant influence on the deflection of the cargo during its transfer and should not be neglected during research on the dynamics of the load. Variations in the trajectory and positioning of the load were also observed depending on the severity of the wind velocity. The greatest deflection of the cargo (when wind velocity is less or equal 6m/s) can be noticed at the free oscillations of the load (t=10s). When wind velocity is equal to 12m/s, the deflection is similar for all kinematic forces (Fig. 9). When the external force $\mathbf{F}_w$ is equal to zero, deflection was noticed only in one direction caused by the kinematics exclusions (Fig. 7).

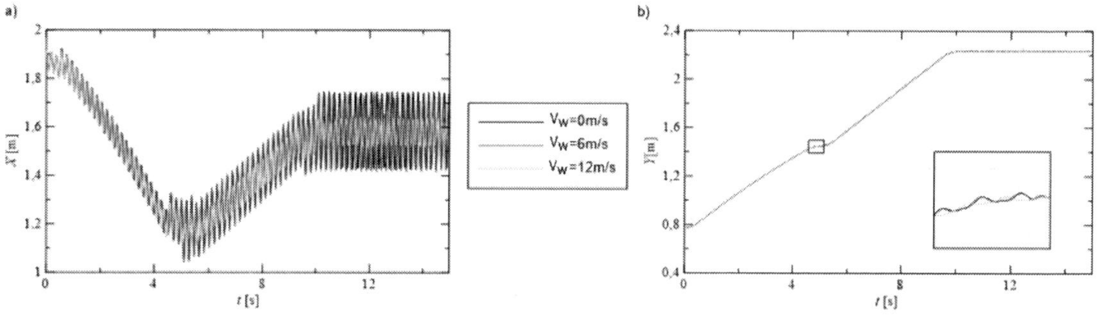

Figure 10. Dependency of wind velocity on trajectory of the load: a) X-coordinate, b) Y-coordinate

5. Conclusion

This paper describes a preliminary theoretical and computational considerations of load movement with wind influence. The cargo was treated as rigid body. First part involved the analysis of a simplified model that was presented as the mathematical pendulum motion. The obtained numerical results were compared with the experimental results. The second part presents the analysis of the mobile crane working cycle, which was the application of the previously determined equation of the load movement and the impact of wind velocity. During the research on the influence of the wind, a simplified model was adopted in which a reference area was assumed as a constant surface of a cuboid. Differences in the trajectory of the load were noted depending on the strength of the wind. The simulation model will be further developed by considering aerodynamic resistance and the friction forces in the joints, and aerodynamic parameters [13]. The presented simulation model may be the basis for the optimization of the load movement and positioning.

Acknowledgements.
The study has been carried out within statutory research of Czestochowa University of Technology

References

[1] Frendo F 2016 *Engineering Failure Analysis* **66** p 479-488
[2] Klinger C 2014 *Engineering Failure Analysis* **43**
[3] Tomczyk J, Cink J and Kosucki A 2014 *Automation in Construction* **42** p 100-111
[4] Cekus D and Kwiatoń P 2018 *Engineering Mechanics* p 137-140
[5] Posiadała B and Tomala M 2011 *Modelling in Engineering* **41** p 323-330 (in Polish)
[6] Kacalak W, Budniak Z and Majewski M 2018 *Int. Journal of Applied Mechanics and Engineering* **23** pp 65-77
[7] Jin Y and Chamorro L P 2017 *Fluid Mech.* **816** p 468-480
[8] Gnatowska R, Sosnowski, M and Uruba V. 2017 CFD modelling and PIV experimental validation of flow fields in urban environments. In *E3S Web of Conferences* Vol. 14, p. 01034
[9] Ahmed W K 2013 *International Journal of Engineerin (IJE)* **7** pp 25-31
[10] Baker C J 1991 *Journal of Fluids and Structures* **5** p 91-111
[11] Sobczyk J, Wodziak W, Gnatowska R, Stempka J and Niegodajew P 2018 Impact of the downstream cylinder displacement speed on the hysteresis limits in a flow around two rectangular objects in tandem–PIV study of the process *Journal of Wind Engineering and Industrial Aerodynamics* **179** p 184-189
[12] Cekus D, Posiadała B and Waryś P 2014 *Archive of Mechanical Engineering* **61** p 57-74
[13] Gnatowska R 2008 Synchronization Phenomena in Systems of Bluff-Bodies. *International Journal of Turbo and Jet Engines*, **25**(2) p 121-128

Aerodynamic assessment of the cycloidal rotor fan

S Dykas[1], M Majkut[1], K Smołka[1], M Strozik[1] , T Staśko[1]

[1] Silesian University of Technology, Institute of Power Engineering and Turbomachinery, Konarskiego 18, 44-100 Gliwice, Poland

E-mail: Krystian.Smolka@polsl.pl

Abstract This paper presents numerical and experimental testing of a cycloidal rotor fan (CRF). The tested cyclorotor is made of four NACA0012 profile blades. CFD calculations using the Ansys CFX® tool and the LDA measurement technique are performed to determine the fan performance for different operating parameters, i.e. rotational speed and the rotor blade maximum swing angle values. One initial position of the blades corresponding to one air-flow direction is considered only. The comparison between the numerical results and the experimental LDA measurements is satisfactory and encouraging further investigation.

Symbols

α – blade actual swing angle
α_0 – blade total swing angle
c – blade chord, m
Δt – time step, s
γ – angle of the blade initial position
R – rotor radius, m
t – time, s
v – velocity, m/s
ω – angular velocity, rad·s^{-1}

1. Introduction

The cycloidal rotor concept is mostly found in lift-based wind turbines (the Darrieus wind turbine) or in propulsion systems of unmanned aircrafts mainly [1-3]. This means that the solution can be applied to major types of turbomachines, such as engines (turbines) or working machines (compressors or pumps). The search for new applications and more and more efficient turbomachines, e.g. fans, encourages implementation of new ideas and designs. One of them is the idea of the cycloidal rotor fan (CRF) characterized by a wide range of rotational speeds and volume flow rates. This type of fan belongs to the family of cross-flow fans, but it has no casing (diffuser). Typical cross-flow fans (CFF's) are widely used in cooling, heating and air conditioning systems, and in automotive as well as industrial applications, e.g. for heating and drying purposes. Most of them have a compact design and generate little noise. CFF's are a particular type of fans having a drum-type rotor with a high span-to-radius ratio and a spiral-type casing. In CFF's the flow field and the performance characteristics are more dependent on the casing shape than the rotor design, which is investigated both experimentally and numerically in [4, 5].

The CRF is an open-rotor type fan and its main features are as follows: an arbitrary change in the flow direction without changing the fan position (only by changing the incident angle of the rotor blades), a big span of the rotor (making it possible to obtain a high volume flow rate or/and a big swept area), a uniform flow along the span (like in most cross-flow fans), a possibility of using the device with any gases (and even liquids, in the pump mode).

The solution can only be analysed through detailed numerical and experimental research. In the first step, a comparison was made between the results of a numerical model created in the Ansys CFX® program and the in-house experimental data. The good agreement between the two methods proves that a numerical model can be used to look for the most efficient solution of the CRF design.

A special script in the Ansys CFX Expression Language (CEL) had to be prepared to develop the numerical model of the air flow in the CRF. The presented experimental investigations draw on the Laser-Doppler Anemometry (LDA) (or the Laser Doppler Velocimetry (LDV)). The technique was developed over 40 years ago and is commonly applied by many researchers around the world in non-intrusive 1D, 2D and 3D point measurements of velocity and turbulence distribution in both free and internal flows. It was implemented in the Institute of Power Engineering and Turbomachinery and has been used there successfully for over 25 years in laboratories intended for the fluid-flow machinery testing, also to measure the flow field in compressors [6, 7].

2. Analysed object

The fan with a cycloidal rotor analysed herein is made of four NACA0012 profile blades with the chord of $c = 50\ mm$. The NACA0012 profile is a fully symmetrical aerofoil with the thickness of 12% of the chord length. The rotor radius totals $R = 70\ mm$ and the initial position angles of the four blades under consideration are $\gamma = 0,\ \pi/2,\ \pi$ and $3\pi/2$, respectively (cf. Fig. 1).

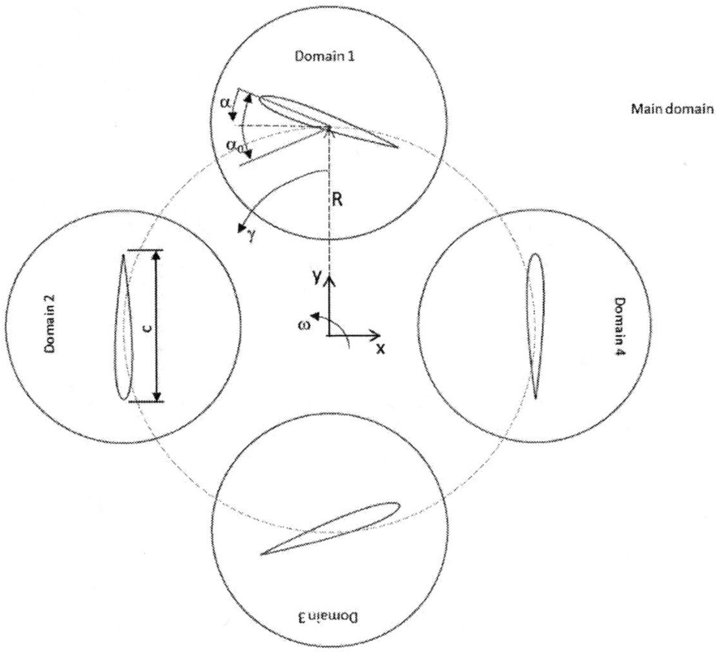

Fig. 1. Configuration of the CRF with four NACA0012 blades

The initial positions of the blades are presented in Fig. 1, where the top blade ($\gamma = 0$) has a positive incident angle and the bottom blade ($\gamma = \pi$) has a negative incident angle. The incident angle of the other two blades (with $\gamma = \pi/2$ and $3\pi/2$) is $0°$. Such initial arrangement of the blades makes it possible to generate the flow from top to bottom, according to the description in Fig. 1. Each blade is swung around the point on its profile camber line in the middle of the chord. For a given rotational speed, the volume flow rate is controlled by the change in the blade total swing angle (α_0). The higher the blade total swing angle, the larger the volume flow should be.

3. Research methodology

3.1. CFD model

The numerical simulation of the unsteady flow in the cyclorotor fan was performed by means of the commercial Ansys CFX® software package. To this end, the whole computational domain consisted of four rotor-blade domains (one domain for one rotor blade) and one big domain surrounding the cyclorotor ensuring an undisturbed flow of the working medium (cf. Fig. 2). The "Opening" boundary condition is assumed for the main domain outer boundary, whereas the "General connection" boundary condition is assumed for interaction of the rotor-blade domains with the main domain. Compressible air is taken into account as the working medium. The k-ω SST turbulence model is used in the computational model with a medium value of the turbulence level at the inlet. Because the CRF prototype was manufactured by means of 3D polymer-based printing methods, 0.1 mm sand-grain roughness is assumed for the blade walls in the numerical model.

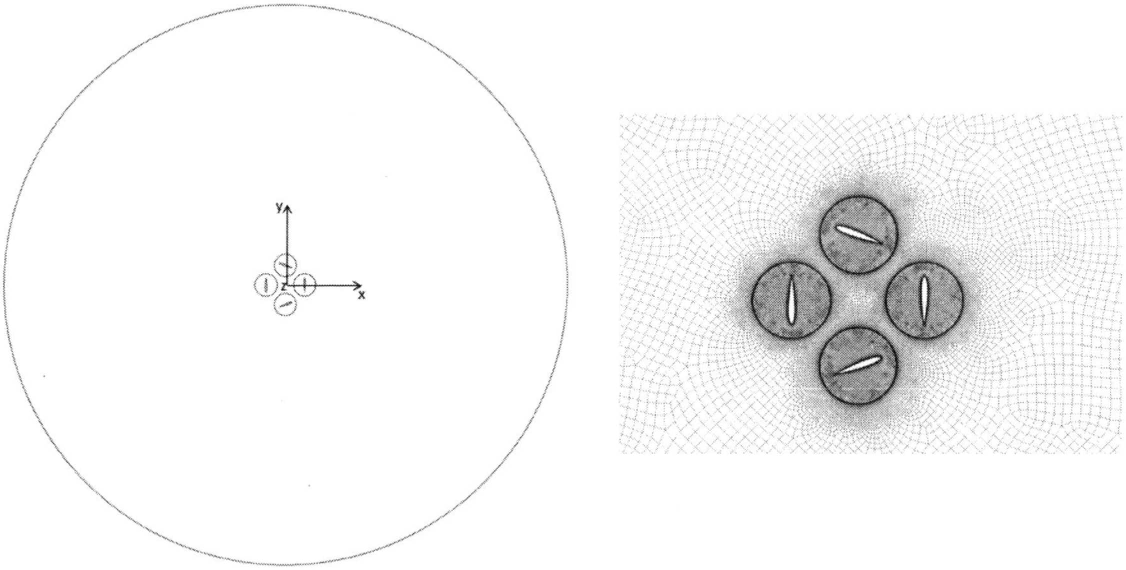

Fig. 2. Computational domain and numerical mesh for the analysed CRF geometry

The total number of the numerical mesh points is ~600k, whereas the main domain consists of ~400k points and each blade domain has ~50k mesh points (cf. Fig. 2). Because only three layers are used in the Z direction, the calculations are treated as two-dimensional.

In the Ansys CFX® software it was impossible to rotate the computational domains simultaneously around more than one coordinate system centre of rotation. For this reason, during the computations the rotor domains (domain 1, 2, 3 and 4 in Fig. 1) had to be moved (transformed) to the centre of the coordinate system, swung by angle $\Delta\alpha$ and then transformed back to the original position and rotated by angle ωt. A CEL script had to be written for this purpose.

The transient calculations were performed with a constant time step calculated based on angular velocity:

$$\Delta t = \frac{2\pi}{\omega} \Big/ n, \tag{1}$$

where n is the constant used for keeping similar time step values for numerical simulations with different values of the cyclorotor angular speed ω.
Gradual increment in the blade swing angle in specific time steps was calculated according to the following relation:

$$\Delta\alpha = \Delta t \cdot \omega \cdot \frac{2\alpha_0}{2\pi} \cdot \frac{\sin(\omega \cdot t+\gamma)}{|\sin(\omega \cdot t+\gamma)|}. \qquad (2)$$

It can be seen from relation (2) that $\Delta\alpha$ may have a positive or a negative value, which ensures the blade return to the initial position after the rotor full rotation (2π).

3.2. Experimental measurements
Additive manufacturing, which is more colloquially referred to as 3D printing, was used to make the cyclorotor fan (CRF) (cf. Fig. 3). All elements of the CRF were made of a polymer material by means of the FFF (fused filament fabrication) technique and roll bearings were used for all rotary connections. An electric motor with an inverter was used to drive the fan to obtain the required rotational speed.

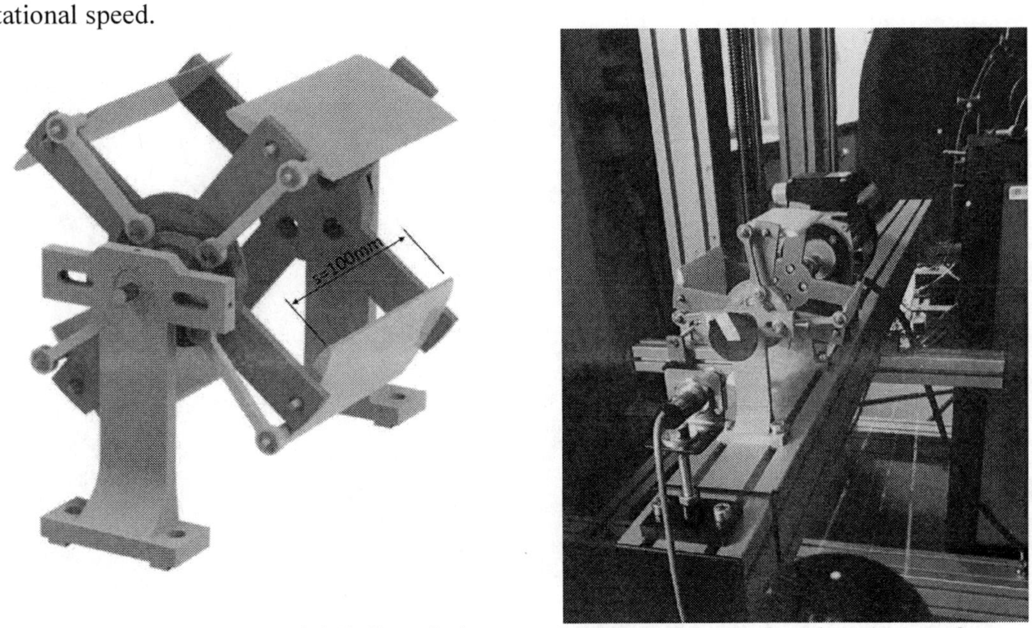

Fig. 3. CAD model (left) and photograph (right) of the analysed cyclorotor fan

The applied LDA system is a two-colour, six-beam, three-dimensional measuring system. It consists of a 4W argon-ion laser, tuned to 488 nm (blue) and 514.5 (green) lines, a transmitter, six beams, a three-dimensional measuring system with two 60 mm 1D and 2D laser probes. The focal length of the LDA probes is 160 mm or 400 mm with additional lenses producing the beam intersection lengths of about 0.66 mm or 4.09 mm, respectively. In order to obtain light impulses strong enough for the LDA, silicon oil particles with a mean diameter of 0.6 – 1.0 micrometre were used.

4. Results
Figures 4 and 5 show the velocity contours obtained from CFD calculations performed by means of the prepared model in the Ansys CFX® software (CRF anticlockwise rotation). The velocity contours are shown in two positions in time for one rotational speed value and for two values of the blade swing angle α_0. It can be seen clearly that the flow velocity rises together with an increase in the blade total swing angle. This proves the ease of controlling the CRF performance: by increasing α_0, it is possible to raise the volume flow rate at the same rotational speed.

The maximum velocity values are observed in the blade vicinity in the position angles between $\gamma = 0$ and $\gamma = \pi/2$. This is related to the non-optimal air flow around the blade for high positive values of the blade incident angle. The flow strong separation on the suction side and on the pressure side close to the trailing edge causes significant flow losses.

This phenomenon can be minimized by using other, more optimal profiles intended for operation at higher incident angle values. More symmetric aerofoils with a different thickness and location have to be tested for this kind of application. The effect can be also minimized by using optimal changes in the blade incident angle (blade swing angle α), better adapted for and synchronized with the CRF angular velocity ω. From the numerical point of view, it would be possible to do that by modifying relation (2), whereas technically, it would be very difficult to implement in a real machine. In such a case, the mechanical way of the incident angle adjustment would have to be replaced by electronic control drives, programmed individually for each blade.

Fig. 4 and Fig. 5 also indicate that for a smaller value of α_0 there is a decrease in aerodynamic losses, which may suggest that to achieve a high volume flow rate, it is better to increase rotational speed rather than the blade total swing angle.

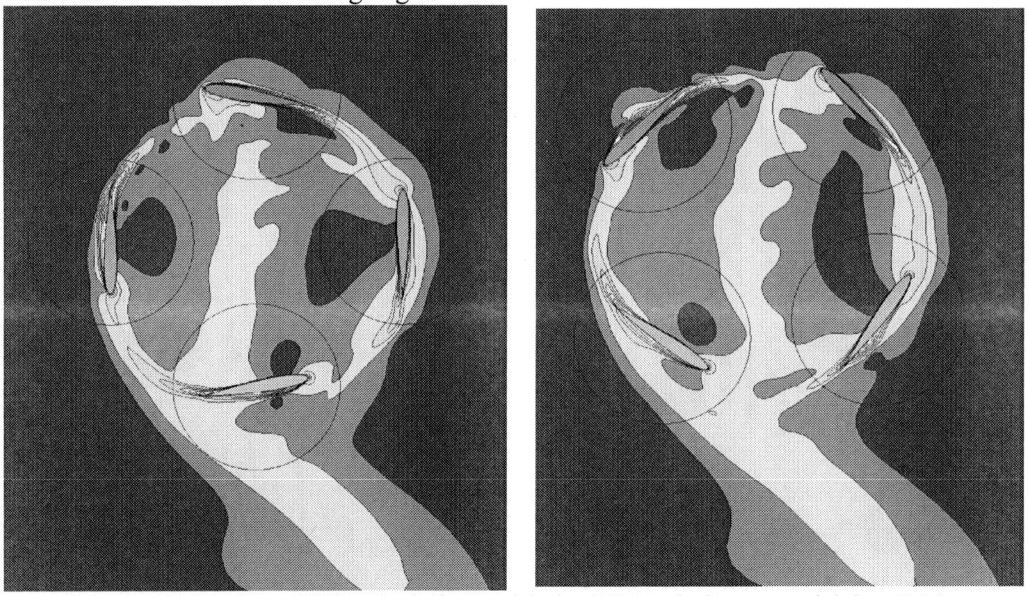

Fig. 4. Velocity fields from CFD calculations with the SST turbulence model for *1000 rpm* and $\alpha_0 = 20°$ ($v = 0 \div 10$ m/s, $\Delta v = 1$ m/s)

Fig. 6 presents the velocity profiles obtained from CFD calculations and from measurements. All the CFD results presented herein are averaged over one full rotation. The velocity profiles are presented at the distance of *140 mm* (*2R*), *240 mm* (4*R*) and *420 mm* (6*R*) downstream from the CRF centre of rotation. The velocity profile obtained from experiments is presented for *31* measuring points with the pitch of *10 mm*. The experimental results are presented at the distance of *140 mm* (*2R*) only, which results from too big a measurement uncertainty in the case of distances *4R* and 6*R*. A significant measurement uncertainty can also be observed for rotational speed lower than *500 rpm,* especially for the lower value of α_0.

Comparing the CFD and the experimental results, it may be concluded that the implemented Ansys CFX® model is able to predict the flow in the CRF with good accuracy in terms of the maximum value of velocity and velocity profiles.

IOP Conf. Series: Journal of Physics: Conf. Series **1101** (2018) 012006 doi:10.1088/1742-6596/1101/1/012006

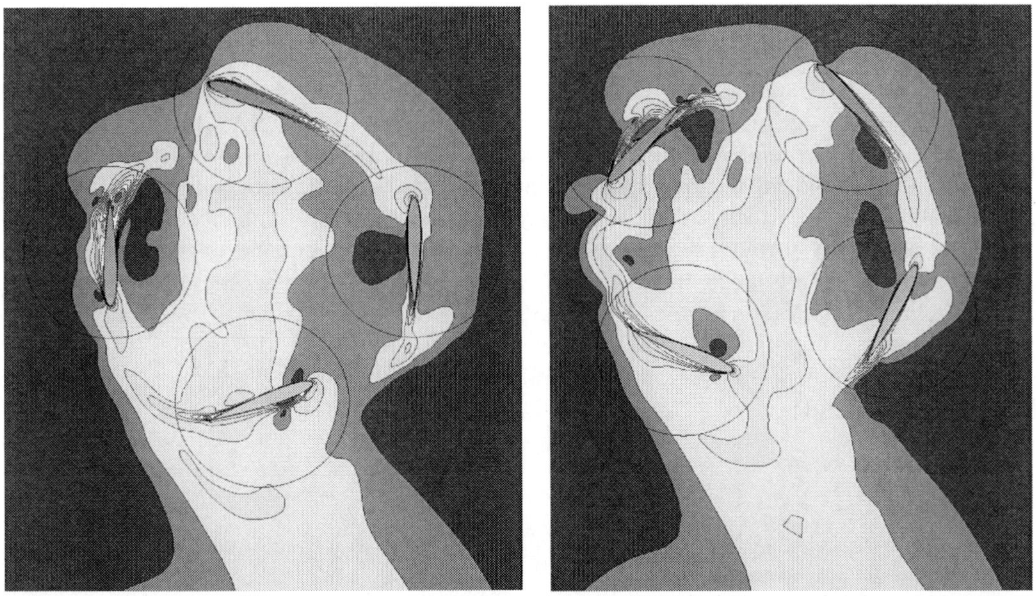

Fig. 5 Velocity fields from CFD calculations with the SST turbulence model for *1000 rpm* and $\alpha_0=40°$ ($v = 0 \div 10$ m/s, $\Delta v = 1$ m/s)

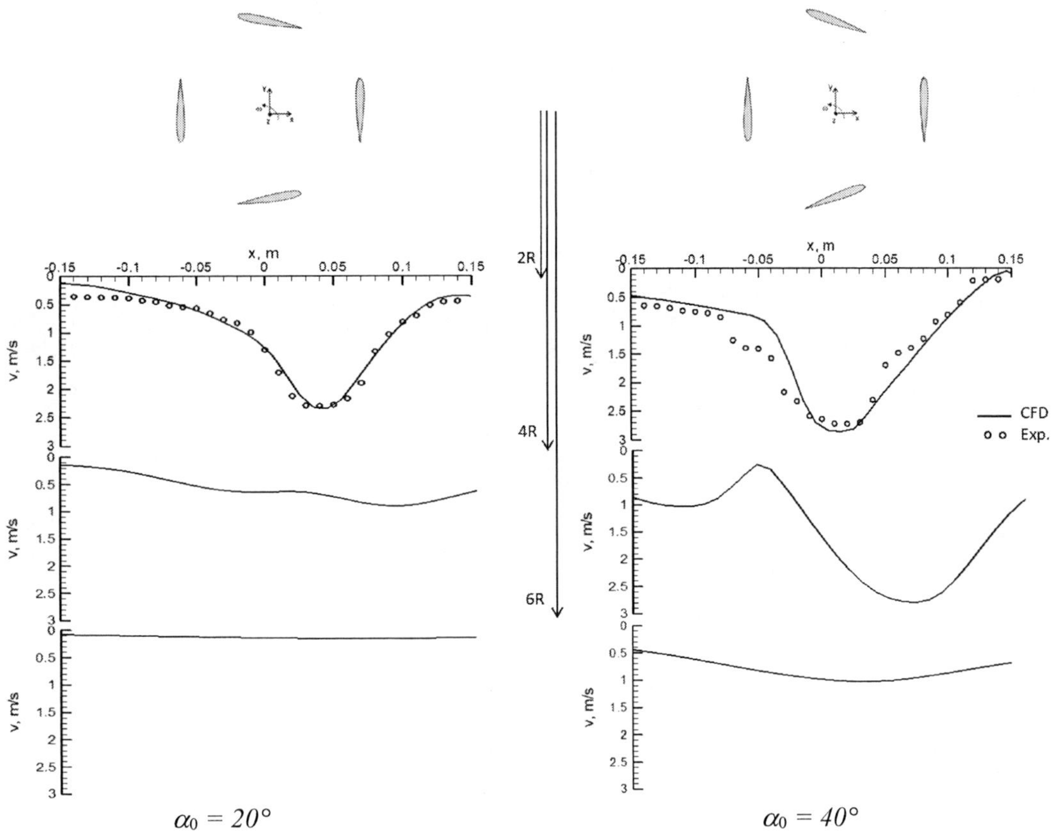

Fig. 6. Velocity profiles obtained from CFD calculations and measurements for *1000 rpm* and two values of total swing angles

The observed differences between the results of CFD calculations and measurements necessitated more testing of the developed numerical model using other turbulence models. The calculations performed for most two-equation turbulence models gave results similar to calculations based on the k-ω SST model. Only the results of calculations performed using the SAS turbulence model differ from the others. Fig. 7 presents a comparison between velocity profiles obtained using the SST and the SST-SAS turbulence models. The maximum value of velocity is similar for both numerical results, but there are small differences in the shape of the velocity profiles. The velocity profile for the results based on the SST-SAS turbulence model is less regular, but it can be expected that for a finer numerical mesh the results with the SAS turbulence model would be smoother. Figure 8 shows momentary velocity flow fields for the CFD results obtained using the SST-SAS turbulence model.

Considering the computational time and the lack of significant differences in the flow field generated by most turbulence models, it is recommended that the k-ω SST turbulence model should be used for future calculations.

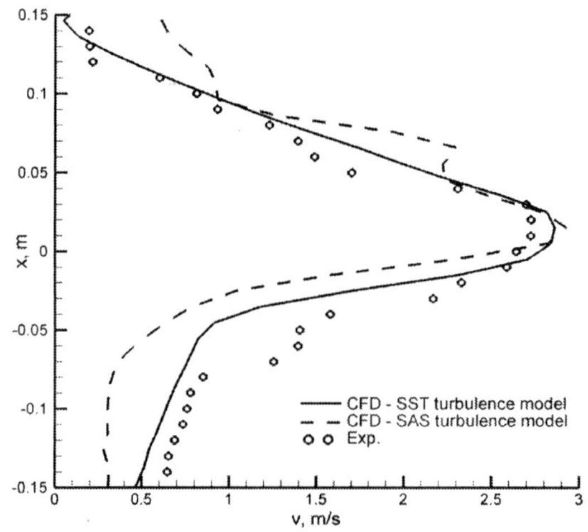

Fig. 7. Velocity profiles for CFD calculations with different turbulence models (*1000 rpm,* α_0 = *40°*)

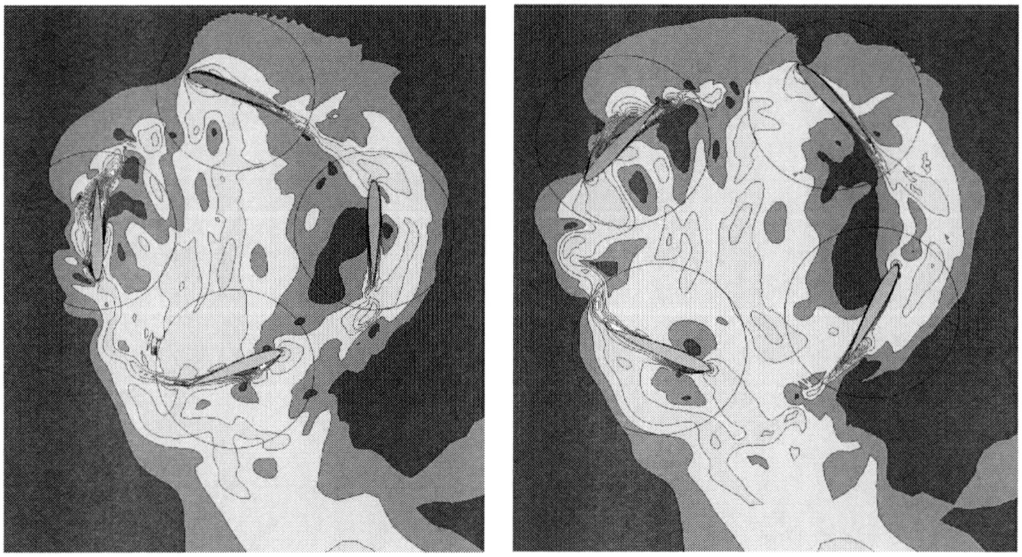

Fig. 8. Velocity fields from CFD calculations with the SST-SAS turbulence model for *1000 rpm* and $\alpha_0 = 40°$ ($v = 0 \div 10$ *m/s*, $\Delta v = 1 m/s$)

5. Summary and future plans

The paper presents an aerodynamic assessment of the cycloidal rotor fan based on CFD calculations and LDA flow field measurements. The numerical and experimental results are compared in terms of velocity profiles. Satisfactory agreement is shown between the numerical simulation results and the experimental data. The observed differences in the velocity flow field and the difficulties in the comparison for low rotational speed values between CFD calculations and measurements may be caused by:

- more significant than expected 3D effects in experimental measurements due to interaction between the side elements,
- small differences in swing angle values (α) between the numerical and the experimental model (the maximum difference was about *1*°),
- considerable difficulties in the measurements using the LDA technique for low values of the flow velocity and high levels of turbulence.

Therefore, future works aiming to obtain a more efficient cyclorotor fan will be carried out using the CFD technique, and the final design will be tested further experimentally.

Future research will require numerical testing of the cyclorotor fan (CRF) design for different configurations of the:

- blade profile (e.g.: NACA0010, NACA0015, N66-012, E168, E169)
- blade chord (c),
- number of blades,
- blade swing centre of rotation,
- cyclorotor radius (R),
- blade initial swing angles (positions)
- value of the blade total swing angle (α_0).

Acknowledgements

The presented results were obtained within Statutory Research Funds of the Institute of Power Engineering and Turbomachinery of the Silesian University of Technology.

References

[1] Hwang S, Min Y, Lee H, and Kim J 2008 Development of a Four-Rotor Cyclocopter *Journal of Aircraft* **45(6)** pp. 2151-2157

[2] Xisto C M, Páscoa J C, Abdollahzadeh M, Leger J A, Schwaiger M and Wills D 2014 PECyT – plasma enhanced cycloidal thruster *Proceedings of the 50th AIAA/ASME/SAE/ASEE Joint Propulsion Conference*

[3] Xisto C M, Páscoa J C, Leger J A, Masarati P, Quaranta G, Morandini M, Gagnon L, Wills D and Schwaiger M 2014 Numerical modeling of geometrical effects in the performance of a cycloidal rotor *Proceedings of the 6th European Conference on Computational Fluid Dynamics*

[4] Gabi M, Dornstetter S and Klemm T 2002 Investigation of the flow field in crossflow fans by particle imaging velocimetry *Proceedings of the 10th Int. Symp. On Flow Visualisation* (Kyoto, Japan)

[5] Gabi M and Klemm T 2004 Numerical and experimental investigations of Cross-flow fans *Journal of Computational and Applied Mechanics* **5(2)** pp. 251-261

[6] Witkowski A and Majkut M 2006 Experimental investigation of inlet guide vane-rotor interaction in a low-speed axial-flow compressor stage *Archive of Mechanical Engineering* **LIII(1)** pp. 23-48

[7] Witkowski A, Ziach M, Majkut M and Strozik M 2008 Experimental investigations of the flow phenomena in the rotor blades of the axial-flow low-speed compressor stage at the unstable part of the overall performance characteristic *Archive of Mechanical Engineering* **LV(4)** pp. 313-330

IOP Conf. Series: Journal of Physics: Conf. Series **1101** (2018) 012007 doi:10.1088/1742-6596/1101/1/012007

Thermal dilatation of solid grains: the LBM approach

A. Grucelski

Institute of Fluid-Flow Machinery, Polish Academy of Sciences, Gdańsk, Poland

E-mail: agrucelski@imp.gda.pl

Abstract. On the way to present the detailed description of the thermal processing of fossil fuels, one has to face the task of modeling physico-chemical phenomena as complex as thermal dilatation of particles of granular bed in a hot stream of gases. The work presents main assumptions and description of modeling tool for simulate of coal grains thermal dilatations with increasing temperature. For this purpose we use the Lattice Boltzmann method with some schemes needed when the process, of coupled phenomena, in a representative element of granular media is modeled.

1. Introduction

Detailed simulation of flow thermomechanics in a complex geometry becomes achievable to perform even on desktop computers. This is due to the rapid increase of computational power and the development of numerical schemes and methods (geometrically detailed for specific phenomena) of the Computational Fluid Dynamics (CFD). Mentioned progress is desirable in particular in the case of problems where physico-chemical phenomena have to be modeled. A very demanding process to describe is the thermal processing of solid fuels where phenomena to be solved include heat and fluid flow with chemical species transport in reactive, random granular media. Great efforts have been taken to numerically model some of the phenomena. The methodology utilized in recent studies is the meso-scale simulation approach. In the case of granular media, only part of the domain is modeled taking into account the grain level with use of micro-scale results, such as chemical kinetics and transport through grain porous structure [1]. Usually the goal is to prepare the simulation case to be used in macroscale modeling. The approach is more and more often used in CFD where we need a detailed description of the process but the scale disparity (micro/macro) in industrial cases make it difficult to obtain.

Meso-scale simulations are usually performed in a selected geometry, usually called the representative element of volume (REV). The REV is chosen in the way to be much smaller than the overall size of the industrial device and much larger then size of the smallest modeled scales. With such non-trivial assumptions regarding geometry, in result some of the simple models of thermal processing of fossil fuel can be proposed. Simplifications usually concern numerical modeling in the micro scale, mentioned above, as well as in the macro scale (influence of the boundaries). Therefore, for the processes characterised by a large span of length and time scales, detailed numerical modeling of industrial scale objects is still intolerably expensive [2].

On the way to a physically sound simulation of the coking process in the meso-scale, one of the important phenomena to model is the thermal dilatation of grains. Aside of fluid flow and heat transfer (including radiation), the list of other phenomena to be modeled contains: chemical kinetics, release of gaseous products and change of physical properties of grains. The last one includes the variations of volume: expansion (in the first stage of heating) and shrinking (in the second stage) accounting for change of shape, with implemented plasticity of the material. In both stages, the process involves also a change of topology of the solid matrix: the granular medium, through plastic deformation, becomes a porous medium.

Due to compact geometry of the granular medium (modeled with the regular lattice as a compromise between accuracy and efficiency) and a need to model physico-chemical phenomena (flow thermomechanics, species transport, plastic deformation, etc.) we choose the Lattice Boltzmann Method (LBM). LBM has received a great interest both from engineers and scientists. Thanks to its meso-scale fundamentals, the method is increasingly extended to model various phenomena; additionally, thanks to its ease of parallelization the method found application over many branches of technological problems.

Although a few papers on the modeling of fluid flow and heat transfer along with chemical species transport in granular medium have been reported [4, 3], the way for modeling thermal dilatation of grains still remains open. After implementation of the method we have to provide closure relationships and describe the process of thermal processing of fossil fuels. The paper, after introduction, contains description of the model and preliminary quantitative results regarding interface diffusion; the paper ends with a short conclusion.

2. Description of the model

2.1. LBM formulation

Here only a brief description of the method along with used schemes is mentioned. Basically, in LBM a given phenomenon is modeled by tracking the evolution of prescribed distribution functions (DF) f_i corresponding to a chosen set of discrete lattice velocities. The main advantage of LBM is the fully local character of advection and collision terms in the evolution of the DF, along with a possibly easy and effective parallelisation (see [1]). The method is usually applied on a uniform mesh and so it is often chosen to model the phenomena in domains characterised by complex geometry and random layout of elements.

The evolution of the DF in LBM over a time increment δt can typically be written as:

$$f_i(\mathbf{x} + \mathbf{c}_i\delta t, t + \delta t) = f_i(\mathbf{x}, t) + \tau^{-1}\left[f_i^{\mathrm{eq}}(\mathbf{x}, t) - f_i(\mathbf{x}, t)\right] + S_i, \tag{1}$$

where τ stands for relaxation time and the equilibrium DF is usually presented in a polynomial form:

$$f_i^{\mathrm{eq}} = \rho\Omega_i\left(A + B\frac{\mathbf{c}_i \cdot \mathbf{u}}{c^2} + C\frac{(\mathbf{c}_i \cdot \mathbf{u})^2}{c^4} + D\frac{\mathbf{u} \cdot \mathbf{u}}{c^2}\right). \tag{2}$$

The form of τ and the coefficients of Eq. (2) depends mainly on modeled phenomenon. So in the case of heat transfer $\tau \sim \alpha$; for the chemical species transport τ depends on the mass diffusivity coefficient. In case of the equilibrium DF, the coefficients and exact form depend mainly on discretisation model (here D2Q9 stands for a two dimensional modeling with nine directions of DF advection). Equation (1), through the presence of source term (S_i), can be used to account for model of additional phenomena; a few to mention are influence of released chemical species on the flow field, change of internal energy due to chemical reactions, convection of the fluid, etc. Here we will not discuss in details the implementation of these as they are presented, along with a selection of results, in [3] and references therein.

In case of multiphase flow (we treat grains as non-newtonian fluid in a given range of temperatures) the model utilized here is presented in detail in [9]. The multiphase modeling can be done with use of two DF, which correspond to the evolution of pressure and phase index; in presented paper prescribed with f and I DF, respectively. Simulation of the fluid flow is done by modifying Eq. (1) with the following terms

$$\text{Eq. (1)}|_f \quad + \quad \frac{2\tau-1}{2\tau}(\mathbf{c}_i-\mathbf{u})\cdot[\Gamma_i(\mathbf{u})(\mathbf{F}_s+\mathbf{G})-(\Gamma_i(\mathbf{u})-\Gamma_i(0))\nabla\Psi(\rho)]\,\delta t \tag{3}$$

$$\text{Eq. (1)}|_I \quad - \quad \frac{(2\tau-1)}{2\tau}\frac{(\mathbf{c}_i-\mathbf{u})\cdot\nabla\Psi(\phi)}{c_s^2}\Gamma_i(\mathbf{u})\delta t\ , \tag{4}$$

for pressure (p) and index DF respectively; Ψ denotes used equation of state described below.

The surface tension and buoyancy forces are calculated from:

$$\mathbf{F}_s=\kappa\phi\nabla\nabla^2\phi,\quad \mathbf{G}=(\rho-\rho_m)\mathbf{g}\ ; \tag{5}$$

where ϕ is the colour index function; the magnitude κ of the surface tension (σ) can be presented:

$$\sigma=\kappa\int\left(\frac{\partial\phi}{\partial\xi}\right)^2 d\xi\simeq\kappa\frac{0.1518(a-a_c)^{1.5}}{1+3.385(a-a_c)^{0.5}}\ ;$$

ξ is the local coordinate in the direction normal to the interface and $\rho_m=(\rho_1-\rho_2)/2$ where ρ_1 and ρ_2 correspond to the densities of considered fluids. In the above equation, a is a parameter describing the miscibility of the fluids; as reported in [9], if $a>a_c$ both fluids are immiscible. The model is complemented by the Carnahan-Starling equation of state (EOS) which describes the process of separation of phases (see [9]):

$$\Psi(\phi)=c_s^2\phi\left[\frac{1+\phi+\phi^2-\phi^3}{(1-\phi)^3}-1\right]-a\phi^2\ , \tag{6}$$

where a determines the strength of molecular interactions. Critical value for Carnahan-Starling equation is $a_c=3.5337$ and in our simulation $a=4$. Then, $\nabla\Psi(\phi)$ describes intermolecular interactions of phases and plays an important role in separating the phases.

Following [9], to calculate differential terms in Eqs. (4)–(5) we use in this work a fourth order compact scheme. Γ_i is used only for pressure and index fields so coefficients are constant:

$$\Gamma_i(\mathbf{u})=\omega_i\left[1+3\frac{\mathbf{e}_i\cdot\mathbf{u}}{c^2}+\frac{9}{2}\frac{(\mathbf{e}_i\cdot\mathbf{u})^2}{c^4}-\frac{3}{2}\frac{\mathbf{u}^2}{c^2}\right];$$

Detailed formulas for macroscopic fields are presented at the end of Section 2.

The resulting density and kinematic viscosity of phases in the one-fluid formulation are calculated with use of index functions ([6]):

$$\rho(\phi)=\rho_2+\frac{\phi-\phi_2}{\phi_1-\phi_2}(\rho_1-\rho_2)\ ,$$

for the viscosity, in [7] it is proposed to account of viscoplasticity:

$$\nu(\phi)=\nu_1\exp\left[\frac{\phi-\phi_2}{\phi_1-\phi_2}\ln\left(\frac{\nu_1}{\nu_2}\right)\right],$$

where ϕ_1 and ϕ_2 correspond to minimum and maximum values of index functions and ν_1, (ρ_1) and ν_2, (ρ_2) to viscosities (densities) of fluids. Here ν_2 stands for the viscosity of coal in a plastic phase; the discussion regarding viscosity of coal in higher temperatures (thermoplastic) is presented in [10].

The developed model has to account for the evolution of temperature and chemical compounds. Most of the works presented till now (for the best knowledge of the Author) concern isothermal multiphase flow, see [6]. As the evolution of temperature field is crucial for the accurate process description, we utilize here the known scheme for Internal Energy Distribution Function (described in [2, 4]). The IEDDF scheme assumes an additional DF (prescribed usually by g) to be solved along with other DFs (like f_i, I_i); the evolution equation for IEDDF is very similar to Eq. (1); equilibrium Eq. (2) differs mostly in implementation (one has to pay attention to the discretisation scheme and DF direction of advection, see [2]).

In case of chemical species transport we utilize a scheme described in [8, 3]. The scheme assumes additional DF (here ϕ_k) for every tracked chemical compound k. The equilibrium distribution functions account for high gradient of chemical compound concentration, this results in equilibrium terms with following modification:

$$\phi^{\mathrm{eq}}_{0,k} = \frac{\rho Y_k}{9}(9 - 7\psi), \qquad \phi^{\mathrm{eq}}_{1,\ldots,6,k} = \frac{\rho Y_k}{9}(\psi + 3\mathbf{e}_i \cdot \mathbf{u}), \qquad \phi^{\mathrm{eq}}_{7,\ldots,14,k} = \frac{\rho Y_k}{72}(\psi + 3\mathbf{e}_i \cdot \mathbf{u}),$$

where $\psi = \rho^*/\rho$ is the ratio of a minimum density and density in a given point. The release of chemical compounds is calculated with use of the model of chemical kinetics utilized in thermal processing of fossil fuels (see [3]).

The macroscopic quantities are calculated from the following equations:

$$p = \sum f_i - 0.5\mathbf{u} \cdot \nabla\Psi(I)\delta t, \qquad \mathbf{u} = \frac{1}{\rho c_s^2}\sum \mathbf{c}_i f_i + \frac{1}{c_s^2}(\mathbf{F}_s + \mathbf{G})/2\delta t,$$

$$\phi = \sum I_i, \quad \theta = \sum_i g_i, \quad Y_k = \sum_i \phi_{i,k}\,.$$

The model assumes simulation of flow thermomechanics with the release of chemical compounds and plastic deformation of grains along with calculation of stresses. The work is in the development stage; an early attempt with temperature-dependent volume of grains was reported earlier. The full set of equations along with the details of the implementation can be found in [2, 4] regarding heat transfer and [8, 3] for chemical compounds evolution.

2.2. Boundary conditions

Like in our previous work, we attempt to model the coupled phenomena within a REV element being in fact a fragment of granular medium; in the present work obstacles cannot intersect the domain boundary and other obstacles surface to simplify the model of thermal dilatation. Grains at lower temperature are treated as solid obstacles and at the increasing temperature we introduce the viscoelasticity.

For the fluid flow, at the opposite walls we use periodic conditions also to simplify description at the early state of thermal dilatation of grains. At the solid boundaries of grains no-slip condition is used and half-way bounce-back scheme is implemented. For the index DF at boundaries of numerical domain we use zero flux (zero derivative) scheme. In case of heat transfer a shift-periodic boundary scheme is implemented at the two opposite walls. The condition introduces constant difference of averaged temperature between chosen points (boundaries) of the computational domain at which the scheme is implemented (for details see references in [5]). The scheme keeps a similar profile of temperature and the continuity of heat flux. At boundaries parallel to the main gradient of temperature, a simple periodic scheme is used.

Utilized boundary conditions result in forced heat transfer, whereas the fluid flow is mainly caused by free convection, thermal dilatation and the release of chemical compounds from the surface of grains. This is still not the ideal case for REV as some pressure drop is expected in the direction perpendicular to the main direction of heat transfer. Also, as the shift-periodic scheme for heat transfer seems to be the right choice, still the difference of temperature (in function of the number of grains) is determined by our needs. Simply we want to have a maximal temperature observed in the process and temperature at which no chemical species release is started.

2.3. Thermal dilatation of grains
Dilatation of grains is mainly caused by release of gaseous products and change of viscoplasticity of the coal. To mimic this situation changes of the appropriate distribution function have to be implemented. Due to the data collected (in the volume of the grain) at every time step, the algorithm have to account (depend on) to the mean temperature of the grain (in contrast to the local temperature in the solid node) and the rate of release of the gaseous products in the current instant (recent results are presented in [3]). Still it is difficult to find a numerical dissertation or experimental results with presented empirical correlation regarding variable viscosity/geometry of fossil coal in function of temperature. Here we propose formulae that intuitively correspond to the thermo-chemical state of the coal grains with increasing temperature.

In the case of index distribution function, we modify Eq. (4) with:

$$Eq.\ (4) + \Omega_i \alpha_s(\theta)\ [\phi_1 - \phi(i,j,k)]$$

where the terms in bracket restrain the highest value of the index function. The α_s stands for expansion function and takes the form:

$$\alpha_s(\theta) = 0.5 + 0.5\tanh(12\theta - 15.6)$$

and models the release of gaseous products. This only mimics results (namely the detailed functional-group or more general one specie model, see [3]) from modeling of product release.

3. Calculation results

At first the interface behaviour is modeled for a single grain (and obstacle) with the density ratio of $\rho_1/\rho_2 = 6$ in the case of no forced convection. The gravity force is added (but with disabled model of buoyancy force).

As it is visible from Fig. 1, with a running simulation the interface of the obstacle undergoes strong diffusion, even for recommended value of $a = 4.0$ (see [7] for details).

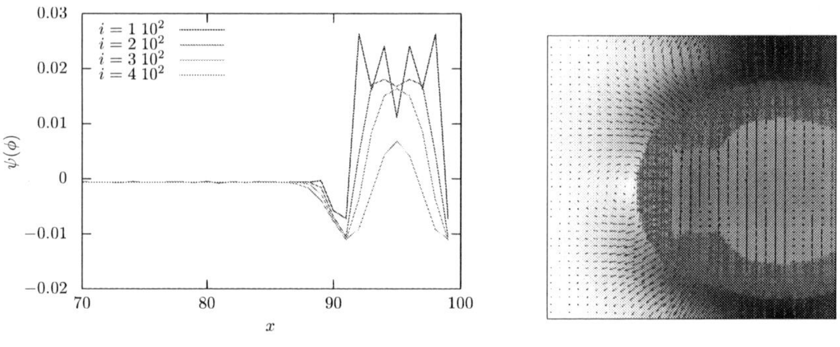

Figure 1. Left: result of the Carnahan-Starling EOS, Eq. (6), in function of iteration number with used $\kappa = 0.001$ and $a = 4.0$ where the value of a is set to be larger than to $a_c = 3.8$; right: velocity magnitude map (grayscale) with vectors and surface nodes as points.

Although the increase of the parameter a improves the immutability of the index function, this also influences the stability of the simulation (not presented here). Thus some decrease of the time step (relaxation parameter) is necessary to hold the stable simulation.

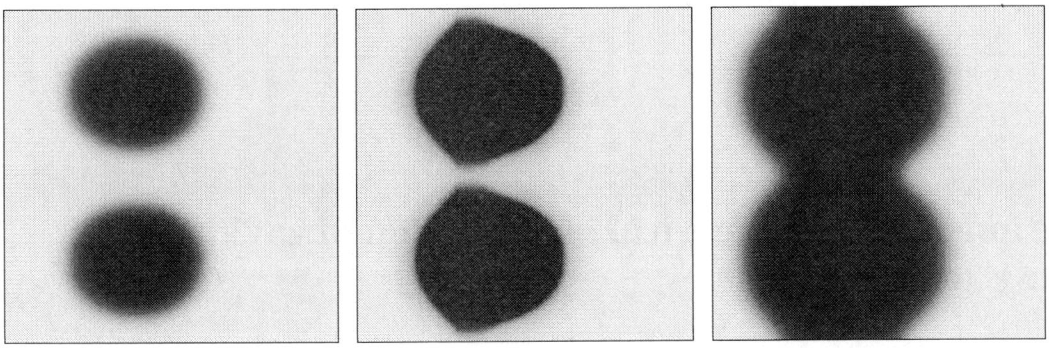

Figure 2. Change of the couple grains volume in instants of time, left boundary is heated.

Figure 2 presents some of qualitative results from simulation of dilatation of obstacles with increasing temperature with implemented models described above (modelling heating by infinitesimally thin wire). The smooth merging of two growing grains (no shear stress) is caused by a lack of a proper condition at the surface of disjoint obstacles.

4. Conclusion and outlook

The presented work is thought to be a brief presentation of the tool used to model the thermal dilatation of grains in reactive flow through a granular medium. In the next step of the implementation of the contact algorithm for modeling the resulting stresses between grains in contact is considered.

Acknowledgement
The Author would like to express his gratitude to prof. Jacek Pozorski for useful comments and help in preparing the manuscript and remarks regarding the model and to Michał Olejnik for discussions on surface tension and Young-Laplace equation.

5. References

[1] C.K. Aidun and J.R. Clausen (2010). Lattice-Boltzmann method for complex flows. *Annu. Rev. Fluid Mech.* **42** 439-472.
[2] J. Wang, M. Wang and Z. Li (2007). A lattice Boltzmann algorithm for fluid–solid conjugate heat transfer. *Int. J. Therm. Sci.* **46** 228-234.
[3] A. Grucelski and J. Pozorski, (2017). Application of Lattice Boltzmann method to meso-scale modelling of coal devolatilisation. *Chem. Eng. Sci.* **172** 503-512.
[4] A. Grucelski and J. Pozorski, (2015) Lattice Boltzmann simulations of heat transfer in flow past a cylinder and in simple porous media. *Int. J. Heat Mass Transf.* **86** 139–148.
[5] A. Grucelski and J. Pozorski, (2016). Shift-periodic boundary condition for heat transfer computations in Lattice Boltzmann Method. *Int. Comm. Heat Mass Transf.* **77** 116122.
[6] Q. Chang, J.I.D. Alexander (2006). Application of the lattice Boltzmann method to two-phase Rayleigh–Benard convection with a deformable interface. *J. Comput. Phys.* **212** 473-489.
[7] P.A. Swain, G. Karapetsas, O.K. Matar and K.C. Sahu, (2015). Numerical simulation of pressure-driven displacement of a viscoplastic material by a Newtonian fluid using the lattice Boltzmann method. *European Journal of Mechanics-B/Fluids* **49** 197-207.
[8] A.F. Di Rienzo, P. Asinari, E. Chiavazzo *etal.* (2012). Lattice Boltzmann model for reactive flow simulations. *Eur. Phys. Lett.* **98** 34001.
[9] P.R. Redapangu, K.Ch. Sahu and S.P. Vanka (2012). A study of pressure-driven displacement flow of two immiscible liquids using a multiphase lattice Boltzmann approach. *Phys. Fluids* **24**, 102110.
[10] S. Nomura, K. Kato, I. Komaki, Y. Fujioka, K. Saito and I. Yamaoka (1999). Viscoelastic properties of coal in the thermoplastic phase. *Fuel* **78**, 1583-1589.

IOP Conf. Series: Journal of Physics: Conf. Series **1101** (2018) 012008 doi:10.1088/1742-6596/1101/1/012008

Experimental and numerical investigation of cavitation on Clark Y foil

D Homa[1, 2], **W Wróblewski**[1], **M Majkut**[1] **and M Strozik**[1]

[1]Institute of Power Engineering and Turbomachinery, Silesian University of Technology, Konarskiego Street 18, Gliwice, Poland
[2]To whom any correspondence should be addressed

Abstract. The results of experimental investigation of cavitation in flow over Clark Y foil are presented. The cavitation test rig included a chamber with a blade profile fixed to the rotary disc that enabled to set different angles of attack. Working fluid was water at temperature 16°C. For different flow conditions (water velocity, pressure) the pictures of cavitating flow were taken and analyzed. The pressure at the inlet and outlet of the chamber were measured, as well as the value of volumetric flow rate. That enabled to determine the cavitation number for each case. The mechanism of cavitation structures appearance, growth and collapse was observed and described. The numerical simulation of cavitating flow was also performed by means of OpenFOAM software with interphaseChangeFoam solver. Kunz cavitation model was chosen for calculations, with k-ω SST turbulence model. The assumption of isothermal, two–phase flow was made. The vapour areas appearance, their shapes and changes in time were described and compared with experimental results. The main features of cavitating flow were caught, but the further adjustment of cavitation model is recommended.

1. Introduction

Cavitation is a complex process that includes both vapour bubbles formation in the flow and their sudden collapses. Cavitation can lead to many unwanted effects such as noise, vibrations, machine's efficiency drop and erosion [1]. There are different types of flow with cavitation, but as far as turbomachinery performance is concerned, the main interest is focused on the flow over a hydrofoil. Many studies have been devoted to experimental investigation of cavitating flow [2-4] which provide substantial information about cavitation inception, development and dynamics. Especially in case of erosion risk assessment, it is necessary to identify locations of bubbles rebounds and collapses [4]. An important part of the experimental investigation is the registration of characteristic features of the cavitating flow, such as induction of reentrant jet or cavitating clouds detachment. Capturing high frequency cyclic changes of the cavitation structures demands high speed video camera and appropriate illumination system. Due to the rapid progress in computer science and computational fluid mechanics, it is recently more popular to simulate cavitating flow by means of CFD programs. Simulation performed in OpenFOAM is also included in this study.

The flow parameter which describes flow susceptibility to cavitation is called cavitation number σ [5]:

$$\sigma = \frac{p_\infty - p_s}{0.5\rho u_\infty^2} \tag{1}$$

where: p_∞ – free stream pressure, Pa; p_s – saturation pressure, Pa; ρ – liquid density, kg/m^3; u_∞ - free stream velocity, m/s.
The lower cavitation number the more intense cavitation should appear in the flow.

2. Materials and methods

2.1. Experimental investigation

To investigate cavitating flow over a foil a cavitation tunnel was built at the Institute of Power Engineering and Turbomachinery at Silesian University of Technology. The test rig included rectangular chamber with the blade fixed to the disc that enabled to set different angles of attack. The investigation concerned Clark Y 11.7 % foil of chord (c) equal to 70 mm. The height of the chamber was 189 mm (2.7c) and the span (width) of the chamber was 70 mm (1c). The foil was placed at the half of the chamber's height, 210 mm downstream from inlet of the chamber. The overall length of the chamber was 700 mm. At the top, bottom and one side wall of the test section transparent acrylic glass was used to enable observations. The scheme of the circuit is shown in figure 1. The 3D model of the test rig is depicted in figure 2. The measurements of static pressure at the inlet and at the outlet of chamber were performed, as well as the temperature at the outlet. Volumetric flow rate was determined by means of electromagnetic flowmeter. On the basis of flowmeter record the free stream velocity was determined. That enabled to calculate the cavitation number for each case. The measurement signals were collected by the computer unit. By means of pump's motor variable frequency drive the rotational speed of pump was changed and the flow rate was increased. These caused the pressure in the chamber to drop and more intense cavitation to occur. During the measurements the movies of cavitating flow were taken by means of high speed camera Phantom Miro C110. The recording speed was set to 1200 frames per second, which was enough to capture the changes during one cycle of cavitation forming and collapse, but also sufficient spatial resolution of frames equal to 1280 x 720 pixels was kept. The series of measurements were performed for three different angles of attack, in the range of cavitation numbers from 1.5 (cavitation inception) to 0.75 (cloud, developed cavitation). The experiment was carried out on water of temperature equal to 16°C. The feeding tank had about 35 m^3 capacity, which enabled to keep water temperature at the constant level during the series of measurements. After experiment, the movies were split into series of frames. The period of changes for each case was estimated. The visualization included two different camera views, so that changes in spanwise and streamwise directions could be observed. It should be noted that only one camera was used in the experiment, but cyclic character of changes enable to match top and side view pictures. The view of recording equipment is shown in figure 3.

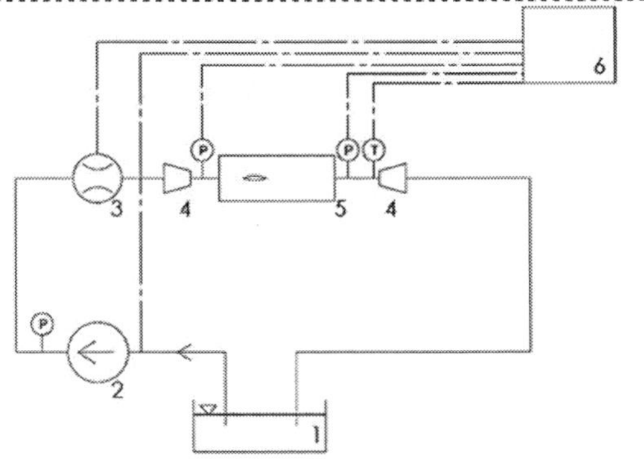

Figure 1. The scheme of the circuit. 1 – water tank, 2 – pump, 3 – flowmeter, 4 – cross section reduction, 5 – chamber with blade 6 – computer unit. P – pressure measurement, T – temperature measurement

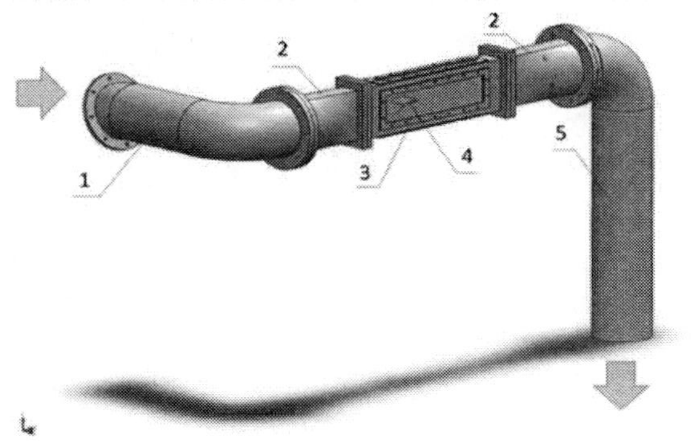

Figure 2. The scheme of the test rig. 1 – inlet pipe, 2 – cross section change, 3 – measurement chamber, 4 – blade, 5 – outlet pipe.

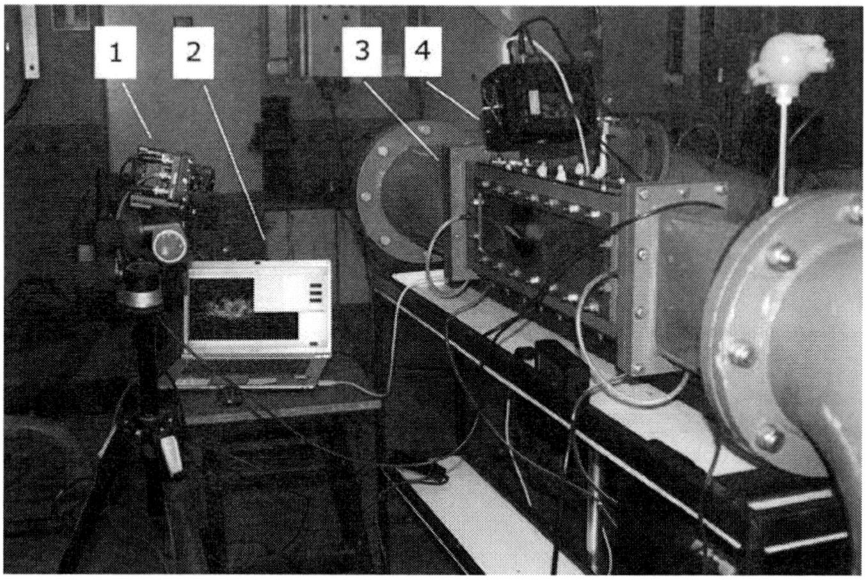

Figure 3. Recording cavitating flow; 1 – Phantom Miro C110 camera, 2 – computer unit, 3 – measurement chamber, 4 – illuminator IL – 105/6X

2.2. Numerical investigation

For modelling the process of cavitation Kunz model was chosen. It was introduced by Kunz et al in [6]. It is a model of one-fluid models group. In the one-fluid model the flow is assumed to be a mixture of two phases and the conservation equations (mass and momentum) for the mixture are to be solved. In this simulation no slip between the phases was assumed. To calculate fraction of gaseous phase mass conservation equation of vapour is solved [7]:

$$\frac{\partial \alpha \rho_v}{\partial t} + \nabla(\alpha \rho_v u) = \begin{cases} R_e & p \le p_s \\ -R_c & p > p_s \end{cases} \tag{2}$$

where: α – vapour volume fraction, -; ρ_v – vapour density, kg·m^{-3}; u – velocity, m·s^{-1}; R_e, R_c – source terms, kg·m^{-3}s^{-1}.

In Kunz model the source terms are derived empirically [6]:

$$R_e = \frac{C_e \rho_v \alpha \min(0, p - p_s)}{0.5 \rho_l u_\infty^2 t_\infty} \tag{3}$$

$$R_c = \frac{C_c \rho_v \alpha^2 (1 - \alpha)}{t_\infty} \tag{4}$$

Where: ρ_l – liquid density, kg·m^{-3}; u_∞ – fee stream velocity, m·s^{-1}; t_∞ –mean flow timescale, s; C_c, C_e – empirical coefficients.

In the domain the foil was placed in the same position as it was during the experiment. The 2D structural grid was generated and extruded in perpendicular direction to the overall width of 0.9 mm (3 layers of 0.3 mm thickness each). The geometry has been divided into 8 blocks and the O-grid was generated around the blade. The blade profile was split into 4 edges: near leading edge, upper side, lower side and trailing edge. On lower and upper side edges 90 grid nodes were set, on the leading and trailing edges – 45 nodes. On the edge normal to the foil 81 elements were set. The whole mesh consisted of 160k hexahedra elements. The overview of the mesh with the zoom of the O-grid region is shown in figure 4.

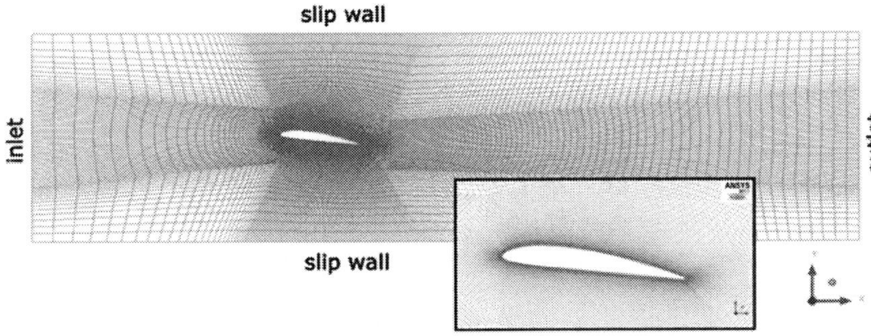

Figure 4. Overview of the mesh, zoom of the O-grid region.

At first calculation with high cavitation number (single phase flow) was performed. The pressure distribution along the foil was monitored and compared with results obtained at Kyushu University, Japan [3] for the same foil. Pressure coefficient C_p is defined as:

$$C_p = \frac{p - p_{inlet}}{0.5 \rho u_\infty^2} \tag{5}$$

The pressure coefficient distribution along the foil is depicted in figure 5. The generated mesh gave good agreement with the experimental results, so it was chosen for further computations.

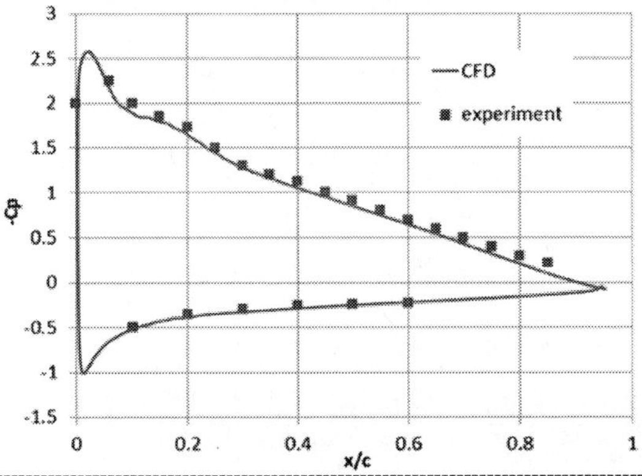

Figure 5. C_p distribution along the foil for cavitation number σ = 3 (one phase flow). Experimental results from [3]

The simulation was performed with inlet velocity and outlet static pressure boundary conditions. The side planes of the domain were set to symmetry conditions, upper and lower planes of the domain - slip wall condition. The pressure at the outlet was consequently lowered and velocity increased, according to the values obtained in the experiment. The cavitating flow was simulated by means of OpenFOAM software. The *interPhaseChangeFoam* solver that was used allows the phase change between two incompressible, isothermal, immiscible fluids. The assumption of vapour incompressibility has been reported in other investigations with satisfying results [8, 9]. Flow was assumed to be isothermal. The turbulence model was k -ω SST with turbulence intensity at the inlet equal to 5%. The transient, first order Euler scheme was chosen. For single phase calculations the time step of value 10^{-5} s was set. For cavitating flow simulations the constant time step was used, set to the value that ensured max Courant number under 0.4. The coefficients C_e and C_c in Kunz model were set to the value of 1000 [3]. During the calculations maximum value of $y+$ parameter on the foil wall was 10 (average value was about 2).

3. Results

3.1. The influence of angle of attack

Four different levels of cavitation number were recorded: 1.3; 1.0; 0.85 and 0.75 with three different

IOP Conf. Series: Journal of Physics: Conf. Series **1101** (2018) 012008 doi:10.1088/1742-6596/1101/1/012008

angles of attack (AoA): 4°, 6°, 8°. The movies were recorded and split into series of frames. On the basis of frame analysis the period of changes was estimated. In figure 6 the case of incipient cavitation is shown (σ = 1.3, α_a = 4°). The estimated shedding frequency for this case was 39 Hz. In figure 6A and 6B the moment of cavitation structure appearance is caught. Cavitation structure was first observed near the contact line of the foil and the rotary disc. It enlarges (6C, 6D, 6E, 6F) and eventually the structure detached from the foil, while a new structure started to grow from the leading edge (6G, 6H). In figure 7 the developed, cloud cavitation is shown with σ = 1.29 and α_a = 8°. It can be observed that angle of attack had significant impact on obtained cavitating flow pattern. The structures became larger and occupied the whole width of the blade. The frequency of changes was about 47 Hz. The estimation of frequency value was difficult due to the fact, that in this case there was no evident moments of structure collapse and growth of a new structure, it happened simultaneously. The example of this is shown in picture 7C and 7D, where the former structure collapses after the detachment, while a new structure reaches about 1/3 of chord length.

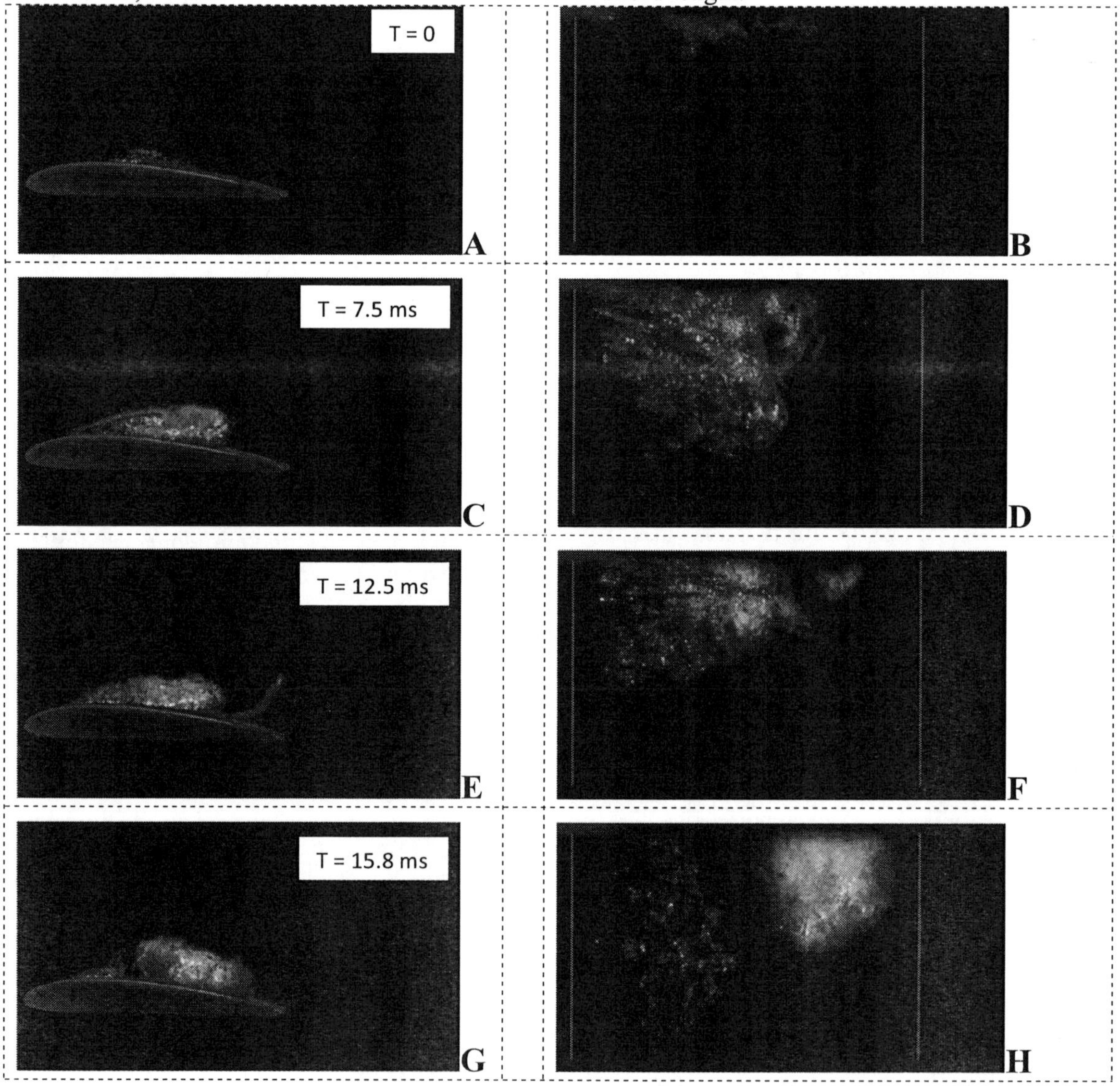

Figure 6. Incipient cavitation, σ = 1.29, α_a = 4°, u_∞ = 9.8 m/s, p_{in} = 63 kPa; left column – side view, right column – top view

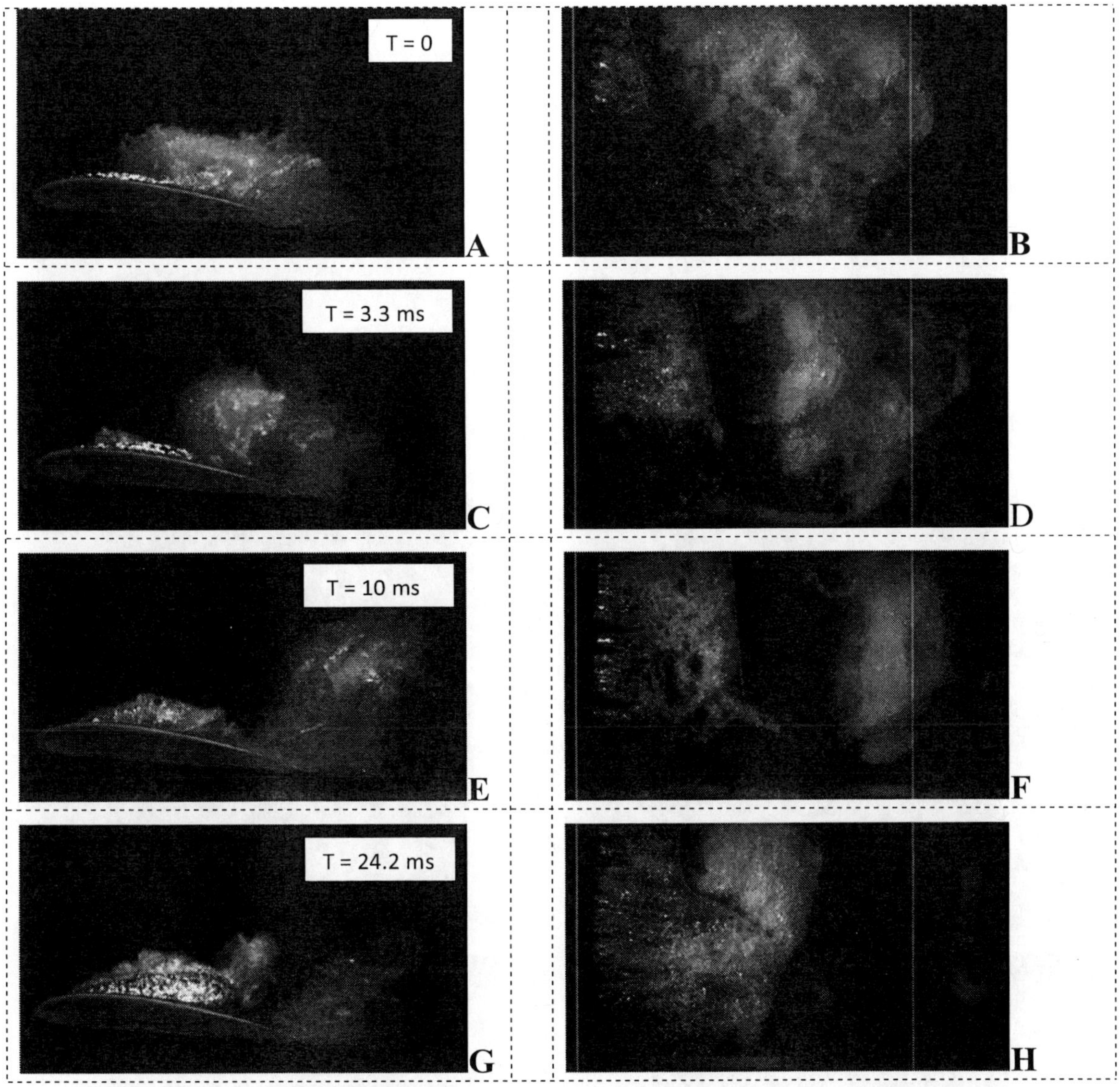

Figure 7. Developed cavitation, $\sigma = 1.37$, $\alpha_a = 8°$, $u_\infty = 9.7$ m/s, $p_{in} = 65.7$ kPa; left column – side view, right column – top view.

The analysis of the top view images showed the different shape of the vapour structures in the streamwise direction. Up to the 20% of chord length was occupied by attached packet of bubbles, then the structures detached from the foil wall and cloud of tiny bubbles can be observed (Fig. 7D). The cloud moved to the rear region of the blade (Fig. 7E, 7F) and eventually collapsed (Fig. 7G, 7H).

3.2. Comparison of experiment and CFD simulation

During the simulations the vapour volume fraction in the computational domain was monitored. Then the frequency of changes was determined and one period selected to images analysis. In figure 8. the change of vapour volume fraction in time is presented for case σ = 1.29 and α_a = 8°, described in chapter 3.1. The frequency of changes was 17 Hz, less than was observed in the experiment. The period selected to visualization is depicted in figure 9 with locations of specific moments shown in figure 10. The comparison of flow pattern obtained in simulations and in the experiment is included in

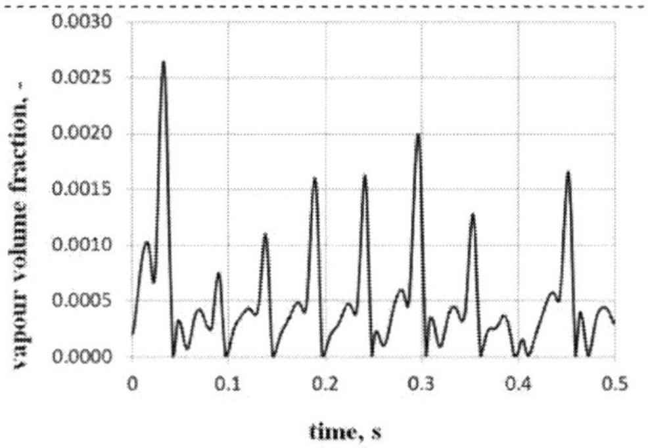

Figure 8. Vapour volume fraction in domain, $\sigma = 1.37$, $\alpha_a = 8°$

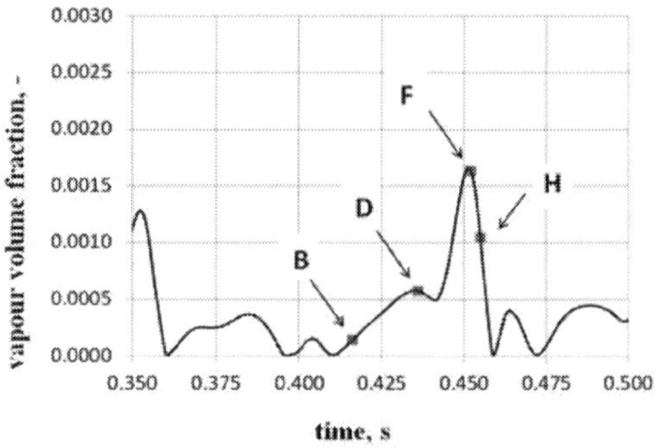

Figure 9. The chosen period of changes for $\sigma = 1.37$, $\alpha_a = 8°$

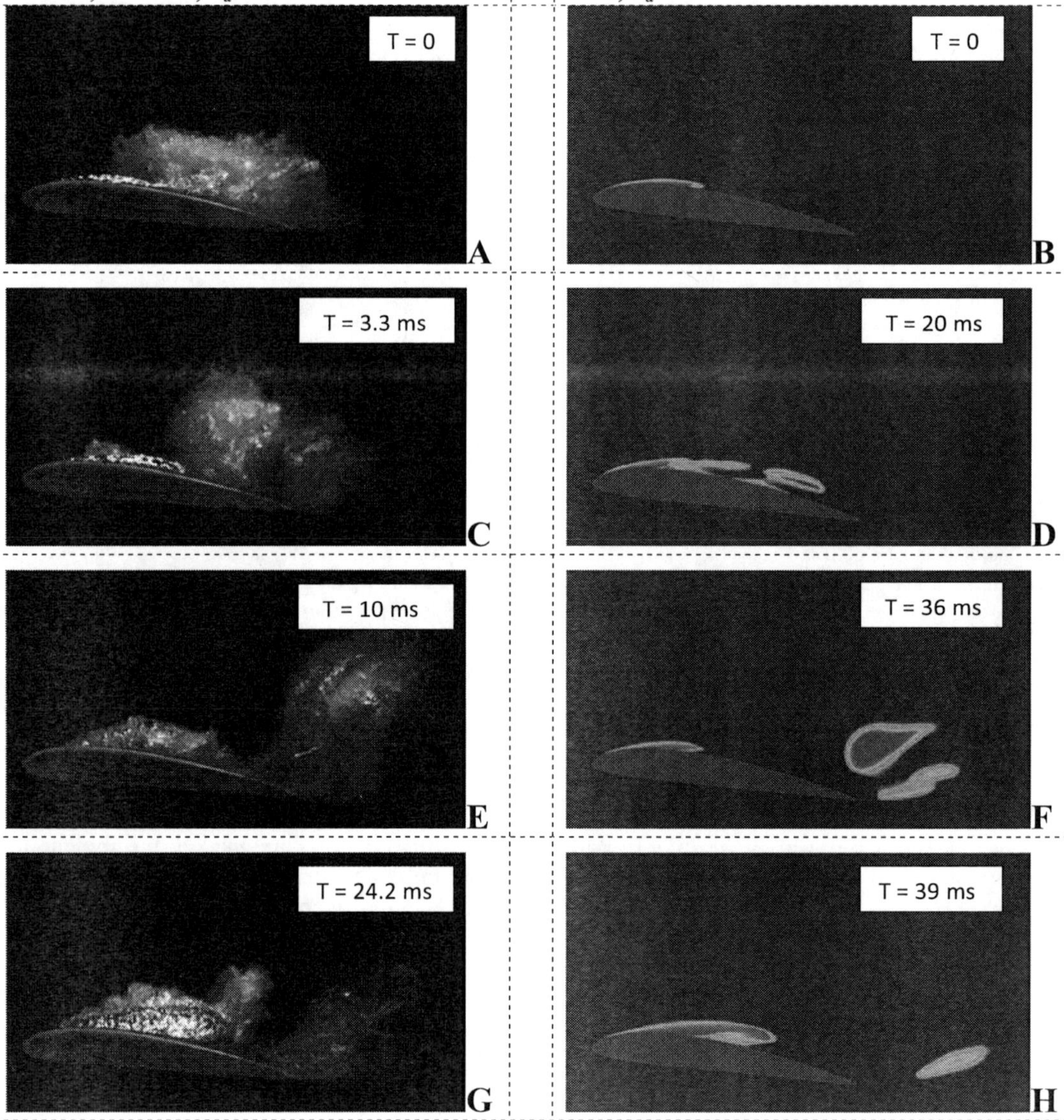

Figure 10. Developed cavitation, $\sigma = 1.37$, $\alpha_a = 8°$, $u_\infty = 9.7$ m/s, $p_{in} = 65.7$ kPa; left column – experiment, right column – numerical simulation

figure 10. Structures in the experiment were higher and longer than the ones obtained in calculations. In numerical simulation the structure grew (Fig. 10B), then split into two clouds (Fig. 10D) – the one closer to leading edge shrank (to about 40% of chord length), the other detached from the foil, moved to the trailing edge and collapsed (Fig. 10F). Meanwhile the structure near the leading edge enlarged to about 60% of chord length (Fig. 10H) and split again. During the experiment the structures appeared simultaneously, which can be seen when comparing images 10A and 10B. During the simulation there were moments when only one cloud existed, contrary to the experiment.

4. Summary and conclusions

In the paper the results of both experimental and numerical investigation of cavitating flow over Clark Y foil are presented. The visualization of cavitation appearing in the flow in both side and top views was done. That provided necessary information about shape and growth of 3D cavitation structures. During the top view frames analysis it was noticed that the structures exhibited non-uniformity in spanwise direction. The longest vapour structures occurred near the contact line of the blade and the disc that it was placed on. It was especially visible in case of low cavitation number and AoA cases. With increase of cavitation number and AoA the vapour clouds covered whole width of the blade. The phenomenon of recurring growth, detachment from foil wall and collapse of the structures near the blade rear region was observed and recorded. On the basis of image analysis also frequency of changes was estimated. For cavitation number equal to about 1.3 it was 39 Hz ($\alpha_a = 4°$) and 47 Hz ($\alpha_a = 8°$). The numerical investigation included simulation of the flow with parameters obtained during the experiment, with Kunz cavitation model. Although some simplifications were assumed, the obtained flow pattern was similar to the one from the experiment, the characteristic features of cavitating flow were noticed. The frequency of changes was lower in the numerical results, it was equal to 17 Hz. That leads to conclusion that selected cavitation model can be used to simulate this type of flow, but some modifications are recommended to improve accuracy of predicted shedding frequency and outreach of vapour structures.

5. References

[1] Brennen C E 1995 *Cavitation and bubble dynamics* (Oxford: Oxford University Press)
[2] Wan G, Senocak I, Shyy W, Ikohago T and Cao S 2001 Dynamics of attached turbulent cavitating flows *Progress in Aerospace Sciences* vol 37 pp 551-81
[3] Matsunari H, Watanabe S, Konishi Y, Suefuji N and Furukawa A 2012 Experimental/Numerical Study on Cavitating Flow around Clark Y 11.7% Hydrofoil *Proceedings of the Eight International Symposium on Cavitation CAV2012*
[4] Sedlar M, Komarek M, Wyroubal M and Muller M 2012 Experimental and numerical analysis of cavitating flow around a hydrofoil *EPJ Web of Conferences* vol 25 01084
[5] Franc JP and Michel JM 2005 *Fluid Mechanics and its Application* vol 76 (Fundamentals of cavitation), ed Kluwe Academic Publishers
[6] Kunz R 2000 A preconditioned Navier-Stokes method for two-phase flows with application to cavitation prediction, Computers & Fluid Vol 29 p. 849-875, 2000
[7] Puffary B 2006 Numerical modelling of cavitation, design and analysis of high speed pumps *Educational Notes RTO-EN-AVT-143* paper 3, pp 1–54
[8] Huang R, Luo X and Ji B 2017 Numerical simulation of the transient turbulent flows around Clark-Y hydrofoil using modified partially averaged N-S methods *Journal of Mechanical Science and Technology* vol 31(6)
[9] Capurso T et al. 2017 Numerical investigation of cavitation on a NACA0015 hydrofoil by means of OpenFOAM *Energy Procedia* 126 pp 794 – 801

Acknowledgments

The presented work was supported by the Polish National Science Centre founds within the project UMO-2016/21/B/ST8/01164.

IOP Conf. Series: Journal of Physics: Conf. Series **1101** (2018) 012009 doi:10.1088/1742-6596/1101/1/012009

Fluid flow and heat transfer analysis of a photovoltaic module under varying environmental conditions

Marek Jaszczur[1], Qusay Hassan[1,2], Mateusz Szubel[1], Ewelina Majewska[1]

[1]AGH University of Science and Technology, Faculty of Energy and Fuels, Poland
[2]University of Diyala, Department of Mechanical Engineering, Diyala, Iraq

e-mail: jaszczur@agh.edu.pl

Abstract. The objective of this work is to evaluate fluid flow and heat transfer from a standard photovoltaic module by means of comprehensive analysis for the polycrystalline photovoltaic modules by using three-dimension CFD numerical modelling and experimental measurement. The analysis shows that PV module temperature is incident solar radiation G_T and ambient temperature T_a dependent but also depends on the wind speed as well as wind direction. It has been additionally shown that the mounting conditions which are not included in any advanced mathematical model also plays a significant role. Base on presented results the already existing models for PV module temperature evaluation can be tuned in order to determine the PV module temperature accurately. The experimental and numerical results that were obtained enable the development of a model for the module operating temperature evaluation under varying real environmental conditions.

1. Introduction

In the last decades, solar energy become considered as an attractive option among renewable energy [1]. Manufacturing of photovoltaic modules uses very modern technologies and keeps a lot of challenge to turn solar radiation into electricity with high efficiency. Recently a considerable effort has been made by researchers to improve photovoltaic (PV) module performance mainly due to maximizing solar absorption and minimizing the optical losses as well as introducing new materials. The photovoltaic cell temperature is one of the critical parameters for assessing the module efficiency and for long-term performance and lifetime. The PV module temperature depends on many parameters such as solar irradiation, installation type and climatic conditions, modules configuration, cells types and thermal properties of materials used in module encapsulation [2] and many more for example dust layer. The effects of the PV module temperature on the power production under environmental condition was studied by Hussain et al. [3]. Authors found that the cell temperature increase, reduced significant the amount of generating electrical produced. The solar cells performance decreases with cell temperature increase, mainly due to increased internal carrier recharge rates, resulting from increased concentrations of carrier [4,5]. In the literature, various authors have developed thermal models relating to the PV module thermal flow and heat transfer mechanisms based on the environmental variables. In the most cases, their prediction due to a large number of arbitrarily selected parameters is poor and new more correlations should be developed. A review of the sensitivity analyses for correlations for PV module thermal process and heat transfer mechanisms has been presented by Coskun et al. [6]. Thermal analysis of wind incident angle on photovoltaic module temperature was studied by Kaplani and Kaplanis [7].

Authors results demonstrated a need for accurate evaluation of the wind velocity and wind incidence angle. It observed that the photovoltaic temperature slightly decreases with increasing wind incidence angle, particularly at high wind velocities. Al-Sabounchi [8] investigate the module tilted angle effect on the PV cell thermal behavior in the countries with hot climate. The investigation takes into account the ambient temperature, solar radiation, module manufacturing properties and other parameters. It was found that module tilted angle has a significant effect on the cell temperature. Jubayer et al. [9] studied the convective heat transfer from a ground mounted stand-alone PV modules. The result shows that the heat transfer rate is higher for solar panels installed on the ground than the solar panel mounted on a tilted surface, mainly due to poor ventilation from the back side. In the research study performed by Lobera and Valkealahti [10] thermal model based on the total energy balance in the PV component has been developed. Similar model for a double-glass multi-crystalline PV module based on the various thermal considerations was developed in [11]. Ju et al. [12] presented an improved method for estimating the temperature of solar cells which can be used for wide range of operating conditions. The photovoltaic module temperature under different wind speeds was analysed by Armstrong and Hurley [13]. Authors shown that there was a parasite difference in convective and radiative heat loss from the module at different wind speeds. Hassan et al. [14] analysed the temperature effect on the efficiency of the PV system and power generation. The results confirmed that the operating temperature of a module plays a significant role in the conversion process and the power output is significantly influenced by the temperature. Based on the thermal efficiency five different analytical models for module temperature evaluation has been verified by Tomar et al. [15]. The environmental effect on the PV temperature has been examined also in other studies [16-19].

The main objective of this research is to model the fluid flow and heat transfer from a standard PV module by using three-dimension CFD numerical modelling and experimental measurement. The obtained results will enable the development of a model for the PV operating temperature evaluation under varying real environmental conditions. By using this type of models, the operating temperature of working PV modules under different actual environmental conditions can be predict. Additionally base on presented results the already available theoretical models can be tuned in order to determine the module temperature with higher accuracy.

2. Experimental set-up

The experimental measurement was conducted using polycrystalline Sharp ND-RJ260 photovoltaic modules (temperature coefficient of power -0.42 %/°C and 1.6 m^2 of surface area, white colour back sheet, tilt angle β=15°) and BrukBet BEP260W (temperature coefficient of power -0.40 %/°C and 1.62 m^2 of surface area, white color back sheet, tilt angle β=35°). The nominal module power is equal to 260W. The azimuth for all PV modules was the same γ=20° West. The main hardware components of the data acquisition system for temperature and solar irradiation measurement consist of 4 Advantech ADAM4018 DAQ, 16-bit, an 8-channel analog input module that provides programmable input ranges on all channels, RS485 transceiver, computer with data acquisition software and data storage.

Figure 1. An experimental set-up with thermocouples arrangement

As temperature sensors, calibrated Copper-Constantan (T-type) thermocouples were used. Each module has been equipped with four thermocouples attached to the back sheet in the recommended by standard arrangement (see Figure 1). Signals from all thermocouples were acquired every second and the five minutes average value has been stored for analysis. The ambient temperature, air speed and solar irradiation (global and diffusive components), were also obtained from the local sensors, at location AGH University of Science and Technology in the city center of Kraków, Poland. The signals from weather station sensors were acquired every second and 5 minutes average value has been stored.

3. Numerical model

In this work, a three-dimensional model of the photovoltaic module with mounting system and surrounding area has been created to analyse fluid flow and heat transfer. The sketch of the analysed geometry and main heat transfer components are presented in Figure 2. The photovoltaic module size under consideration was 1.654 x 0.989 x 0.04m (length, width and thickness). However, it should be noticed that glass thickness was only 5mm. Due to the small thickness of layers, the averaged heat conduction coefficient was calculated. The size of the computational domain was equal to 4.2 x 2.8 x 2.2m, and an unstructured hex-dominant mesh was generated for the domain. The fluid flow domain contains boundary layer with prismatic cells for the photovoltaic module near-wall region. The area with the high impact of turbulence has a locally much finer mesh. Before a set of final CFD analysis were performed different meshes have been tested to obtain mesh independent solution. Based on tests mesh with 300,000 - 500,000 high-order elements has been found to give mesh-independent solution.

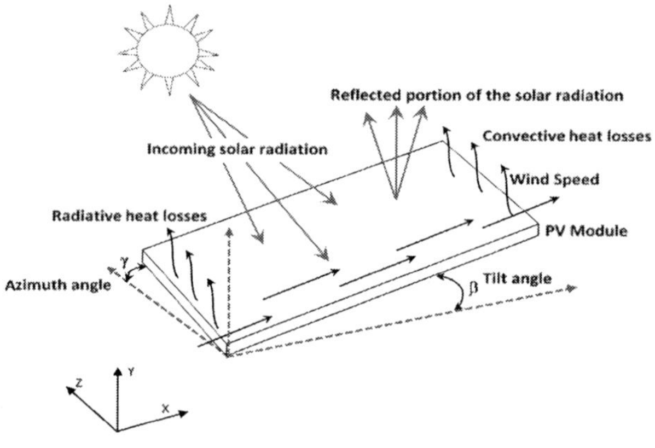

Figure 2. An analysed geometry and main heat transfer components

In this work, it assumed that temperature is a passive scalar and have no influence on the air flow. The considerations for the mathematical model eq. (1)-(3) are: steady state, newtonian fluid, negligible viscosity dissipation and the use of the Boussinesq approximation. The time averaged governing equations for continuity, momentum and energy in tensor notation, are as follows:

$$\frac{\partial \overline{U}_i}{\partial x_i} = 0 \tag{1}$$

$$\overline{U}_j \frac{\partial \overline{U}_i}{\partial x_j} = -\frac{1}{\rho}\frac{\partial \overline{P}}{\partial x_i} + \nu \frac{\partial^2 \overline{U}_i}{\partial x_j^2} - \frac{\partial}{\partial x_j}\left(\overline{u_i u_j}\right) \tag{2}$$

$$\overline{U}_j \frac{\partial \overline{T}}{\partial x_j} = \kappa \frac{\partial^2 \overline{T}}{\partial x_j^2} - \frac{\partial}{\partial x_j}\left(\overline{u_j \theta}\right) \tag{3}$$

where x_i, x_j are the Cartesian coordinates of the system, $\overline{U}, \overline{P}, \overline{T}$ are the mean velocity, dynamic pressure and temperature; ν, κ are fluid viscosity and thermal diffusivity. In order to model turbulent flow the realizable *k-ε* model which is a modification of the standard *k-ε* model was used. Model is a semi-empirical 2-equation eddy viscosity model, which is based on the Boussinesq hypothesis - Reynolds stresses can be expressed in terms of mean velocity gradients and that the turbulent eddy viscosity is related to the turbulent kinetic energy and the dissipation rate of turbulent kinetic energy. The k–ε turbulence models incorporate, on average, the influence of turbulent eddies (Reynolds stresses) through turbulent viscosity, which is related to the fluid rate of strain according to:

$$-\rho\overline{u_i u_j} = \tau_{ij} = \mu_T\left(\frac{\partial \overline{U}_i}{\partial x_j} + \frac{\partial \overline{U}_j}{\partial x_i}\right) = 2\mu_T S_{ij} \quad ; \mu_T = \rho C_\mu \frac{k^2}{\varepsilon}; \qquad \overline{u_i\theta} = -\left(\kappa + \frac{\nu_T}{\mathrm{Pr}_T}\right)\frac{\partial \overline{T}}{\partial x_i} \tag{4}$$

Realizable k-ε model contains a improved formulation for the turbulent viscosity. The model equation for the turbulent kinetic energy and for the dissipation rate can be written as follows:

$$\frac{\partial}{\partial x_i}\left(\overline{U}_i k\right) = \frac{\partial}{\partial x_i}\left(\frac{\nu_{eff}}{\sigma_k}\frac{\partial k}{\partial x_i}\right) + P_k - \varepsilon \tag{5}$$

$$\frac{\partial}{\partial x_i}\left(\overline{U}_i \varepsilon\right) = \frac{\partial}{\partial x_i}\left(\frac{\nu_{eff}}{\sigma_\varepsilon}\frac{\partial \varepsilon}{\partial x_i}\right) + C_1 S\varepsilon - \frac{C_2\varepsilon^2}{k + \sqrt{\nu\varepsilon}} \tag{6}$$

where k and ε are turbulence kinetic energy and dissipation rate, P_k represents the turbulent kinetic energy production as function of turbulent kinematic viscosity, $\mu_{eff} = \mu + \mu_T$ and $\nu_{eff} = \mu_{eff}/\rho$ are turbulent effective viscosity. The turbulent model constants are as follows:

$$C_1 = max\left[0.43, \frac{\eta}{\eta+5}\right]; \ \eta = S\frac{k}{\varepsilon}; \ S = \sqrt{2S_{ij}S_{ij}}; \ C_\mu = \frac{1}{A_0 + A_S\frac{k\sqrt{S_{ij}S_{ij}+\tilde{\Omega}_{ij}\tilde{\Omega}_{ij}}}{\varepsilon}}; \ S_{ij} = \frac{1}{2}\left(\frac{\partial \overline{U}_j}{\partial x_i} + \frac{\partial \overline{U}_i}{\partial x_j}\right)$$

$$\tilde{\Omega}_{ij} = \Omega_{ij} - 2\varepsilon_{ijk}\omega_k; \ \Omega_{ij} = \overline{\Omega}_{ij} - 2\varepsilon_{ijk}\omega_k; \ A_0 = 4.04; \ A_S = \sqrt{6}cos\phi; \ \phi = \frac{1}{3}cos^{-1}(\sqrt{6}W)$$

$$C_{\varepsilon 1} = 1.44; \ ; \ \sigma_k = 1.0; \ \sigma_\varepsilon = 1.2; \ W = \frac{S_{ij}S_{jk}S_{ki}}{\tilde{S}^3}$$

Presented heat transfer model includes solar model and radiation heat transfer model. For the radiative heat transfer Surface-to-Surface (S2S) radiation model has been used:

$$q_{out,k} = \varepsilon_k \sigma T_k^4 + \rho_k q_{in,k} \tag{7}$$

$$A_k q_{in,k} = \sum_{j=1}^{N} A_j \, q_{out,k} F_{jk} \tag{8}$$

where $q_{out,k}$ and $q_{in,k}$ are energy flux leaving the surface and energy flux incident on the surface from surroundings; A_k is an area of k-surface, F_{jk} is view factor between k-surface and j-surface. As a numerical solver Fluent v.18 was used with fluid and solid properties presented in Table 1.

Table 1. Fluid and solid properties.

Material name	Specific heat (J/kgK)	Density (kg/m^3)	Viscosity/ Emissivity (kg/ms) / (-)	Thermal conductivity (W/mK)
Air	1006.43	1.225	1.78*10^{-5}	0.0242
Panel	700	2700	0.92	197
Soil	1600	1600	0.95	2

Mean wind speed and turbulence intensity profiles for aerodynamics roughness length 0.03 m are employed at the domain entrance as an inlet boundary conditions. Five different wind speeds in the range 0.5-5m/s are selected. At the domain exit, a uniform zero gauge pressure boundary condition is used. Side walls of the computational domain have pressure outlet boundary condition applied. The bottom wall of the domain has been treated as a rough wall with no-slip boundary condition. For the thermal boundary conditions, at the inlet plane of the domain the air temperature T_a equal to the values obtained from measurements have been applied in the range 2-16°C (spring season). Solar radiation is applied to the thermal radiation model as an incident solar radiation in the range 100-1000W/m^2.

4. Results and discussion

Presented results show the effect of solar radiation, air temperature, air velocity and direction as well as module inclination and mounting conditions on the module operating temperature T_c. The incident solar irradiation G_T was in the range 100-1000 W/m^2 while the air velocity V_w was between 0.0-5.0m/s. The mathematical model and numerical schemes were validated against experimental measurement and the results of the analysis can be found in [20]. In this paper only selected experimental results for the module tilt angle β=15° are presented in Figure 6.

The flow field around the PV module for the air velocity V_w = 3.1 m/s is shown in Figure 3(a)-3(d). For the configuration presented in Figure 3(a), 3(c)-3(d) the wind direction is opposite to the X-axis direction while in Figure 3(b) air direction is opposite to the Z-direction (is from the module left side). It is worth to notice that for the configuration Figure 3(d) module tilt angle is β=0° and the space under the module is closed (the air flow can't flow under the module) which reduces cooling conditions. The module temperature for the configuration presented in Figure 3(a)-3(d) and for air temperature equal to T_a=8°C is shown in Figure 4(a)-4(d). Depend on configuration temperature distribution and mean module temperature varied significantly. Due to poor cooling conditions, the highest temperature has been obtained for module mounted horizontal (β=0°). In contrast, the lowest temperature was obtained for the case 4(b) with β=15° and side cooling (air flow from the side).

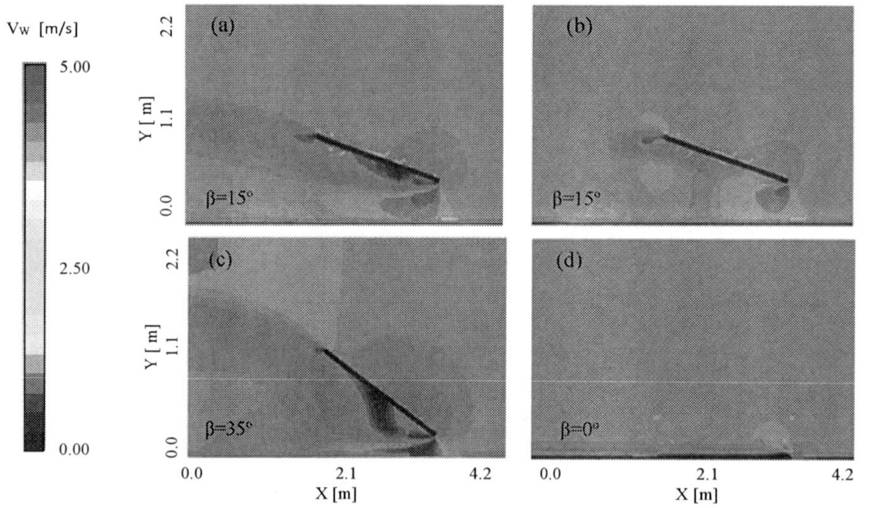

Figure 3. Flow field at different module inclination β and air flow direction.

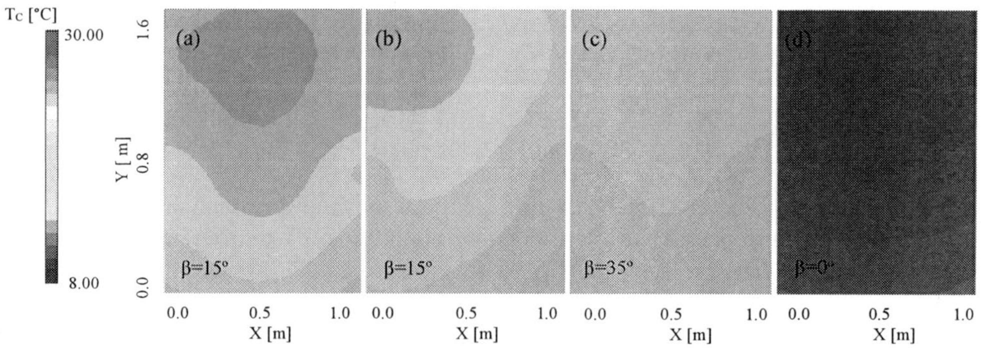

Figure 4. Temperature distribution on the module surface.

As can be seen the temperature distribution at module surface is highly non-uniform and vary significantly during the day depends on wind speed and the Sun position. The effect of the Sun position on the module temperature distribution for incident solar irradiation G_T = 470 W/m^2, wind speed V_w = 1.0 m/s, ambient temperature T_a=8 °C, inclination β = 0°, 15° and for different time (at day 12.04.2018) is presented in Figure 5.

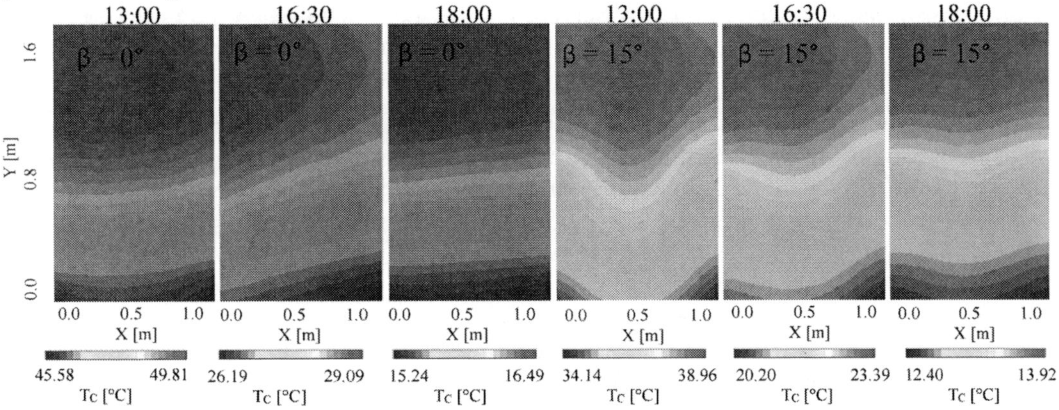

Figure 5. The PV module surface temperature during the day (G_T = 470 W/m^2, V_w = 1.0 m/s, T_a=8°C).

Two of the three essential parameters which influence module temperature are the wind speed V_w and ambient temperature T_a. The effect of those parameters on the mean module temperature for β = 15°, 30° are presented in Figure 6(a)-6(b). It can be seen that for low wind speed heat loss is weak and the module temperature becomes very high which decrease module efficiency about 20%. The presented data compared with the experimental measurement.

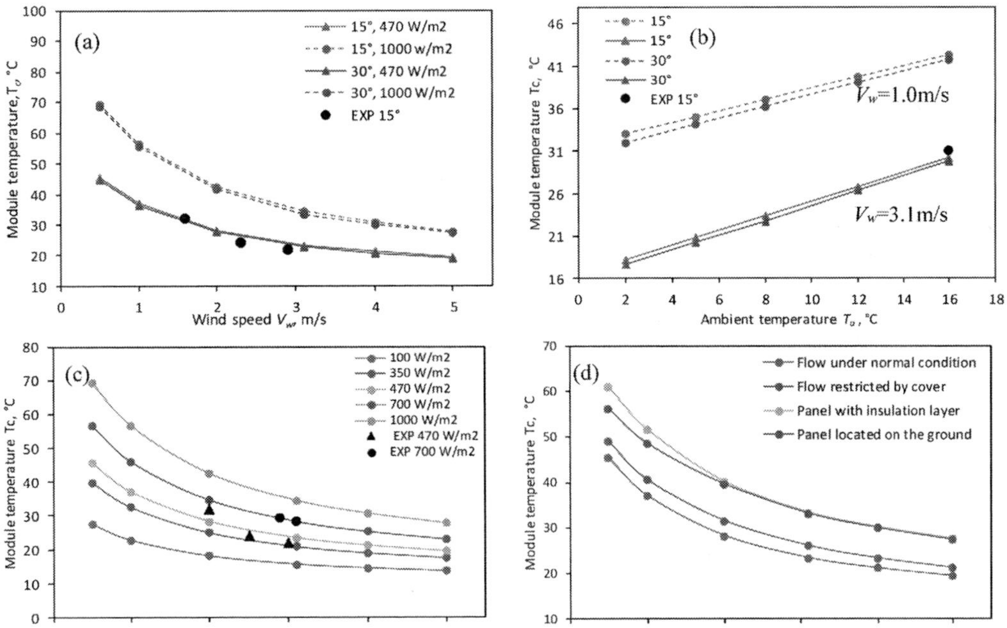

Figure 6. The effect of wind speed V_w (a), ambient temperature T_a (b), solar irradiation G_T (c) and mounting configuration (d) on the average module temperature T_c.

The third important parameter which directly influenced PV module temperature is incident solar irradiation G_T. The effect of solar irradiation on module temperature for different wind speed V_w and at two module angles $\beta = 15°, 30°$ is shown in Figures 6(c). The last (fourth) important parameter is related to the module mounting configuration. Usually module front surface is well cooled, but the module back side heat transfer can very significant. For the cases when the modules back surface is under good cooling conditions air flow from two sides can be nearly the same and the module temperature become lowest. However, when cooling conditions become poor the temperature an increase about 2.6-3.6°C in reference to the normal flow condition is observed. On the other hand when the module is mounted horizontally poor cooling condition cause additional module temperature increase about 10.0-11.2°C (see Figure 6(d)).

In the literature, several correlations have been proposed to evaluate PV module temperature. However, most of them use two or three parameters only. In the present work modification of the one of most popular correlation for the module T_c operating temperature evaluation will be used [21]:

$$T_C = T_a + kG_T \tag{9}$$

As can be seen in this linear expression no wind speed V_w is included. The mounting conditions are taken into account and depend on the single dimensional parameter k, known as the Ross coefficient. Reported values for k are in the range 0.02–0.04 Km^2/W [21]. In the literature, a few models can be found which include the effect of wind speed but without mounting conditions consideration [22]:

$$T_C = 0.943T_a + 0.028G_T - 1.528V_w + 4.3 \tag{10}$$

In the present paper with the use of numerical data and experimental data new correlations is proposed which takes into account four - all key - parameters (solar irradiation, ambient temperature, wind speed and mounting configuration) as the follows:

$$T_C = T_a + kG_T + e^{(3.75-0.75V_w)} \tag{11}$$

where T_a and G_T are ambient temperatures and incident solar radiation, V_w is the wind velocity, k, is the Ross coefficient. The equation coefficients were evaluated using the least squares method. This correlation gives a possibility to take into account ambient temperature, solar radiation, wind speed and mounting scheme or another PV configuration arrangement. The correlation eq.(11) accuracies have been assessed using: root mean squire error (RMSE), the mean bias error (MBE), the coefficient of determination (R^2) presented in Table 2.

Table 2. The models accuracies statistical indicators.

Model	MBE	RMS	R^2
Model Eq. (9)	1,52	4,57	0,613
Model Eq. (10)	3,06	5,34	0,467
Model current Eq. (11)	0,72	2,71	0,933

5. Conclusions

The temperature of the photovoltaic panel is an important issue and it is directly related to the PV system performance. The present paper shows the fluid flow and heat transfer from a PV module by means of experimental measurement and three-dimensional simulation for determining the temperature of the module. Presented results contribute to a better knowledge of the module behaviour influenced by the various environmental conditions as well as mounting conditions. The analysis shows that PV module temperature depends on: incident solar radiation G_T, ambient temperature T_a wind speed V_w and wind direction. It has shown that the mounting conditions (which are not included in any advance model) play a significant role and may cause an additional module temperature increase in the range 10.0-15.5°C. Such temperature additional increase can cause mono and polycrystalline module efficiency decrease about 3.8-6.5%. The presented experimental and numerical results enable the development of an improved model for the PV module temperature related with the environmental and mounting conditions. The presented model is much sought after by the computer software, malfunction detection tools, and system designers.

6. Acknowledgement
The present work was supported by the Polish Ministry of Science (Grant AGH No. 11.11.210.312).

References

[1] Sawin L Sverrisson F Seyboth, K Adib R Murdock E Lins C Edwards I Hullin M Nguyen H Prillianto S Satzinger K 2017 *Renewables 2017 Global Status Report*
[2] Hassan Q Jaszczur M Przenzak E Abdulateef J 2016 The PV cell temperature effect on the energy production and module efficiency *Cont. Problems of Power Eng. and Env. Pro.* p 33
[3] Hussein H Muta I Hoshino T Osakada M 1995 Maximum photovoltaic power tracking: an algorithm for rapidly changing atmospheric conditions *Pro. Gen. Tra. and Dis.* **142** pp 59-64
[4] Zhao B Chen W Hu J Qiu Z Qu Y Ge B 2015 A thermal model for amorphous silicon photovoltaic integrated in ETFE cushion roofs *Energy con. and management* **100** pp 440-448
[5] Hammad B AlAbed M AlGhandoor A AlSardeah A AlBashir A 2017 Modeling and analysis of dust and temperature effects on photovoltaic systems *Ren. and Sus. Ene. Reviews*
[6] Coskun C Toygar U Sarpdag O Oktay Z 2017 Sensitivity analysis of implicit correlations for photovoltaic module temperature: A review *Journal of Cleaner Production* **164** pp1474-1485
[7] Kaplani E Kaplanis S 2014 Thermal modelling and experimental assessment of the dependence of PV module temperature on wind velocity and direction *Solar Energy* **107** pp 443-460
[8] Al-Sabounchi M 1998 Effect of ambient temperature on the demanded energy of solar cells at different inclinations *Renewable Energy* **14** pp 149-155
[9] Jubayer M Siddiqui K Hangan H 2016 CFD analysis of convective heat transfer from ground mounted solar panels *Solar Energy* **133** pp 556-566
[10] Lobera T Valkealahti S 2013 Dynamic thermal model of solar PV systems under varying climatic conditions *Solar energy* **93** pp 183-194
[11] Notton G Cristofari C Mattei M Poggi P 2005 Modelling of a double-glass photovoltaic module using finite differences *Applied Thermal Engineering* **25** pp 2854-2877
[12] Ju X Vossier A Wang Z Dollet A Flamant G 2013 An improved temperature estimation method for solar cells operating at high concentrations *Solar Energy* **93** pp 80-89
[13] Armstrong S Hurley G 2010 A thermal model for photovoltaic panels under varying atmospheric conditions *Applied Thermal Engineering* **30** pp1488-1495
[14] Hassan Q Jaszczur M Przenzak E 2017 Mathematical model for the power generation from arbitrarily oriented photovoltaic panel *In E3S Web of Conferences* **14** p 01028
[15] Tomar V Tiwari N Bhatti S Norton B 2018 Thermal modeling and experimental evaluation of five different photovoltaic modules integrated on prototype test *Ene. Con. Man.* **165** pp 219-235
[16] Styszko K Jaszczur M Teneta J Hassan Q Burzyńska P Marcinek E Łopian N Samek L 2018 An analysis of the dust deposition on solar photovoltaic modules *Environmental Science and Pollution Research* pp 1-9
[17] Jaszczur M Teneta J Styszko K Hassan Q Burzyńska P Marcinek E Łopian N 2018 The field experiments and model of the natural dust deposition effects on photovoltaic module efficiency *Environmental Science and Pollution Research* pp 1-9
[18] Mattei M Notton G Cristofari C Muselli M Poggi P 2006 Calculation of the polycrystalline PV module temperature using a simple method of energy balance *Renewable energy* **31** pp 553-567
[19] Bardhi M Grandi G Tina M 2012 Comparison of PV cell temperature estimation by different thermal power exchange calculation methods *ICREPQ'12 Santiago de Compostela* pp 28-30
[20] Jaszczur M Teneta J Hassan Q Styszko K Skiba A Majewska E 2018 An experimental and numerical investigation of photovoltaic module temperature under varying environmental conditions ICCHMT2018, 21-24 May 2018, Krakow, Poland
[21] Ross RG 1976 Interface design considerations for terrestrial solar cell modules *In 12th Photovoltaic Specialists Conference* pp 801-806
[22] Chenni R Makhlouf M Kerbache T Bouzid A 2007 A detailed modeling method for photovoltaic cells *Energy* **32** pp 1724-1730

IOP Conf. Series: Journal of Physics: Conf. Series **1101** (2018) 012010 doi:10.1088/1742-6596/1101/1/012010

Influence of deflocculants on shear stress in hydromixture flow

Beata Jaworska and Artur Bartosik[1]

[1]Faculty of Management and Production Engineering, Kielce University of Technology, Kielce, Poland

E-mail:b.jaworska@tu.kielce.pl, artur.bartosik@tu.kielce.pl

Abstract. The paper presents experimental studies on influence of two arbitrary chosen additives on wall shear stress in hydromixture, which consists solid particles of averaged diameter 0.05 mm. Experiments were performed for two mass concentration of solids equal to 20% and 43%. Measurements were performed for varied doses of deflocculants in three different proportions in wide range of shear rates. Experiments confirmed influence of chosen additives on decreasing hydromixture viscosity and as a consequence decreasing of shear stresses. The analysed process is complex and strongly depends on doses of deflocculants and solids concentration. However, the rheological property of such mineral suspension is complex and there is no single parameter that can solely explain it. Physical and chemical properties affect on the level of interparticle interaction or aggregation and therefore it is useful to control them in industrial process such as transportation of slurries.

1. Introduction

Process of lime production causes that wasted minerals appears in large amount during rinsing process of lime stones. Wasted minerals are continuously transported to selected landfill using water as a carrier liquid. Unfortunately, the solids are transported with low solid mass concentration. That circumstances cause that such process requires much amount of water and energy to feed centrifugal pumps. To improve properties of hydromixture during its flow in pipeline, mechanical and chemical methods are considered to deal with. Mechanical methods are focused on especially designed grooves or mechanically deformed pipes. Chemical methods are using deflocculants mainly in order to decrease of hydromixture viscosity and as a consequence decrease shear stress. Presence of solid particles in a carrier liquid could increase or attenuate turbulence and depends on solid sizes, flow conditions and carrier liquid properties [1-5].

The paper presents method of increasing solids concentration of wasted minerals during its transportation by adding two components to hydromixture in order to reduce the wall shear stress in a flow.

2. Hydromixture flow

2.1. *Types of hydromixture flow in pipeline installation*

Horizontal flow of hydromixture consisting of fine solid particles and water, as a carrier phase, can occur in several ways. Due to several factors, like interaction between particles, particle diameter, critical velocity of a hydromixture at which the bottom sediment forms, and the rate of settling of solid particles, four different flow regimes can be distinguished [6]:

- homogeneous or pseudo-homogeneous flow, characterized by a homogeneous distribution of solid particles in a flow cross section,
- heterogeneous flow in which all particles remain in suspension or heterogeneous with bottom sediment moving by leaps,
- flow with moving bottom sediment (sliding bed),
- flow with stationary bottom sediment (fixed bed).

For hydromixture consisting of liquid and fine dispersed solid phase with a moderate ratio of particles density to the liquid phase density and at corresponding high flow rates, homogeneity of flow is usually assumed [7]. Such hydromixture can be classified as monophasic liquid with increased viscosity and density. When hydromixture flow with moderate velocity is considering we usually deal with pseudo-homogenous flow. At sufficiently high velocities, the momentum forces dominate, while the share of other forces acting on particles is negligibly small. In this case, solids contained in hydromixture are evenly distributed in the flow cross-section [1]. For low flow velocities, bottom sediment appears, which can be fixed or sliding. The formation of bottom sediment is an undesirable phenomenon leading to a drop in flow rate or even clogging of pipeline.

Figure 1 presents the velocity distribution in horizontal pipeline of transported hydromixture in cross-section; average volume concentration of solid phase in the flow cross-section and transport concentration of solid phase in the unit of cross-section for mentioned flow types.

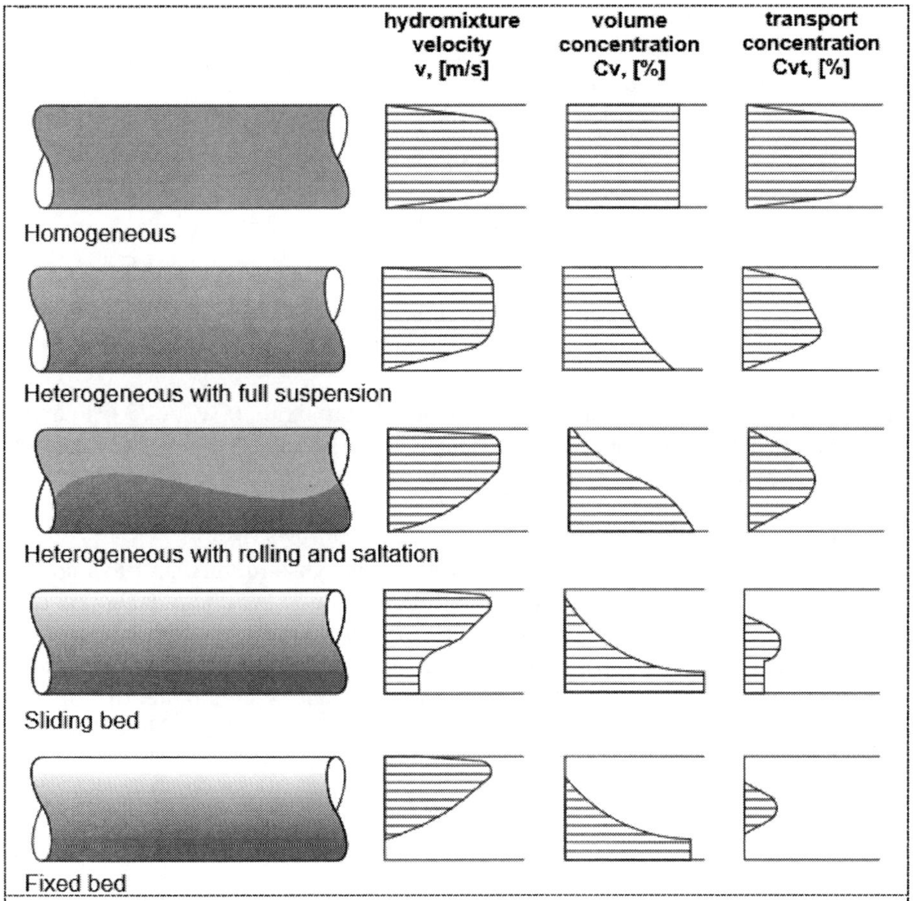

Figure 1. Types of hydromixture flow – horizontal pipeline [7].

2.2. *Hydromixture flow in lime production process*

Pipeline transport of fine-grained solid particles and water, named as hydromixture, appears in a lime production process. In the final stage, hydromixture is subjected to the sedimentary reservoir located at a considerable distance from the plant. Its further transport by pipeline system is difficult because of its rheological properties. The research is focused on applying deflocculants with a defined composition and serve them to transported hydromixture in order to be able to increase solids concentration and making process economically efficient. This may also an effect on increasing the flow rate without increasing the energy consumption of the pumps or pumps system. Another advantage of proposed solution would be reduction of water consumption, which has been used as a carrier liquid.

2.3. *Physical properties of hydromixture*

Sample of hydromixture used in experiments was procured from lime production process. In order to determine mass concentration of tested material, three samples of 100 ml each were collected and dried to obtain averaged values. Laboratory tests were carried out for two different mass concentrations of hydromixture equal to 20% and 43%. Mass concentration determines the percentage content of solids in total weight of hydromixture.

Granulometric analysis was carried out on the particle size analyser, named Mastersizer 3000, in the range from 0.01 to 3500 μm. Measured particle sizes of hydromixture were in the range from 0.5 to 163.5 μm. The averaged particle diameter was 45.5 μm. Table 1 shows percentage content of grain sizes with different particle diameter in tested hydromixture.

Table 1. Percentage content of grain fraction in hydromixture. Measured by using Mastersizer 3000.

Grain size (μm)	Content (%)
0 ÷ 2	6.5
2 ÷ 50	82.4
50 ÷ 163.5	11.1

As the result of analysis, three types of fractions have been separated. Dust fraction with diameter in the range of 2-50 μm is the highest percentage in the sample (82.4%). Sand fraction with an average grain diameter greater than 50 μm is 11.1%, while the phase of clay with grain diameter smaller than 2 μm is 6.5% of all particles.

Chemically, tested hydromixture consists mostly of calcium oxide (CaO – 73.6%) and silicon oxide (SiO_2 – 13%). Other chemicals included in sample are: MgO – 0.6%, Fe_2O_3 – 0.3%, Al_2O_3 – 1.1% and SO_3 – 0.3%. Others, unidentified substances constitute about 11.1% of the sample.

2.4. *Decreasing shear stress in hydromixture flow by using deflocculants*

Deflocculants are used in order to reduce shear stresses in hydromixture and to improve conditions encountered during its flow in a pipeline. Adding thinners or any other chemical compounds to flowing medium is chemical processing. It is vital that quantity and quality of added substances should be adapted to physical and chemical properties of hydromixture. Very important is also character of interactions of deflocculants with hydromixture. Added substances should stabilize hydromixture and improve its ability to disperse solid particles in order to increase its fluency. However, the main task is to reduce viscosity of hydromixture and therefore the shear stress. In describing case, value of viscosity and shear stress increases as the result of saturation by dust particles derived from the processed raw material.

3. Experimental studies

The experiments were performed in Laboratory of Rheology at the Kielce University of Technology. To determine the effect of tested deflocculants on shear stress in hydromixture for adopted shear rates, a rotational viscometer Anton Paar MCR 302 was used. Measurements were carried out in systems of coaxial cylinders with measuring gap equals to 1.1 mm for samples of 18 ml volume in temperature of 20°C. Experimental studies included investigations of the influence of varied doses of deflocculants on rheological properties of hydromixture. Additives were added in concentrations of 0.1%, 0.2%, 0.3%, 0.4% and 0.5%, according to dry mass of solid particles in three different ratio of each substance. Studies were carried out for hydromixture with two mass concentrations: 20% and 43% after 10 seconds of mixing with $\partial U/\partial y=200$ [1/s]. Measurements of shear stresses versus shear rates in the range from 1 to 1000 [1/s] were performed.

3.1. *Characteristic of deflocculants used in the experiment*

Compounds used in the experiments are calcareous groats and sodium water glass. These two substances are used as deflocculants – chemical additives intended to push away-suspended particles from each other and increase fluency of hydromixture.

Calcareous groats is a side-product of the lime production process. It is a remnant of lime slaking process by which the hydrated lime is formed. In the final step of lime slaking the hydration of lime and water mixture in suitable mixing chamber takes place. Waste materials that have not been slaked and unburned, sink at the bottom of the chamber. Next, they are periodically removed outside by the trigger aperture located in the lower chamber of the mixer. Obtained hydrated lime is subjected to two-stage separation process in order to separate it from contaminants. Hydrated lime is selling as a finished product, while the remains of the separations are called calcareous groats [2].

Figure 2 and 3 presents image of the surface of calcareous groats made by scanning electron microscope. Figure 3 presents micro-scale features of tested sample made with 12000 magnifications after application of 10kV electron-accelerating voltage.

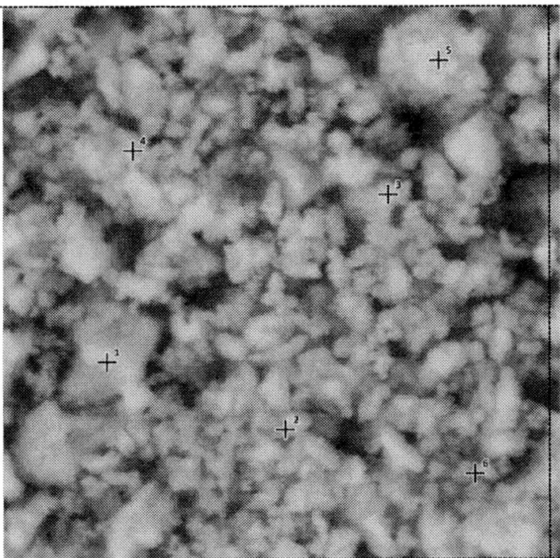

Figure 2. Image of the surface of calcareous groats made by using Scanning Electron Microscope Phenom Pro X after application of 15kV electron-accelerating voltage.

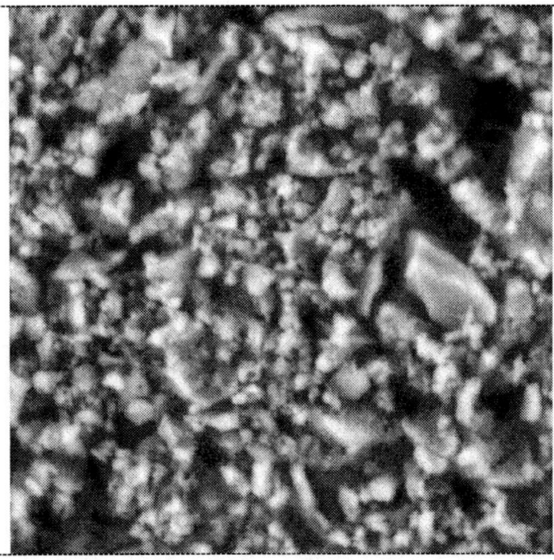

Figure 3. Image of the surface of calcareous groats made in 12000 zoom by using Scanning Electron Microscope Phenom Pro X.

Table 2. Chemical composition of calcareous groats made by Energy Dispersive X-Ray Spectroscopy.

Element Number	Element Symbol	*Element Name*	Atomic Concentration	Error
20	Ca	*Calcium*	13.1	0.0
8	O	*Oxygen*	79.7	0.1
6	C	*Carbon*	2.2	0.9
53	I	*Iodine*	0.2	0.0
7	N	*Nitrogen*	4.3	2.1
12	Mg	*Magnesium*	0.5	0.8

Table 2 includes results of measurements of chemical microanalysis made by Energy Dispersive X-Ray Spectroscopy in conjunction with SEM.

Another deflocculant used in experimental studies is well known dispersant sodium water glass (also liquid glass). Aqueous solution of sodium silicate is characterised by high viscosity and density even 20-70% more than the water. Table 3 presents main properties of this substance.

Table 3. Physical and chemical properties of sodium water glass.

Properties	
Chemical formula	Na_2SiO_3
Appearance	thick liquid
Density	2.61 g/cm^3
Specific gravity	1.39 g/cm^3 (20°C)
Melting point	1.088°C
Solubility in water	22.2 g/100ml (25°C)
pH	approximately 11.3

A mixture of calcareous groats and sodium water glass was added to hydromixture and used as deflocculants in experimental studies.

3.2. *Results of measurements*

The rheological theory of concentrated hydromixture is strongly related to the viscosity-dependence on the particle size. From this point of view, the investigation of rheological behaviour of tested raw material with respect to particle size and distribution is necessary.

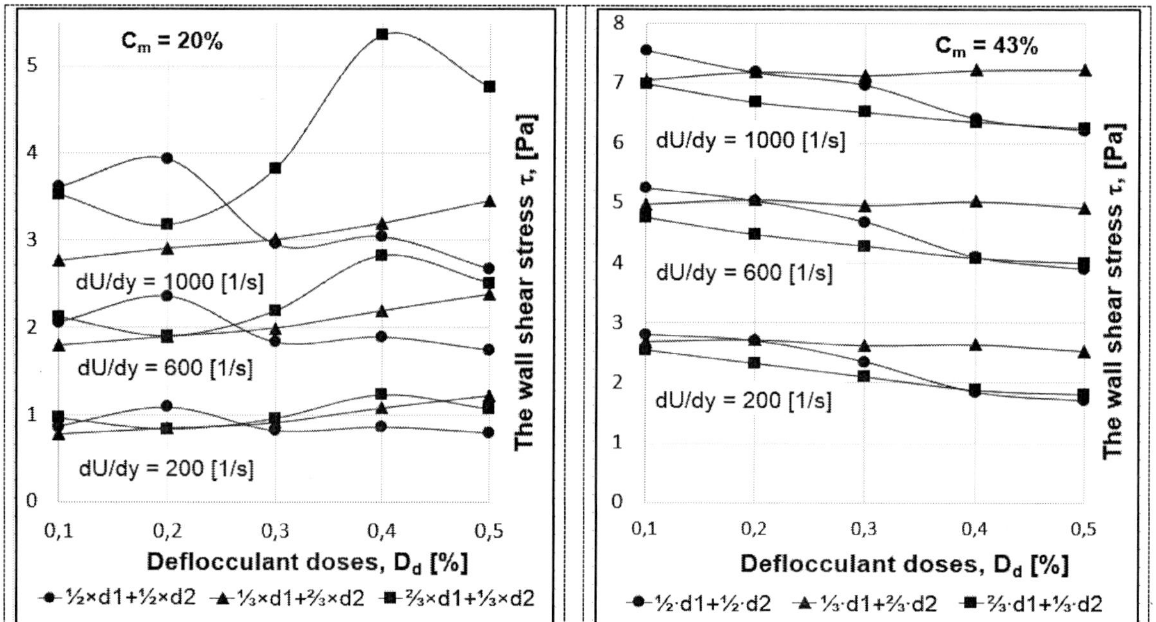

Figure 4. Dependence of the wall shear stress on different samples of deflocculants and its doses for three arbitrary chosen values of shear rates in hydromixture with two different mass concentrations.

The paper presents experiments on influence of two arbitrary chosen deflocculants, which are sodium water glass, denoted as d1, and calcareous groats, denoted as d2, on wall shear stress in tested hydromixture. An average diameter of solids included in hydromixture was 0.05 mm. Solids concentrations in hydromixture was equal to 20% and 43% by mass. Measurements of wall shear stress for different shear rates in hydromixture with varied doses of mixed deflocculants were performed in the range of shear rates up to 1000 [1/s].

Figure 4 presents dependence of the wall shear stress on different samples of deflocculants for hydromixture with mass concentration 20% and 43%. Samples were made in proportions: (1/3·d1+2/3·d2), (1/2·d1+1/2·d2) and (1/3·d1+2/3·d2). Its doses were in the range from 0.1% to 0.5% for three arbitrary chosen values of shear rates, equal to (200; 600; 1000) [1/s].

On the bases of presented results of experiments we can observe that influence of chosen additives on shear stress is complex and depends on solids mass concentrations of tested hydromixture and doses of deflocculant samples. As we can see, for hydromixture with 43% of solid mass concentration, the lowest wall shear stress refers to additive of deflocculants in proportions of (2/3·d1+1/3·d2) for doses in the range of 0.1% to 0.4% and (1/2·d1+1/2·d2) for deflocculants doses up to 0.5%. Simultaneously, for the mass concentration equals to 20% the similar effect are obtained with deflocculants in proportions of (1/3·d1+2/3·d2) for doses in the range of 0.1% to about 0.3% and (1/2·d1+1/2·d2) for deflocculants doses up to 0.5%. This phenomenon is the same for all chosen shear rates in both cases.

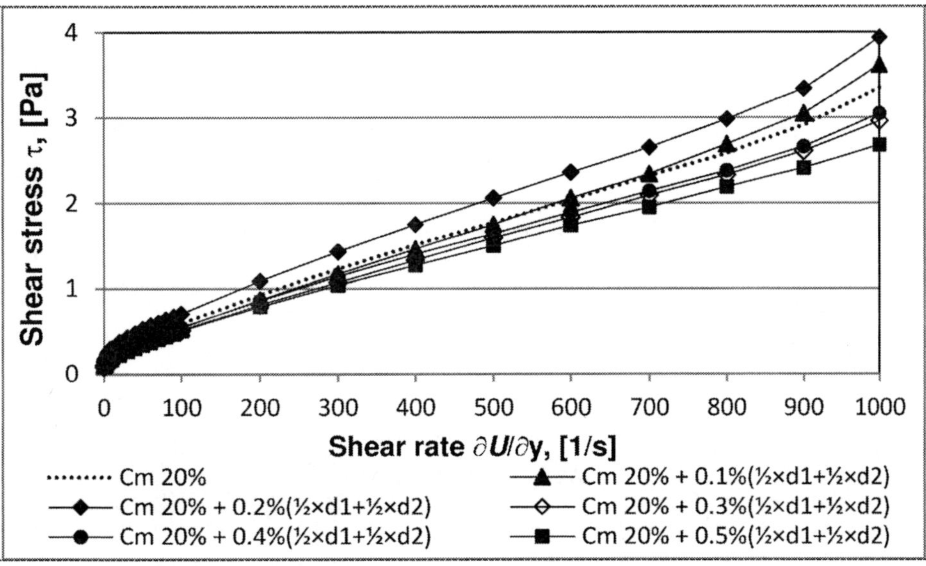

Figure 5. Flow curve of hydromixture with mass concentration C_m=20% with additive of 0.1...0.5% of deflocculants measured after 10 seconds of mixing at $\partial U/\partial y$=200 [1/s].

Figure 5 shows shear stress versus shear rate for raw hydromixture with deflocculants in proportions of (1/2·d1+1/2·d2). As we can see, the addition of combination of chosen chemicals at dose 0.1% and 0.2% in relation to dry mass included in hydromixture with C_m=20% do not improve its rheological properties but even worsen it. Addition of higher doses of deflocculants affect on reduction of shear stress arisen in hydromixture in wide range of shear rate. The largest decrease of shear stress was obtained after application of 0.5% of deflocculants into hydromixture. Results of measurements of generated shear stress and viscosity values for analyzed hydromixture are presented in Table 4.

Table 4. Measured values of shear stress and viscosity generated in hydromixture with mass concentration of 20%.

Shear Rate [1/s]	C_m 20%		C_m 20% + 0.5%·(1/2·d1+1/2·d2)	
	Shear Stress [Pa]	Viscosity [Pa·s]	Shear Stress [Pa]	Viscosity [Pa·s]
1000	3.350	0.00335	2.680	0.00268

900	2.910	0.00323	2.410	0.00267
800	2.580	0.00322	2.190	0.00274
700	2.320	0.00331	1.950	0.00279
600	2.040	0.00340	1.740	0.00290
500	1.770	0.00354	1.500	0.00301
400	1.510	0.00378	1.270	0.00318
300	1.230	0.00410	1.040	0.00346
200	0.925	0.00463	0.787	0.00394
100	0.593	0.00593	0.505	0.00505

If higher mass concentration of hydromixture is considered (C_m=43%) we can observe clear decrease of shear stress for all doses of deflocculants in full range of shear rates. The best influence on decrease the shear stress we can notice for dose of 0.5% of sodium water glass and calcareous groats in relation to dry mass of hydromixture. Data presented in Figure 6 demonstrate shear stresses in hydromixture, which includes deflocculants in proportions of (1/2·d1+1/2·d2).

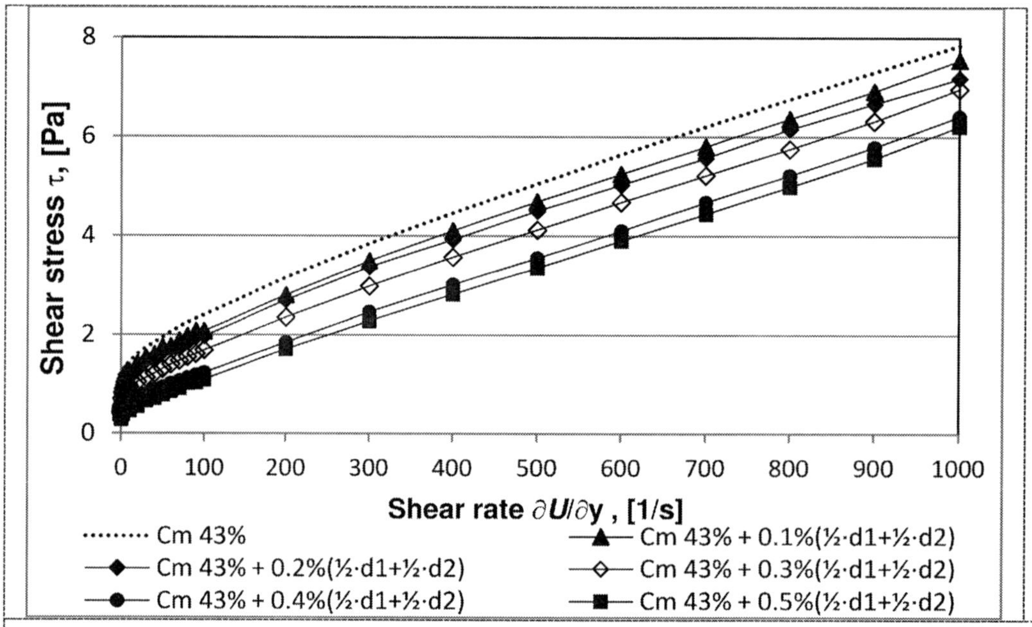

Figure 6. Flow curve of hydromixture with mass concentration C_m=43% with additive of deflocculants from 0.1% to 0.5% measured after 10 seconds of mixing at $\partial U/\partial y$=200 [1/s].

Experimental data presented in Table 5 show observable decrease of viscosity and shear stress using dose of deflocculants equals to 0.5% in proportions of (1/2·d1+1/2·d2).

Table 5. Measured values of shear stress and viscosity generated in hydromixture with mass concentration of 43%.

Shear Rate [1/s]	**C_m 43%**		**C_m 43% + 0.5%·(1/2·d1+1/2·d2)**	
	Shear Stress [Pa]	Viscosity [Pa·s]	Shear Stress [Pa]	Viscosity [Pa·s]
1000	7.850	0.00785	6.220	0.00622
900	7.310	0.00812	5.560	0.00618
800	6.760	0.00845	4.990	0.00624
700	6.210	0.00888	4.440	0.00635
600	5.640	0.00940	3.900	0.00651

500	5.050	0.01010	3.350	0.00670
400	4.460	0.01110	2.820	0.00705
300	3.840	0.01280	2.280	0.00761
200	3.150	0.01580	1.710	0.00857
100	2.400	0.02400	1.100	0.01100

Results of experiments confirmed that influence of additives on shear stress is strongly depending on proportion and doses of deflocculant in samples. In order to decrease viscosity in hydromixture, and as a consequence the shear stress, proper proportion of additives related to dry mass of hydromixture has to be applied.

4. Conclusions

Influence of solids concentration of hydromixture on rheological properties is significant since various ranges of solids concentration can lead to different types of flow curves. Rheological behaviour of tested hydromixture is pseudo-plastic (shear-thinning). The degree of its pseudo-plasticity increases with increasing solids concentration. Results of research proved that addition of small amount of suitable thinner can improve flow-ability of hydromixture by decreasing its viscosity. This can enhance the energy efficiency of hydromixture transport.

Presented analysis of the influence of various doses of sodium water glass and calcareous groats gave response to a question regarding the amount of substances needed to apply in hydromixture with mass concentrations equal to 20% and 43% in order to reduce flow resistance and then to increase solids concentration. Generally, after addition of 0.5% of deflocculants in proportions of (1/2·d1+1/2·d2) by mass, the viscosity of hydromixture is lowest in both mass concentrations. The gradual increase of deflocculant from 0.1% up to 0.5% leads to reduction of shear stress in all investigated ranges of shear rates.

Additives used in the research are easy to access and are commercially cheap. Proposed solution can decrease shear stresses in hydromixture flow for about 20% for both mass concentrations. Such reduction causes that it is possible to decrease energy consumption for further transportation of hydromixture to natural reservoir, and decrease volume of water serving as a carrier liquid.

References

[1] Bartosik A 2009 Simulation and experimental studies of axially-symmetrical flow of fine- and coarse-dispersive slurry in delivery pipelines M11 (Kielce: Kielce University of Technology) p 257

[2] Bartosik A and Jaworska B 2014 *Proc. J. of Physics* The influence of additives on rheological properties of limestone slurry vol 530

[3] Caenn R, Darley H and Gray G 2017 *Composition and properties of drilling and completion fluids* Elsevier

[4] Spelay R 2007 Solids transport in laminar, open channel flow of non-Newtonian slurries (PhD Thesis) University of Saskatchewan

[5] Korhonen M 2015 Flocculation/deflocculation of cellulosic materials and mineral particles by polyelectrolyte complexes and nanocelluloses (DhD) Aalto University 83/2015

[6] Miedema S and Ramsdell R 2015 Slurry Transport Fundamentals. A Historical Overview & The Delft Head Loss & Limit Deposit Velocity Framework Delft University of Technology p 24-26

[7] Chung T 1996 *Applied continuum mechanics* Cambridge University Press

IOP Conf. Series: Journal of Physics: Conf. Series **1101** (2018) 012011 doi:10.1088/1742-6596/1101/1/012011

An investigation of secondary flow features in a low pressure turbine

P. Jonak[1], T. Borzecki[2], S. Kubacki[1]

[1]Institute of Aeronautics and Applied Mechanics, Warsaw University of Technology, Nowowiejska 24, 00-665, Warsaw, Poland; pjonak@meil.pw.edu.pl
[2]Avio Aero, Grazynskiego 141, 43-300 Bielsko-Biala, Poland; Tomasz.Borzecki@avioaero.it

Abstract. A flow simulation through a four-stage low pressure turbine (LPT) of aero-engine, using eddy-viscosity based RANS (Reynolds-averaged Navier-Stokes) model, is performed. The numerical results are compared with experimental data obtained in Polonia Aero Lab in Zielonka, Poland. Good agreement between measured and predicted global flow features and the pressure coefficient on surface of inlet strut is reported. The development of vortex structures in inlet duct of the low pressure turbine (LPT), named a turbine central frame (TCF), is analysed. An analysis of secondary flow losses by means of entropy generation rate coefficient is provided.

1. Introduction

A complex three-dimensional flow in endwall region near the base of turbine blades and vanes has an important impact on the level of aerodynamic losses. In the low pressure turbine (LPT) the three main sources of aerodynamic losses might be reported: the profile losses, the flow leakage losses and the endwall losses. The growth of the boundary layers on the airfoil surface is responsible for the profile losses. The flow leakage losses are generated by a leakage of the flow over tips of the rotor blades and stator vanes. The endwall losses appear as result of enhanced mixing caused by the secondary vortex structures developing near the base of the rotor blades or stator vanes. In spite of the fact that all three types of losses have their own origin, they almost never occur separately. For instance, the flow over the blade and vane surfaces is strongly affected by the secondary flow characteristics along the endwall.

The oncoming flow impacts the leading edge of the blade and the horseshoe vortex is formed. Farther downstream, the horseshoe vortex develops in the form of streamwise-oriented vortices called pressure- and suction-side legs of the horseshoe vortex. The pressure-side leg of the horseshoe vortex moves towards the suction-side of adjacent blade, due to strong transverse pressure gradient, and becomes a part of the passage vortex. The suction-side leg of the horseshoe vortex may develop above the passage vortex, it might be entrained into the passage vortex or dissipate rapidly downstream the leading edge depending on the case. We refer to work by Dunham [1], Sieverding [2], Langston [3] and Ligrani et al. [4] for a discussion of secondary flow losses in axial turbines.

Denton [5] provides an extensive analysis of the loss generation mechanism in turbomachines. The physical sources of losses like viscous and turbulence mixing effects in the boundary layers, heat transfer and influence of shock waves were discussed. The formulas for loss coefficient based on entropy generation were defined. The experimental study of the stagnation pressure losses through the passage in a linear turbine blades cascade was made by Harrison [6]. Endwall shear stress, measured by hot film probe, was used to define the loss coefficient. It was shown that the wake mixing is relevant for accurate prediction of the loss coefficient downstream of the trailing edge of the blade. Experimental study of the leakage losses in a linear turbine cascade with tip clearance was performed by Yamamoto [7]. The clearance gap size and the angle of incidence were claimed to strongly influence the measured leakage and total pressure losses. The interaction between the leakage flow vortices and the passage vortex was reported. Benner et al. [8] [9] performed a two-part study on empirical method for prediction of the secondary flow losses. New loss breakdown scheme with improved correlation for the spanwise penetration depth of the passage vortex was presented in [8]. A novel empirical prediction model for secondary losses was reported in [9]. A considerably large database was used for comparison of predicted and measured cascade data. Analysis revealed that a general scaling factor is required for the new secondary loss correlation in order to give reasonable results for wide range cases. Recently, numerical study of secondary flow in the linear cascade with endwall using LES was made by Cui et al. [10].

The present work reports the simulation results of main flow path (flow without taking into account the flow through the blade tip and inter-stage labyrinth seals) through the four-stage low pressure turbine (LPT). Numerical results are compared with the experimental data from the test rig. The origin of the vortex structures which develop in the vicinity of the shroud and hub in the entrance duct, called a turbine central frame (TCF), is discussed. The coefficient based on entropy generation is used to describe the aerodynamic losses in the passage. In our earlier work [11], the three different turbulence models were tested for simulation of flow through linear cascade and the k–ω SST was found to give the best correspondence with experiment. The present simulations are performed with the k–ω SST model.

2. Test rig

The low pressure turbine consists of 11 conceptual parts: swirler, turbine central frame (TCF), four turbine stages (each consists of vane and blade rows) and exhaust vane turbine rear frame (TRF). The main flow inside the LPT was measured in Polonia Aero Lab in Zielonka (Poland). The measurements of the static pressure on the surface of inlet strut and the global flow characteristics were performed.

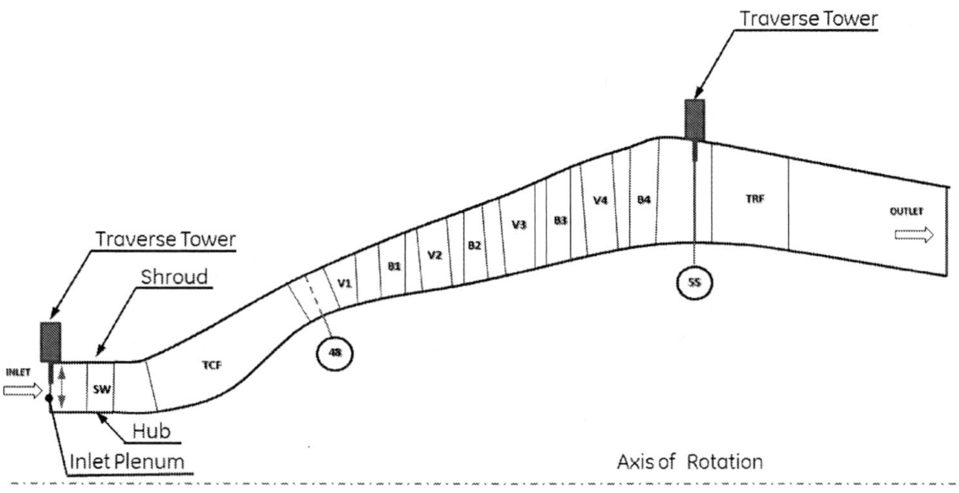

Figure 1. Schematic of LPT and localisation of measuring traverse towers.

Figure 1 shows the cross-section through the LPT and positioning of measuring traverse towers. The three traverse systems, spaced 120°, were installed at the inlet to the LPT. The five-hole probe that was installed in the traverse tower can rotate ±15° relative to its central axis. The measurement of temperature, pressure and fluid flow angles were possible by the radial movement of the probe along the span. One traverse passes the total cycle in about 2 hours. The same test system was installed at the outlet from LPT. The pressure taps, were placed at 20, 50 and 80% of a radial span on the TCF vane. The mean distance between the taps was about 9% of the TCF axial chord length.

3. Computational details

The Moving Reference Frame (MRF) method was used to take into account for the rotating flow dynamics. The air was treated as an ideal gas. The energy equation was used with the viscous work term active. High resolution scheme of ANSYS CFX solver was used for discretization of momentum, continuity and transport equations. Normalized residuals drop down to 10^{-4} for all equations. Constant values of the total pressure, total temperature and turbulence intensity were defined at the inlet to the LPT placed $2.6c_x$ upstream of the leading edges of the swirler vanes (c_x refers to the swirler vane axial chord). The mass-averaged inlet values of total pressure and total temperature were set according to experiment. The flow velocity vectors were set normal to the inlet plane. The value of the static pressure at the outlet from the LPT was set according to experiments. The periodic boundary conditions were applied on side boundaries. The shroud, hub, vane and blade surfaces were treated as no-slip, adiabatic walls. The wall function technique was employed near to walls. The Stage and Frozen Rotor approaches (see description below) of the ANSYS CFX solver were used at the interfaces between consecutive vane- and blade-rows in different simulations. Only connection between the swirler and TCF domain, located in the front part of the turbine, was defined without the use of any interface technique. The pitch of the swirler domain was adjusted somewhat, in order to obtain 1:1 area contact between the swirler and the TCF zone.

The Stage technique employs the circumferential averaging of fluxes in bands at the outlet from a given part (vane and blade domains) and later transmits the averaged fluxes to the downstream part. The model is useful for large pitch ratios between the consecutive domains (vane and blade domains). This way, a single blade domain is used for simplified representation of the whole blade-row and a single vane domain is used for representation of whole vane-row. This reduces significantly the computational costs related to simulation of flow through the whole LPT. In present work, the Stage interface model has been applied for the grid sensitivity study related to the TCF region (see discussion below).

The Frozen Rotor technique produces a steady state solution to the local frame of reference on each side of the interface. The two frames of reference are connected in such a way that they have a fixed relative position during the calculation. If the frame changes the appropriate transformation of the equations is introduced. With the Frozen Rotor technique the pitch change between the consecutive domains has to be avoided. But if the pitch changes, the fluxes are scaled by the pitch change coefficient. In current work, the largest pitch was reported for the TCF domain. The single vane and blade domains downstream of TCF domain were multiplied several (few) times to match the pitch of the TCF domain. The computational grid for simulation with the Frozen Rotor technique is therefore much larger than for simulation with the Stage model. An analysis of secondary flow details in TCF region and partly also an experimental validation of the LPT results was performed using the Frozen Rotor interface model.

Block-structured mesh was generated for simulation of the flow through the low pressure turbine (LPT). The grid in boundary layer region in every part of the LPT contained 25 nodes and was characterized by 1.15 cell growth ratio. The thickness of the boundary layer mesh zone (O-grid block) was defined to be about 20% larger than the turbulent boundary layer thickness estimated from the correlation. The vane and blade grids consisted of about 1 million nodes. The swirler, TCF and TRF meshes had about 1.3, 2 and 3 million nodes, respectively. The mesh refinement was introduced in the TCF region only due to the fact that the mesh resolution in this part of the LPT domain was found to have a strong influence on the numerical results. About 25, 35 and 45 nodes were assigned inside the boundary layer for coarse, medium and fine grids in TCF region, respectively. The thickness of O-grid mesh and the growth ratio was kept the same in all three cases. The maximal values of y^+ on the strut surface in TCF region were equal to 21, 11 and 6 on coarse, medium and fine meshes, respectively. The maximal value of y^+ at other walls was less than 24. Table 1 summarizes the total number of points applied in the present work for simulations using the Stage (basic, medium and fine grids applied in TCF region) and the Frozen Rotor techniques.

IOP Conf. Series: Journal of Physics: Conf. Series **1101** (2018) 012011 doi:10.1088/1742-6596/1101/1/012011

In present work the reduced-order approach, namely the eddy-viscosity RANS model, was employed for prediction of the mean flow characteristics. Note that the grids applied by Cui et al [10] for LES of flow through a single blade cascade with endwall are about 40 times denser than the current single blade grid applied with the Stage model. In present work, the whole grid for simulation of flow through the LPT using the Frozen Rotor technique consists of about 100 million od nodes (Table 1). An application of LES will require much denser grid (refined more than two-order of magnitude with respect to current one) and integration of governing equations in time using a low value of the CFL number. We cannot afford such prohibitive costs of application of LES in the present study.

The grid sensitivity study was performed by means of the pressure coefficient

$$C_p = \frac{p-p_1}{0.5\rho_1 U_1^2}\,, \qquad (1)$$

where p is the static pressure on the blade surface and p_1, ρ_1, and U_1 are the mean static pressure, density and freestream velocity at midspan at the inlet to LPT, respectively. The second quantity was the normalized turbulent kinetic energy

$$q = 2k/(U_1^2)\,, \qquad (2)$$

where k denotes the turbulent kinetic energy. The C_p distribution along 50% of span of TCF vane on different density grids is presented in figure 2(a). Hardly any difference between coarse, medium and fine grids is visible. Figure 2(b) shows the circumferentially averaged profiles of normalized turbulent kinetic energy, q, at the outlet from TCF. Despite some difference are reported in the predicted turbulent kinetic energy profiles, between 5 and 80% of span, they had almost no influence on the mean flow characteristics at the outlet from TCF region (results not shown). As shown in Table 1 there are also only small differences reported between predicted mean flow characteristics on different density meshes.

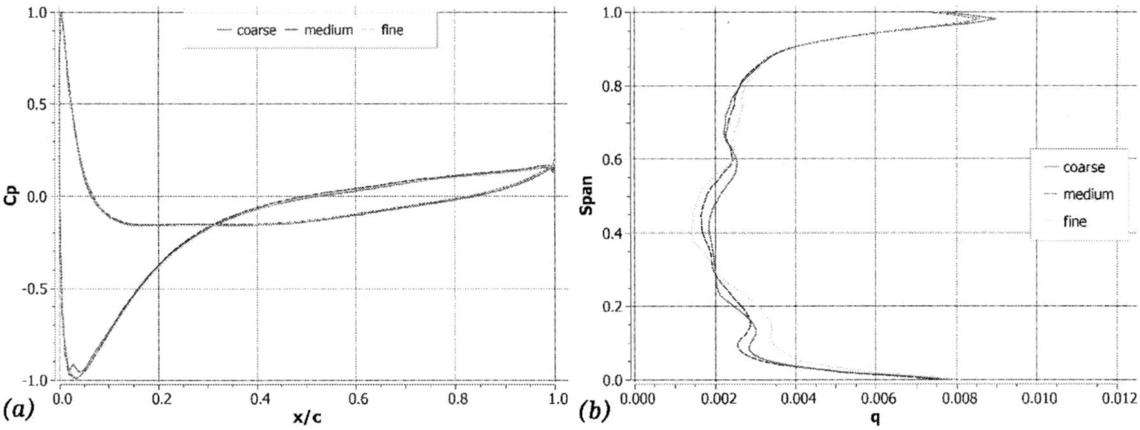

Figure 2. Pressure coefficient along TCF vane at 50% of span (a) and profiles of circumferentially averaged normalized turbulence kinetic energy at outlet from TCF (b) for different density grids.

4. Results

4.1. Comparison of numerical results with experiment

Table 1 shows a comparison between measured and predicted global flow characteristics on different density meshes, using the two different approaches at interfaces between domains (Stage vs. Frozen Rotor technique). The comparison is made by means of the relative difference

$$D = \frac{\varphi_{num} - \varphi_{exp}}{\varphi_{exp}} \times 100\% , \quad (3)$$

where φ_{num} and φ_{exp} are the numerical and experimental values of given variable. The mass flow rate (mf), total pressure (p_t) and total temperature (T_t) are analyzed. The global flow parameters were measured in two sections one at the outlet from TCF (Section 48) and one after the Blade04 (Section 55). Their locations are indicated in figure 1. The lack of tip blade and inter-stage sealing cavities in numerical model might be responsible for small differences between the predicted and measured mass flow rate (difference less than 1.5%). But the differences between the measured and predicted average values of the total pressures and total temperatures are small. They do not exceed 0.4% in all analyzed cases. So the agreement between measured and predicted global flow quantities is encouraging.

Numerical results of the pressure coefficient on the TCF vane at two spanwise locations, namely 20% and 80% from the hub surface, obtained using the Frozen Rotor technique (Case 4 in Table 1) are compared with experimental data in figure 3. Measurement uncertainty of the pressure sensors is ±0.05%. One can notice some under-prediction of C_p value at x/c = 0.2 on the suction and pressure side of the vane in figure 3(b). The satisfactory level of agreement between measured and predicted pressure coefficient at TCF vane is obtained.

Table 1. Total number of nodes is simulation using the Stage and Frozen Rotor techniques and relative differences between measured and predicted global flow parameters at sections 48 and 55. Figure 1 shows the location of measuring sections 48 and 55.

Case	TCF grid	Interface model	LPT grid size [nodes]	Section 48			Section 55	
				mf [%]	p_t [%]	T_t [%]	p_t [%]	T_t [%]
Experiment	–	–	–	ref	ref	ref	ref	ref
Case 1	coarse	Stage	20.8 M	1.2431	-0.2173	0.2195	-0.2240	-0.2949
Case 2	medium	Stage	22.7 M	1.3812	-0.1284	0.2195	-0.2489	-0.3042
Case 3	fine	Stage	26.8 M	1.3352	-0.1581	0.2195	-0.2489	-0.2949
Case 4	fine	Frozen Rotor	98.7 M	1.4733	-0.1680	0.2170	-0.3733	-0.3840

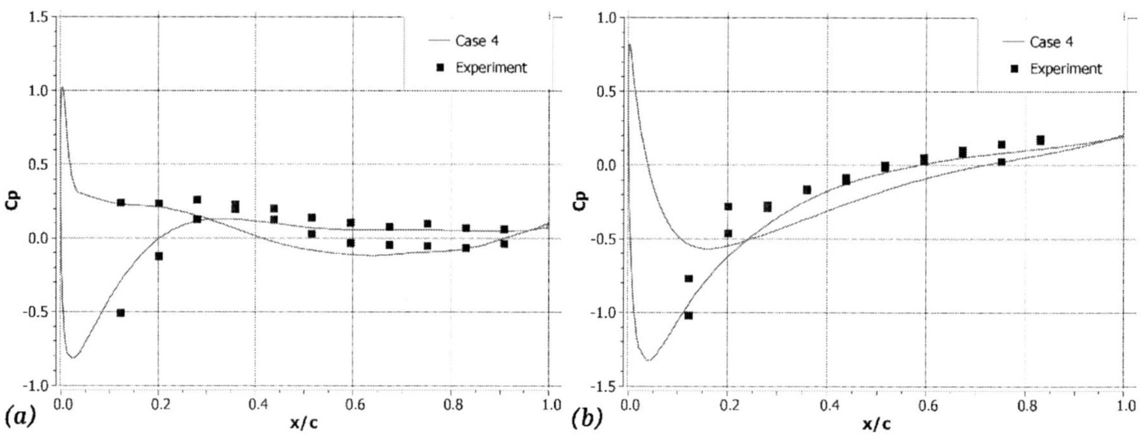

Figure 3. Pressure coefficient distribution on the TCF vane at 20% (a) and 80% (b) of span.

4.2. Secondary flow losses

In present section, the secondary flow structures in TCF region are analyzed. Figure 4 presents the visualization of the vortex structures by means of the q-criterion (level 0.001) and the mean flow pathlines in vicinity of the shroud and hub. The blue pathlines, which have been released in the vicinity of the shroud at the leading edges of swirler vanes, are denoted by A – E symbols (figure 4(a)). The blue pathlines, which have been released near the hub at the leading edges of the swirler vanes, are denoted by H – L symbols (figure 4(b)). Figure 4(a) shows also the pressure- and suction-side legs of the horseshoe vortex at the leading edge of the strut. They are denoted by G and F symbols, respectively. The interaction of the vortex structures, originating from the endwalls of the swirler passages, with the endwall boundary layers in TCF region results in a formation of the streamwise-oriented vortex structures (figure 4). Some vortices impact onto the axially-oriented grooves on the shroud and hub in TCF region (notice that the axially-oriented groves are present due to the technological and assembly reasons). A red circle in figure 4(a) indicates a perturbation of the vortex structure nearby TCF shroud by one of the grooves. The small separation flow region is formed near the shroud as result of this interaction. The suction-side leg of the horseshoe vortex (F symbol) and the streamwise-oriented vortex structures denoted by A and B symbols interact with each other at the outlet from TCF. Somewhat weaker interaction is reported between the pressure-side leg of the horseshoe vortex (denoted by G symbol) at the other streamwise-oriented vortices.

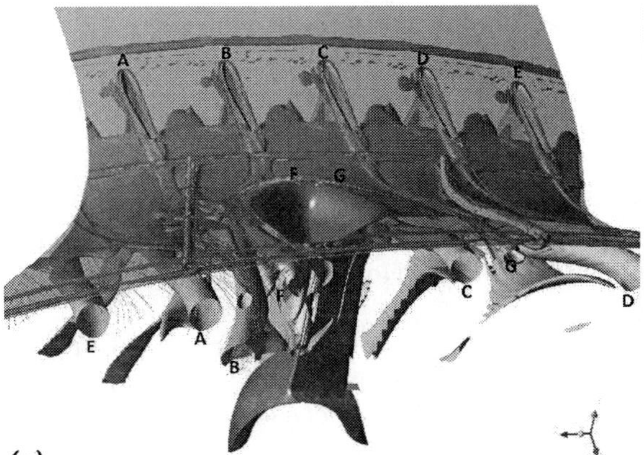

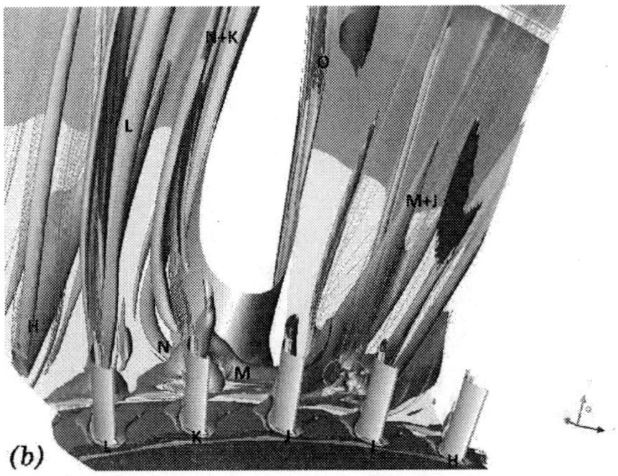

Figure 4. Vortex structures in the TCF region near shroud with the main flow directed outward the paper (a) and near hub with the main flow directed towards the paper (b).

Figure 4(b) shows the vortex structures near to the hub. The oncoming boundary layer impacts the leading edge of the TCF vane and starts to roll-up. The pressure- (N symbol) and suction-side (M symbol) legs of the horseshoe vortex are formed. The suction-side leg of the horseshoe vortex (M symbol) is pushed away from the TCF surface. Subsequently, it interacts with the one of the swirler wakes denoted by J symbol. A flow perturbation in the vicinity of one of the grooves on TCF hub, connected with the local flow separation is denoted by orange circle in figure 4(b). Finally, the J and M structures merge and form a single vortex denoted by M+J symbol. The pressure-side leg of the TCF horseshoe vortex (symbol N) interacts with the vortex structure generated behind one of the swirler vane (symbol K). The structures move nearby the pressure side of the strut. After reaching the trailing edge they merge into (N+K) structure which can be observed at the outlet form TCF region. The formation of a strong counter vortex (O symbol) is observed on the suction side of the TCF vane. The structure, aligned with the red pathlines, develops in the TCF wake region. The swirler wakes denoted by L and H symbols travel through the TCF passage without interaction with the strut boundary layer.

The secondary flow losses in the LPT passage is evaluated by means of the loss coefficient expressed by:

$$Y_e' = \frac{2h_{in}}{\rho_{in} U_{in}^3} TP_S \,, \tag{4}$$

where h is span, ρ is mean density, U is mean velocity, T is temperature and P_S denotes to the rate of entropy generation per volume unit. Subscript *in* corresponds to the values at the inlet to TCF domain. The entropy generation rate in equation (4) is defined by Adeyinka and Naterer [12]:

$$TP_S = \left(\frac{k+k_t}{T}\left(\frac{\partial T}{\partial x_i}\right)^2 + \left(1+\frac{\mu_t}{\mu}\right)\tau_{ij}\frac{\partial u_i}{\partial x_j}\right), \tag{5}$$

where k and k_t are thermal molecular and turbulent conductivities, μ and μ_t are dynamic molecular and turbulent viscosities and τ_{ij} is the viscous stress tensor.

Figure 5 presents the contour plots of loss coefficient (normalized entropy generation rate) at outlet from TCF (a) and at outlet from the first vane-row (b), denoted by V1 in figure 1. The lines at constant radial distances correspond to 25 and 75% of the span. The tilting of wakes coming from the swirler vanes can be observed in central part of the figure 5 (a). The wake structure which is formed behind the middle swirler vane at the shroud (C symbol in figure 4(b)) is visible on the pressure side of the TCF vane (same C symbol in figure 5(a)). In contrast, the wake structure which is formed behind the same middle swirler vane at the hub (J symbol in figure 4(b)) can be found on the suction side of the TCF vane (M+J symbol in figure 5(a)). A redistribution of the middle swirler wake in circumferential direction is caused by the backward skewed tips of the swirler vanes. An interaction of the suction-side leg of the horseshoe vortex (F symbol) and two vortex structures generated behind the swirler wakes (A and B symbols) results in a high loss production at the trailing edge of the strut, visible in central part of the figure 5(a). The increased value of the loss coefficient which results from interaction of the pressure-side leg of the horseshoe vortex (N symbol) and one of the swirler wakes (K symbol) and the counter vortex (O symbol) can also be noticed at about 25% of span. Note that the high entropy losses occur mostly near to the endwalls.

The secondary flow losses distribution at the outlet from first vane-row (downstream of the TCF outlet) are presented in figure 5(b). The increased losses production is observed in the bended wakes behind the vanes of the first stage. A large patch of high entropy production can also be observed in the central part of the figure between the third and fifth wake from the right. The origin of the patch should be searched in the downstream development of the wake behind the strut.

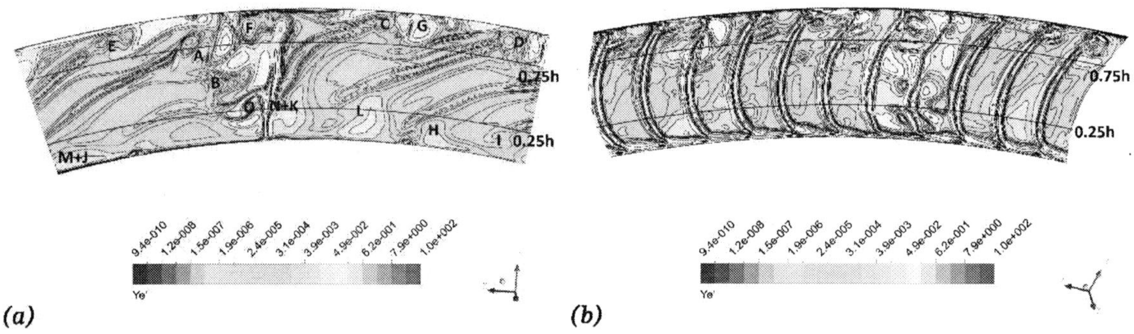

Figure 5. Contour plots of the loss coefficient (eq. 4) at outlet from turbine central frame, TCF (a) and outlet from first vane-row (b).

5. Conclusions

The flow through the four-stage low pressure turbine was simulated using the RANS technique with the moving reference frame approach. Satisfactory agreement between experimental and numerical data in terms of the pressure coefficient on surface of the strut at entrance to LPT was reported. Good agreement between measured and predicted global flow characteristics (1.5% difference in the mass flow rate) was obtained.

The identification of the vortex structures in the entrance duct (turbine central frame, TCF) to the low pressure turbine was made. The interaction of the streamwise-oriented vortex structures, forming at the endwalls of the swirler passages, with the endwall boundary layers in the TCF region was reported. Analysis revealed that the streamwise-oriented vortex structures interacted strongly with the suction- and pressure-side legs of the horseshoe vortex developing around the inlet strut. The local separation of the endwall boundary layer, caused by impact of the streamwise-oriented vortex structures onto the axially-oriented grooves on hub and shroud in TCF region, was also reported.

The secondary flow losses were evaluated by means of the entropy generation rate coefficient. Spots with high loss production were identified at the outlet from turbine central frame (TCF) and also farther downstream at outlet from first vane-row. A great part of losses were generated in the vicinity of the endwalls. The losses were later redistributed and dissipated across the passage due to interaction of the streamwise-oriented vortex structures with the endwall boundary layers. The interaction of the streamwise-oriented vortices, originating from the swirler passages, with the vortex system along TCF vane resulted in formation of the complex vortex pattern in the wake region behind the TCF vane.

Acknowledgments
Authors acknowledge a support from the research project COOPERNIK financed partly by the Polish National Centre for Research and Development (INNOLOT/I/11/NCBR/2014) and partly by Avio Polska Sp. z o.o. The work was partly supported by the Interdisciplinary Centre for Mathematical and Computational Modelling (ICM) of the University of Warsaw (grant ID GA73-24).

References
[1] Dunham J 1970 A review of cascade data on secondary losses in turbines *Journal of Mechanical Engineering Science* **12** 48–58
[2] Sieverding C H 1985 Recent progress in the understanding of basic aspects of secondary flows in turbine blade passages, *Journal of Engineering for Gas Turbines and Power* **107** 248–257
[3] Langston L S 2001 Secondary flows in axial turbines – a review, *Annals of the New York Academy of Sciences* **934** 11–26
[4] Ligrani P, Potts G and Fatemi A 2017 Endwall aerodynamic losses from turbine components within gas turbine engines, *Propulsion and Power Research* **6** 1–14
[5] Denton J D 1993 Loss mechanisms in turbomachines *Journal of Turbomachinery* **115**, 621–656
[6] Harrison S 1990 Secondary loss generation in a linear cascade of high-turning turbine blades *Journal of Turbomachinery* **112** 618–624
[7] Yamamoto A 1989 Endwall flow/loss mechanisms in a linear turbine cascade with blade tip clearance *Journal of Turbomachinery* **111** 264–275
[8] Benner M W, Sjolander S A and Moustapha S H 2005 An empirical prediction method for secondary losses in turbines – part I: A new loss breakdown scheme and *Journal of Turbomachinery* **128** 273–280
[9] Benner M W, Sjolander S A and Moustapha S H 2005 An empirical prediction method for secondary losses in turbines – part II: A new secondary loss correlation *Journal of Turbomachinery* **128** 281–291
[10] Cui J, Nagabhushana Rao V and Tucker P 2016 Numerical investigation of secondary flows in a high-lift low pressure turbine *International Journal of Heat and Fluid Flow* **63** 149–157
[11] Jonak P, Borzęcki T, Konopa M and Kubacki S 2018 Prediction of secondary flow features in a low pressure turbine GT2018-75673 *Proc. of ASME Turbo Expo 2018: Turbomachinery Technical Conference and Exposition GT2018* June 11-15 Oslo Norway 12 pages
[12] Adeyinka O B and Naterer G F 2004 Modeling of entropy production in turbulent flows *Journal of Fluids Engineering* **126** 893–899

IOP Conf. Series: Journal of Physics: Conf. Series **1101** (2018) 012012 doi:10.1088/1742-6596/1101/1/012012

Leakage flow reduction in different configuration of labyrinth seal on a turbine blade tip

Piotr Kaczyński, Ryszard Szwaba, Filip Wasilczuk, Paweł Flaszyński, Piotr Doerffer

Institute of Fluid-Flow Machinery, Polish Academy of Sciences

E-mail: piotr.kaczynski@imp.gda.pl

Abstract. This paper contains experimental and numerical investigation of the leakage flow over the blade tip in turbine stage. Experiments were conducted in non-rotating linear channel. Test rig was intended to model the geometry of the labyrinth seal of the blades of the lower pressure turbine stage. Investigated model contained two different geometries of labyrinth fins. Moreover smooth and honeycomb stator landing were tested. Experimental measurements have been supported by CFD simulation which gives valuable information of flow structure in labyrinth seal and show complex flow physics in the investigated model.

1. Introduction

Aircraft engine manufacturers are continuously working on reducing the overall engine weight and efficiency increase which result in reduction of specific fuel consumption. One of the way to reach these goals is to decrease a losses generated in turbine stage. The sources of those losses can be divided into three main category: profile losses, end-wall losses and leakage losses [1]. The great importance for overall performance of turbine stage have a losses generated between turbine casting and the tip of a blade. Leakage flow in sealing region can reach up to one of third of total losses in turbine cascade [2].

The shrouded blades with labyrinth seal equipped with fins are a commonplace in turbomachinery [3]. It is intended to reduce leakage flow by dissipating the kinetic energy of fluid which flows through the clearance or to increase the flow resistance by modifying a sealing geometry. The most used types of labyrinth seals are the straight, interlocking, slanted, stepped ones and their combinations [4].

On one hand it is desire to keep clearance as small as possible to avoid additional losses. However the different operating conditions of turbine stage significant influences on thermal and mechanical displacement of stationary and rotating parts of the sealing. Therefore the operational gap size changes during the work and it can influence on leakage mass flow rate. In order to minimalize tip clearance the wear-in materials in stationary landing are used. An example of abradable coating is a honeycomb structured material. Such a solution allows to transient contact between labyrinth fins and landing. Moreover that type of seal provides to reduce the operating tip clearance [5]. Despite of benefits of honeycomb land, it enlarges effective gap size and increases leakage flow rate compare to the same configuration with smooth wall [6].

The paper contains the experimental investigations and numerical calculation which have been conducted in non-rotating linear channel. The test rig is intended to model the geometry of the labyrinth seal of the blades of the lower pressure turbine stage. Labyrinth seal was equipped with two angled fins which can be replaced by the other fins geometry. The stator land has two configurations, first with smooth wall while the second is material with honeycomb structure. Additionally, four different gap size have been tested in the range from 0.6 mm up to 1.2 mm. The geometry used in numerical calculation corresponds to experimental one. The computational structured mesh has around 6 million cells. The solver which have been used in numerical calculation is Ansys/Fluent with k-w SST turbulence model.

The research focuses on impact of the clearance size and labyrinth seal configuration on leakage flow. Experimental measurements have been supported by CFD simulation which gives valuable information of flow structure in labyrinth seal and show complex flow physics in the investigated model.

2. Test stand

The real view of the measurement test stand is shown in figure 1. Downstream of the inlet (on the left) a laminar mass flow meter with automatic control valve is located. The test stand outlet is connected to the vacuum tank throughout the control valve, so the air flows through the measurement section driven by the pressure difference between the ambient and very low pressure in the vacuum tank. Measurement section contains a model of two fins labyrinth sealing and appropriate scaled to its size the narrow inlet and outlet channel. The test stand has a possibility to control the inlet and outlet pressure independently. By means of two valves (at inlet and outlet) the mass flow and adequate pressure drop in the test section is controlled.

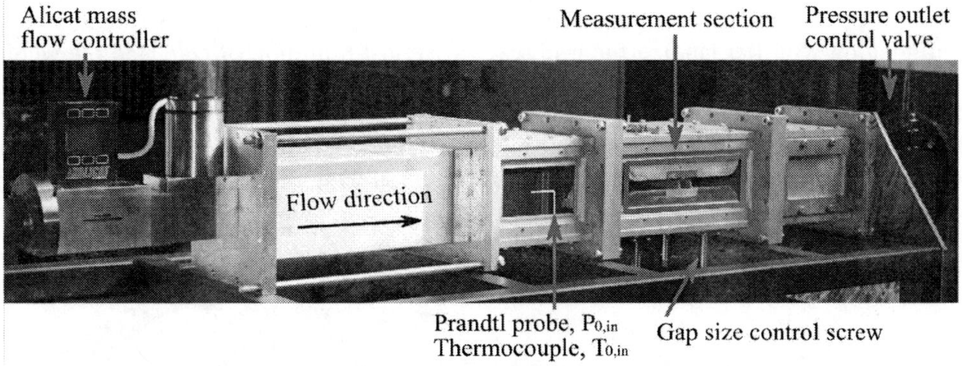

Figure 1. View of the test stand.

The schematic view of the measurement section is shown in figure 2. The main part of this section is the bottom insert with two fins which are 10 mm high and it can be replaced by the other type of fins. In reference case the width of fin tip is 0.8 mm. Modified fins are thickened and their width of tip has 2.6 mm. Just above the fins removable top insert is located. In reference configuration it is flat (smooth) wall and it can be replaced by honeycomb landing. The clearance between the fins and top landing is adjustable by means of four control micro screws (see figure 1) which are working independently allowing to set even clearance above two fins and also parallel in spanwise direction. Size of inlet and outlet channel is 11 mm for the gap of 1 mm and changing with the different clearance size. Measurement section was equipped with the Prandtl Probe and thermocouple located 135 mm upstream of the main measurement channel and 200 mm upstream of the finned bottom insert. Moreover 11 pressure taps were located at the bottom wall and bottom insert (red vertical lines in figure 2). The inlet total pressure was set at 48 kPa

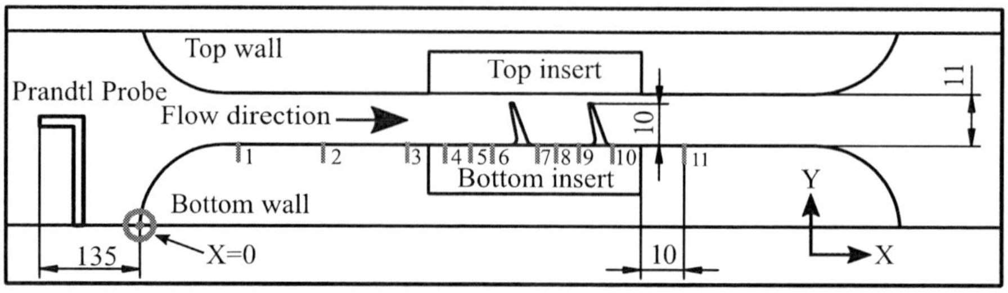

Figure 2. Scheme of the measurement section.

3. Numerical model

Inlet conditions as a stagnation pressure was set on 48 kPa, the same as in experimental measurements. Other boundary conditions at the inlet were agreed as follows: temperature on 297 K, turbulent viscosity ratio on 4.5 and turbulence intensity on 0.5%. As outlet boundary condition the static pressure was used, with range from 25 to 46 kPa. Moreover, boundary condition for the walls in numerical model are non-slip and adiabatic. Inlet cavern and labyrinth seal region were connected using non-matching interface. It was used in order to decrease number of cells of the model. Both sides of non-matching connection in the boundary layer mesh was kept identical to eliminate any distortion in these areas.

In configuration with flat landing the grid reached almost 6 million of cells and the whole channel width (figure 3, top) was simulated. In case of honeycomb landing (figure 3 bottom) the width of the model was limited to the approximately 5 mm in order to reduce amount of cells which was 6.4 million. Mesh in honeycomb region was prepared as an unstructured grid and connected with main model by non-matching connection technique. The other parameters set in numerical calculations are presented in Tab. 1.

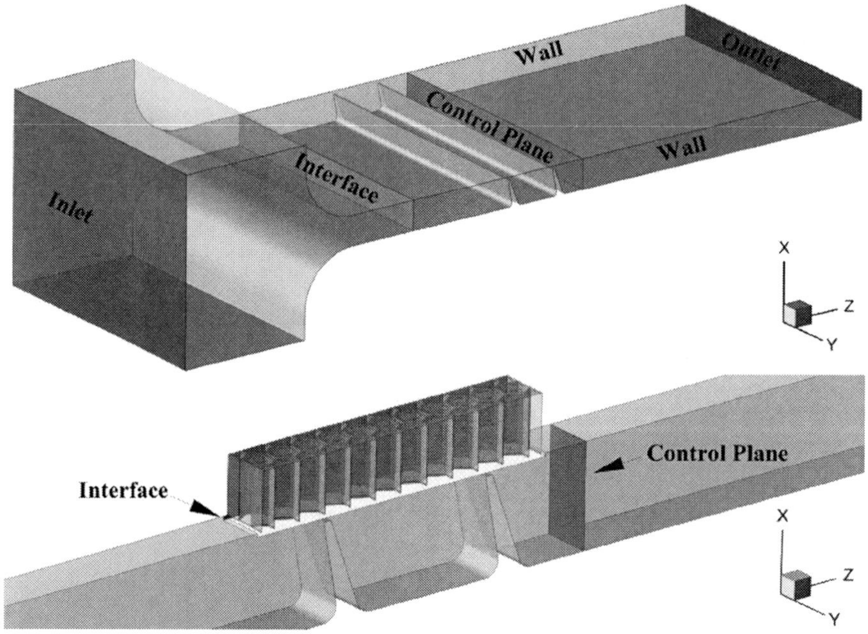

Figure 3. Boundary conditions for the numerical calculations.

IOP Conf. Series: Journal of Physics: Conf. Series **1101** (2018) 012012 doi:10.1088/1742-6596/1101/1/012012

Table 1. Numerical features

Solver	Ansys Fluent, Finite volume, pressure based
Type of grid	Structured and Unstructured
Physical model	Reynolds-average Navier-Stokes (RANS)
Turbulence model	k-ω SST
Gas model	Perfect gas
Numerical scheme	SIMPLE
Spatial discretization	Upwind scheme – 2nd order
Viscosity	According to Sutherland's law

4. Flow parameter determination

In order to compare data from different configuration the dimensionless approach was applied. Operational pressure ratio π and discharge coefficient C_D were used. The pressure ratio is calculated dividing inlet total pressure ($p_{0,in}$) by the static pressure ($p_{s,out}$) measured at the outlet of the measurement section. Inlet total pressure was taken from Prandl Probe. Static pressure at the outlet was measured by means tap located 10 mm downstream of the second fin (pressure tap number 11). In case of numerical calculation $p_{s,out}$ was calculated using average pressure at the plane corresponding to pressure tap (Control plane, see figure 3). It is defined by the following relation

$$\pi = \frac{p_{0,in}}{p_{s,out}} \tag{1}$$

The mass flow rate parameter is often insufficient to comprehensively compare and evaluate leakage reduction in different labyrinth seal configuration. It depends from many factors such as geometry of the sealing or the total inlet parameters, which are variable in time due to the characteristic of the installation. Therefore frequently met in the literature dimensionless flow coefficients is used [7]. Discharge coefficient allows to compare the level of leakage flow through the different sealing shapes and sizes including varying values of fluid parameters. The C_D coefficient is defined as a ratio of the mass flow rate through the model of the seal to that of an isentropic nozzle with the same surface area.

$$C_d = \frac{\dot{m}}{\dot{m}_{ideal}} \tag{2}$$

The mass flow rate under the perfect fluid condition is defined as:

$$\dot{m}_{ideal} = \frac{p_{0,in}A}{\sqrt{T_{0,in}}}\sqrt{\frac{2\kappa}{R(\kappa-1)}\left[\left(\frac{1}{\pi}\right)^{\frac{2}{\kappa}} - \left(\frac{1}{\pi}\right)^{\frac{\kappa+1}{\kappa}}\right]} \tag{3}$$

where: $p_{0,in}$ is inlet total pressure, $T_{0,in}$ – inlet total temperature, A – clearance area in spanwise direction, R – individual gas constant, κ – isentropic exponent, π – operational pressure ratio.

5. Results

The figures from 4 to 6 have the same pattern and consists of two charts. On the left hand side discharge coefficient in function of operational pressure ratio is presented while on the right hand side mass flow rate for different operating pressures is drawn. Figures contains different fins and landing configuration and also four level of gap sizes (0.6 mm, 0.75 mm, 1 mm and 1.2 mm). Red line represents reference thin fin geometry (REF, 0.8 mm) and black line thickened fin (TF, 2.6 mm) one, for flat landing (FW, figure 4) and honeycomb landing (HC, figure 5) with gap size 0.6 mm. Results for tip clearance of 0.75

mm are shown by blue (thin fin) and green (thickened fin) colour lines on the same figures. Chart in figure 6 is drawn in the same pattern as previous while it is dedicated for gap size 1 mm and 1.2 mm.

Figure 4 shows the comparison of reference and thickened fins for flat landing and tip clearance of 0.6 and 0.75 mm. In this figure one can observes that thickened fins (tip width of 2.6 mm) contribute to the increase in mass flow rate regarding to reference type (tip width of 0.8 mm) in case of flat landing. The distinction between reference and thickened fins increases as the gap size is growing. There is no noticeable discrepancies in result between the discharge coefficient and mass flow rate values in terms of leakage reduction.

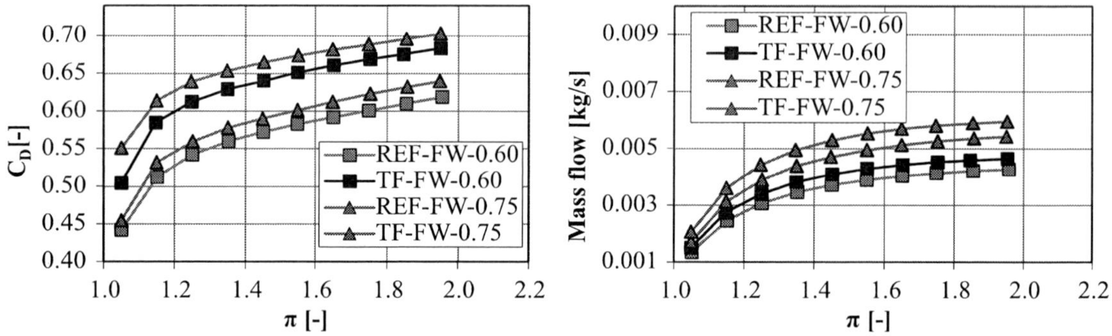

Figure 4. C_D coefficient and mass flow rate for reference and thickened fins with smooth land, gap size: 0.6 mm and 0.75 mm.

In the contrary to the flat type of landing, the thickened fins in the honeycomb configuration (figure 5) reduces the leakage flow in relation to the thinner fins. The largest differences are observed for gap size of 0.6 mm. As the clearance size increases the differences in mass flow between reference and thickened fins gradually vanishes (figure 6). In case of the highest gap size (1.2 mm) leakage flow rate for both type of fins is almost similar.

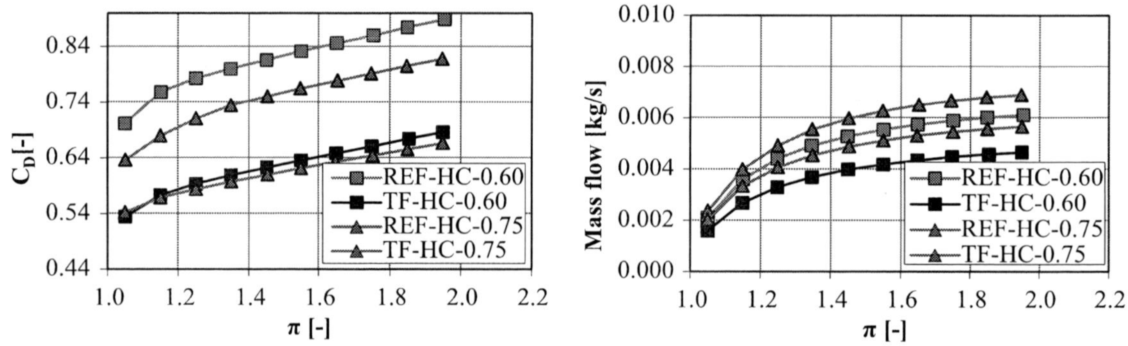

Figure 5. C_D coefficient and mass flow rate for reference and thickened fins with honeycomb land, gap size: 0.6 mm and 0.75 mm.

The higher mass flow rate through labyrinth seal in configuration with thickened type of fins with flat landing can be explained by the numerical simulation and it results from different flow pattern in the gap region. The flow above the reference fins (tip width of 0.8 mm) starts to separate at the leading edge (figure 7) and the reattachment on such short distance is not observed. It results in decrease of effective gap size and thereby reduces the leakage flow. For the thickened fins (tip width of 2.6 mm) a slightly different flow structure in the gap region is observed. Initially the flow separation at the leading edge of fin also appears, afterwards one can observe the reattachment of the flow (figure 8). It results in increasing of effective gap size and higher mass flow rate.

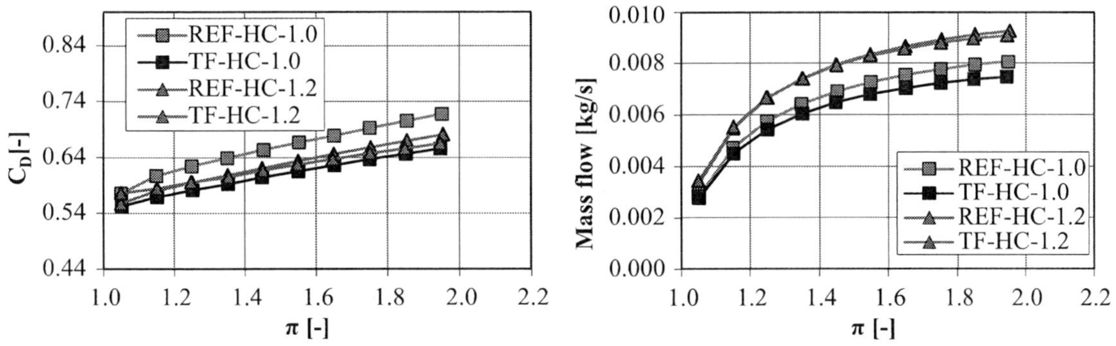

Figure 6. C_D coefficient and mass flow rate for reference and thickened fins with honeycomb land, gap: 1.0 mm and 1.2 mm.

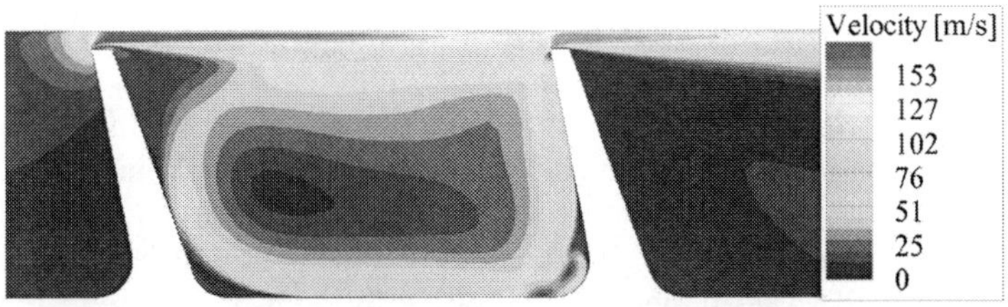

Figure 7. Velocity contour for reference fins with smooth land.

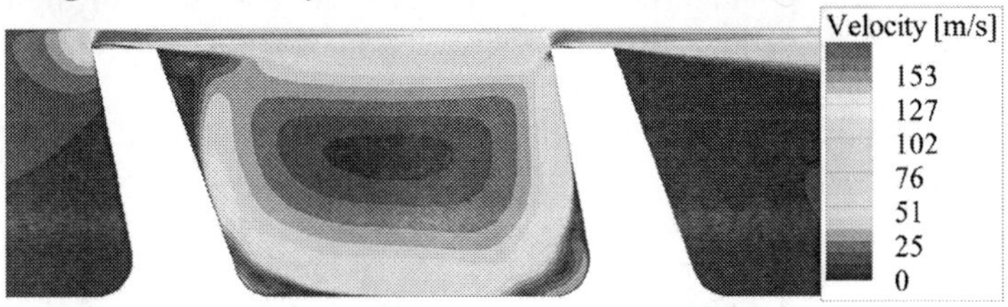

Figure 8. Velocity contour for thickened fins with smooth land.

The presence of the honeycomb landing in the labyrinth sealing considerably modifies the flow structure in the tip clearance region. As it is shown on velocity contours in the figure 9 and 10, part of the flow can use additional available space in honeycomb cells above fins. It happens especially in case of reference thin fins (figure 9) where the leakage flow tries to omit the fin by the space of honeycomb cell and in consequence the effective gap size increases. Thickened fins which covers higher area of honeycomb cell are the larger obstacle and such geometry puts forward a higher resistance to the flow. It results in reduction of effective clearance size in relation to the thinner fins and finally the mass flow rate decreases. As the gap size is increasing, honeycomb cells have less influence on leakage flow and the differences in the mass flow rate between reference and thicken fins gradually diminishes.

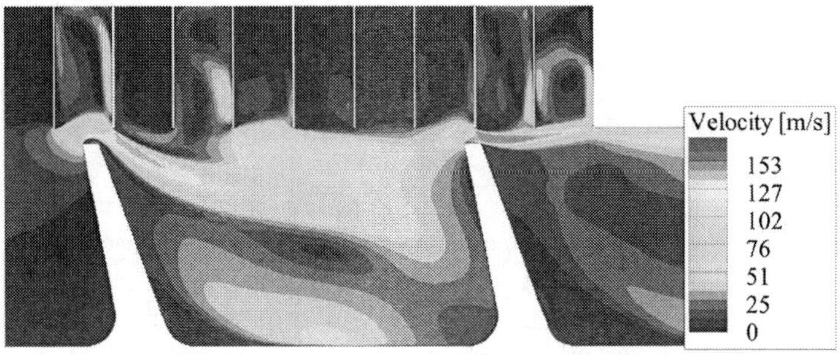

Figure 9. Velocity contour for reference fins with honeycomb land.

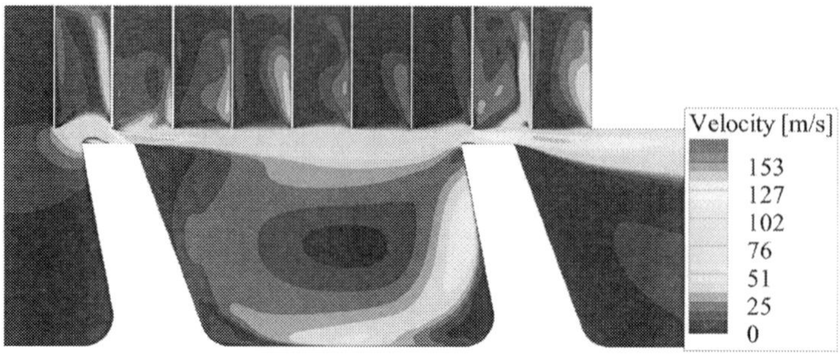

Figure 10. Velocity contour for thickened fins with honeycomb land.

The plot in figure 11 shows an example of normalized static pressure distribution along the measurement section. The stagnation pressure $p_{0,in}$ was equal to 48 kPa and pressure ratio $\pi = 1.25$. The coordinates X in range of 96.5mm to 103.5mm are located between the fins. Red and black line represents the flat landing with reference (tip width of 0.8 mm) and thickened fins (tip width of 2.6 mm), respectively. Blue line shows the distribution for honeycomb landing configuration with the reference fins and the green colour line represents the thickened fins. It is worth to noticed, that the pressure distribution on the lower wall varies depend on which sealing configuration is applied. In the figure 11 one can notice constant pressure and velocity region at the channel inlet of the measurement section. Next there is a large pressure drop over first fin, between the fins the pressure is more or less constant and finally downstream of the second fin a further pressure drop is observed. The pressure at the outlet slightly increases (velocity decreases) as a stabilisation process downstream of the vortex pattern created behind the labyrinth seal.

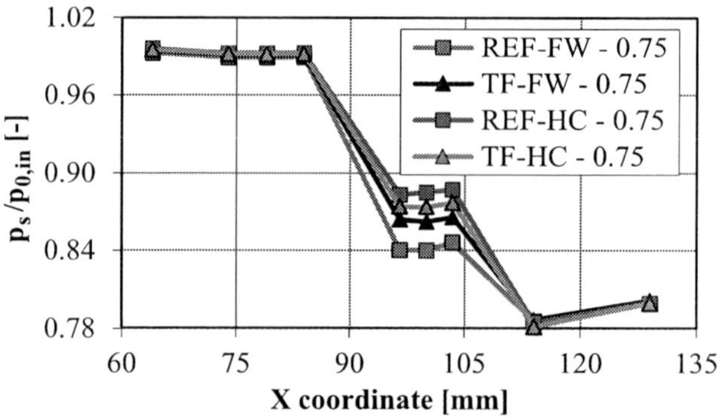

Figure 11. Normalized pressure distribution on bottom wall of measurement section.

The range of pressure jump on particular fins is different depend on configuration of labyrinth seal. As the total pressure drop is the same for all flow cases it can be observed that large pressure drop on the first fin is accompanied with small one on the second fin. It means that depends on sealing geometry, the loading of first and second fin is different. The higher difference one can notice especially in case of thinner reference fins for flat and honeycomb landing. It is due to various inflow angle upstream of the second fin resulting from different flow pattern between the fins in these flow cases, see figure 7 and 9. The stream deflection downstream of the first fin is totally different in case of honeycomb landing (figure 9). For thicker fins the difference in flow pattern is not so much visible, therefore the pressure drops on particular fins in case of flat and honeycomb landing are more approached.

6. Conclusions

An experimental investigations has been performed to investigate the possibilities of the mass flow reduction in the blades labyrinth seal. During the research there was tested geometries which comprised modified labyrinth seal with different landings (flat and honeycomb). The seal geometries of thickened fins, besides the valuable results showing its work in different sealing configurations, is basis for future investigation of passive flow control (air curtain) solution performance in the shrouded blades. The results of this leakage reduction method was compared against the reference geometry (fins, 0.8 mm thick) in terms of seal aerodynamic performance. Reynolds number based on the fin height was varied between 1500 and 9500 depend on flow condition, meanwhile for such Re the flow velocity in the channel inlet was changing from U=2.7 to 15.5 m/s. The following conclusions can be drawn from the carried out experimental investigations:

- The pressure drop in the channel on particular fins depends on configuration of labyrinth seal. The larger pressure drop on the first fin is accompanied with smaller one on the second fin, which means that the loading of first and second fin is different. The higher difference is observed in case of comparison of thinner reference geometry in flat and honeycomb landing. It is due to various inflow angle upstream of the second fin. For thicker fins the difference in flow pattern between both type of landings is not so noticeable, therefore the pressure drops on particular fins are more approached.
- In the channel with flat landing, the fins thickening causes the mass flow increase in the sealing in relation to the reference flow (thinner fins) in the range of 5÷20%. The different flow behaviour is observed in the channel with honeycomb landing. The mass flow reduction (up to 25%) for geometry with thicken fins is observed, but the tendency towards leakage reducing with the growing gap size is decreasing.
- There is the other possible solution to reduce the leakage in sealing with honeycomb, i.e. by decreasing the honeycomb cell size, what in turbine applications can be easier than the thickening of sealing fins. However this proposal should be verified in further research.

References

[1] Denton J D 1993 Loss mechanisms in turbomachines *J. Turbomach.* **115** 621-656
[2] Hamik M and Willinger R 2008 An innovative passive tip leakage control method for axial turbines: linear cascade wind tunnel results *ASME Turbo Expo 2008: Power for Land, Sea, and Air (Berlin)* vol 6 927-938
[3] Ghaffari P, Willinger R, Bauinger S and Marn A 2015 Impact of passive tip-injection on tip-leakage flow in axial low pressure tubine stage *ASME Turbo Expo 2015: Turbine Technical Conf. and Exposition (Montreal)* vol 2A V02AT38A002.
[4] Chupp R E, Hendricks R C, Lattime S B and Strinetz B M 2006 Sealing in turbomachinery *J. Propulsion and Power* **22** 313-349
[5] Burcham R E and Keller R B Jr 1978 Liquid rocket engine turbopump rotating-shaft seals *NASA SP-8121*
[6] Schramm V, Willenborg K, Kim S and Wittig S 2000 Influence of a honeycomb facing on the flow through a stepped labyrinth seal *ASME J. Eng. Gas Turbines Power* **124** 140-146
[7] Braun E, Dullenkopf K and Bauer H J 2012 Optimization of labyrinth seal performance combining experimental, numerical and data mining methods *ASME Turbo Expo 2012: Turbine Technical Conf. and Exposition (Copenhagen)* vol 4 1847-54

IOP Conf. Series: Journal of Physics: Conf. Series **1101** (2018) 012013 doi:10.1088/1742-6596/1101/1/012013

Diffuse interface models for two-phase flows in artificial compressibility approach

Adam Kajzer and Jacek Pozorski

Institute of Fluid-Flow Machinery, Polish Academy of Sciences, ul.Fiszera 14, 80-231 Gdańsk

E-mail: akajzer@imp.gda.pl

Abstract. We present an approach to numerical simulation of two-phase flows in the so-called one-fluid formulation, combining the Entropically Damped Artificial Compressibility (EDAC) method for flow solution and a chosen variant of the Phase Field Method (PFM) that belongs to a wider family of the Diffuse Interface (DI) methods of interface capturing. The resulting governing equations express a set of conservation laws and are of the advection–diffusion type, convenient in numerical handling. These equations, rewritten in the semi-discrete form, can be efficiently solved on parallel devices using the method of lines. We applied the conservative finite difference method for the spatial discretisation and a variant of the Runge–Kutta method to advance the solution in time. The results presented in this work contain the analysis of spurious currents in the case of stationary droplet together with the pressure jump across its surface, as well as the deformation of a droplet subjected to the laminar Couette flow. The presented EDAC–DI approach, offering very high computational efficiency, gives results comparable to other well-established methods for the simulations of the interfacial flows.

1. Introduction

The present work aims to assess the potential of the artificial compressibility (AC) approach for simulations of two-phase interfacial flows based on a chosen one-fluid formulation. The AC methods are attractive due to their "elliptic-free" character which matches the advection (possibly advection-diffusion) type of the equations governing the dynamics of fluid interfaces arising in the interface capturing methods like Volume Of Fluid, Level-Set and Phase Field Method (PFM). The systems of equations of the same type are easier in numerical and algorithmic handling and offer a potential of very efficient solution algorithm using the Single Instruction Multiple Threads (SIMT) approach on parallel devices (e.g. Graphics Processing Units, GPU). On the other hand, the explicit integration methods preferred for the time marching problems suffer from the time step size restrictions that in our case stem from the acoustic effects and/or strongly diffusive behaviour.

In the following we first present the chosen model of the two-phase flow and describe the discretisation of the governing equations. Next we show results of the standard benchmark tests – the stationary droplet and deformation of droplet under the Couette flow. Last but not least, we discuss the numerical and computational issues stemming from the chosen modelling approach and its usefulness at particular flow regimes.

2. Two-phase flow model

The flow is modelled by means of the Entropically Damped Artificial Compressibility (EDAC) method introduced by Clausen [3], Eqs. (1) and (2) below, while the dynamics of the fluid-fluid interface is governed by a variant of the Phase Field Method, Eq. (3), proposed by Chiu and Lin [2]:

$$\partial_t \mathbf{u} = -\nabla \cdot (\mathbf{u}\mathbf{u} + p\mathbf{I} - \mathrm{Re}^{-1}\nabla\mathbf{u}) + \mathbf{g} + \mathbf{F}_{st}, \tag{1}$$

$$\partial_t p = -\nabla \cdot (p\mathbf{u} + \mathrm{M}^{-2}\mathbf{u} - \mathrm{Re}^{-1}\nabla p), \tag{2}$$

$$\partial_t \Phi = -\nabla \cdot (\Phi\mathbf{u} + \gamma\Phi(1-\Phi)\mathbf{n} - \gamma\epsilon\nabla\Phi). \tag{3}$$

In the above equations, $\mathbf{u} = (u, v, w)$ is the velocity field, p is the pressure, $\Phi : \Omega \mapsto [0,1]$ is the phase indicator function and $\mathbf{n} = \nabla\Phi/|\nabla\Phi|$ is the unit vector normal to the interface; $\mathbf{I}$ stands for the unit tensor; $\mathbf{g}$ and $\mathbf{F}_{st}$ denote the body force and the contribution of the surface tension, respectively. M and Re are the Mach and Reynolds numbers, defined by a suitably chosen velocity scale U, length scale L and the artificial speed of sound c_s.

Let us now take a closer look at Eq. (3). This PFM model is based on the Conservative Level Set (CLS) method of Olsson and Kreiss [8] who proposed to capture the interface by advecting the level set function Φ and performing its re-initialisation in artificial time τ separately:

$$\partial_t \Phi = -\nabla \cdot (\Phi\mathbf{u}), \tag{4}$$

$$\partial_\tau \Phi = -\nabla \cdot (\Phi(1-\Phi)\mathbf{n} - \epsilon\nabla\Phi). \tag{5}$$

Equation (5) can be considered as a modification of the viscous Burgers equation: the first term on the right hand side sharpens the interface profile (creates a "shock") and bounds $\Phi \in [0,1]$; the second term introduces diffusion (therefore the name Diffuse Interface). In the equilibrium state the interface profile (for clarity in one spatial dimension) is given by $\Phi(x) = \frac{1}{2}(1+\tanh\frac{x}{2\epsilon})$. It is clear that the width of the interface is of the order of ϵ. The constant γ in Eq. (3) controls the "strength" of the interface profile reshaping. The advantage of the considered PFM model over the CLS is mainly algorithmical – all equations are solved simultaneously and there is no need to check the convergence of the integration process in artificial time. A detailed discussion of different one-fluid formulations of multiphase flow models can be found in [6].

The term $\mathbf{F}_{st}$ couples the interfacial effects and the momentum of the system and is defined according to the classical Continuum Surface Force (CSF) model of Brackbill et al. [1], which consists in smoothing the surface tension across the interfacial region:

$$\mathbf{F}_{st} = -\mathrm{We}^{-1}\kappa\mathbf{n}\delta \approx -\mathrm{We}^{-1}\kappa\nabla\Phi, \tag{6}$$

where $\mathrm{We} = \rho U^2 L/\sigma$ is the Weber number (σ being the surface tension coefficient) and ρ is the reference density. The interface curvature $\kappa = \nabla \cdot \mathbf{n}$. Clearly, the CSF utilises the fact that $|\nabla\Phi| \approx \delta$, where δ is the surface Dirac delta function.

The boundary conditions (BC) for the solid walls (oriented parallely to the $x - z$ plane) are as follows:

$$\mathbf{u} = \mathbf{0}, \qquad \partial_y p = \mathrm{Re}^{-1}\partial_{yy} w, \qquad \partial_y \Phi = 0, \tag{7}$$

i.e. the velocity vanishes on the solid walls, the pressure is computed from the momentum equation, there is no flux of the phase indicator function through the walls and the contact angle (when relevant) is 90 degrees.

We emphasize here that the presented model is expressed as a set of conservation laws, in contrast to standard Level Set method and some variants of the PFM (see e.g. [9]). On the other hand, the CSF approach can not be rewritten in the divergence form and therefore minor imperfections in the momentum conservation can be expected. This disadvantage can be overcome by the application of the Continuum Surface Stress (CSS) model of the surface tension effects, see [12]. However, for the preliminary tests we decided to use CSF since the CSS implementation is more complicated.

Another important feature of the presented model is the purely parabolic type of the equations. Application of an explicit time integration method eliminates the need of solving linear algebraic system, which is an inherent part of modelling the truly incompressible flows due to the divergence–free velocity field restriction which in turn is no longer valid in the EDAC approach. The diffusive terms present in the EDAC equations as well as the considered PFM equation allow for the use of central finite difference schemes in spatial discretisation. In our recent work we have shown that the above features allow for easy, effective and efficient implementation of the single-phase EDAC flow solver on parallel devices [4]. We have also disscussed there the incompressible flow limit with the EDAC model.

3. Discretisation

As mentioned earlier, we use the method of lines to numericaly solve Eqs. (1)–(3) by rewritting the system in semi-discrete form (the indices i, j, k number the grid nodes in the x, y, z directions, respectively):

$$\frac{\mathrm{d}}{\mathrm{d}t}\mathbf{U}_{i,j,k} = -\frac{(\mathcal{F}_{i+\frac{1}{2},j,k} - \mathcal{F}_{i-\frac{1}{2},j,k})}{\Delta x} - \frac{(\mathcal{G}_{i,j+\frac{1}{2},k} - \mathcal{G}_{i,j-\frac{1}{2},k})}{\Delta y} - \frac{(\mathcal{H}_{i,j,k+\frac{1}{2}} - \mathcal{H}_{i,j,k-\frac{1}{2}})}{\Delta z}, \tag{8}$$

where $U = (u, v, w, p, \Phi)$ is the vector of the dependent variables and $\mathcal{F}, \mathcal{G}, \mathcal{H}$ are the discrete fluxes. Using parallel devices the solution is advanced in time at each node by a separate thread (SIMT). All the terms are discretised with standard central 2nd order finite difference (FD) scheme excluding the convective terms for which the central skew-symmetric form is applied (for details see [4]). For the x flux component we have (other fluxes are defined by analogy):

$$\mathcal{F}_{i+1/2,j,k} = \begin{bmatrix} \mathcal{C}(pu) + \mathrm{M}^{-2}\mathcal{D}(u) - \mathrm{Re}^{-1}\mathcal{L}(p) \\ \mathcal{C}(uu) + \mathcal{D}(p) - \mathrm{Re}^{-1}\mathcal{L}(u) \\ \mathcal{C}(uv) - \mathrm{Re}^{-1}\mathcal{L}(v) \\ \mathcal{C}(uw) - \mathrm{Re}^{-1}\mathcal{L}(w) \\ \mathcal{C}(u\Phi) + \gamma\mathcal{D}(\Phi(1-\Phi)n_x) - \gamma\epsilon\mathcal{L}(\Phi) \end{bmatrix}_{i+1/2,j,k}, \tag{9}$$

where (for brevity, we omit the j, k indices from the notation):

$$\mathcal{C}(u^\alpha u^\beta)_{i+1/2} = \frac{1}{4}(u_i^\alpha + u_{i+1}^\alpha)(u_i^\beta + u_{i+1}^\beta), \tag{10}$$

$$\mathcal{D}(f)_{i+1/2} = (f_i + f_{i+1})/2, \tag{11}$$

$$\mathcal{L}(f)_{i+1/2} = (f_{i+1} - f_i)/\Delta x\ . \tag{12}$$

The normal vectors $\mathbf{n}$ and the interface curvature κ are also computed with 2-nd order centered FD schemes. For example the x - component of $\mathbf{n}$ is given by:

$$n_{x,i} = \frac{(\Phi_{i+1} - \Phi_{i-1})}{2\Delta x}\frac{1}{|\nabla\Phi|} \tag{13}$$

and for the curvature we have

$$\kappa_i = \frac{(n_{x,i+1} - n_{x,i-1})}{2\Delta x} + \frac{(n_{y,j+1} - n_{y,j-1})}{2\Delta y} + \frac{(n_{z,k+1} - n_{z,k-1})}{2\Delta z}. \tag{14}$$

We use the ghost node technique to impose the boundary conditions in the conservative manner. Let the wall be placed in the middle between the nodes indexed by $j = 0$ and $j = 1$.

The BCs are discretised as follows:

$$u_{i,0,k} = -u_{i,1,k} \qquad v_{i,0,k} = -v_{i,1,k} \qquad w_{i,0,k} = -w_{i,1,k}, \tag{15}$$

$$p_{i,0,k} = p_{i,1,k} - \mathrm{Re}^{-1} \frac{\mathrm{d}^2}{\mathrm{d}y^2} \left[L_v \left(\frac{y_0 + y_1}{2} \right) \right] \Delta y, \tag{16}$$

where $L_v(y)$ is the Lagrange interpolation polynomial based on the values $v_{i,0,k}, v_{i,1,k}, v_{i,2,k}, v_{i,3,k}$ at y_0, y_1, y_2, y_3 and

$$\Phi_{i,0,k} = \Phi_{i,1,k}. \tag{17}$$

The normal vectors and curvature at the ghost nodes are computed using one–sided FD stencils.

To advance the solution in time we use the explicit 3-rd order Runge-Kutta method in the low-storage variant proposed by Williamson [10]. The low-storage property is important since we are going to execute the EDAC-DI solver on GPUs. For the details of implementation and issues according to execution on the GPUs we refer the reader to our paper [4]. In 3-dimensional case, the time step size constraints arising from the acoustic effects (CFL condition), viscous terms, surface tension and the diffusion of the interface are (Δ is the minimal grid spacing):

$$\Delta t_{CFL} \leq 1.6 \frac{\Delta}{u_{max}(1 + \mathrm{M}^{-1})}, \quad \Delta t_{\nu} \leq \frac{2.78\Delta^2}{12\mathrm{Re}^{-1}}, \quad \Delta t_{sf} \leq \sqrt{\frac{\Delta^3}{8\pi \mathrm{We}^{-1}}}, \quad \Delta t_{PFM} \leq \frac{2.78\Delta^2}{12\gamma\epsilon}. \tag{18}$$

The time step is chosen according to the most severe of the four restrictions above. Many flows in physics and engineering are convection dominated and therefore the CFL condition is the most restrictive. However, in the next section we present also the diffusion dominated case. Fortunately, the constraints resulting from the PFM are usually weaker (or of similar order) than the CFL condition and in AC methods it is commonly accepted to set M≤ 0.1. For reasonable Weber numbers, the constraint stemming from the surface tension is usually the least restricitve.

4. Tests

All the test have been performed on uniform grids in 3D. The coefficients in the PFM equation were set according to the work of Mirjalili et al. [7]: we take $\gamma = 1.5$ and $\epsilon = 5\Delta/6$. This setting allows to avoid over- and undershoots of Φ which can occur when central FD schemes are used and the interface is resolved using a few grid nodes. In all the cases we assume the density ratio of the fluids of either side of the interface equal 1; the same assumption is made on the viscosity.

4.1. Stationary droplet

As a first test we performed the simulation of the stationary (initially spherical) droplet in a cubic domain. Since there is no velocity scale U (physically, the fluid is at rest), to stay in the non-dimensional setup we treat the Reynolds number as a free paramter that determines the diffusion of the momentum. The domain size is $2 \times 2 \times 2$ and the droplet diameter $D = 1$. The spatial resolution was 128^3. The aim of this test is to check the agreement of the pressure drop through the interface with the analytical result (Young-Laplace law) $\Delta p = 4/\mathrm{We}$. We also quantify the numerical artifacts, the so-called parasitic (or spurious) currents. The phenomenon is fully described by one nondimensional parameter – the Laplace number $\mathrm{La} = \mathrm{Re}^2/\mathrm{We}$. The measurements of parasitic currents for different values of La are gathered in Table 1 (left panel). They are expressed in terms of the capillary number $\mathrm{Ca} = \overline{U}\mathrm{We}/\mathrm{Re}$, where $\overline{U}$ is the maximum observed velocity magnitude in the quasi-steady state. The results are in good agreement with other reports, see e.g. [12]. We emphasize that we observed only minor sensitivity of results to M for a wide range of the Mach numbers (based on $\overline{U}$) tested, from 0.1 to 0.001.

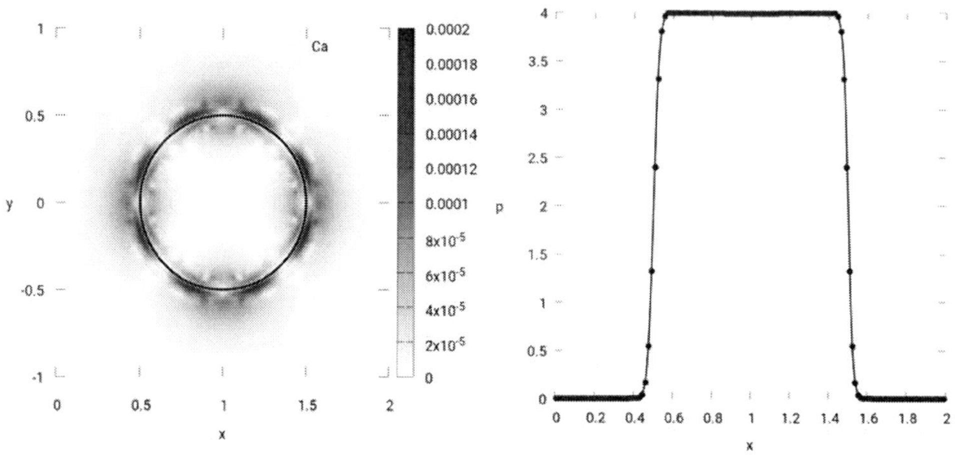

Figure 1. Stationary droplet at We=1 and La=3×10^5. Left: parasitic currents, circle marks the interface $\Phi = 0.5$. Right: the pressure profile; the Young-Laplace law predicts $\Delta p = 4$.

In the left panel of Fig. 1 the domain cross-section in $x - y$ plane is shown and the grayscale map denotes the magnitude of the velocity at La=3×10^5. A pattern similar to other reports is visible. In the right panel of Fig. 1 the pressure profile across the droplet is shown. The pressure jump agrees well with the theoretical prediction.

4.2. Droplet deformation in Couette flow

In this case we placed the droplet between two walls sliding in the opposite directions with speed U. Under the shear flow the droplet deforms (towards an ellipsoid) and a steady state is achieved if the surface tension is high enough to balance the viscous forces. The deformation of the droplet depends on the capillary number defined now as Ca = We/Re. The domain size is set to $4 \times 2 \times 2$ and the droplet diameter is $D = 0.5$. The spatial resolution is $256 \times 128 \times 128$. The Reynolds number is Re = 1. Thus, the capillary number value is changed by modifying the Weber number only. In Fig. 2 the final droplet shape is shown for Ca=0.1. Let l and s denote the large and small axes of the ellipsoid. Assuming creeping flow, Taylor [11] has shown that:

$$E = \frac{l - s}{l + s} = \frac{35}{32}\mathrm{Ca}. \qquad (19)$$

Table 1. Left: steady droplet, spurious currents. Right: droplet deformation in Couette flow.

La	Ca
10^2	1.2×10^{-2}
10^4	5.6×10^{-4}
3×10^5	1.7×10^{-4}

Ca	E_{comp}	E
0.1	0.1072	0.1094
0.2	0.2228	0.2188

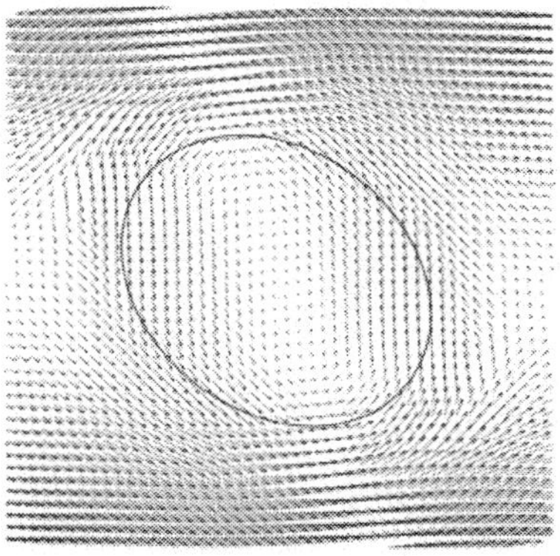

Figure 2. The deformation of a droplet in the Couette flow at Ca=0.1. The line marks the interface; the velocity field in the $x - y$ plane passing through the droplet centre is also shown.

In Table 1 (right panel) the results of our computations are gathered. A good agreement with the theoretical correlation, Eq. (19), is visible and the errors are ~2% which is as accurate as the phase field model based on the Cahn-Hilliard equation [5]. In the above simulations we set Ma=0.1 and therefore the time step is restricted by the viscous effects, resulting in the Courant number ~ 0.2 so the explicit discretisation is far from optimal choice in this case.

5. Conclusions and additional remarks

The proposed complete approach for the simulation of two-phase flows with interfaces gives good results for standard benchmark tests. In our opinion, the important advantages are the simplicity of the method and very high effciency that can be obtained on the desktop computers. Thanks to the GPU implementation the execution time in all the presented cases did not exceed 20 minutes. However, the method is not the best option for the diffusion dominated problems, like droplet in the Couette flow, since the explicit time integration is applied. In such cases implicit integration methods should be used, especially when very high spatial resolution (or long integration time) is required. This, however, undermines the advantage of the EDAC approach over the well-established truly incompressible flow solvers. Additionaly, we checked that the GPU we used would make it possible (due to the memory resources) to use grids of $\mathcal{O}(10^8)$ nodes what allows us to think about applications beyond the academic cases.

In a future work we intend to validate the EDAC-DI model for dynamic cases involving bigger interface deformations which remains the main challenge for existing models of interfacial flows.

References

[1] Brackbill J U, Kothe D B and Zemach C 1992 *J. Comp. Phys.* **100**, 335–354
[2] Chiu P-H and Lin Y-T 2013 *J. Comp. Phys.* **230** 185–204
[3] Clausen J R 2013 *Phys. Rev. E* **87** 013309
[4] Kajzer A and Pozorski J 2018 *Comp. Math. App.* article in press
[5] Kim J 2005 *J. Comp. Phys.* **204** 784–804
[6] Mirjalili S, Jain S S and Dodd M S A 2017 *CTR Ann. Res. Briefs* 117–135
[7] Mirjalili S, Ivey C B and Mani A 2018 *arXiv:1803.01262v2*
[8] Olsson E and Kreiss G 2005 *J. Comp. Phys.* **210** 225–246
[9] Sun Y and Beckermann C 2007 *J. Comp. Phys.* **220** 626–653
[10] Williamson J H 1980 *J. Comp. Phys.* **35** 48–56
[11] Taylor G I 1934 *Proc. Roy. Soc. Lond. A* **146** 501–523
[12] Tryggvason G, Scardovelli R and Zaleski S 2011 *Direct Numerical Simulations of Gas-Liquid Multiphase Flows* (New York: Cambridge University Press)

IOP Conf. Series: Journal of Physics: Conf. Series **1101** (2018) 012014 doi:10.1088/1742-6596/1101/1/012014

Validation of numerical models for flow simulation and wave propagation along human aorta

N Kizilova[1] and J Mizerski[2]

[1] Institute of Aeronautics and Applied Mechanics, Warsaw University of Technology, Warsaw, Poland
[2] Interdisciplinary Centre for Mathematical and Computational Modelling, Warsaw University, Warsaw, Poland

E-mail: n.kizilova@gmail.com

Abstract: Numerical computations on 2d model of blood flow along human aorta based on incompressible axisymmetric Navier-Stokes equations for blood and momentum equations for incompressible viscoelastic arterial wall are compared to the blood flow oscillation curves measured in vivo by Doppler ultrasound in the larger systemic arteries of healthy volunteers. The analysis of the wave propagation and reflection along the geometrical 91-tube model of aorta as a tube with side branches is based on Lighthill's theory of waves in arteries. It is shown due to individual geometry positive wave reflections at some aortic branches and aortic bifurcation may lead to increase of the pressure amplitudes, high pressure and wall shear stress oscillations that may lead to wall damage and development of aortic aneurism and stenoses of branched arteries. The numerical results obtained on the linearized 2d and nonlinear 1d models are compared to the in vivo pressure and flow measurements along the aorta and in its branches. A good qualitative correspondence is obtained. The model can be used for determination of the individual parameters for patient-specific cardiovascular models.

1. Introduction

Mathematical modeling of blood flow and pulse wave (PW) propagation in arterial vasculatures and novel methods of analysis of the pressure p(t) and flow v(t) waveforms have become important constituents of contemporary medical diagnostics of the cardiovascular diseases [1]. Such approaches related to the computational medicine based on the vast detailed numerical computations on individual patient specific mathematical models are essential for early medical diagnostics, *in silico* planning of surgery, therapy and rehabilitation procedures. Traditional contour and spectral analyses of the waveforms as well as novel wave-intensity analysis, fractal analysis and other mathematical methods revealed important information on the state of arteries, vessel wall, microcirculation, general and specific diseases and pathologies. The most part of numerical computations of the blood flow in human systemic arteries have been carried out on the 55-tube model that had been developed on cadaveric dissections [2]. The lengths L_j and diameters d_j of the largest arteries of human trunk, head, upper and low extremities have been measured in the state of their maximal vasodilatation. In this study blood flow and wave propagation along aorta are studied on the 84-tube model of aorta (Fig.1a) based on the cadaveric and *in vivo* ultrasound measurements when the arteries are in their natural state [3].

Aorta is the largest blood vessel that directs the blood flow [4] and distributes it between the organs, muscles and other tissues via its branches. Due to its high elasticity the aorta converts the pulsatile flow produced by heart to almost steady flow in peripheral vessels (the Windkessel effect) [1]. Periodic heart contractions generate small excitations of blood pressure and velocity which propagate as PWs along the aorta with the pulse wave velocity (PWV) c = 5-10 m/s which significantly exceed the average blood flow velocity u =0.5-0.8 m/s [1]. The propagating PWs experience numerous reflections at arterial branching, narrowing (stenosis), enhancements (aneurysms), regions with low or high wall compliance (atherosclerotic plaques or age-related wall degeneration). Therefore, PW propagation and reflection in the aorta are of great interest [5]. Wave reflections influence the proximal aortic pressure and play a role in systolic hypertension [6]. Significant wave reflections originate due to narrowing between the thoracic and abdominal aorta [7].

PWV value is an important index of the arterial stiffness which well correlates with a risk of cardiovascular disease and has been shown to be an independent predictor of cardiovascular outcome [8]. Aortic, brachial and femoral PWVs determine probability of cardiovascular survival of patients with kidney failure [9] and patients with Marfan syndrome [10], ageing and hypertension [11]. 3D CFD modeling was found useful for detection of rigid inclusions in the aortic wall by the PWV values [12].

In this paper the most detailed model of human aorta as a tapered tube composed from 34 segments with 57 branches is developed and validated by the *in vivo* ultrasound measurements of the blood pressure $p(t)$ and flow $u(t)$ signals, and compared to the numerical data obtained in literature on different simplified geometries.

2. Mathematical modeling of the pulse waves in arteries

According to the theory of pulse waves in arteries as fluid filled compliant tubes developed by Sir. J. Lighthill [13], the wave propagation and reflection at the arterial branching are determined by local properties of the parent and daughter arteries, i.e. their characteristic wave admittances

$$Y_0 = \frac{\pi d^2}{\rho_b c_0}, \qquad c_0 = \sqrt{\frac{Eh}{\rho_b d}}, \tag{1}$$

where c_0 is the pulse wave speed by the Moens-Korteweg formula, ρ_b is the density of blood, d, E, h are diameter, Young modulus and wall thickness of the artery.

In that way, the wave reflection coefficient $\Gamma = p_r / p_i$ which is the ratio of the amplitudes p_r and p_i of the reflected and incident waves is determined as [13]

$$\Gamma = \frac{Y_0^{(0)} - \sum_{j=1}^{m} Y_0^{(j)}}{Y_0^{(0)} + \sum_{j=1}^{m} Y_0^{(j)}}, \tag{2}$$

where the superscript indicates the number of the artery in the branching; (0) is for the parent artery, *j=1,2...m* are for the daughter arteries; m=2 for the bifurcations, m=3 for the trifurcations, etc.

Accounting for the blood and arterial wall viscosities, the Womersley model [1] of the axisymmetric pulsatile flow of a viscous fluid along the viscoelastic tube can be used for generalization of the pulse wave theory. Then the wave speed instead of (1) is a complex value

$$c = c_r + Ic_i = c_0 \sqrt{\frac{1 - F_{01}}{1 - \sigma^2} e^{I\theta}}, \tag{3}$$

where c_0 is determined by (1), $F_{01} = 2J_1(\beta)/(\beta J_0(\beta))$ is the Womersley function, J_0, J_1 are the Bessel functions of the first kind and orders 0 and 1, accordingly, $\beta = \alpha(I)^{3/2}$, $I = \sqrt{-1}$, $\alpha = d\sqrt{\omega\rho_b/4\mu}$ is the Womersley number, ω is the wave angular frequency, σ is the Poisson ratio of the wall material, θ is the phase shift between the pressure and the wall displacement due to the wall viscosity [14].

The pressures $P^{(j)}(t, x_j)$ and volumetric flow rates $Q^{(j)}(t, x_i)$ in the *j-th* artery can be computed as superpositions of the incident and reflected running waves in the form [1]

$$\begin{aligned} P^{(j)}(t, x_j) &= P_0^{(j)}(e^{I\omega(t - x_j/c_j)} + \Gamma_j e^{I\omega(t + (x_j - 2L_j)/c_j)}), \\ Q^{(j)}(t, x_i) &= Y_0^{(j)} P_0^{(j)}(e^{I\omega(t - x_j/c_j)} - \Gamma_j e^{I\omega(t + (x_j - 2L_j)/c_j)}), \end{aligned} \tag{4}$$

where j is the number of the vessel, $P_i^0 = P_i|_{x_i=0}$ is the wave amplitude at the inlet of the artery, $x_j \in [0, L_j]$ is the longitudinal coordinate computed from the inlet, L_j is the length of the artery.

According to (6), any positive wave reflection ($\Gamma > 0$) leads to an increase in the pressure amplitude and decrease in the flow rate because any reflected wave carries some fluid against the flow. When the downstream wave admittance is bigger than the upstream one and $\Gamma < 0$, the so called negative wave reflection is observed. Then the reflected wave propagates downstream and brings additional amount of fluid that leads to the decrease in the pressure at the outlet and, therefore, in the increase of the flow rate by the increased pressure drop along the artery (suction effect). Negative wave reflections have been found in the pulmonary circulation [15], coronary vasculature [16,17], and some others [3,18].

3. Materials and methods

The values of $\{L_j, d_j\}_{j=1}^{91}$ of the aortic segments and its branches (Fig.1) have been measured on five cadavers during the post-mortem examination. The $p(t)$ and $v(t)$ patterns in different locations along the aorta (marked in Fig.1 by capital letters A-G) have been measured on five healthy volunteers (ages from 18 to 42) by triplex scanner 'Sonolaine Elegra Advanced' (Siemens). The $p(t)$ and $v(t)$ curves in the main coronary arteries have been measured by transesophageal ultrasound. The measured curves have been smoothed by Bayesian filter (Fig.2) and analyzed. In total 150 pairs of $v(t)$ and $p(t)$ curves have been recorded and treated (30 measurement sites per an individual). The measurement data are presented in Table 1. The Roman numeration corresponds to the direction from the aortic root to bifurcation. The averaged non-scaled data (model I) and scaled to h=175 cm, and inlet diameter d=35 mm (model II) are marked by $<L>, <d>$.

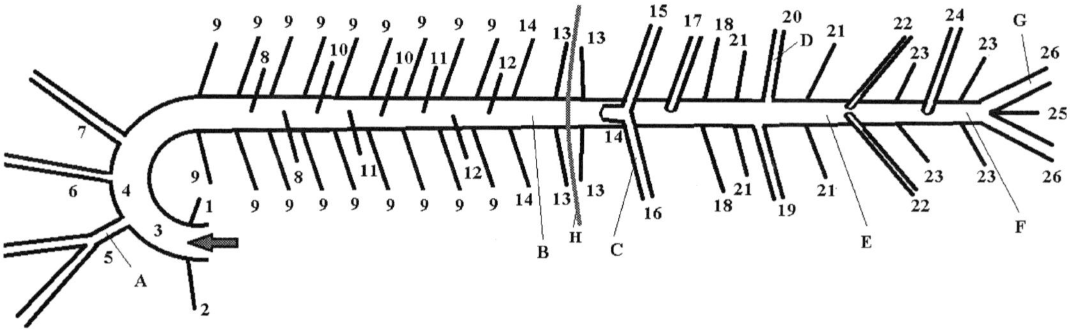

Figure 1. Model of aorta: left (1) and right (2) coronary, aortic root (3) and arc (4), brachiocephalic trunk (5), left carotid (6), subclavian (7), bronchial (8), intercostal (9), esophageal (10), mediastinal (11), pericardial (12), diaphragmal (13), subcostal (14), cervical trunk (14), splenic (15), hepatic (16), upper mesenchymal (17), suprarenal (18), left (19) and right (20) renal, upper lumbar (21), testicular (22), lower lumbar (23), lower mesenchymal (24), sacral (25) and iliac (26) arteries; the arrow indicates blood inflow; letters A-G correspond to the sites of *in vivo* measurements, H - diaphragm

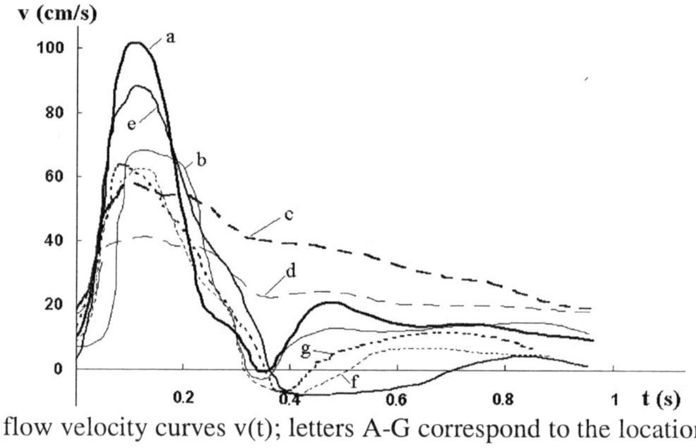

Figure 2. Digitized flow velocity curves v(t); letters A-G correspond to the locations marked in Fig.1

IOP Conf. Series: Journal of Physics: Conf. Series **1101** (2018) 012014 doi:10.1088/1742-6596/1101/1/012014

Table 1. Diameters (d) and lengths (L) of the arterial segments in five humans h1, h2, h3, h4, h5 (height/age), males (m) and females (f), model I and model II averaged values

N	name	h1(m) d(mm)	172/60 L(mm)	h2(m) d(mm)	177/58 L(mm)	h3(f) d(mm)	164/38 L(mm)	h4(f) d(mm)	155/43 L(mm)	h5(m) d(mm)	172/62 L(mm)	<d> (mm) (model I)	<L> (mm) (model I)	<d> (mm) (model II)	<L> (mm) (model II)
1	ascending aorta I	33	3	33.3	3	32	3	29	3	35	3	32.5	3.0	35.0	3.1
2	ascending aorta II	33	91	26.7	85	32	58	29	60	35	85	31.1	75.8	33.6	78.5
3	aortic arch I	32	23	32	23	30	15	29	4	35	6	31.6	14.2	34.1	14.6
4	aortic arch II	32	14	32	14	30	14.5	29	5	35	17	31.6	12.9	34.1	13.3
5	thoracic aorta I	31	1	23.5	5	30	2	29	6	29	24	28.5	7.6	30.9	7.9
6	thoracic aorta II	31	2	23	51	30	4	29	7	29	8.5	28.4	14.5	30.8	14.7
7	thoracic aorta III	31	2.5	23	2	30	3	29	2	29	1.5	28.4	2.2	30.8	2.3
8	thoracic aorta IV	31	21	22	22	30	23	29	12	29	9	28.2	17.4	30.6	18.1
9	thoracic aorta V	31	18.5	22	17	30	19	29	8	29	4	28.2	13.3	30.6	13.8
10	thoracic aorta VI	30.5	32	21.5	10	30	10	28	12	29	25	27.8	17.8	30.1	18.4
11	thoracic aorta VII	30.5	15	21.5	5	29.5	8	28	10	29	13	27.7	10.2	30.0	10.7
12	thoracic aorta VIII	30.5	17	21.5	15	29	18	28	22	29	12	27.6	16.8	29.9	17.7
13	thoracic aorta IX	30.5	32	21.5	15	29	18	28	22	29	25	27.6	22.4	29.9	23.4
14	thoracic aorta X	30	20	21.5	11	27	18	28	14	29	25	27.1	17.6	29.3	18.3
15	thoracic aorta XI	30	12	21.5	4	27	11	28	8	29	11	27.1	9.2	29.3	9.6
16	thoracic aorta XII	30	32	21	15	27	7	28	22	29	14	27.0	18.0	29.2	18.8
17	thoracic aorta XIII	29	32	21	15	25	18	28	22	29	25	26.4	22.4	28.6	23.4
18	thoracic aorta XIV	29	16	21	10	25	9	28	22	29	25	26.4	16.4	28.6	17.2
19	thoracic aorta XV	28.5	16	21	5	24	9.5	28	10	29	12	26.1	10.5	28.3	11.0
20	thoracic aorta XVI	28	32	21	15	23	17.5	28	12	29	13	25.8	17.9	27.9	18.6
21	thoracic aorta XVII	28	32	20.5	15	23	18	28	22	29	25	25.7	22.4	27.8	23.4
22	thoracic aorta XVIII	26	18	19.5	17	20	6.5	23	4	26	14.7	22.9	12.0	24.7	12.3
23	thoracic aorta XIX	25.7	20	19	18	20	8.5	23	4	26	27	22.7	15.5	24.6	15.8
24	abdominal aorta I	24	9	18	8	20	6	22	11	26	13	22.0	9.4	23.8	9.8
25	abdominal aorta II	23.6	33	17	44.5	19.5	12	22	13	25	34	21.4	27.3	23.2	27.9
26	abdominal aorta III	23	1	19	8.5	19	10	20	4	24	6	21.0	5.9	22.7	6.1
27	abdominal aorta IV	23	1	19	9.5	19	11	22	8	24	5	21.4	6.9	23.1	7.3
28	abdominal aorta V	23	1	20	14	19	12	20.5	12	24	17	21.3	11.2	23.0	11.7
29	abdominal aorta VI	22.5	19	20.5	12.5	19	10	20.5	7.5	23.5	14	21.2	12.6	22.9	13.0
30	abdominal aorta VII	22.5	2	20.5	2.5	19	2	20	1.5	23	3	21.0	2.2	22.7	2.3
31	abdominal aorta VIII	22.5	8	21	6	18.5	10	20	7	23	14	21.0	9.0	22.7	9.4
32	abdominal aorta IX	22	41	20	28	18	21	20	21	23	30	20.6	28.2	22.2	29.2
33	abdominal aorta X	22	18	20	8	18	13	20	18	23	16	20.6	14.6	22.2	15.3
34	abdominal aorta XI	22	23	19	22	17.5	11	19	12	21	16	19.7	16.8	21.3	17.3
35	abdominal aorta XII	22	24	19	26	17	20	19	21	21	30	19.6	24.2	21.2	25.1

36	abdominal aorta XIII	22	23	20	24	17	19	19	17	21	28	19.8	22.2	21.4	23.0
37	right coronary	3.1	37	3.5	38	6.3	26	3	48	2.7	35	3.7	36.8	4.0	38.6
38	left coronary	3.9	29	4.9	29	6.5	24	3.8	11	2.9	15	4.4	21.6	4.8	22.3
39	brachiocephalic trunk	19	42	11	36	13.5	35	11	36	16	31	14.1	36.0	15.2	37.6
40	left common carotid	8.8	109	7.3	116	8.5	110	7.2	117	10.5	130	8.5	116.4	9.1	121.5
41	left subclavian	12	64	13.7	44	8	26	7	29	10.5	35	10.2	39.6	11.0	40.9
42	bronchial	1.8	36	1.4	35	1.2	30	1	23	1.3	28	1.3	30.4	1.4	31.5
43	esophageal	2.4	28	1.8	27	1.6	25	0.8	19	1.2	26	1.6	25.0	1.7	26.0
44	mediastinal	1.5	2	1.1	25	1.2	23	1.2	16	1.4	24	1.3	18.0	1.4	18.8
45	pericardial	1.2	21	1	25.5	1.2	20.5	1.1	15	1.2	22	1.1	20.8	1.2	21.6
46	intercostal I	3.2	23	2	22	2.9	12	2.5	9	2.5	14	2.6	16.0	2.8	16.5
47	intercostal II	3.1	23	2.5	22	2.9	11	2.3	9	2.2	15	2.6	16.0	2.8	16.5
48	intercostal III	3	23.5	2.7	22	2.9	10.5	2.3	8	2.2	15	2.6	15.8	2.8	16.2
49	intercostal IV	3	23	2.5	22.5	2.9	11	2.3	8	2.2	14	2.6	15.7	2.8	16.1
50	intercostal V	3	23	2.5	22	2.9	11.5	2.3	8	2.1	16	2.6	16.1	2.8	16.5
51	intercostal VI	3.1	22.5	2.4	22	2.9	10.5	2.2	8	2.2	15	2.6	15.6	2.8	16.0
52	intercostal VII	3	23	2.5	22	2.9	11	2.2	8	2.2	15	2.6	15.8	2.8	16.2
53	intercostal VIII	3	23	2.5	22	2.9	11	2.3	8	2.2	16	2.6	16.0	2.8	16.4
54	intercostal IX	3	23	2.4	21.5	2.9	1.5	2.3	7	2.1	16	2.5	13.8	2.8	14.1
55	subcostal	3	23	2.5	22	2.9	11	2.3	7	2.1	14	2.6	15.4	2.8	15.8
56	phrenica superior	2	131	1.8	19	2.1	10	1.4	32	1.5	176	1.8	73.6	1.9	75.6
57	phrenica inferior	2	131	1.8	18	2.2	9	1.5	36	1.6	155	1.8	69.8	2.0	71.8
58	celiac trunk	8.9	11	7.5	11	8	17	8.7	14	10	15	8.6	13.6	9.3	14.3
59	upper mesenteric	8	67	8.5	19	9.8	47	13	17	7	17	9.3	33.4	10.2	34.7
60	suprarenalis	1	28	1.5	22	1.1	19	1	23	1.4	26	1.2	23.6	1.3	24.6
61	lumbar I	2	46	1.8	158	3	56	1	23	2.2	38	2.0	64.2	2.1	65.5
62	left kidney	6	41	4.3	53	8.5	87	5.6	37	4.4	68	5.8	57.2	6.3	59.6
63	right kidney	5.4	28	4.2	69	8.5	45	6.7	47	4.1	54	5.8	48.6	6.3	50.5
64	gonadal	1.5	310	1.3	340	1.8	300	1	223	2	250	1.5	284.6	1.6	295.6
65	lumbar II	2	46	1.7	158	3	56	2.6	50	2.2	38	2.3	69.6	2.5	71.6
66	lumbar III	2	46	1.7	158	3	56	2.6	50	2.2	38	2.3	69.6	2.5	71.6
67	lower mesenteric	5.5	50	3.5	26	3.5	28.5	3	43	4	52	3.9	39.9	4.2	41.7
68	lumbar IV	2	46	1.7	158	3	56	2.6	50	2.2	38	2.3	69.6	2.5	71.6
69	left iliac	13.5	69	11	38	10	52	10	70	10	57	10.9	57.2	11.8	60.1
70	lumbar V	2	46	2.1	158	3	56	2.6	50	2.2	38	2.4	69.6	2.6	71.6
71	right iliac	13.3	67.5	11	53	10	52	8.3	53	12	53	10.9	55.7	11.7	58.1
72	sacral	0.7	87	1.4	97	1.4	70	1.4	67	1.5	75	1.3	79.2	1.4	82.2

4. FSI mathematical model

Here 2D fluid-structure interaction axisymmetric model [1] of the pulsatile blood flow along a thick wall viscoelastic tube has been used for detailed computer simulations. The model is based on axisymmetric incompressible Navier-Stokes equations for blood flow

$$\nabla\cdot\vec{v}=0,\quad \rho_f\left(\frac{\partial\vec{v}}{\partial t}+\vec{v}\cdot\nabla\vec{v}\right)=-\nabla p+\mu_f\Delta\vec{v}\,, \tag{5}$$

where $\vec{v}$ is the flow velocity, ρ_f, μ_f are the fluid mass density and viscosity, p is the hydrostatic pressure, and the mass and momentum balance equations for the incompressible wall

$$\nabla\cdot\vec{u}=0,\quad \rho_w\frac{\partial^2\vec{u}}{\partial t^2}=-\nabla p_w+\nabla\cdot\hat{\sigma}\,, \tag{6}$$

where $\vec{u}$, ρ_w, p_w, $\hat{\sigma}$ are the displacement, density, hydrostatic pressure and stress tensor of the wall.

The Kelvin-Voigt rheological model is taken for the viscoelastic arterial wall

$$\tau_w\frac{\partial\sigma_i}{\partial t}+\sigma_i=A_{ik}\varepsilon_k+\mu_w\frac{\partial}{\partial t}\varepsilon_i \tag{7}$$

where A_{ik} is the matrix of the elasticity coefficients, μ_w is the wall viscosity, τ_w is the stress relaxation time, $\bar{\sigma}$ and $\bar{\varepsilon}$ are the stress and strain vectors in the form $\bar{a}=\{a_{11},a_{22},a_{33},a_{23},a_{13},a_{12}\}$.

The continuity conditions for pressures and velocities at the fluid–solid interface in each vessel

$$r=0:\qquad v_r=0, \tag{8}$$

$$r=R:\qquad \frac{\partial\vec{u}}{\partial t}=\vec{v},\quad -p+\mu\frac{\partial v_r}{\partial r}=-p_w+\sigma_{rr},\quad \mu\left(\frac{\partial v_x}{\partial r}+\frac{\partial v_r}{\partial x}\right)=\sigma_{rx}\,, \tag{9}$$

the attachment condition of the outer surface of the wall to the surrounding rigid tissues

$$r=R+h:\qquad \vec{u}=0\,, \tag{10}$$

have been taken as the boundary conditions for (5)-(7).

The inlet pressure wave into the aortic root has been taken from the in vivo measurement data [1] as a Fourier expansion with 6 harmonics

$$x=0:\qquad p^{(1)}(t)=\sum_{k=0}^{6}P_k^{(1)}e^{i\omega_k t} \tag{11}$$

where $P_k^{(1)}$ are amplitudes of the harmonics with angular velocities ω_k.

The arterial wall is considered as composed of an incompressible isotropic material with radius dependent Young modules $E^{(j)}(R_j)$ [1]. The approximations for $E^{(j)}(R_j)$ and for the term $E^{(j)}(R_j)h^{(j)}(R_j)$ needed for the PWV calculations on (1) have been taken from [19], though in some models of aorta the constant Young modulus E = 225 kPa has been used [20]. In the 55-tube model [2] higher wall rigidity have been used: E = 0.4 MPa for aortic wall, E = 0.8 MPa and E = 1.6 MPa for its larger and smaller branches accordingly. The approximation for c_0 was accepted from the approximation $(Eh/\rho_b d)=15.18\cdot d^{-1.0514}$ with the coefficient of determination $R^2=0.8296$ [2]. The approximation for the aortic wall thickness has been taken from [20] in the form

$h(d)=R_0(0.2802\cdot\exp(-5.053R_0)+0.1324\exp(-0.1114R_0)$,

where the local unperturbed radius $R_0(x)$ in every arterial segment is determined from linear interpolation between the inlet R_{in} and outlet R_{out} radii : $R_0(x)=R_{in}-(R_{in}-R_{out})x/L$.

Solution of (5)-(7) has been found in the form of the running waves

$$f(t,r,x)=\sum_{j=0}^{\infty}f_j(r)e^{I\omega_j(t-x/c_j)}\ \text{for}\ f=\{v_x,v_r,u_x,u_r,p,p_w\}. \tag{12}$$

The numerical procedure for the FSI problem is described in detail and validated in [21]. The following material parameters have been used for numerical simulations: $\rho_b = 1.05$ kg/m^3, $\mu_b = 4 \cdot 10^{-3}$Pa·s, ρ_w=1.12 kg/m^3, $\tau = 0.045$ c, $\tau_2 = 0.035$s, $\sigma = 0.5$. The wall thickness and elasticity have been approximated as it is explained before. The structured tree data $Y_{tj}^{(k)}(\omega_j, \phi_j(d^{(k)}), \psi_j(d^{(k)}))$ have been taken from [20].

5. Results and discussions

The pressure and flow waveforms have been computed in the middle cross section of each segment in the aortic model (Fig.1) and the correspondent dependencies measured *in vivo* at different locations have been used for validation. The digitized signals (Fig.2) has been presented as Fourier series and compared to the correspondent harmonics of the computed signals in the form (12). The first harmonics in the solution (12) has been determined by the individual heart beat rate taken from the ultrasound data.

It was found, small variations in the parameters ϕ_j, ψ_j that determine the terminal structured outflow trees influenced the pressure and flow oscillations in the closest tube only, while the signals in other tubes remain almost the same, with minor differences. It may be connected to fast attenuation of the reflected waves and small input of the multiple reflections. In that way, when the measured inlet pressure profile has been used in the form (16), the first computations of the pressure and flow signals in the sites A-G exhibited noticeable differences with the measured signals when the initial $\phi_j = 1, \psi_j = 1$ values has been used (Fig.4). When the calibration procedure for ϕ_j, ψ_j values has been conducted by the least square method, the computed curves became quite close to the measured signals (Fig.3). In that way the structured tree outflow boundary conditions at the terminuses provide better approximation giving different wave impedances for different harmonics [19]. In the 1D model the terminal admittances are also determined by two parameters of the Windkessel model, e.g. frequency independent resistivity and capacity [1]. Despite the 1D model, the model (5)-(11) describes wave dispersion and wave propagation in the fluid and viscoelastic walls.

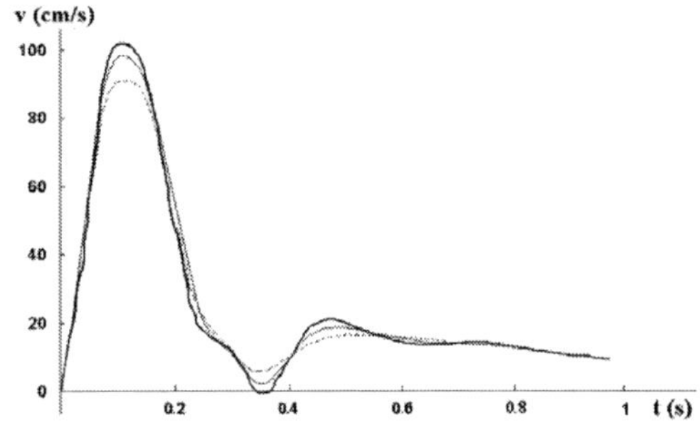

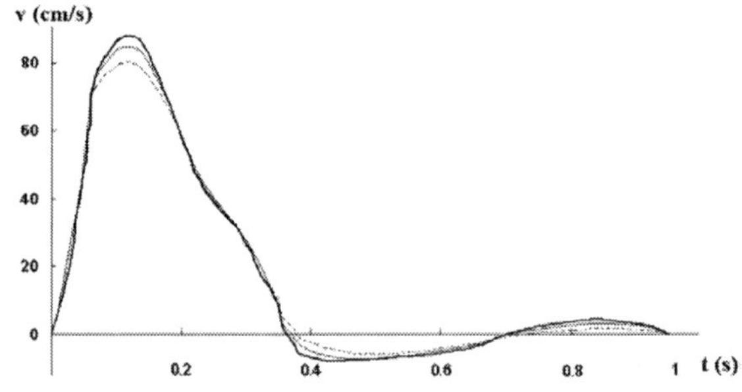

a b

Figure 3. Blood flow velocity measured (thick solid) and computed before (dashed line) and after (thin solid line) the corrections, at the locations A (a) and E (b) (see Fig.1)

After validation, the averaged values $<\phi_j>, <\psi_j>$ have been computed for the averaged geometrical models I (Fig.4) and II (Fig.5) dimensions (Table 1). An example of numerical computations on the averaged model is presented in Fig.4. The main features of the pulse wave propagation along aorta i.e. increase in the pressure amplitude and steepening, appearance of the dicrotic wave due to wave reflections, decrease of the flow amplitude and absence of the reverse flow at the end systole, are presented in the computed curves. The same signals computed for the model II scaled to the height h=175 cm give more smooth curves for the same input data. This may be connected with big differences in the lengths of the aortic segments of the taller individuals (h=172, 177cm) and shorter ones (h=155, 164 cm).

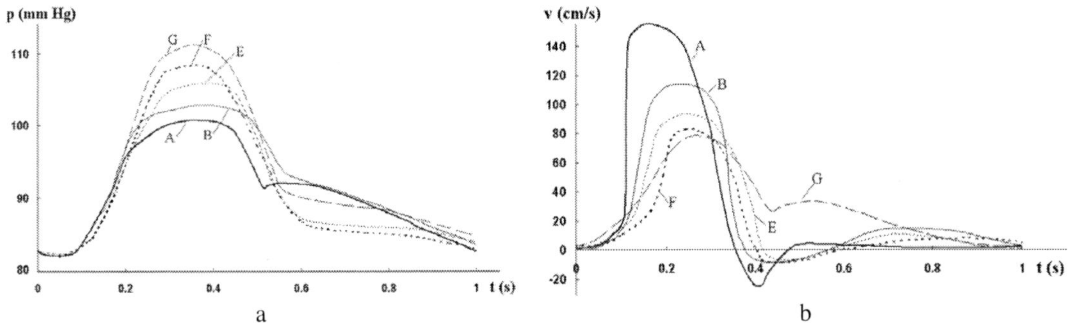

Figure 4. Blood pressure (a) and flow (b) waveforms in the locations A,B,E,F,G of the averaged model

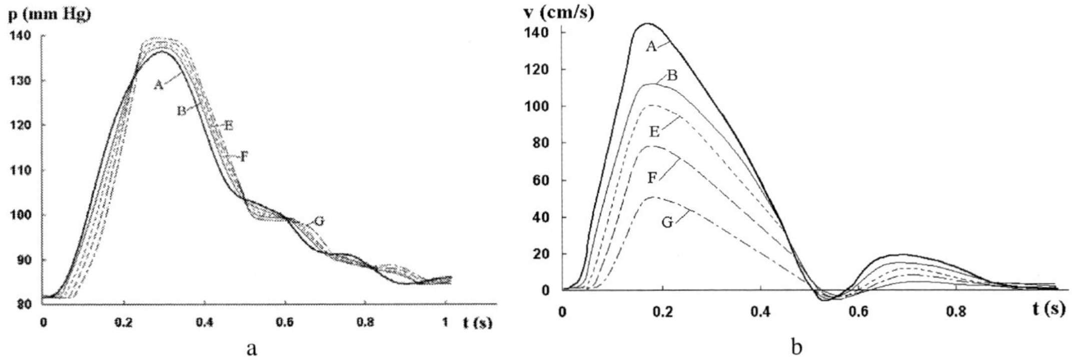

Figure 5. Blood pressure (a) and flow (b) waveforms in the locations A,B,E,F,G of the scaled models

The data on the arterial diameters are more uniform than on the lengths that produce significant differences in the phases of the waves reflected at the same branches in different individuals. For instance, the aortic length between the inlet and aortic bifurcations varied between 58-72cm. When the individual dimensions are scaled by the personal height, and then averaged and scaled to a fixed height, it produces fewer differences between personal specific coordinates of the same branches along the aorta.

6. Summary

Based on personal MRI of CT-extracted individual geometry of the larger blood vessels, the corresponding 3D structure can be restored and the hemodynamic parameters p(t), v(t), PWV and WSS can be computed. The smallest vessels remain invisible and the proper boundary conditions are needed for the CFD simulations. Here the most detailed aortic model composed from 34 straight segments with 57 side branches is developed on the measurement data. It was shown, the average geometrical model rescaled by individual height of the individual exhibit steady flow and wave propagation properties similar to those known from literature. Most of the side branches promote wave reflections along the flow, not upstream, that produces lower pressure at the outlets of the aortic segments and flow acceleration due to the suction effect.

The presented mathematical model gives more realistic results than those obtained on 3d geometries with rigid walls. The test computations revealed physically consistent pressure wave peaking and steepening along aorta, and blood flow decrease in the amplitude, negative flow phase at the end systole and dicrotic wave appearance and development due to the wave reflections. The major input in the dicrotic wave is produced by aortic bifurcation [1].

The developed mathematical model allows fast numerical computations for different geometries. The scaled model II can be 'adjusted' for any individual when the inlet and outlet diameters of the aorta and height are measured. The boundary conditions can be reconstructed from p(t) and v(t) curves measured on 6-10 arteries by US. The model can predict high reflection sites that are dangerous for wall damage. It can be also useful for modeling the hydrodynamic outcomes of aortic stenosis, aneurism and other pathologies, as well as early diagnoses of them.

The authors are grateful to organizers and participants of the XXIII Fluid Mechanics Conference KKMP2018 (Zawiercie, Poland, 9 -12 September 2018).

References

[1] Nichols W and O'Rourke M 2005 *McDonald's, blood flow in arteries. Theoretical, experimental and clinical principles* (Oxford: Oxford Univ. Press)
[2] Westerhof N, Bosman F, de Vries C J and Noordegraaf A 1969 *J. Biomech* **2** pp 121-143
[3] Kizilova N, Philippova H and Zenin O 2010 *Mechanika w Medycynie* vol 10 ed Korzynski M and Cwanka J (Rzeszow: Rzeszow Univ. Press) pp 103-118
[4] Daves J E, Parker K H, Francis D P, et al. 2008 *Heart* **94** pp 1545-1547
[5] Nguyen K T, Clark Ch D, Chancellor Th J and Papavassiliou D V 2008 *J. Biomech.* **41** p 11-19
[6] Westerhof B E, Guelen I, Westerhof N, et al. 2006 *Hypertension* **48** pp 595-601
[7] Latham R D, Westerhof N, Sipkema P, et al. 1985 *Circulation* **72** pp 1257-1269
[8] Vappou J, Luo J, Okajima K, et al. 2011 *Artery Res.* **5** pp 65-71
[9] London G M and Pannier B 2010 *Nephrol. Dial Transplant.* **25** pp 3815–3823
[10] van der Palen R L F, Barker A J, Bollache E, et al. 2017 *J. Cardiovasc. Magn. Resonance* **19** 10.1186/s12968-017-0345-7
[11] London G M and Guerin A P 1999 *Am. Heart J.* **138** pp 220-224
[12] Shahmirzadi D and Konofagou E E 2012 *Artery Res.* **6** pp 114-123
[13] Lighthill M J 1978 *Waves in fluids* (Cambridge: Cambridge Univ. Press)
[14] Bergel D H 1961 *J. Physiol.* **156** pp 458–469
[15] Hollander E H, Wang J J, Dobson G M, et al. 2001 *Am. J. Physiol.* **281** pp 895-902
[16] Sun Y-H, Anderson T J, Parker K H and Tyberg J V 2000 *J. Appl. Physiol.* **89** pp 1636–1644
[17] Zenin O K, Kizilova N N, Philippova E N 2007 *Biophysics* **52** pp 924-930
[18] Kizilova N N 2007 *Biophysics* **52** pp 131-136
[19] Olufsen M S 1999 *Am. J. Physiol.* **276** pp 257-268
[20] Blanco P J, Watanabe S M, Passos M A, et al. 2015 *IEEE Trans. Biomed. Eng.* **62** pp 736–753
[21] Hamadiche M and Kizilova N N 2005 *Intern. J. Dynam. Fluids* **1** pp 1-23

Pre-swirl energy saving device in marine application

P Król[1,2] and K Tesch[1]

[1]Faculty of Mechanical Engineering, Gdańsk University of Technology, Gdańsk, Poland
[2]Ship Design and Research Centre, Gdańsk, Poland

E-mail: przemyslaw.krol@cto.gda.pl

Abstract. This paper covers topics of energy saving device (ESD) with application to marine propulsors. The form of ESD, considered in this paper, consists of fixed lifting foils mounted in front of the screw propeller (the pre-swirl stator/guide vanes). An algorithm for designing propulsion systems, consisting of guide vanes and screw propeller, is presented. The proposed method relies on hybrid lifting line (guide vanes)-lifting surface (screw propeller) vortex model. Additionally, an analysis method for such systems is also presented. It utilises lifting surface model for guide vanes as well as for screw propeller blades. Accuracy of the proposed method is notably lower in comparison with CFD calculations, however, the former is substantially faster and at the same time capable of providing accuracy sufficient for practical application. Further, the entire process of design and analysis is presented on the example of Nawigator XXI ship. The designed propulsion system, consisting of propeller CP753 and pre-swirl stator ST002, was manufactured and tested in model scale. Model tests revealed increased efficiency by 4% in the design point in comparison with with the Nawigator's original propeller CP469. Finaly, model scale experimental results, conducted in Ship Hydrodynamics Division of CTO Gdańsk, are compared with numerical predictions.

1. Introduction

The pre-swirl stator is an energy saving device in the form of fixed stator blades mounted in front of the operating propeller. ESD are widely applied in the field of energy generation industry, however, their application in marine propulsion systems is rather limited. Although such devices are known in the literature [11], modern design methods of pre-swirl stators involve multiple CFD calculations of subsequent configurations [8, 9, 11]. Methods based on inviscid flow theory are much faster but at the same time they are less accurate than CFD [4, 11]. However, carefully conducted inviscid calculations are even capable of realistic reproduction of unsteady cavitation phenomena [1]. Such a method would find an immediate practical application if it were thoroughly verified.

The vortex lattice method for designing propeller-pre-swirl stator systems was developed in CTO Gdańsk [5, 6]. This paper presents its improved version and its application. The propeller-stator system was designed for Nawigator XXI ship by means of the hybrid lifting line-lifting surface algorithm and analysed both numerically (lifting surface model CFD) and experimentally.

The discussed method takes advantage of the lifting line-lifting surface hybrid approach within the design-analysis loop. Despite simplicity of the adopted model, main characteristics of velocity fields are reproduced correctly. This is because an iterative procedure for relaxing free vortices shedding from the propeller and stator blades is applied. The geometry of free

IOP Conf. Series: Journal of Physics: Conf. Series **1101** (2018) 012015 doi:10.1088/1742-6596/1101/1/012015

vortices segments is rearranged in each time step in such way to satisfy force free condition on free vortex filaments [2]. Convergence is usually achieved after less than 20 iterations. This approach makes it possible to obtain reliable results with reasonable effort and time. There is no need for extensive computational power.

2. Practical application

The propulsion system consisting of controllable pitch propeller CP753 and pre-swirl stator ST002 was designed for Nawigator XXI vessel. Nawigator XXI was chosen because it was subject to numerous tests and a variety of full scale data is available. The original propulsion system consists of the controllable pitch propeller CP469 operating at constant RPM without any stator. Standard towing tank model tests were carried out in CTO Ship Hydromechanics Division. Open water characteristics for both propellers were determined – stator ST002 was not included in these tests as it was regarded as the hull appendage. Subsequently, self-propulsion tests with CP469, CP753 and with CP753+ST002 propulsors were conducted. This made it possible to confirm the main theoretical conclusions on stator's influence mechanism and operating features. Full scale power and rotational speed prediction indicate that the new propulsion system will increase efficiency by 4% in the design point. This can be considered as significant progress, since in case of huge marine engines reducing the demanded power even by a single percent leads to notable energy saving.

2.1. Design algorithm

An algorithm based on a lifting surface vortex model is used in order to design a screw propeller and the pre-swirl stator vanes. First of all, the number of stator blades, their length and angular positions as well as standard data for propeller design must be known in advance. These include the desired thrust force, number of blades, diameter, angular velocity, actual velocity field round the propeller and the design velocity of the ship. The main steps of the algorithm are shown in figure 1.

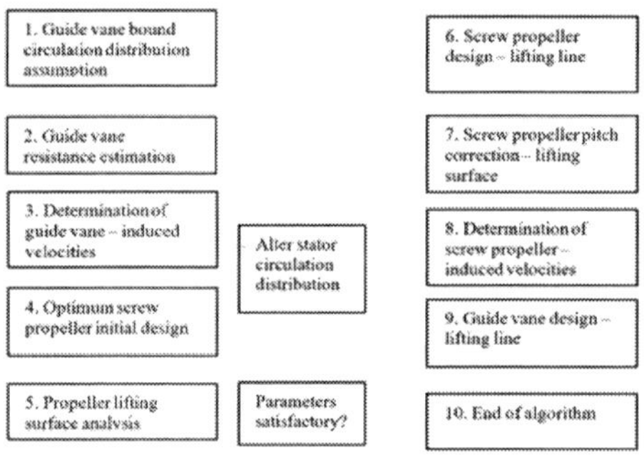

Figure 1. Block scheme of the screw propeller+guide vane design algorithm.

The main problem related to step 1 is to assume the bound circulation magnitudes and its distribution on each stator foil. The general idea here is to apply higher foil loading at positions where velocity field in propeller plane can by improved by the stator presence. From the lifting line theory, the lifting foil accelerates the velocity at its suction side and decelerates it at its pressure side, while total change in axial velocity is zero (if viscous effects are neglected). This effect can be used to make sharp velocity gradients smoother. The bound circulation distribution has important influence on the system operation as it influences directly the velocity field at the propeller location and the stator drag.

Further, the drag of the guide vanes in step 2 is estimated by using simplified formula [5]

$$D = 1.5\frac{L^2}{\pi\frac{1}{2}\rho V^2\lambda^2} \tag{1}$$

where D is a single stator foil resistance, L – the lifting force generated by this foil, ρ – water density, V is the design speed and λ is the respective foil length. This value is used later for initial evaluation of propulsor system efficiency. Equation (1) is based on the induced drag of lightly loaded hydrofoils with an elliptical loading distribution, operating in ideal fluid. It is increased here by a factor of 50% in order to take account for deviations from these assumptions, foils mutual interaction, viscous drag component, etc.

Lifting line representation of the stator foils is used to determine their induced velocities in step 3. These are calculated for the location of the propeller disc. Each foil is replaced with a set of straight line vortex segments with bound circulation variable spanwise. The geometry of the trailing vortex wake of the stator foils is calculated with an iterative scheme [5, 6]. What is more, the screw propeller presence is neglected during these calculations.

The optimal screw propeller design is prepared in step 4. This is achieved by means of the classical lifting line algorithm. The only difference is to take the stator-induced velocities into account, which is obviously not the case for the screw propeller operating alone. This propeller is then analysed further in step 5 by using a lifting surface software to estimate thrust and torque values at the design point. The next point is simply determination of the system efficiency. As the stator generates drag on itself it is included at this stage as the thrust decrease. It can be assumed, based upon CTO experience, that stator's drag is about 2-5% of the total resistance of the hull with operating propeller.

After that, the system efficiency is calculated according to the following formula

$$\eta = \frac{T - D}{2\pi n Q} \tag{2}$$

where η is the system efficiency, T – screw propeller thrust force, n – screw propeller rpm and Q is the screw propeller torque. If the calculated parameters are not satisfactory then the stator circulation distribution should be modified and all above-mentioned calculations repeated, starting from step 2. When the stator loading distribution, preserving satisfactory operation of the system, is found the design process can be continued.

The screw propeller geometry is prepared with more detail in step 6. First of all, the cavitation analysis is conducted with use of software whose description can be found in [1]. Moreover, the pitch distribution is corrected in step 7 by means of a lifting surface software. The propeller is analysed in a velocity field, taking both non-uniform wake field and stator-induced velocities into account. The hull presence alters the velocity field in which the propeller operates through its displacement and boundary layer. The velocity field measured past the hull without a propeller is referred to as a nominal wake. However, the propeller actually operates in the so called effective wake – the wake virtually experienced by the propeller. According to potential flow theory, the effective wake would be produced by a hull singularity and boundary layer system, altered by the propeller presence. What is more, the effective wake is merely a mathematical concept, being impossible to measure, because it takes no propeller induced velocities into account. Both concepts, i.e. nominal and effective wakes are well known in ship hydrodynamics. Many well established methods exist for recalculating measured nominal wakes into effective wakes [3]. In vortex calculations mean value of the effective wake is applied for each radial position. Further, it is assumed constant along the axial coordinate and taken as 'undisturbed' flow during the calculations. In viscous approach realised by means of CFD the effective wake concept is redundant, as the hull is directly present within the simulation.

The unit-span thrust force is calculated for each radii according to

$$\tau_i = \frac{T_i}{r_{i+1} - r_i} \tag{3}$$

where τ_i is the unit-span propeller blade thrust force at analysed radius i, T_i is the thrust force generated by the propeller blade bound vortices starting from analysed radius i and r_i are respective radii. Next, the unit-span thrust force values are compared with the ones determined within a screw propeller design process at respective radii. If these differ less than 1% over the whole propeller blade then the geometry is accepted. If, however, the difference is higher then the blade pitch angle is corrected at respective radius by means of the following formula

$$\Delta\varphi = \frac{\tau_D - \tau_C}{2\pi} \tag{4}$$

where φ is the screw propeller blade pitch angle, τ_D – demanded τ value at analysed radius and τ_C is the value calculated with the use of lifting surface model. After completing the screw propeller design, the geometry is finally analysed by using a lifting surface software. This time, however, not only the screw propeller dynamic characteristics are determined, but also its induced velocities in the stator plane. These serve as an input for a stator design in step 9. Since the propeller rotates with respect to its axis and the stator remains fixed, the average values for each radius are applied.

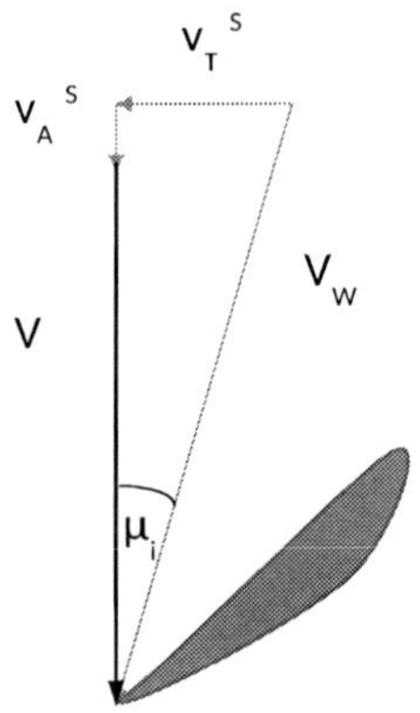

Figure 2. Velocity components on the guide vane foil element.

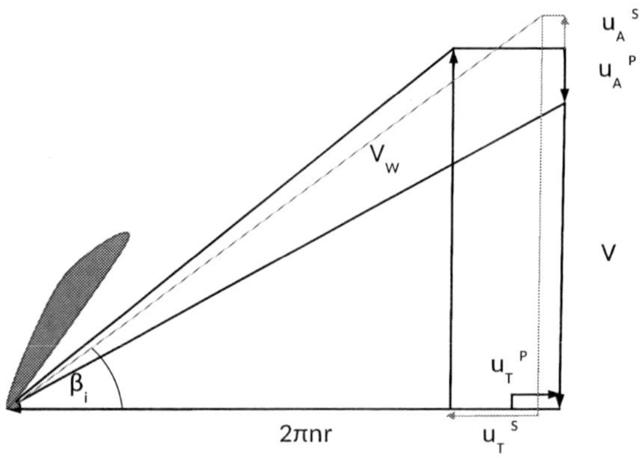

Figure 3. Velocity components on the propeller blade element.

Subsequently, the design of guide vane foils in step 9 is based on the algorithm that is very similar to the classical one used for the screw propellers in steps 4 and 6. Main differences are in the way of determination of local inflow angles, which are for a screw propeller given by

$$\beta_i = \arctan \frac{V + u_A^P - u_A^S}{2\pi nr - u_T^P + u_T^S} \tag{5}$$

where u is the velocity induced at the screw propeller disc, see figures 2 and 3. Subindex A means axial component and T denotes tangential. Superindex P denotes velocity induced by the propeller and S induced by the stator. Typically, the induced velocities u_T^P and u_A^S are negative. Furthermore, the velocities induced by the stator are taken as mean values for an individual radius. Respectively, the local inflow angle in case of the guide vane section profile is

$$\mu_i = \arctan \frac{v_T^S}{V + v_A^P + v_A^S} \tag{6}$$

where v is the velocity induced at the stator plane. Indices are the same as in the case of formula (5). Propeller-induced tangential velocity is not taken into account in equation (6) as due to the vortex theory its mean value is zero in any plane in front of it. Velocities induced by the propeller are taken as mean values for respective radius.

Moreover, as the guide vane foils have relatively high aspect ratio, no corrections for lifting surface are applied for determination of the angle of attack and profile camber. Pure lifting line model is used.

2.2. Designed system

The propulsor system consisting of a controllable pitch propeller CP753 and guide vanes ST002 was designed by means of the algorithm described above. The vessel used as an example was Nawigator XXI – training ship belonging to the the Szczecin Maritime University. The design assumptions for guide vanes + screw propeller propulsor are listed in table 1.

Table 1. Design input parameters.

Quantity	Symbol	Unit	Value
Total thrust	T	kN	125
Ship speed	V	knots	13
Number of propeller blades	Z	-	5
Propeller diameter	D	m	2.26
Rotation speed	n	rpm	259.7
Propeller delivered power	P_D	kW	1024
Number of guide vane foils	N	-	3
Guide vane foils length	λ	m	1.3

The propulsion system was manufactured and tested in the model scale (scale ratio 1:10). Further, the propulsor model mounted on the hull model is shown in figure 4. Self-propulsion test was conducted by means of continental method [14]. Full scale predictions was prepared according to the ITTC Recommendations and Guidelines [15].

Figure 4. Hull model equipped with propeller CP753 and stator ST002.

The most important design values of the designed screw propeller are the number of propeller blades $Z = 5$, propeller diameter $D = 2.26\,\text{m}$, pitch ratio at radius 0.7 $P_{0.7}/D = 0.828$ and expanded area ratio EAR $= 0.7592$.

Guide vane design parameters are given for whole stator. Particular foils differ slightly among each other, however, these differences are so small that mean values provide sufficient information about the guide vane. The guide vane design values are: angular interval $\Delta\gamma = 30°$, foil length $\lambda = 1.3\,\mathrm{m}$, mean angle of attack $\alpha_{\mathrm{mean}} = 18.91°$ and mean profile chord ratio $b_{\mathrm{mean}}/\lambda = 0.174$.

Full scale predictions based upon the model tests are in good agreement with design input. This can be inspected in table 2.

Table 2. Full scale prediction versus design input.

Quantity	Symbol	Unit	Des. value	Exp. value	Difference %
Propeller delivered power	P_D	kW	1024	1028	0.4
Propeller rate of revolution	n	rpm	259.7	265	2
Total thrust	T	N	125	126	0.8

Table 3. Model scale measurements in self-propulsion test.

Pt.	V_{M} m s^{-1}	n_{M} s^{-1}	T_{M} N	Q_{M} Nm
1	1.95	12.0	91.50	2.67
2	1.95	11.4	80.26	2.36
3	2.03	12.4	96.04	2.82
4	2.03	13.0	108.69	3.17
5	2.11	13.3	111.46	3.27
6	2.11	13.9	124.86	3.65
7	2.19	15.2	152.01	4.44
8	2.19	14.6	137.08	4.02
9	2.28	16.3	176.99	5.17
10	2.28	15.7	160.30	4.79

The self-propulsion test was conducted by means of continental method, which involves two measuring points for each tested ship speed. Model scale results are given in table 3. V_{M} is the ship model speed, n_{M} – propeller model rate of revolution, T_{M} – propeller model thrust and Q_{M} – propeller model torque.

3. Numerical methods

3.1. Lifting Surface method

The Lifting Surface software was developed as an auxiliary tool for the design process. It is based on well-known idea of replacing lifting foil with a set of discrete bound and free vortices, modelling loading effects and straight line source filaments, modelling foils finite thickness. The theoretical background of this method can be found in [1, 2, 4, 5, 6] or in the majority of papers on vortex lattice computations. Only simulation settings and assumptions are discussed here.

Both propeller blades and stator foils are discretised by 20 spanwise and 20 chordwise sections. Spanwise elements are concentrated around the tip by means of sine formula

$$r_i = r_0 + (1 - r_0)\sin\frac{\pi i}{2A - 2} \tag{7}$$

IOP Conf. Series: Journal of Physics: Conf. Series **1101** (2018) 012015 doi:10.1088/1742-6596/1101/1/012015

where r_i is a discrete radius i and A is the total number of discrete radii (in this case 20). Moreover, the chordwise spacing of discrete chordwise segments is achieved with constant steps both in the case of screw propeller blades and the guide vane foils. Analogical vortex grid was investigated in [7]. Consequently, the vortex grid convergence given in this reference is also representative for calculations with a pre-swirl stator.

The guide vanes vortex wake was assumed to remain flat and parallel to direction of ship advance speed during the simulation. It may appear to be significant simplification, however, only mean values of induced velocities are taken into account for propeller-stator interaction, as the propeller rotates, unlike the stator. According to well-known Stokes' theorem these mean values depend only on the calculation point radius, free vortices radial position and circulation. Stator-induced axial velocities in the propeller plane are much lower than tangential ones.

The screw propeller vortex wake is iteratively relaxed by a convective scheme. At each end of a free vortex discrete segment the total velocity is calculated – as a sum of undisturbed flow velocity, propeller-induced velocities and mean values of guide vane-induced velocities, proper for considered axial and radial position with respect to the guide vane. In order to preserve stability of a convective scheme the discrete vortex segment endpoints are not allowed to move more than $0.05D$. Moreover, the total velocity in each time step is taken as a weighted average of current (20%) and previous (80%) time steps. The time step value is taken such that propeller revolution angle is 5° within it. It takes around 15 iterations for a calculation point to get stable values of calculated propeller thrust and torque.

3.2. CFD

The Reynolds-Averaged Navier-Stokes approach to turbulence modelling is chosen to solve for averaged quantities. A closed system of equations for incompressible fluids can be formulated consisting of the continuity and Reynolds equation. Additional equations for the two-equation k-ω SST model [12, 16] are equation for transport of kinetic energy of velocity fluctuations k and equation for transport of turbulence frequency ω.

The system of equations is further discretised by means of the Finite Volume Method [10]. Convection terms and other diffusive terms take advantage of Gaussian integration. As for the discretised convection term, centroid variables are interpolated to the cell faces variables by second order limited linear upwind scheme. This make it possible to limit towards first-order upwind in regions of rapidly changing gradients. Diffusive terms involve surface normal gradients. An additional non-orthogonal correction is considered here to maintain second order accuracy for non-orthogonal meshes. The same applies to Laplacian schemes. Moreover, gradient terms, such as pressure gradient, are discretised by means of Gaussian integration. Central differencing is used here in order to interpolate values from cell centres to face centres. Additionally, cell limiting is also introduced in order to improve boundedness and stability [13].

The discretised governing equations are solved by means of the open source software for CFD OpenFOAM [13]. The SIMPLE algorithm is used in order to solve pressure-velocity coupling. The pressure equation is solved by means of GAMG solver with DIC smoother [13]. For the velocity fields as well as turbulent quantities standard solver using a Gauss-Seidel smoother is used. Additionally, under-relaxation factors are used in order to improve stability of a solution.

The flow domain, shown in figure 5, consists of two parts, namely the rotating propeller and the steady hull (figure 6) inside a towing tank. Breadth and depth of the tank correspond to the real dimensions of the Ship Design and Research Centre towing tank, which dimensions length × breadth × depth are 270 × 12 × 6 m respectively. Both domains are discretised separately and merged by means of the arbitrary mesh interfaces. Computational grids can be classified as Cartesian. The merged mesh consists of 8 961 314 nodes and 8 408 464 volumes where 7 536 421 are hexahedra. The maximal y^+ value on the blades does not exceed 5. Figure 7 displays mesh convergence by showing the influence of number of nodes on the T and Q. It can be seen that

IOP Conf. Series: Journal of Physics: Conf. Series **1101** (2018) 012015 doi:10.1088/1742-6596/1101/1/012015

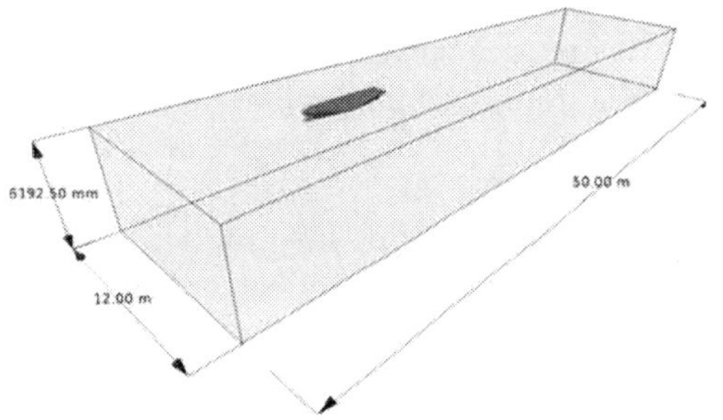

Figure 5. Flow domain.

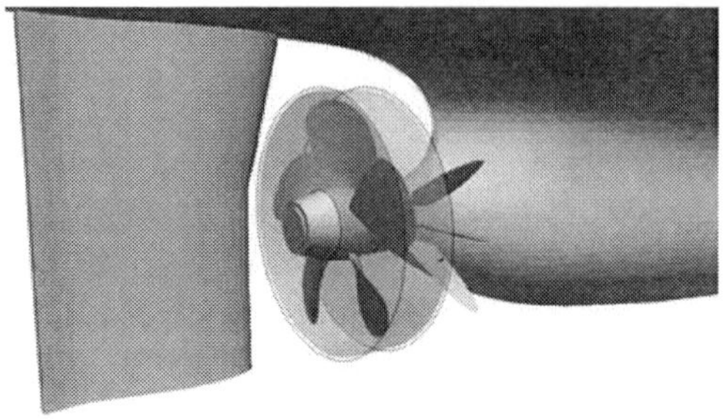

Figure 6. Rotating propeller and steady hull.

increasing the number of the mesh nodes above 9×10^6 has negligible effects on T and Q in particular. Mesh convergence test was conducted for a single point 5 in table 3.

The main boundary conditions include inlet, outlet, walls (side and bottom walls, hull, rudder and shaft) and top symmetry surface. As for the inlet, the fixed velocity perpendicular to inlet surface together with zero normal gradient pressure is specified. Low turbulence intensity case is also assumed. Furthermore, the constant pressure distribution is assumed at the outlet accompanied with zero gradient velocity. This can be justified simply because the outlet surface is located far from the propeller. The impermeability and adhesion condition are specified on the walls, i.e. the no-slip condition. The scalable wall function approach is used in the near wall region. Finally, the so called cyclic arbitrary mesh interface is considered allowing for coupling between stationary domain and rotating propeller. Most importantly, a steady state approach is used also referred to as a multiple reference frame simulation. This saves computational time considerably in comparison with the full transient rotor-stator interaction.

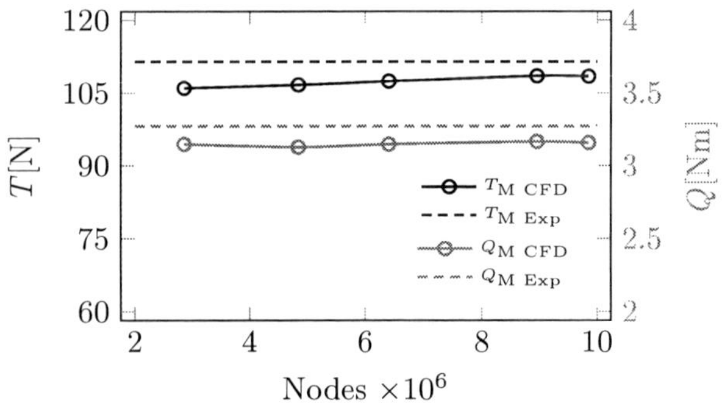

Figure 7. Mesh convergence.

4. Results

The model scale measurements were reproduced by two numerical methods, i.e. lifting surface analysis (in-house code) and CFD (OpenFOAM). The model scale speed V_{M} and propeller rate of revolution n_{M} were regarded as input values and the propeller model thrust T_{M} and torque Q_{M} were calculated during the simulations. Results are presented in table 4 where 'Exp' denotes experimental value, 'LS' Lifting Surface method results and 'CFD' results from CFD calculations. The relative error is denoted here as ΔT or ΔQ.

It can be easily seen that in general the error of CFD predictions is notably lower and more stable than in Lifting Surface Method results. On the other hand a single calculation point with LS method requires around 30 minutes (quad-core processor i5-4590, 3.30 GHz – 1 core engaged) while CFD requires 180-200 minutes (six-core processor i7-6850K, 3.60 GHz – 6 cores engaged). Keeping in mind that the six-core i7 processor is several times faster than quad-core i5, it becomes obvious that CFD approach is about 20 times slower. This makes the vortex methods still attractive since it is capable of providing accuracy sufficient for practical application.

Table 4. Model scale calculations in self-propulsion test.

Pt.	$T_{\mathrm{M\,Exp}}$ N	$T_{\mathrm{M\,LS}}$ N	$T_{\mathrm{M\,CFD}}$ N	$\Delta T_{\mathrm{M\,LS}}$ %	$\Delta T_{\mathrm{M\,CFD}}$ %	$Q_{\mathrm{M\,Exp}}$ Nm	$Q_{\mathrm{M\,LF}}$ Nm	$Q_{\mathrm{M\,CFD}}$ Nm	$\Delta Q_{\mathrm{M\,LS}}$ %	$\Delta Q_{\mathrm{M\,CFD}}$ %
1	91.50	87.53	87.10	4.34	4.80	2.67	2.67	2.55	0.19	4.67
2	80.26	82.96	75.90	3.36	5.42	2.36	2.48	2.23	4.95	5.48
3	96.04	96.99	92.51	0.99	3.67	2.82	2.94	2.70	4.95	4.01
4	108.69	111.10	104.87	2.22	3.50	3.17	3.31	3.05	4.95	3.71
5	111.46	114.95	108.53	3.13	2.63	3.27	3.13	3.16	4.40	3.22
6	124.86	129.24	121.94	3.51	2.33	3.65	3.85	3.54	5.54	2.84
7	152.01	159.84	150.43	5.15	1.03	4.44	4.74	4.35	6.68	2.06
8	137.08	143.75	135.47	4.87	1.17	4.02	4.28	3.93	6.47	2.16
9	176.99	156.21	176.30	11.74	0.38	5.17	4.78	5.08	7.47	1.55
10	160.30	151.37	160.02	5.57	0.17	4.70	4.56	4.62	2.94	1.45

Acknowledgments
The experimental results and design algorithm presented and discussed in the paper were developed within the CTO Internal Research Found.

References
[1] Bugalski T and Szantyr J 2014 Numerical Analysis of the Unsteady Propeller Performance in the Ship Wake Modified By Different Wake Improvement Devices, *Polish Maritime Research* 3(83), 32
[2] Greeley D and Kerwin J 1982 Numerical Methods for Propeller Design and Analysis in Steady Flow, *Transactions of Society of Naval Architects & Marine Engineers* **90**, 415
[3] Jarzyna H 1993 Wzajemne oddzialywanie kadluba i pednika statku (Wroclaw: Ossolineum)
[4] Kerwin J, Coney W and Hsin C 1988 Hydrodynamic Aspects of Propeller/Stator Design, *Propellers' 88 Symposium 5th (Virginia Beach)*
[5] Król P and Bugalski T 2017 Application of vortex flow model in propeller-stator system design and analysis, *Polish Maritime Research* **1**(97) 35
[6] Król P, Bugalski T and Wawrzusiszyn M 2017 Development of numerical methods for marine propeller – pre-swirl stator system design and analysis *Fifth International Symposium on Marine Propulsion (Espoo)*
[7] Król P and Tesch K 2018 Experimental and numerical validation of the improved vortex method applied to CP745 marine propeller model, *Polish Maritime Research* **2**(98) 56
[8] Kim J, Choi J, Choi B, Chung S and Seo H 2015 Development of energy-saving devices for a full slow-speed ship through improving propulsion performance *International Journal of Naval Architecture and Ocean Engineering* **7**(2) 390
[9] Chen K, He W, Qin J and Wang L 2013 On the Design of Energy Saving Shaft Bracket for a Twin Screw Vessel *2013 Proceedings of the Twenty-third International Offshore and Polar Engineering Conference (Anchorage)*
[10] Ferziger J H and Perić M 2002 Computational methods for fluid dynamics (Berlin: Springer-Verlag)
[11] Kim M, Chun H and Kang Y 2004 Design and Experimental Study on a New Concept of Preswirl Stator as an Efficient Energy-Saving Device for Slow Speed Full Body Ship *Transactions-Society of Naval Architects and Marine Engineers* **112** 111
[12] Menter F R 1994 Two-equations eddy-viscosity turbulence models for engineering applications *AIAA-Journal* **32**(8)
[13] OpenFOAM user guide 2015, OpenFOAM Foundation Ltd.

[14] Propulsion Committee of the 28th ITTC Propulsion/Bollard Pull Test ITTC Quality System Manual Recommended Procedures and Guidelines 2017
[15] Propulsion Committee of the 28th ITTC 1978 ITTC Performance Prediction Method ITTC Quality System Manual Recommended Procedures and Guidelines 2017
[16] Wilcox D C 1994 Turbulence modeling for CFD (California: DCW Industries)

IOP Conf. Series: Journal of Physics: Conf. Series **1101** (2018) 012016 doi:10.1088/1742-6596/1101/1/012016

Prediction of endwall losses in a low pressure turbine cascade with an algebraic intermittency model

S Kubacki[1], P Jonak[1] and E Dick[2]

[1]Institute of Aeronautics and Applied Mechanics, Warsaw University of Technology, Nowowiejska 24, 00-665, Warsaw, Poland
[2]Department of Flow, Heat and Combustion Mechanics, Ghent University, St.-Pietersnieuwstraat 41, B-9000, Ghent, Belgium

e-mail: slawomir.kubacki@meil.pw.edu.pl

Abstract. The predictive qualities of a newly developed algebraic intermittency model are analysed by simulation of the flow through a linear cascade of low pressure turbine (LPT) blades with endwalls. Both steady RANS (Reynolds-averaged Navier-Stokes) and time-accurate RANS (URANS) simulations are performed. The results are compared with reference LES (Large Eddy Simulation) by Cui et al. (2017, Numerical investigation of secondary flows in a high-lift low pressure turbine, Int. J. of Heat and Fluid Flow, vol. 63) for a turbulent endwall boundary layer (TBL) at the entrance to the cascade. Good agreement is obtained between simulations with the algebraic model and the reference LES for the evolution of the mass-averaged total pressure loss coefficient and for profiles of pitchwise-averaged turbulent kinetic energy.

1. Introduction

Endwall losses may contribute up to 30% to the total aerodynamic losses of a turbine cascade. Therefore, proper accounting of secondary flow features at the endwalls is of particular importance in the design of an axial turbine. An oncoming endwall boundary layer impacts the leading edge of the blades and separates. The separated boundary layer rolls up and forms the streamwise-oriented pressure-side (HSVPS) and suction-side (HSVSS) legs of a horseshoe vortex. The pressure-side leg travels, due to a strong transverse pressure gradient, towards the suction side of the neighbouring blade and becomes a part of the passage vortex (PV). The suction-side leg develops near the suction side of the blade, but there are several scenarios reported in the literature on this development, depending on the cascade load and load distribution. It may travel above the passage vortex, wrap around the passage vortex or dissipate quickly downstream of the leading edge. Goldstein and Spores [1] found the suction-side leg of the horseshoe vortex travelling above the passage vortex until the trailing edge of the cascade. They observed formation of two counter-rotating vortices in the corner of the suction side of the blade and the endwall at the intersection point of the horseshoe vortex separation line in the endwall plane and the blade surface. Generation of a corner vortex was also observed on the blade pressure side. Gross et al. [2] studied experimentally and numerically the secondary flow of a cascade with front-loaded L2F-blades. The experiments were performed at a high inlet turbulence level, Tu=3%, whereas a zero turbulence level was taken in the LES. A separated boundary layer was reproduced on the blade surface away from the endwall using LES. The boundary layer did not separate on the blade surface in the experiment owing to the high freestream turbulence level. Interestingly, the difference in blade surface boundary layer state has only a minor effect on the pressure losses and the secondary flow structure at the endwall. Gross et al. [2] observed fast dissipation of the suction-side leg of the horseshoe vortex at the onset of the adverse pressure gradient region on the blade suction side. They also observed a corner separation vortex (SSCSV) on the blade suction side with the same sense of rotation as the passage vortex. The LES results showed the formation of a shed vortex (SV) at the trailing edge, with the rotation sense opposite to that of the passage vortex. Bear et al. [3] studied experimentally the total pressure loss and the loss production mechanism for the same cascade with front-loaded L2F-blades, with and without a suction side corner fillet. The flow was subjected to a high freestream turbulence level, Tu=3%. They found that the passage vortex primarily generates loss within the blade passage, while the suction surface corner separation vortex (SSCSV) generates loss within the passage and downstream of it. The shed vortex (SV) generates loss downstream of the trailing edge. Cui et al. [4] studied the secondary flow through a linear cascade of T106A-blades with endwalls, using LES. The simulations were performed with three sets of inlet conditions: laminar and turbulent endwall boundary layers and laminar endwall boundary layers combined with free-stream inflow perturbed by wakes. A low freestream turbulence level, Tu=0.5%, was specified in the simulations with the laminar and turbulent boundary layers. The results obtained in the present paper with the algebraic model are compared with the reference LES by Cui et al. [4] for turbulent endwall boundary layers (TBL) at the cascade entrance.

2. Algebraic transition model

The algebraic intermittency model [5,6] is combined with the newest version of the k-ω RANS turbulence model by Wilcox [7]. It takes into account two effects in an attached pre-transitional boundary layer: damping of short-wavelength disturbances induced by the free stream and breakdown of near-wall disturbances into fine-scale turbulence.

The transport equations for turbulent kinetic energy k and specific dissipation ω rate are

$$\frac{Dk}{Dt}=\gamma P_k+(1-\gamma)P_{sep}-\beta^* k\omega+\frac{\partial}{\partial x_j}\left[\left(\nu+\sigma^*\frac{k}{\omega}\right)\frac{\partial k}{\partial x_j}\right], \tag{1}$$

$$\frac{D\omega}{Dt}=\alpha\frac{\omega}{k}P_k-\beta\omega^2+\frac{\partial}{\partial x_j}\left[\left(\nu+\sigma\frac{k}{\omega}\right)\frac{\partial\omega}{\partial x_j}\right]+\frac{\sigma_d}{\omega}\frac{\partial k}{\partial x_j}\frac{\partial\omega}{\partial x_j}. \tag{2}$$

The intermittency function γ is a multiplier of the production term P_k in the k-equation. The production term is $P_k=\nu_s S^2$, where ν_s is the small-scale eddy viscosity, which is part of the full eddy viscosity ν_T. S is the magnitude of the shear rate tensor. In the laminar part of a boundary layer, γ is set to zero. There is then no production of k, but turbulent kinetic energy enters by diffusion out of the free stream. In the laminar part of a boundary layer, the ω-equation stays active. The term $(1-\gamma)P_{sep}$ models turbulence production by instability and breakdown of the laminar free shear layer generated by boundary layer separation in a low turbulence level background flow.

The turbulent kinetic energy k is split, based on the laminar-fluctuation kinetic energy transition model by Walters and Cokljat [8], into a small-scale part and a large-scale part:

$$k_s=f_{SS}k, \quad k_l=k-k_s. \tag{3}$$

The splitting by the factor f_{SS} expresses the shear-sheltering effect in a pre-transitional boundary layer, which is the damping of high-frequency components of the turbulence coming from the free stream. Shear sheltering depends on the ratio of two timescales in a laminar layer: the timescale of convection of disturbances relative to an observer inside the layer and the timescale of diffusion in the wall-normal direction. In present work, it is modelled by [6]:

$$f_{SS}=exp\left(-\left(\frac{C_{SS}\nu}{\sqrt{k}y}\right)^2\right), \tag{4}$$

where y is the distance to the wall and $C_{SS}=C_S(1+C_A f_W \psi)$ is a flow-dependent coefficient. C_S and C_A are model constants (Table 1). The ψ and f_W functions are:

$$\psi = tanh\left(\frac{-\Omega(S-\Omega)}{C_\psi\left(\beta^*\omega\right)^2}\right), \quad f_W = 1 - tanh\left(\frac{k}{C_W \nu\omega}\right), \tag{5}$$

with Ω the magnitude of the rotation rate tensor. The role of the flow-dependent term $f_W\psi$ is increasing C_{SS} (larger shear sheltering) in accelerating flow ($f_W\psi>0$), and reducing C_{SS} (smaller shear sheltering) in decelerating flow ($f_W\psi<0$), for acceleration or deceleration due to streamline curvature. The f_W function limits the correction to the border zone between the laminar and turbulent parts in a pre-transitional boundary layer.

The eddy viscosity associated to small scales is calculated in the same way as the eddy viscosity of the original turbulence model by replacing k by k_s :

$$\nu_s = \frac{k_s}{\tilde{\omega}} \quad with \quad \tilde{\omega} = max\left[\omega, \frac{C_{lim}S}{a_s}\right]. \tag{6}$$

The large-scale eddy viscosity, is, similarly defined with k_l :

$$\nu_l = \frac{k_l}{\tilde{\omega}} \quad with \quad \tilde{\omega} = max\left[\omega, \frac{C_{lim}S}{a_l}\right]. \tag{7}$$

The resulting eddy viscosity, used in the Navier-Stokes equations, is $\nu_T=\nu_s+\nu_l$.

The intermittency function γ determines when a flow region is laminar or turbulent. In the present work, γ is prescribed algebraically as a function of the distance to the wall by [6]:

$$\gamma = min\left(max\left(\frac{\sqrt{k}y}{A_\gamma \nu} - 1, 0\right), 1\right), \tag{8}$$

where A_γ is a constant.

The model includes turbulence production due to breakdown of a laminar separated boundary layer at a low free-stream turbulence level by the term $(1-\gamma)P_{sep}$. We adopt a term with the same purpose in the newest intermittency-transport transition model by Menter et al. [9]:

$$P_{sep} = C_{sep}F_{sep}\nu S^2, \quad F_{sep} = min\left(max\left(\frac{R_V}{2.2A_V} - 1, 0\right), 1\right), \tag{9}$$

with $R_V=y^2S/\nu$. The closure coefficients and other terms of the underlying turbulence model are given by Wilcox [7]. In the current model, we change the model function f_β to $f_{\beta S}=f_{d\gamma}f_\beta+(1-f_{d\gamma})$, with $f_{d\gamma}=min(\gamma, f_d)$, where f_d is the modified DES model shielding function by Spalart et al. [10]:

$$f_d = 1 - \tanh\left(\left(8r_d\right)^n\right), \quad r_d = \frac{k/\omega + \nu}{\kappa^2 y^2 \sqrt{S^2 + \Omega^2}}. \tag{10}$$

The exponent n is set to 1, instead of 3 as in the original formula [10], to allow for more gradual variation of the shielding function near a wall. This change enhances the performance of the model in simulations with 3D RANS. Table 1 lists the constants of the algebraic intermittency model (C_{lim}=7/8 and a_s=0.3 keep the standard values). The boundary conditions recommended by Wilcox [7] were specified at the no-slip walls: k=0 at the wall and $\omega=6\nu/(\beta_0(\Delta y)^2)$ in the first near-wall point, with Δy the distance of this first point to the wall.

Table 1. Transition model constants.

A_γ	C_S	C_A	C_ψ	C_W	C_{sep}	A_V	a_l
12.0	21.0	1.0	10.0	10.0	2.0	550.0	0.6

3. Computational setup

Figure 1 (a) shows the geometry of the computational domain and the types of boundary conditions. The inlet to the computational domain was placed at $0.4C_x$ upstream of the leading edge of the blades (x=0). The cascade outlet was specified at x=$1.6C_x$. A 2D grid was made, composed of a structured part near to the blade surface and an unstructured part around the structured boundary layer grid. This 2D grid was extruded in the spanwise direction. The simulations were performed on a grid with 7.2 million cells. The y^+ value in the first grid point near the blade surface and the endwall was below 1.5. The reference LES by Cui et al. [4] were obtained on a grid with 40 million cells. The second order upwind scheme was applied for the convective terms in the momentum and transport equations. The URANS simulations were performed with the second-order backward discretisation scheme in time with a CFL number less than 2. The normalised residuals for the momentum and transport equations fell below 1e-5 in the RANS simulations. Typically, about 15 inner iteration steps were required to obtain a converged solution at each time step in URANS, with residuals dropped down to 1e-5 at each time step. In the present work, a turbulent endwall boundary layer was imposed at the inlet to the cascade, according to Cui et al. [4]. The inlet profiles of the mean velocity, turbulent kinetic energy and the specific dissipation rate were obtained from a precursor simulation of a fully developed turbulent boundary layer flow. Next, the velocity profiles were projected into the inlet plane of the cascade. The angle of attack (in x-y plane) was set to α=39.7deg according to [4]. The freestream turbulence level was set to Tu=0.5%. Figure 1 (b) shows the mean velocity magnitude and the turbulent kinetic energy profiles at the cascade inlet by the RANS and LES. The agreement is good.

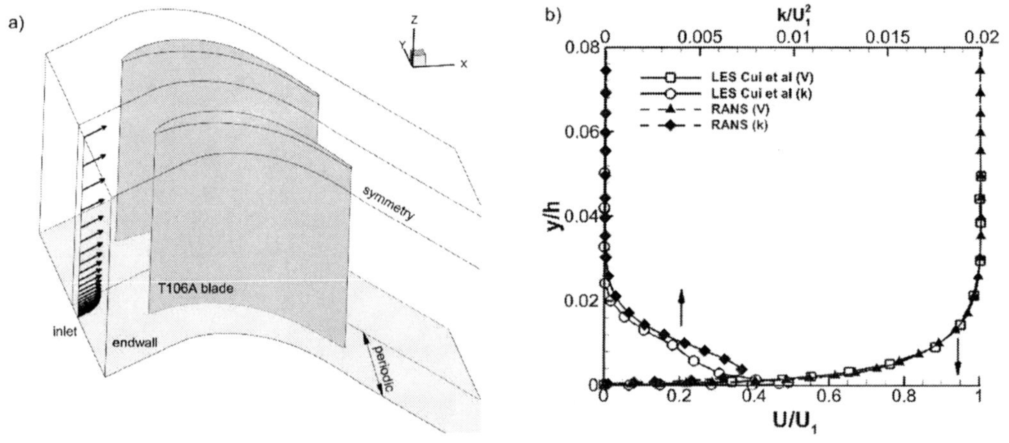

Figure 1. a) Computational domain (copied once in y-direction) and types of boundary conditions, b) Mean velocity and turbulent kinetic energy profiles at inlet to the domain.

4. Results

Flow simulations were performed using the algebraic intermittency model, employing both RANS and URANS techniques, and the standard k-ω model (fully turbulent flow). The latter are for the purpose of comparison. With the algebraic intermittency model, the boundary layer separates in the rear part of the blade, away from the endwall, due to the low freestream turbulence level (Tu=0.5%), transitions to turbulent state and reattaches near to the trailing edge. The separation is not obtained by the fully turbulent simulation. Figure 2 shows the comparison of the pitchwise-averaged turbulent kinetic energy profiles at the streamwise distances, C_x=0.5, 0.8, 1.1 and 1.3. The results are compared with the reference LES by Cui et al. [4]. The endwall is located at z/h=0 (h-is the blade span). Figure 2 (a) shows a peak of k at z/h=0.02, caused by increased turbulence production due to the passage vortex. The peak value is underpredicted by all models. The fully turbulent model produces too high k values away from the endwall (z/h>0.1), owing to the turbulent character of the boundary layers on the blade surface. The near-wall peak value is better captured with the algebraic model farther downstream the cascade, at

x/C_x=0.8 (Fig. 2 b). The fully turbulent simulation shows again a too high value of k for z/h>0.16. Surprisingly, the fully turbulent model reproduces a too narrow turbulent patch at z/h=0.12-0.15, approaching the cascade outlet (Fig. 2 c and d). The algebraic model shows better agreement with LES. The very small difference between the total (k_{tot}) and the modelled (k_{mod}) turbulent kinetic energy profiles with URANS in figure 2 is remarkable. It means that most of the turbulence resists in the modelled part. Some differences between the steady RANS and the time-accurate RANS (URANS) techniques are noted at x/C_x=1.1 and 1.3. This is caused by increased activity of the large-scale unsteadiness in the separated boundary layer along the blade surface away from the endwall, at the streamwise distance x/C_x=0.9 (not shown). The good agreement between the results obtained by the algebraic intermittency model and the reference LES, in terms of the pitchwise-averaged turbulent kinetic energy profiles, allows for trustworthy analysis of the secondary flow features.

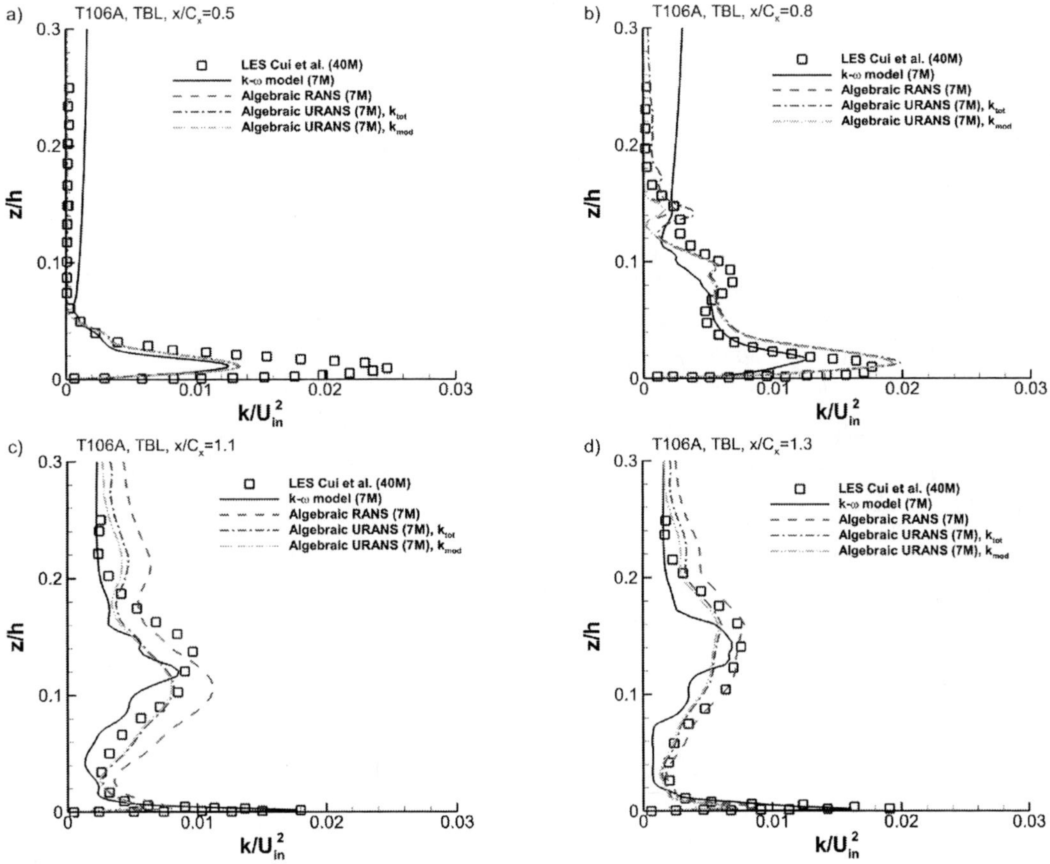

Figure 2. Pitchwise averaged turbulent kinetic energy profiles at distances a) x/C_x=0.5, b) x/C_x=0.8, c) x/C_x=1.1, d) x/C_x=1.3.

Figure 3 shows the visualisation of the secondary flow by means of iso-surfaces of the q-criterion (q=10,000), coloured by the static pressure. For further visualisation, a primary flow is defined as the time-averaged flow at midspan, extruded in spanwise direction (thus, a 2D flow). Secondary velocity vectors are then defined by the difference between the local time-averaged velocity (3D flow) and the primary flow velocity (2D flow). Secondary flow velocity vectors are shown in selected y-z planes (in every 15th cell), projected into these planes. In Figure 2 (b), also limiting streamlines are shown. With the secondary flow velocity vectors and the limiting streamlines, the rotation sense of the secondary flow vortices can be determined. The URANS results are discussed here, but the steady RANS results show similar secondary flow characteristics.

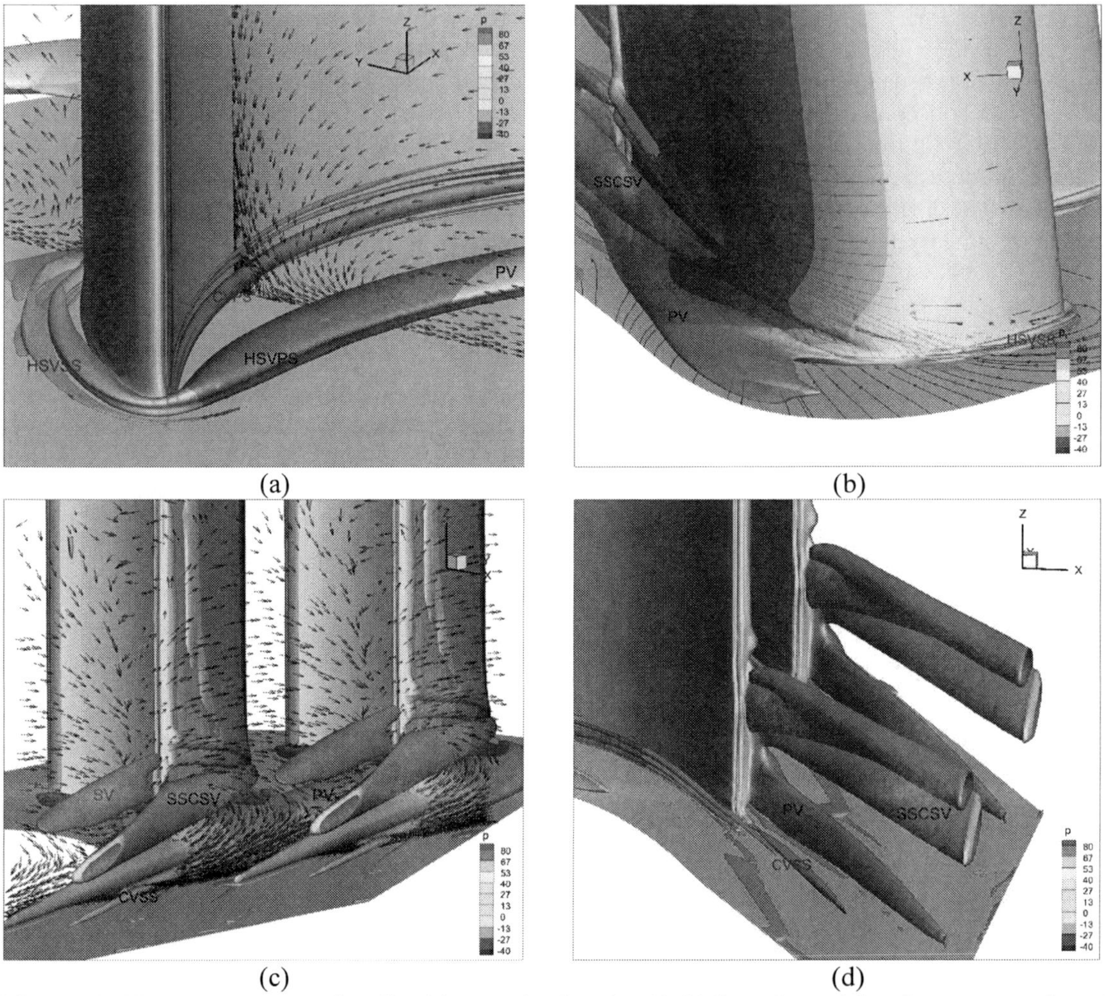

Figure 3. Vortex structures visualised by q-criterion (q=10,000), coloured by time-averaged static pressure and secondary velocity vectors in y-z planes (a, c and d), projected into these planes: a) view near the blade leading edge (x/C_x=0.2), b) in the front part of the blade (suction side), and near the trailing edge c) view from the suction side (x/C_x=1.1) and d) view from the pressure side.

The pressure- and suction-side legs of the horseshoe vortex are clearly visible in figure 3 (a). The pressure-side leg (HSVPS) develops into the passage vortex (PV). The rotation sense of the passage vortex is counter-clockwise when viewed from the cascade inlet. A small corner vortex (CVPS) is formed on the blade pressure side (Fig. 3 a), rotating in the opposite sense to the passage vortex. The suction-side leg of the horseshoe vortex (HSVSS) is pushed towards the midspan of the blade by the corner separation of the boundary layers on the blade suction side and the endwall, which starts around x/C_x=0.2, as is visible in figure 3 (b). As the flow is turned through the passage, it is subjected to a centrifugal force due to streamline curvature. The centrifugal force is balanced by the pressure difference between the pressure and suction sides of neighbouring blades. This centrifugal force is reduced, however, inside the endwall boundary layer due to reduced velocity. Since pressure is approximately constant in the wall-normal direction across a boundary layer, an imbalance occurs in the endwall region. This imbalance causes the passage vortex and the associated cross flow near the endwall from the pressure side to the suction side, as is visible in figure 3 (b). This cross-flow is responsible for the formation of a pair of corner vortices in the separation zone downstream of x/C_x=0.2 (Fig. 3 b). It consist of a small vortex deeply in the corner (CVSS in Fig. 3 c and d), with rotation sense against the passage vortex and a much stronger vortex more towards the midspan, with the same rotation sense as the passage vortex, called the suction side corner separation vortex (SSCSV in Fig. 3 c and d), by Gross et al. [2] and Bear et al. [3]. It gains in strength as it travels downstream the passage. At the trailing edge, a shed vortex is formed (SV in Fig. 3 c and d), with rotation sense against the passage vortex. The SV and HSVSS have the same rotation sense and merge downstream of the trailing edge, forming what universally is called the counter vortex (CV). The vortex system, produced by the algebraic intermittency model, agrees very well with the vortex system produced by the LES of Cui et al. [4] and observed by Goldstein and Spores [1]. In our interpretation, it accords also with the vortex system observed by Gross et al. [2] and Bear et al. [3]. The difference is that these researchers claim that the HSVSS dissipates. But they do not stress that, in their results, immediately downstream of the dissipation position, a new rather weak vortex is generated with the same rotation sense as the HSVSS. This new vortex can be interpreted as the continuation of the HSVSS.

Figure 4 shows contours of the shape factor H_{12} over the blade suction side and the endwall, obtained with the unmodified k-ω model (Fig. 4 a) and with the algebraic intermittency model (Fig. 4 b). The blanked zones correspond to H_{12}>3.5 and indicate boundary layer separation. No separation is obtained on the suction side of the blade near to the midspan (z/h=0.5) with the unmodified k-ω model. With the algebraic model (Fig. 4 b), the blade boundary layer separates at the rear part of the blade, transitions to turbulent state and reattaches as fully turbulent near the trailing edge. Another difference between the result by the algebraic model and the fully turbulent model is reproduction of a flow region prone to separation (indicated by the red arrow in Fig. 4 b) at the outlet from the cascade. The contours of the shape factor at the endwall reproduced with the algebraic model agree well with the contours obtained with the reference LES [4].

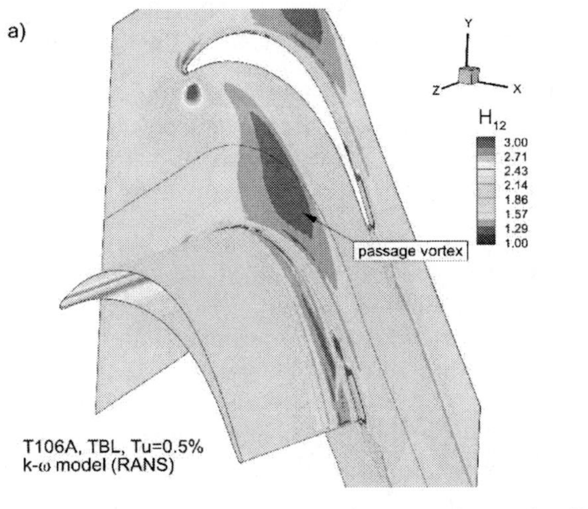

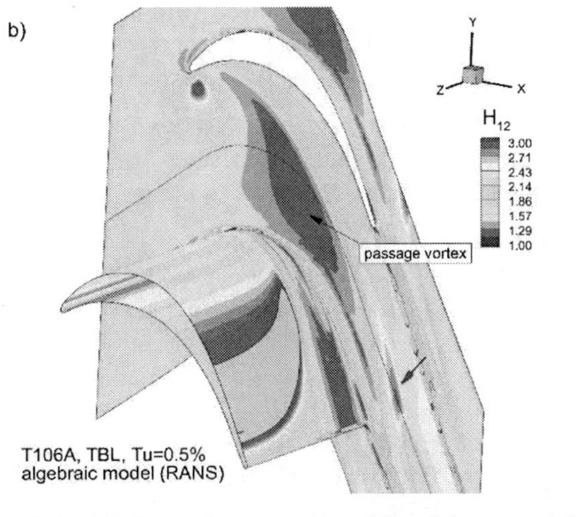

Figure 4. Contours of shape factor at the endwall and the blade surfaces predicted by a) k-ω model (fully turbulent) and b) the algebraic intermittency model. The blanked zone shows the separated flow region in which H_{12}>3.5.

Figure 5 shows the evolution of the mass-averaged total pressure loss coefficients by the unmodified k-ω turbulence model, the algebraic transition model (steady RANS and URANS) and the reference LES. Very good agreement is obtained between predictions by the algebraic model using URANS and the LES. The accuracy of the algebraic model is somewhat less with the steady RANS, approaching the cascade outlet. The difference in the calculated mass-averaged total pressure loss using the unmodified k-ω model and LES becomes only high downstream of x/C_x=0.5. This is partly due to the fully turbulent endwall boundary layer at the inlet to the domain. But one has to note that the endwall flow features, which are responsible for the major part of the total pressure losses through the cascade, are only to small extent influenced by the boundary layer status on the blade surface [3].

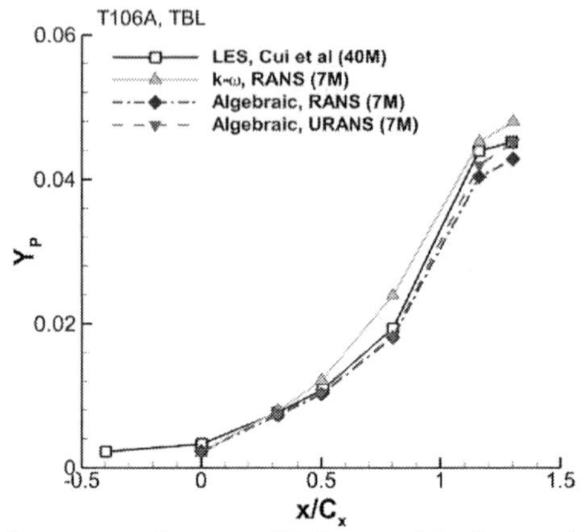

Figure 5. Mass-averaged total pressure loss coefficients with the models and the reference LES.

5. Conclusions

An algebraic intermittency model was employed for calculation by RANS and URANS of the three-dimensional transitional flow through a linear cascade of T106A blades with an endwall. Results were compared to these by a reference LES. Compared to the LES, the flow separation and reattachment over the suction side of the blades and the secondary flow features near the endwall are properly captured. Good agreement is obtained for the evolution through the cascade of the pitchwise-averaged profiles of turbulent kinetic energy and the mass-averaged values of total pressure loss. The results with URANS are somewhat better than these with RANS. Simulations were also done by RANS without an added transition model. The separation and reattachment on the suction side of the blades is then not captured and the total pressure loss is overestimated.

Acknowledgements. The first two authors acknowledge support from the research project COOPERNIK financed partly by the Polish National Centre for Research and Development (INNOLOT/I/11/NCBR/2014) and partly by Avio Polska Sp. z o.o.

References

[1] Goldstein RJ and Spores RA 1988 Turbulent transport on the endwall in the region between adjacent turbine blades. *Journal Heat Transfer* **110** 862-869.

[2] Gross A, Marks CR, Sondergaard R, Bear PS and Wolff JM 2018 Experimental and numerical characterization of flow through highly loaded low-pressure turbine cascade. *J. Propulsion Power* **34** 27-39.

[3] Bear P, Wolff M, Gross A, Marks CR, Sondergaard R 2018 Experimental investigation of total pressure loss development in a highly loaded low-pressure turbine cascade. *J. Turbomachinery* **140** 031003/1-9.

[4] Cui J, Rao VN and Tucker PG 2017 Numerical investigation of secondary flows in a high-lift low pressure turbine. *Int. J. Heat Fluid Flow* **63** 149-157.

[5] Kubacki S and Dick E 2016 An algebraic model for bypass transition in turbomachinery boundary layer flows. *Int. J. Heat Fluid Flow* **58** 68-83.

[6] Kubacki S and Dick E 2016 An algebraic intermittency model for bypass, separation-induced and wake-induced transition. *Int. J. Heat Fluid Flow* **62** 344-361.

[7] Wilcox DC 2008 Formulation of the k-ω turbulence model revisited. *AIAA J.* **46** 2823-2838.

[8] Walters DK and Cokljat D 2008 A three-equation eddy-viscosity model for Reynolds-averaged Navier-Stokes simulations of transitional flow. *J. Fluids Eng.* **130** 121401/1-14.

[9] Menter FR, Smirnov PE, Liu T and Avancha R 2015 A one-equation local correlation-based transition model. *Flow Turbul.Combust.* **95** 583-619.

[10] Spalart PR, Deck S, Shur ML, Squires KD, Strelets M Kh and Travin A 2006 A new version of detached-eddy simulation, resistant to ambiguous grids. *Theor.Comput. Fluid Dyn.* **20** 181-195.

IOP Conf. Series: Journal of Physics: Conf. Series **1101** (2018) 012017 doi:10.1088/1742-6596/1101/1/012017

Numerical analysis of centrifugal compressor operating in near-surge conditions

M Kulak, F Grapow and G Liśkiewicz

Institute of Turbomachinery, Lodz University of Technology, Łódź, Poland

E-mail: michal.kulak@p.lodz.pl

Abstract. In order to reproduce the near-surge operation of centrifugal compressor a series of numerical simulations in ANSYS CFX have been conducted. Different settings of boundary conditions have been investigated in purpose of providing a combination allowing to model surge without non-physical flow constraints. This, in turn, would be a step towards modelling of anti-surge system. As the conclusions, it was observed that there is no possibility of simulating near-surge working regime without mass flow rate definition as a boundary condition. However, it is achievable to determine a throttle characteristic, which can be used as a function defining the outlet pressure and indirectly the mass flow rate. In case of transient simulations, it would allow to analyse both steady and unsteady regimes of the flow in a centrifugal compressor.

1. Introduction

1.1 The surge in centrifugal compressors

The most dangerous flow instability occurring in centrifugal compressors is the surge. It occurs when the mass flow rate is decreased below its critical value. The surge is known as low-frequency fluctuations of the pressure and the mass flow rate that in most severe cases might even lead to reverse flows within the compression system [1]. Surge always leads to significant decrease of the machine performance but is also capable of destroying compressor in just few seconds. This is due to the fact that flow oscillations generate high thermal and mechanical loads acting on machine structure and also increase blades vibration level which can be reason of a fatigue fracture [2]. Moreover, the surge is a global instability affecting whole compression system including other components of industrial rig like pipes, reactors etc. [3]. Undisturbed operation of the compressor is a key in generating incomes in most of industries and this is why avoiding or suppressing of this phenomenon is essential. The setting of so-called the surge margin is the most widely used method of prevention from the surge. Compressor operating range is limited by automatized system which keeps operational point sufficiently far from the surge onset. It is usually set somewhere close to 10%-15% away from the surge limit. This results in narrowing the operation range of the machine.

It is also uneasy to investigate surge on real machines because of its destructive nature and the surge limit is in many cases set based on theoretical calculations. This leaves a space for CFD analysis that might be more suitable for that application. Numerical analysis allows to investigate the surge margin without causing risk of machine failure. Nevertheless, it is very prone to non-physical conditions such as the turbulence model, timescale setting and many other simulation parameters. It is also not clear which boundary conditions are best in reproducing the conditions present in the compression system affected by a global flow instability such as the surge.

1.2 CFD analysis of an unstable centrifugal compressor operation
In literature, several papers considering CFD analysis of an unstable, near-surge compressor operation can be found. Most commonly it is an unsteady RANS (URANS) simulation implemented in commercial codes like ANSYS CFX or Fluent, however, the inlet and the outlet boundary conditions vary in wide range. In [4] set of boundaries recommended by CFX modelling guide [5] has been used. Those settings include constant mass flow rate on the inlet and constant static pressure on the outlet of the compressor. Advantages of such approach are robustness and full control of operational point while they unnecessarily stiffen the simulation and may introduce uncertainties. Those settings have been partially reversed in [6] where the mass flow rate is set at the outlet and at the inlet the total pressure is set to preserve real energy level of inflowing air. In this case the mass flow may still reveal some unnatural flow structures but the inlet boundary ensures better energy conservation. The simulations without direct fixing of the mass flow rate have been described in the literature. For example, in [7], the total pressure is set at the inlet, while the static pressure is set at the outlet. In [8] the opening boundary with pressure dependent of plenum mathematical model is used at the outlet in pair with total pressure boundary at the inlet. Despite the fact that there is no direct setting of the mass flow rate in the CFX model, the constant mass flow rate is defined as an outlet boundary in equations describing the plenum.

Another direction of researching near-surge operating conditions is via in-house made codes like those presented in [9–12] . In all of those papers the total pressure is set as the inlet condition. In [9,11] the static pressure is defined at the outlet while in [10,12] more sophisticated pressure function dependent on plenum model has been used.

In neither of mentioned papers the outlet plenum has been modeled and in only few of them the inlet pipe was added. Very often the boundary is set not even at the end of volute but at the outlet of diffuser. Only one of those studies contained surge analysis [8] and the remaining ones - the near-surge unsteady flows. It is caused by the fact that it is difficult to enter the surge region without the mass flow definition at inlet/outlet boundary. This is why those simulations do not provide clear information whether they are suitable for the surge investigation.

Next to RANS simulations also few LES simulations have been performed [13–15]. All of them had mass flow rate at inlet and static pressure at outlet as boundaries. Also inlet [13–15] and outlet [13,15] pipe was included in simulations. The precision of this type of simulation is considered as a considerable advantage over other types of simulations but they are highly time and resource consuming which makes them unpractical for everyday use.

1.3 Aim of study
In this paper CFD simulations of the near-surge compressor operation using commercial ANSYS CFX solver have been conducted. Six settings of boundary conditions have been investigated in purpose of providing a combination allowing to model surge in centrifugal compressor without non-physical flow constraints. Findings of such a study would be very useful in development of numerical models aiming at fault-free detection of the surge margin.

2. Method
The geometry of the investigated centrifugal compressor is a virtual model of an experimental test stand located at the Institute of Turbomachinery, Lodz University of Technology, Poland. The test rig is constructed using impellers and casing of a large industrial compressor VRK-3. The computational domain consists of six elements, which are presented in Figure 1.

Flow enters the inlet pipe (1) with a defined inlet boundary condition, and throughout the converging nozzle (2) it enters an impeller (3). Its geometry was reconstructed from a 3D laser scan of compressor stage mounted on the test rig. From the impeller exit, the air goes through the parallel wall, a vaneless diffuser (4), afterwards being collected by a spiral volute (5). Afterwards, the straight pipe with a contraction is modelled (6) at the end of which the outlet boundary condition is defined.

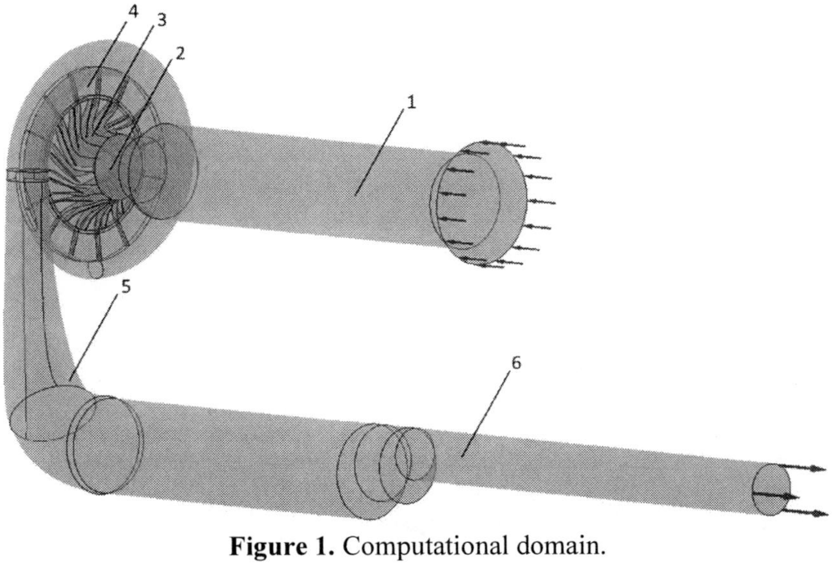

Figure 1. Computational domain.

The CFD simulations have been performed with ANSYS CFX 18.2. The computational domain consisted of around 2.5 million elements (hexahedral in inlet and outlet pipes as well as rotor and tetrahedral in remaining regions). A mesh independence study was performed according to [16]. GCI (Grid Convergence Index) value for majority of monitored parameters was in a satisfactory level (<1%) even for the coarse mesh, therefore it was chosen as a good compromise between results quality and computation time. Shear Stress Transport turbulence model was applied, along with high resolution advection scheme and turbulence equations spatial differencing. As for the heat transfer, the kinetic energy effects were of importance as well as transport of enthalpy, therefore the total energy approach was used – a static, ambient temperature was set at 293 K. The rotor domain was set to rotate at 15855 rev/min. At the inlet pipe/rotor and rotor/diffuser connections frozen rotor interfaces were applied including a GGI (General Grid Interface) mesh connection method.

In this study six sets of boundary conditions combinations were analysed – all of them are collected and presented in Table 1. All sets will be described in details in next section.

Table 1. Comparison of applied boundary conditions.

Set no.	Simulation type	Inlet boundary condition	Outlet boundary condition
1	Steady State	Total pressure	Mass flow rate
2	Steady State	Total pressure	Opening
3	Steady State	Total pressure	Static pressure
4	Steady State	Mass flow rate	Static pressure
5	Steady State	Total pressure	Static pressure as a function
6	Transient	Total pressure	Static pressure as a function

3. Results

Amongst combinations of inlet and outlet boundary conditions commonly used in ANSYS CFX, velocity/mass flow at inlet and static pressure at outlet is considered as the most robust option [5]. Nevertheless, the simulation of the surge is much different from the classic simulation at the nominal point. In the case of global instability one may expect that the pressure and mass flow rate will fluctuate to a large extent not only within the compressor, but also in the plenum and the inlet pipe. Therefore, the inlet and outlet boundary condition would be influenced by this fluctuation. The response of a boundary condition to those changes is crucial in obtaining a physical model.

It is a considerable simplification to assume the fixed value of mass flow entering the computational domain. Therefore, a set of different boundary conditions combinations (see Table 1) was tested in search of a one, allowing to model unstable working conditions in near-surge region.

Set 1 incorporated total pressure at the domain inlet and the mass flow condition at the outlet. It is considered as rather robust boundary conditions combination but may introduce non-physical phenomena when the flow should be reversed. In set 2 the mass flow outlet condition was replaced by pressure condition of opening (entrainment), enabling the fluid to both enter and leave the domain. Velocity distribution at the plane in the axis of outlet pipe is presented in Fig. 2. Additionally, the streamlines in whole computational domain are included.

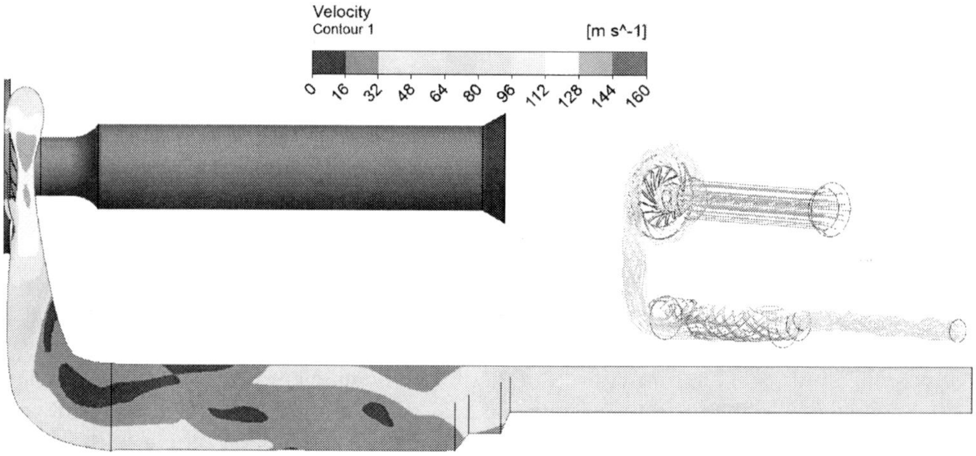

Figure 2. Velocity distribution in the outlet pipe, flow streamlines; BC set 2.

Set 3 included static pressure definition at the domain outlet. As it is described in [5] this combination is valid, with a mass flow as a part of solution. However, the solution is sensitive to initial conditions. Again, the velocity distribution and streamlines are presented in Fig.3. At that point, the decision was made to introduce a function for modelling the value of static pressure at the outlet, resulting from the value of the mass flow (equivalent to mimicking the throttle performance curve). In order to apply this approach, an additional step was necessary - boundary combination presented in set 4 was introduced. It consisted of mass flow defined at the inlet and static pressure at the outlet. This simulation (although considered as over-stabilized) provided a values of pressure p_{pre} and mass flow rate $\dot{m}_{pre}$. They were used to prepare a definition of throttle function for calculation of p_{sim} on the basis of value of the mass flow rate $\dot{m}_{sim}$. A function in the following form was introduced at the outlet BC in a pre-final set 5:

$$p_{sim} = \frac{p_{pre}}{\dot{m}_{pre}{}^2}\dot{m}_{sim}{}^2$$

Throughout the solution, monitored parameters such as mass flow rate or pressure ratio presented an oscillations of considerable amplitude, therefore an influence of various values of timescale applied locally was also studied. ANSYS CFX enables different time scales to be applied in different regions of the calculation domain – the smaller time scales are applied to regions where flow has a higher velocity [5]. Regardless of the local timescale choice, the average mass flow rate value was close to the expected one (see Fig. 4).

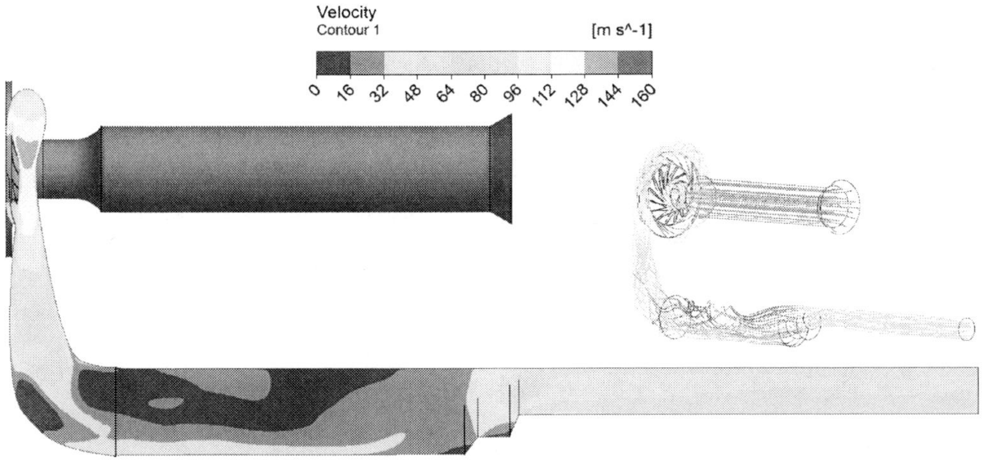

Figure 3. Velocity distribution in the outlet pipe, flow streamlines; BC set 3.

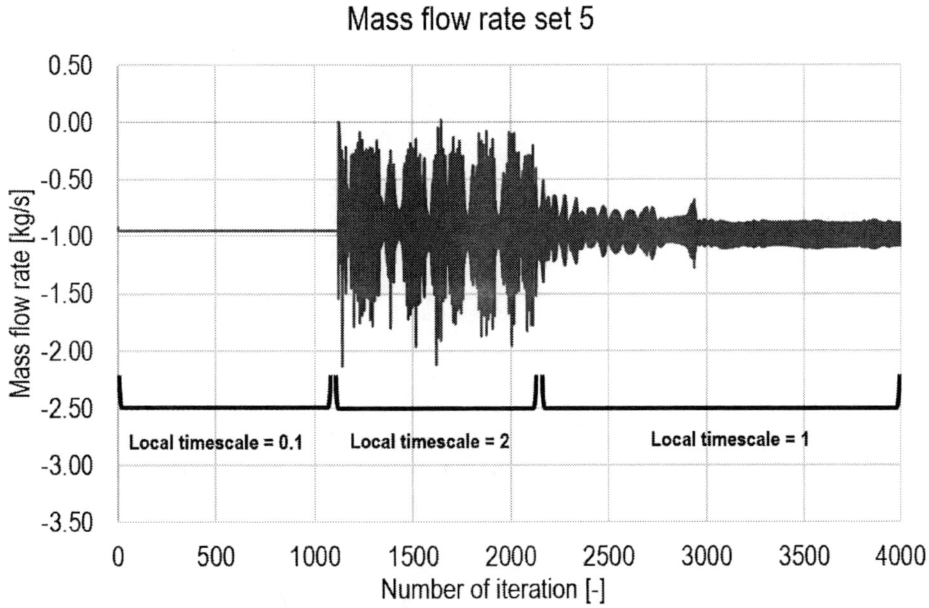

Figure 4. Mass flow rate fluctuations with different local timescales.

Finally, the outlet pressure defining function was incorporated into transient solution (set 6). The amount of iterations per one blade passage was set to 20. The flow parameters presented considerably lower fluctuations but at the cost of significant simulation time increase.

Figure 5 collects above mentioned results (excluding set 1 – non convergent solutions and set 4 – used only as a data source) and combines them with the results of preliminary simulations (series of numerical investigations performed on a simplified geometry including only rotor, diffuser and volute, in order to study a performance of a centrifugal compressor in different working conditions). Presented trendline is a fourth degree polynomial.

4. Discussion

Taking into consideration set 1, one can conclude that settling the amount of the medium exiting the domain leads to non-convergence of numerical solution in case of the centrifugal compressor. In reality the mass flow is regulated by the valve at some distance downstream the volute outlet, what could be utilized by introducing a throttle function (working point at the intersection of throttling valve curve and compressor characteristic).

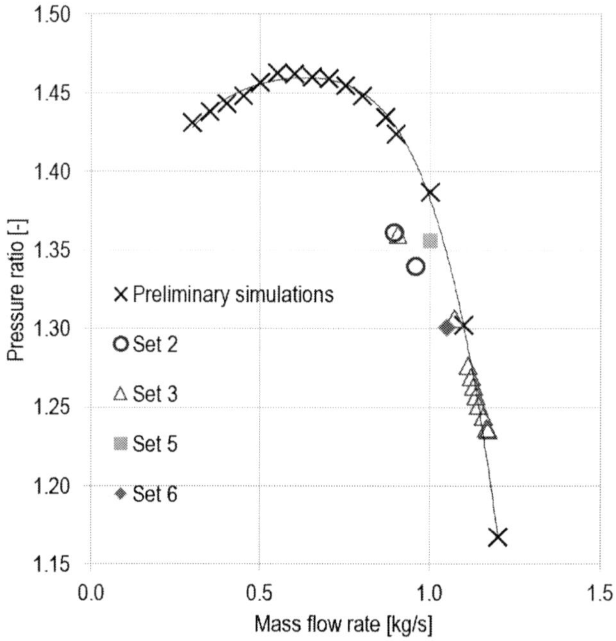

Figure 5. Pressure ratio as a function of mass flow rate – simulations results.

Introducing an opening condition at the outlet (set 2), enabling the fluid to flow in and out of the domain, allowed to relax this boundary, resembling experimental or real life conditions. As this approach resulted in correct flow representation only for stable working regime (mass flow in the range of 0.9 – 1.0 kg/s) it was decided to test yet another approach.

For set 3, similarly to set 2, it was possible to simulate correctly the working conditions for stable flow regime. For static pressure at the outlet higher than ca. 135 000 Pa, simulations did not converge (with a sudden mass flow rate drop to 0 and flow blockage). This fact confirmed that the steady state solutions without varying pressure condition at the outlet are able to provide reliable results only for stable work region. Steady state simulations without direct mass flow (set 2 and 3) were not capable of reaching peak of the compressor curve so were far from even near-surge conditions. Similar observation was made in [7]. In mentioned paper a transient simulation with pressure at both the inlet and the outlet boundary condition were conducted to reach the near-surge operation. In this article this step was skipped because definition of pressures on both inlet and outlet cannot be used to enter the surge region.

The only discussed paper with surge simulation is [8]. In this paper an indirect mass flow definition was used in form of outlet pressure function based on mathematical plenum model in which mass flow rate is one of the boundary conditions. Similar operation was made in sets 5 and 6 were pressure is defined as function of the mass flow rate depending on the throttle characteristic. This allows to abandon unphysical mass flow condition and use pressure function instead. Those simulations, however, are very sensitive to the time scale and pseudo time scale since they are dependent of mass flow rate from the previous iteration. Too big time step leads to oscillatory character of simulation.

Set 5 with steady state simulation and throttle characteristic set at the outlet can be used to simulate stable and near surge iteration but only when the mean value is needed. Its oscillatory character may be misleading, especially when one needs to analyse a flow field or a particular physical quantity at the particular iteration.

Set 6 is a transient version of the set 5. It is most universal and complete variant useful in stable and surge simulations. As it was mentioned before it is highly sensitive to the time step which has to be small enough. In literature for stable and unstable compressor operation with constant values of the boundary conditions, value of time step is chosen in a way which guarantees 15 to 20 iterations per blade passage [7,17,18]. It is time and resource consuming for a rotor with many blades or with high rotational speed but it also guarantees that function based on values taken from previous time step will operate with satisfactory stability.

One can notice that according to Fig. 5 simulation of compressor with and without inlet and outlet pipe differs from each other significantly. It may be the effect of pipe contraction located close to volute outlet. During simulations this region was highly unstable and generated local areas of stagnation or even led to errors in simulation. Additional study of this phenomenon is needed because contraction in the vicinity of volute outlet has not been modelled in the compressor unstable operation in the state-of-the-art publications. What is more, very often CFD simulations overestimate pressure rise with respect to the experimental values. This might be an independent confirmation of the fact, that the inlet and the outlet pipes should be included in computational models of the near-surge simulations.

Many researches of anti-surge systems are based on calculations using Greitzer model and similar one-dimensional mathematical models [19–21]. Despite the fact that surge is considered to be one dimensional instability, the influence of anti-surge system may lead to developing local two or three dimensional flow structures which cannot be observed in simple mathematical model. In contrast to mentioned models, CFD gives almost unlimited possibilities of implementing different anti-surge system actuators at different locations. Furthermore, the three dimensional flow analysis in any region of domain can be conducted. That can be considered as significant advantage of CFD approach to designing anti-surge systems requiring wide variety of possible modifications and easy observation methods. It is also important to point out the fact, that a given capability and level of accuracy are acceptable for designing process – the industry still accepts shortcomings of RANS (in contrary to cost, time and user skills necessary for scale-resolving computations) [22].

5. Conclusions

The presented research is a part of a project aiming at design and construction of anti-surge systems of different types. It includes both: experimental studies on an industrial compressor and numerical analysis of its virtual model. Eventually, it is expected to introduce the anti-surge systems concepts into the numerical simulations of centrifugal compressors. However, it is crucial to be able to obtain the surge or the near-surge working regime beforehand. In terms of analysing combinations of boundary conditions allowing to do that one can conclude as follows:

- it is difficult to enter the surge region without the mass flow definition;
- the direct mass flow rate definition as the outlet boundary condition may lead to a non-convergence of the steady-state solution;
- both opening pressure (entrainment) and static pressure condition at the outlet allows the simulations to converge, however only in the range of the stable flow regime;
- it is possible to define a throttle characteristic, which can be used as a function defining the outlet pressure; it requires the knowledge about the mass flow rate and equivalent pressure at single operational point, which can be obtained in a disassociated simulation (with a most robust boundary conditions combination – mass flow rate at inlet and static pressure at outlet)
- the outlet pressure function allows to indirectly define the mass flow rate, not as a fixed value, enabling the flow fluctuations to occur;
- transient simulation with an application of described outlet pressure function is most universal because it allows to analyse both steady and unsteady regimes of the flow in a centrifugal compressor;

Acknowledgements
This work was funded by the National Centre for Research and Development (LIDER/447/L-6/NCBR/2015).

References

[1] Willems F and Jager B de 1998 Modeling and control of rotating stall and surge: an overview *IEEE Int. Conf. Control Appl.* 1 331–5
[2] Meuleman C, Willems F, de Lange R and de Jager B 1998 Surge in a low-speed radial compressor *Int. Gas Turbine Aeroengine Congr.*
[3] de Jager B 1995 Rotating stall and surge control: A survey *Decision and Control, 1995., Proceedings of the 34th IEEE Conference on* pp 1857–62
[4] Shahin I, Gadala M, Alqaradawi M and Badr O 2014 Unsteady CFD Simulation for High Speed Centrifugal Compressor Operating Near Surge *Volume 2D: Turbomachinery* (ASME)
[5] ANSYS 2016 CFX-Solver Modelling Guide--Release 17.0
[6] Ding M Y, Groth C, Kacker S and Roberts D 2005 CFD Analysis of Off-design Centrifugal Compressor Operation and Performance *2006-Int-ANSYS-Conf* 252
[7] Dickmann H-P, Secall Wimmel T, Szwedowicz J, Filsinger D and Roduner C H 2006 Unsteady Flow in a Turbocharger Centrifugal Compressor: Three-Dimensional Computational Fluid Dynamics Simulation and Numerical and Experimental Analysis of Impeller Blade Vibration *J. Turbomach.* 128 455
[8] Guo S, Chen H, Zhu X and Du Z 2011 Numerical Simulation of Surge in Turbocharger Centrifugal Compressor - Influence of Downstream Plenum *Proc. ASME Turbo Expo* GT2011-451 1–12
[9] Niazi S, Stein A and Sankar L 1998 Development and Application of a CFD Solver to the Simulation of Centrifugal Compressors *AIAA Pap.* 934
[10] Stein A, Niazi S and Sankar L N 2000 Numerical analysis of stall and surge in a high-speed centrifugal compressor *38th Aerospace Sciences Meeting and Exhibit* p 226
[11] Stein A, Niazi S and Sankar L N 2001 Computational Analysis of Centrifugal Compressor Surge Control Using Air Injection *J. Aircr.* 38 513–20
[12] Stein A, Niazi S and Sankar L 2000 Numerical studies of stall and surge alleviation in a high-speed transonic fan rotor *38th Aerospace Sciences Meeting and Exhibit* (Reston, Virigina: American Institute of Aeronautics and Astronautics)
[13] Hellstrom F, Gutmark E and Fuchs L 2012 Large Eddy Simulation of the Unsteady Flow in a Radial Compressor Operating Near Surge *J. Turbomach.* 134 051006
[14] Shahin I, Gadala M, Alqaradawi M and Badr O 2015 Large Eddy Simulation for a Deep Surge Cycle in a High-Speed Centrifugal Compressor With Vaned Diffuser *J. Turbomach.* 137
[15] Semlitsch B, Jyothishkumar V, Mihaescu M, Fuchs L and Gutmark E J 2013 Investigation of the surge phenomena in a centrifugal compressor using large eddy simulation *ASME International Mechanical Engineering Congress and Exposition, Proceedings (IMECE)* vol 7 A pp 1–10
[16] Celik I B, Ghia U and Roache P J 2008 Procedure for estimation and reporting of uncertainty due to discretization in CFD applications *J. fluids Eng. ASME* 130
[17] Ljevar S 2007 *Rotating Stall in Wide Vaneless Diffusers* (Eindhoven University of Technology) Doctoral dissertation
[18] ANSYS 2015 Fluent User's Guide
[19] Backi C J, Gravdahl J T and Skogestad S 2016 Robust control of a two-state Greitzer compressor model by state-feedback linearization 1226–31
[20] Willems F and de Jager B 1998 Active compressor surge control using a one-sided controlled bleed/recycle valve *IEEE Conference on Decision and Control* vol 3 pp 2546–51
[21] Greitzer E 1976 Surge and rotating stall in axial flow compressors—Part I: Theoretical compression system model *J. Eng. Gas Turbines Power* 98 190-198
[22] Duraisamy K 2017 Status , Emerging Ideas and Future Directions of Turbulence Modeling Research in Aeronautics *NASA Tech. Memo.* 1–9

IOP Conf. Series: Journal of Physics: Conf. Series **1101** (2018) 012018 doi:10.1088/1742-6596/1101/1/012018

Local Nusselt number evaluation in the case of jet impingement

Tomasz Kura[1], Elżbieta Fornalik-Wajs[1], Jan Wajs[2] and Sasa Kenjeres[3]

[1]AGH University of Science and Technology, Department of Fundamental Research in Energy Engineering, al. Mickiewicza 30, 30-059 Krakow, Poland
[2]Gdansk University of Technology, Faculty of Mechanical Engineering, Department of Energy and Industrial Apparatus, ul. Narutowicza 11/12, 80-233 Gdansk, Poland
[3]Delft University of Technology, Faculty of Applied Sciences, Department of Chemical Engineering, van der Maasweg 9, 2629 HZ, Delft, The Netherlands

kura@agh.edu.pl

Abstract. Jet impingement still is one of demanding cases regarding computational fluid dynamics, due to its highly turbulent behaviour, with occurrence of turbulent-laminar transition. Even recently developed methods exhibit some drawbacks – RANS based simulations lack accuracy, LES and DNS based ones require too much computational time. Hybrid methods also exist, but their development and validation is in progress. Nevertheless, CFD application can play major role in the investigation of jet impingement phenomena. While the flat surface impingement is widely discussed in the literature, there is lack of data regarding non-flat surfaces – the ones that might exist for example in the heat exchangers. In the following paper, the numerical simulation of both flat and non-flat surfaces single jet impingement is presented, with the aim of precise description of the turbulence models impact on the thermal and hydrodynamic results. Choice of turbulence model is crucial for sufficient calculation outcome. Only the complex analyses, shown in the article, including the turbulence and momentum budgets comparison between particular models, can reveal significant and meaningful differences.

1. Introduction

Jet impingement phenomena is applied for the heat transfer enhancement for many years – for example in turbine blade cooling or metallurgy. Wajs et al. [1-3] proposed and verified another implementation of this method. They presented a prototype of cylindrical heat exchanger, in which numerous orifices generate turbulent jets. In contact with the heat exchanging surface they interfere with boundary layer, causing significantly higher heat rates transferred between the cold and hot fluid sides. The experimental results were very promising, but only general parameters, such as overall heat transfer coefficient, total pressure losses etc., were available. The reason of such positive effect on performance, regarding non-flat surfaces, was an interesting problem to be investigated, especially in the light of future optimization. This issue is poorly described in the literature and it demonstrates lack of universal correlations describing heat transfer phenomena. Therefore the numerical methods, especially their ability to perform multiscale analyses, can lead to better understanding of phenomena occurring in the near-wall zones of heat exchangers. Because of relatively short computation time the RANS methods were used as the first ones. They can be important for the analyses of the single jet, but crucial when the jet arrays will be considered with additional crossflow – the conditions which exist in the abovementioned heat exchanger. The selection of particular RANS model is necessary for the reliable results. Unfortunately, according to Zuckerman [4], the unresolved issue of heat transfer prediction is common for almost all available RANS models. The one recognized as presenting the best performance is Durbin's v^2-f [5]. Its theoretical superior performance was reported for example by Behnia [6, 7]. In the following paper, this model was validated for single jet impingement on flat and non-flat surfaces. It is important from the future, more detailed analyses of real heat exchangers, point of view. Its more recent, modified version, ζ-f by Hanjalic et al. [8], is also taken into account. Calculations were conducted with utilization of the OpenFOAM software, in which the first model is implemented by default, while the second one was implemented by the Authors. The ultimate goal is to choose the RANS model that would be used in the future for the custom hybrid RANS/LES simulations – in Authors opinion the best one to investigate such complex devices.

2. Mathematical model, geometry and numerical procedure

The analyses required consideration of mass (1), momentum (2) and energy (3) conservation laws [9]. Simulations, performed with open-source OpenFOAM software, were steady state and incompressible, performed using 2D geometries. For the single jets impingement, steady state is possible to be used, due to chosen boundary conditions. Reynolds averaging approach was utilized. Classic SIMPLE method was included, and all numerical schemes were of second-order types. The air was used as the working medium.

$$\frac{\partial \overline{u_i}}{\partial x_i} = 0, \tag{1}$$

$$\rho \frac{\partial}{\partial x_j}\left(\overline{u_j}\,\overline{u_i}\right) = -\frac{\partial \overline{p}}{\partial x_j} + \frac{\partial}{\partial x_j}\left(2\mu S_{ij} - \rho \overline{u_i' u_j'}\right), \tag{2}$$

$$\frac{\partial}{\partial x_j}\left(\overline{u_j}\Theta\right) = \frac{\partial}{\partial x_j}\left(\alpha \frac{\partial \Theta}{\partial x_j} - \overline{u_j \theta}\right), \tag{3}$$

where $u_{i,j}$ are the velocity components, m/s; ρ is the density, kg/m^3; p is the pressure, Pa; μ is the dynamic viscosity, Pa·s; S_{ij} is the strain rate tensor, 1/s; $\overline{u_i' u_j'}$ is the Reynolds stress term, m^2/s^2; Θ is the mean temperature, K; α is the thermal diffusivity, m^2/s and $\overline{u_j \theta}$ is the turbulent heat flux. The symbol "‾" represents averaged quantity.

In figure 1, the geometries of flat and non-flat cases are shown. Presented dimensions correspond with the ones included in the ERCOFTAC association database [10] – they assure reliable validation of chosen numerical approach. As it can be seen, axisymmetric geometries were investigated. In the non-flat type cases, dimension H is the same as in the flat case. One additional parameter – curvature radius R, was introduced. Its value, the same for convex and concave types, was chosen on the basis of Authors' previous publication [11]. Moreover, it corresponds with the radius from the heat exchanger [1-3].

In table 1, list of boundary conditions and their values is presented. Flow at the orifice exit was fully developed, which was achieved with the boundary mapping technique. Conditions were chosen to correspond with the ERCOFTAC [10] reference data. Apart from validation with reference case, the computational space division was validated with mesh independence tests.

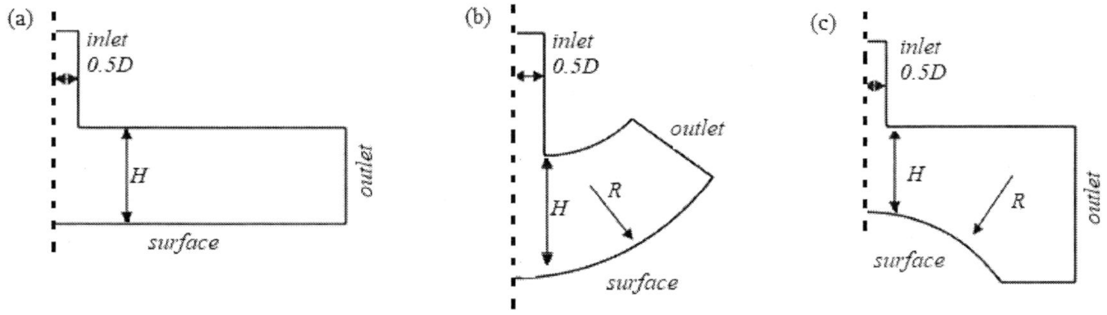

Figure 1. Types of analyzed geometries, (a) flat, (b) concave and (c) convex.

Table 1. Boundary conditions

Bulk Reynolds number, orifice exit	Temperature, orifice exit, K	Heat flux density at the surface, W/m^2	*H/D*	*R/D*
23000	293	1000	2	4

3. Results

Comparison of obtained data was performed in the basis of on the local velocity profiles, local Nusselt number values, local turbulence and momentum budgets. Therefore it was important to establish the method for flat and non-flat geometries comparison. Figure 2 presents, how the reference points were determined. For analyzed cases, the differences of the chord length and corresponding segment connecting those points and stagnation points were negligible – as a results, the non-dimensional distance *x/D* was used as sufficient and reliable parameter.

Figure 2. Determination of the reference point, (a) flat and (b) non-flat cases.

3.1. Velocity profiles

At first, the local velocity profiles normal to the heated surface, obtained with v^2-f and ζ-f models for the flat geometry were compared with the reference data, to check their accuracy. Results are presented in figure 3. As it can be seen, the shape is very similar, however slightly better agreement with the experimental data for the ζ-f model could be found.

Since the ζ-f represented good matching with the experimental data, it was selected to obtain the local velocity profiles for non-flat surfaces. Comparison of the results with flat case is shown in figure 4. The most visible differences could be observed in the region close to the stagnation point.

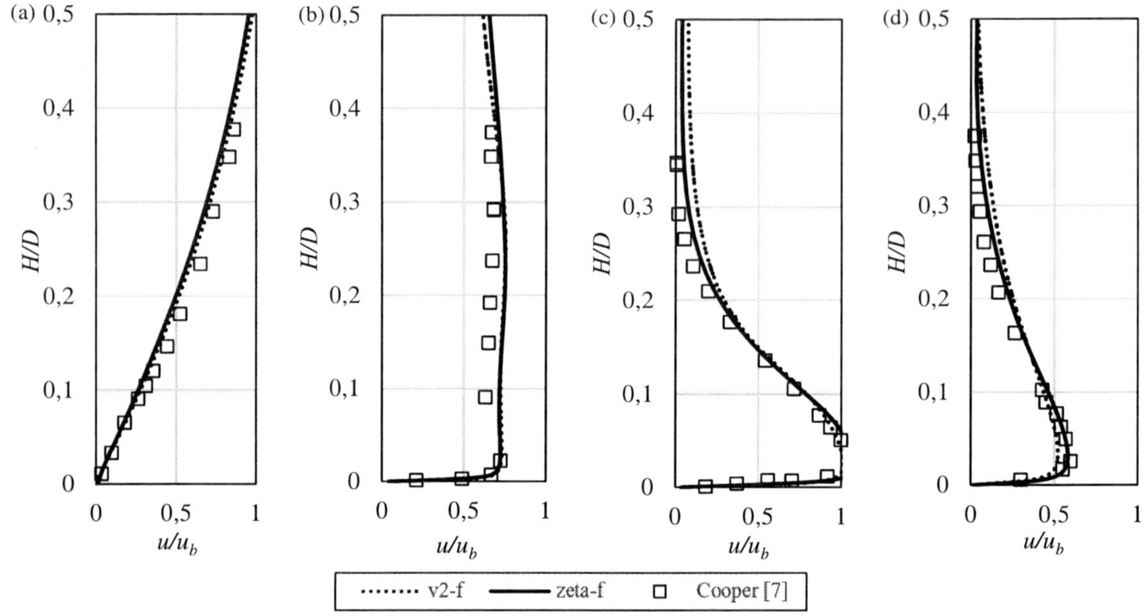

Figure 3. Local velocity profiles at selected radial locations, normalized by bulk velocity in the orifice exit u_b, (a) $x/D = 0$, (b) $x/D = 0.5$, (c) $x/D = 1$, (d) $x/D = 2.5$. Flat surface, v^2-f vs ζ-f.

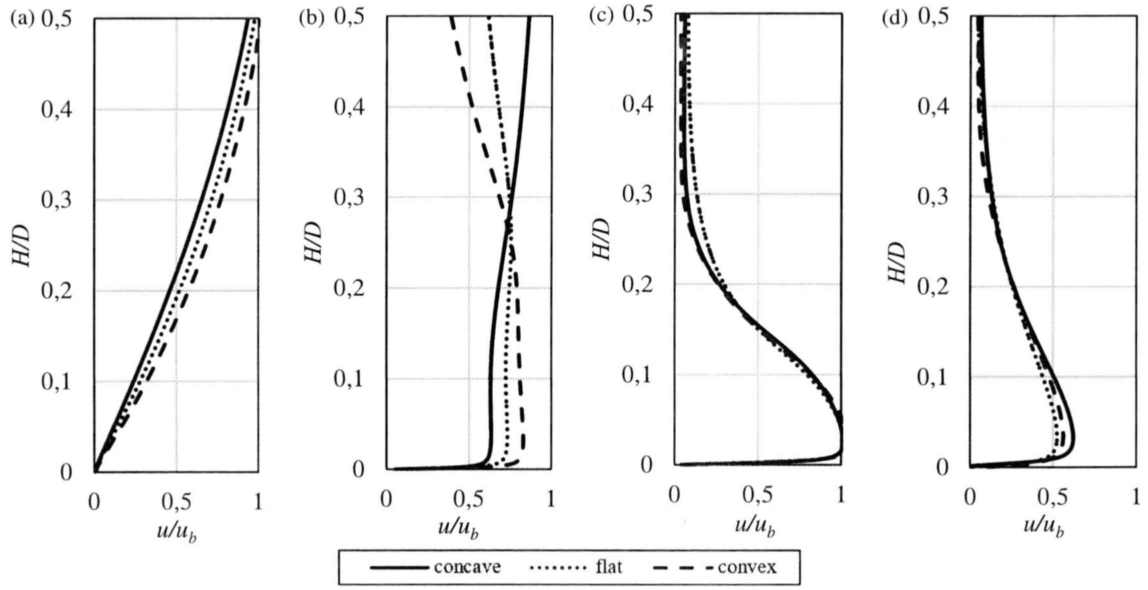

Figure 4. Local velocity profiles at selected radial locations, normalized by bulk velocity in the orifice exit u_b, (a) $x/D = 0$, (b) $x/D = 0.5$, (c) $x/D = 1$, (d) $x/D = 2.5$. All surface types, ζ-f model.

3.2. Local Nusselt number

Another comparison was done for the local Nusselt number distribution along the heated surface, defined by equation (4):

$$\mathrm{Nu} = \frac{h_{ef} D}{\lambda}, \tag{4}$$

h_{ef} is the effective heat transfer coefficient, W/(K·m^2), obtained locally with the use of the software and λ is the thermal conductivity, W/(K·m). This parameters is crucial for the analysis of heat exchangers.

In figure 5(a) the local Nusselt number values, obtained with both considered models, are presented and compared with the experimental and numerical data from ERCOFRACT database. Surprisingly, results by Behnia [6, 7], who used v^2-f model, are in poor agreement with the ones obtained by Authors using the same model in the stagnation zone. On the other hand the results obtained with ζ-f model, very adequate in the stagnation zone, are slightly over-predicting the Nusselt number values far from it. However the most important feature of the phenomena – secondary peak, demonstrated by the experimental results and located close to the the x/D value equal to 2, is successfully predicted by ζ-f model. The reason of such noticeable discrepancy in both v^2-f results comes from the fact, that the model commonly implemented in CFD software is not the original one [12]. Changes applied by Lien and Kalitzin [13] made it more robust, but at the same time less precise. Hanjalic [8], who also modified the original v^2-f, proposed different approach, which led to better accuracy, as presented here. In figure 5(b), comparison of the local Nusselt number parameter, for various surface types, is presented. The differences between the results are not significant at the stagnation zone and near it, but increase at longer distance. Importance of this discrepancy may increase when instead of single jet, whole array of impinging jets would be analyzed.

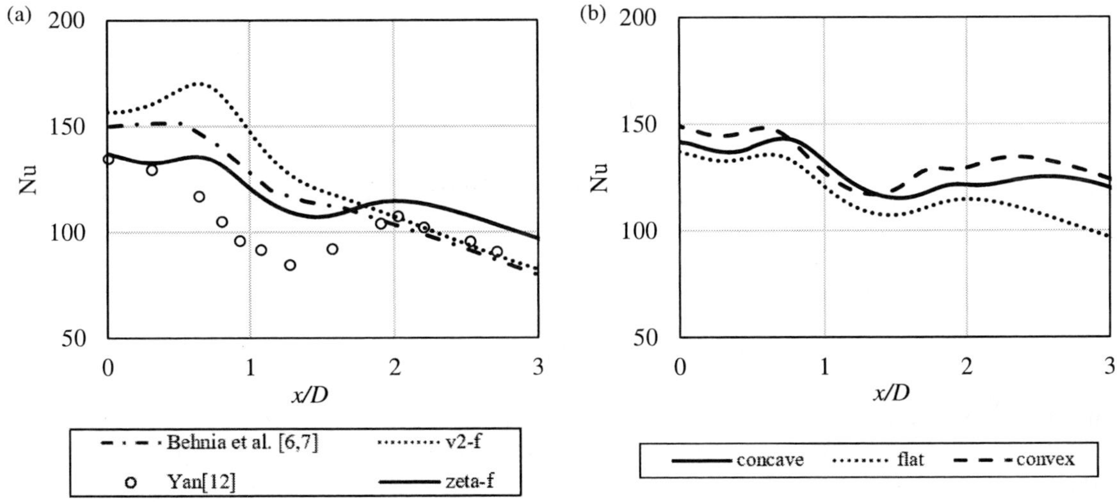

Figure 5. Local Nusselt number distribution along the heated surface, (a) comparison of the turbulence models results, flat surface and (b) comparison between analyzed surface types.

3.3. Turbulence budgets

While the velocity profiles do not differ much between both investigated turbulence models, the uncertainties coming from the local Nusselt number results require identification of their source. One possible way is to check the turbulence kinetic energy (k) budget, as the turbulent character of the flow is expected to play major role in the heat transfer intensification in the heat exchanger [1-3]. The terms of kinetic energy of turbulence were determined in accordance with their representation in [9]. In figure 6, all the terms representing particular parts of the turbulence kinetic energy budget are shown. For this particular analysis, their values were obtained on the curve equally distant from the heated surface. Distance between them was chosen to correspond with the distance between the maximum value of turbulence kinetic energy and the surface. As it can be concluded from the results shown in figure 6, the overall tendency is the same for all types of surfaces, the only differences refer to the maximal values and their positon x/D. However, this statement is checked only for analyses of non-flat surfaces, in which R/D ratio was equal to 4. Its universality have to be verified for more configurations.

IOP Conf. Series: Journal of Physics: Conf. Series **1101** (2018) 012018 doi:10.1088/1742-6596/1101/1/012018

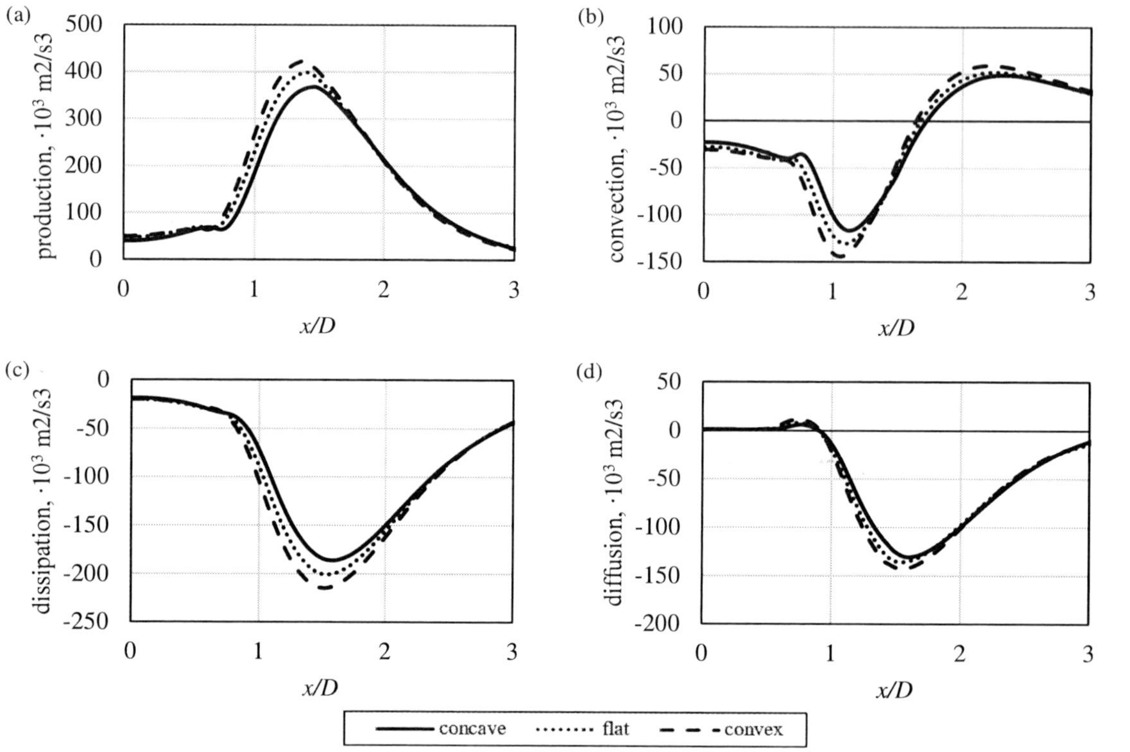

Figure 6. Kinetic energy of turbulence budget, (a) production, (b) convection, (c) dissipation and (d) diffusion. ζ-f model.

It is important to note, that in figure 6, the results obtained with Hanjalic model are presented. In figure 7(a), the production term for all simulated geometries is shown, obtained when using Lien and Kalitzin v^2-f model. This time, significant differences between each case exist in the stagnation zone. Noticeable peak of production term in the stagnation zone, not present when ζ-f model was used, is the main reason of heat transfer over-prediction, mentioned in the section 3.2. It has to be compensated in the stagnation zone by other terms, here represented for example by the higher values of dissipation term, shown in figure 7(b). In general, particular curves are characterized by higher values for v^2-f model.

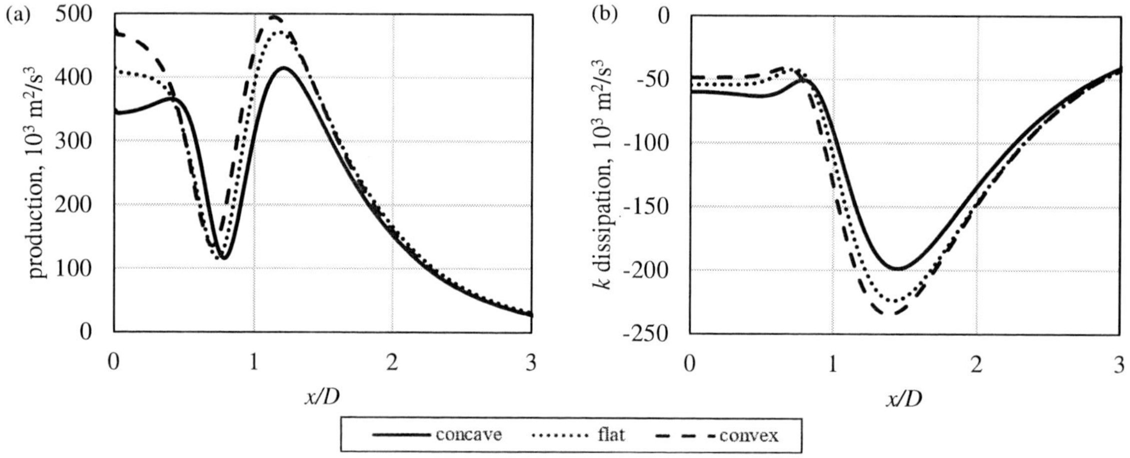

Figure 7. (a) Production term and (b) dissipation term of kinetic energy of turbulence, v^2-f model.

3.4. Momentum budgets

The differences in the local Nusselt number values presented in figure 5(a) and the questions raised by them led also to comparison of particular terms of the momentum budget. In figures 8 and 9, the results calculated with two analyzed models, at the location of *x/D* equal to 0.5 and 2 are presented. The first location represents the place, where stagnation zone starts to vanish, the second location represents the place, where secondary Nusselt peak occurred.

One noticeable feature of v^2-f model is the significantly higher diffusion of momentum near the wall, visible in figures 8(a) and 9(a) – probably it plays a role in turbulence limitation near the wall, which is the main issue with v^2-f, as described in section 3.3.

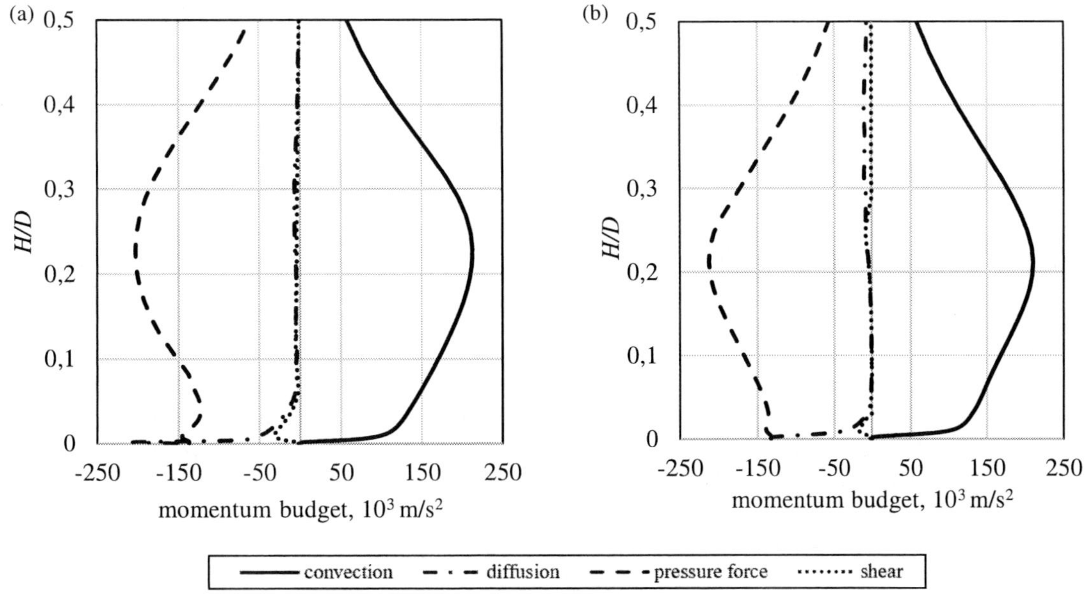

Figure 8. Momentum budget terms, $x/D = 0.5$. (a) v^2-f model and (b) ζ-f model.

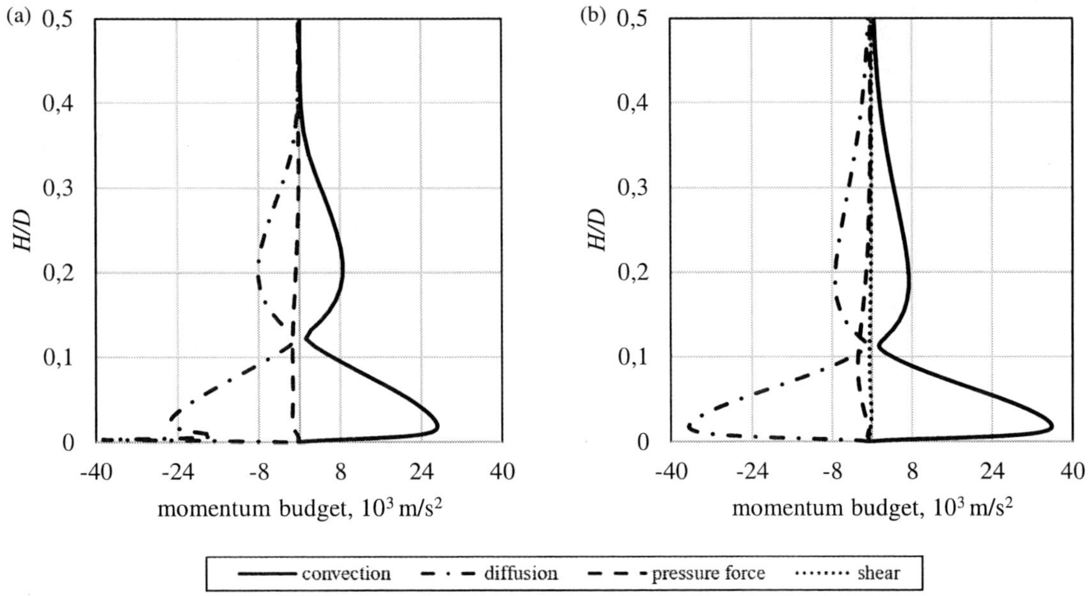

Figure 9. Momentum budget terms, $x/D = 2$. (a) v^2-f model and (b) ζ-f model.

4. Summary

This paper discussed the performance of two RANS turbulence models, v^2-f and ζ-f, implemented in the OpenFOAM software, in the case of flat and non-flat surfaces jet impingement. The main goal was to determine the model that can be further used in preparation of hybrid RANS/LES model to perform complex analysis of phenomena occurring in the heat exchanger [1-3].

Both models were selected in the basis of literature overview and their ability to simulate cases of jet impingement. However, as revealed in the following work, both models exhibit some disadvantages, that need to be overcome before performing more advanced analysis. It can be stated, that Hanjalic ζ-f [8] model is more suitable for the future purposes. It is able to better predict the heat transfer rates, moreover it does not over-predicts turbulence generation in the stagnation regions. In addition, it turns out to be numerically stable. It is worth to mention that to make the presented results more reliable, also more recent, 2012 Billard and Laurence [12] RANS model, in opinion of its creators the most accurate 4 equations v^2-f based model, was included in the research. It turned out however to be very unstable with higher order numerical schemes, also obtained heat transfer rates were not satisfactory. The results therefore were not included in this paper.

To sum up, only very detailed analysis of particular turbulence model can explain the reasons of its better or worse representation of real problem. With understanding of particular terms and variables, its tuning can be performed, which can lead to improvement of the simulation results.

5. References

[1] Wajs J Mikielewicz D Fornalik-Wajs E 2013 *Cylindrical jet heat exchanger dedicated to heat recovery, especially from low temperature waste sources* (Patent PL224494, in Polish)
[2] Wajs J Mikielewicz D Fornalik-Wajs E Bajor M 2015 *Arch. of Thermodynamic* **36** 48-63
[3] Wajs J Mikielewicz D Fornalik-Wajs E Bajor M 2018 *Heat Tr. Eng.* doi:10.1080/01457632.2018.1442369
[4] Zuckerman N Lior N 2005 *J. Heat Transfer* **127** 544-552
[5] Durbin P A 1995 *AIAA Journal* **33** 659-664
[6] Behnia M Parneix S Durbin P A 1998 *Int. J Heat Mass Transfer* **41** 1845-1855
[7] Behnia M Parneix S Shabany Y Durbin P A 1999 *Int. J. Heat and Fluid Flow* **20** 1-9
[8] Hanjalic K Popovac M Hadziabdic M 2004 *Int. J. Heat Fluid Flow* **25** 1047–1051
[9] Holzmann T *Mathematics, Numerics, Derivations and OpenFOAM* (online book)
[10] Cooper D Jackson D C Launder B E Liao G X 1993 *Int. J. Heat Mass Transfer* **36** 2675-2684
[11] Kura T Fornalik-Wajs E Wajs J 2018 *Arch. of Thermodynamics* **39** 147-166
[12] Billard F Laurence D 2012 *Int. J. Heat Fluid Flow* **33** 45-58
[13] Lien F S Kalitzin G 2001 *Int. J. Heat Fluid Flow* **22** 53-61

Acknowledgments

The present work was supported by the Polish Ministry of Science and Higher Education and by the PLGrid Infrastructure.

IOP Conf. Series: Journal of Physics: Conf. Series **1101** (2018) 012019 doi:10.1088/1742-6596/1101/1/012019

A system for measuring environmental data in full-scale

Tomasz Lipecki, Paulina Jamińska-Gadomska, Andrzej Sumorek, Ewa Błazik-Borowa, Jarosław Bęc

Department of Structural Mechanics, Faculty of Civil Engineering and Architecture, Lublin University of Technology, Nadbystrzycka 40, 20-618, Lublin, Poland
e-mail: t.lipecki@pollub.pl

Abstract. The paper presents a system created for measuring wind speed and its direction in full-scale conditions. The system consists of devices used to measure environmental parameters in one point at a time and a complex weather station. On the basis of the results obtained from the first part of the system, the static effect of wind action on structures or climatic comfort can be approximated, while the second part of the system is used to determine the dynamic wind action on objects. The use of the designed system is illustrated by in-situ studies.

1. Introduction

There are three main ways to conduct research in wind engineering: 1. wind tunnel tests, 2. in-situ measurements, 3. CFD simulations. Nowadays, CFD is used more and more frequently to solve wind engineering problems, however, there is still a need to validate CFD results with wind tunnel data (eg. [1-2]) or even better with full-scale data (eg. [3]). Moreover, the use of CFD for wind engineering purposes often requires very huge computing power and produces several problems with scaling, and related to them issues with maintaining Reynolds number.

From the point of view of wind engineer and civil engineer there are not too many data about in-situ measurements of wind loading on different structures. There are a few buildings which throughout the years were measured in details and now the data collected on them are the base for wind tunnel tests and CFD simulations as well. Such structures are Commonwealth Advisory Aeronautical Council high-rise building, Texas Tech Building, and finally Silsoe Cube, all of them are thoroughly described in the literature. Recently, the more common monitoring of long span bridges, high-rise buildings etc. provides invaluable database of information. From the other hand, the urbanization of cities makes the flow more complex and estimation of the real wind action becomes almost impossible without extensive full-scale, model-scale or CFD investigations.

Various equipment designed for the measurements of wind speed, pressure, deflections of structures under wind loading is commonly used in full scale tests in wind engineering. Measurements usually concern the static and dynamic response of buildings to wind action (eg. [4-6]), chimneys (eg. [7-8]), and other structures. Many data come from measurements of the behavior of high-rise buildings during cyclones (eg. [9-10]) or investigations on the vertical wind profile in cities (eg. [11]) and spatial features of strong winds near ground (eg. [12-13]). Simple devices (such as KIMO) that are also a component of the measuring system described in this paper are less frequently used in wind engineering, and are mainly designed to investigate ventilation systems (eg. [14-17]).

This work presents application of a developed system designed for measuring climate parameters, mainly the parameters of the wind speed and its direction in a full-scale. The system consists of two parts: 1. two devices, each of them allows to measure the environmental parameters in one point in a space, 2. expanded weather station. The operation of the system is illustrated by examples.

IOP Conf. Series: Journal of Physics: Conf. Series **1101** (2018) 012019 doi:10.1088/1742-6596/1101/1/012019

2. Description of measuring devices

Basic information about environmental parameters are obtained with KIMO multifunction instrument AMI 310, held in hands or mounted in a given point of the structure. The described system uses two KIMO devices, each of them have the following probes: telescopic vane probe for wind speed measurements in one direction (Fig. 1a, Fig. 2) and other exchangeable modules, allowing measuring atmospheric pressure, humidity, temperature, and illuminance. The advantage of this device is its portability and the ability to set up in virtually any condition, as well as simultaneously recorded measurement data being saved directly on the device's memory card. The main disadvantage is the relatively low resolution and low measurement accuracy. Usually such devices can be used for climatic comfort experiments, both inside and outside the building. In this paper an estimation of static wind action on the facade scaffolding is presented as the example. The detailed description of two different analyses, based on the data from the measurements performed with use of KIMO devices is presented in [18-19].

The second part of the measuring system – an extensive weather station – allows for simultaneous measurements of the wind speed and its direction in 6 points in the space (5x2D anemometers, and 1x3D anemometer – Fig. 1b,c), atmospheric pressure, temperature and humidity in one point and differential pressure in 16 points (Fig. 1d). The data are archived with the use of National Instruments equipment, which has been specially designed to manage these measurements (Fig. 1e). A description of the designed system, with the emphasis put on the cooperation of software and hardware is presented in [20] and the detailed description of sample measurements in [21]. Here, in this paper the short example which illustrates the capability of the system to estimate wind action on a scaffolding is enclosed.

a) b) c) d)

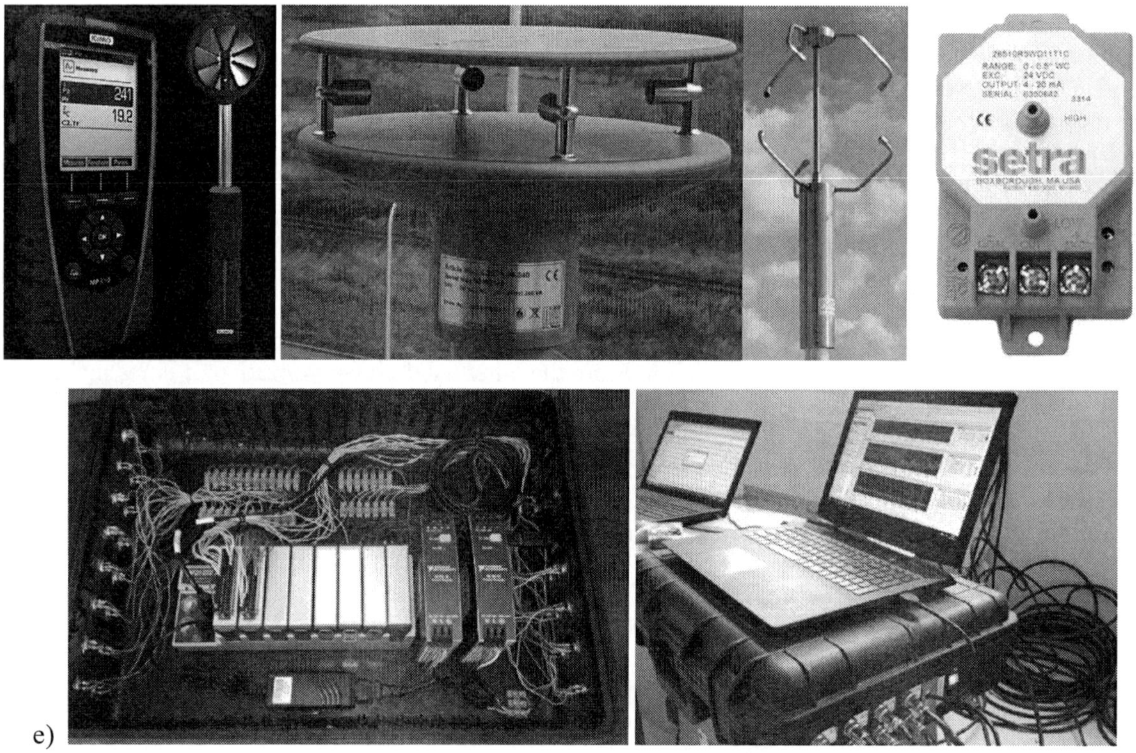

e)

Figure 1. Components of the system.

3. Examples of measurements

3.1. Static wind action

The use of the measuring system for estimation of the static wind action is based on KIMO data which are averaged over a given time, in the example below over 1-min. The operation of the system is illustrated by the example of its use on the façade scaffoldings (Fig. 2). Figure 3a shows the wind speed time history registered with the KIMO device around left top corner of the scaffolding showed in Fig. 2. The data were recorded during 1 minute every 1 second. Individual curves indicate two directions of the wind flow velocity, v_1, v_2, respectively perpendicular and parallel to the façade of the building. Moreover, there are h – humidity and T – temperature 60-sec. time histories in Fig. 3. To get an overview of overall wind action on the scaffolding, similar measurements were repeated in several points of the structure. The results based on one week course of measurements are presented in Fig. 4. There are statistical characteristics of the wind speed in both directions, and mean values of humidity and temperature for several repetitions of measurements (days and hours). The results concern point of measurements located at the left top corner of the scaffolding from Fig. 2.

More complex sample results based on the data measured on 50 scaffoldings in several points located at every object (from 4 to 12 locations) containing maximum and minimum values of the mean wind speed, standard deviation and gusts for v_1 (perpendicular component) are presented in Fig. 5. These data concern the whole one week course of measurements carried out on every structure and statistical values calculated from all points together. Measurements carried out at various locations on the scaffolding allow for estimation of the approximate value of the wind static load (more information in [18]). This approach is burdened with an error introduced by non-simultaneous measurements of the wind speed, what does not occur while using the second part of the system.

Figure 2. Examples of measurements on scaffolding.

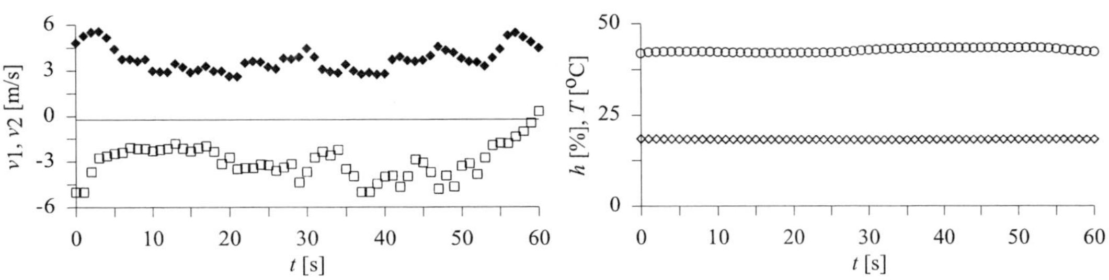

Figure 3. 60-sec measurements of climatic parameters: v_1 – ◆, v_2 – □, h – ○, T – ◇.

a)

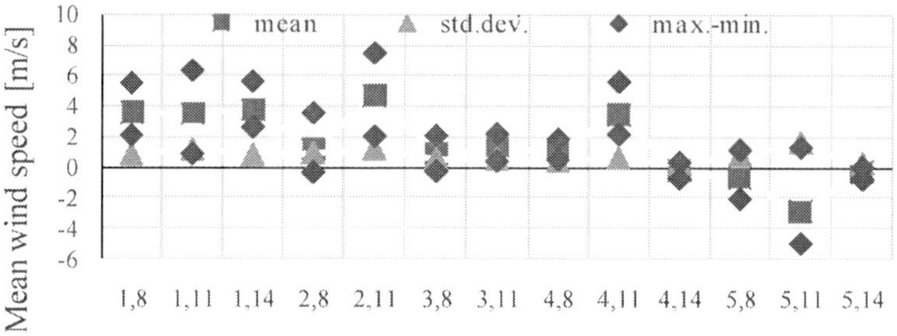

b)

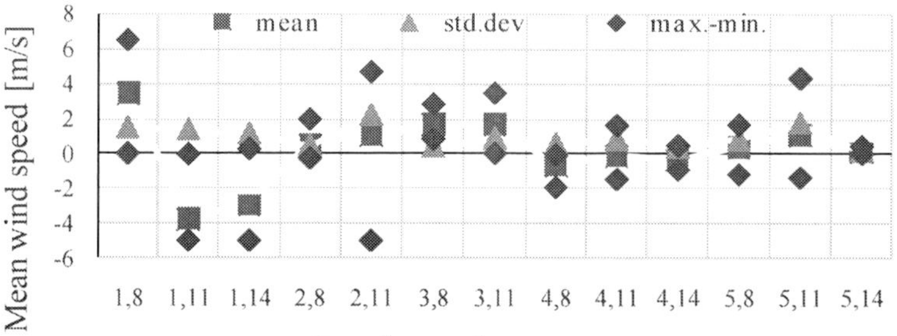

c)

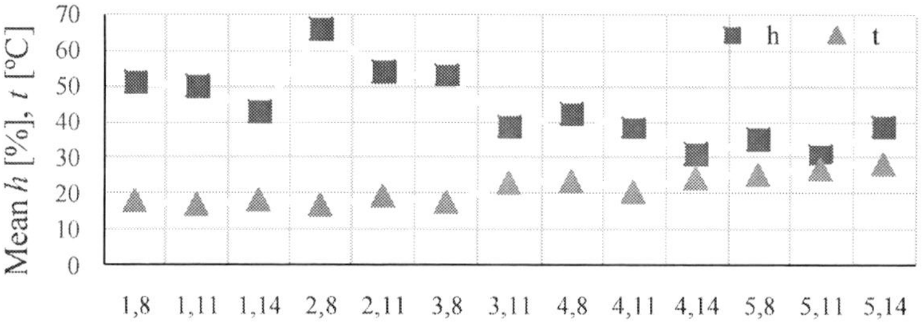

Figure 4. 60-sec statistical characteristics of climatic parameters: a) v_1, b) v_2, c) h and T.

a)

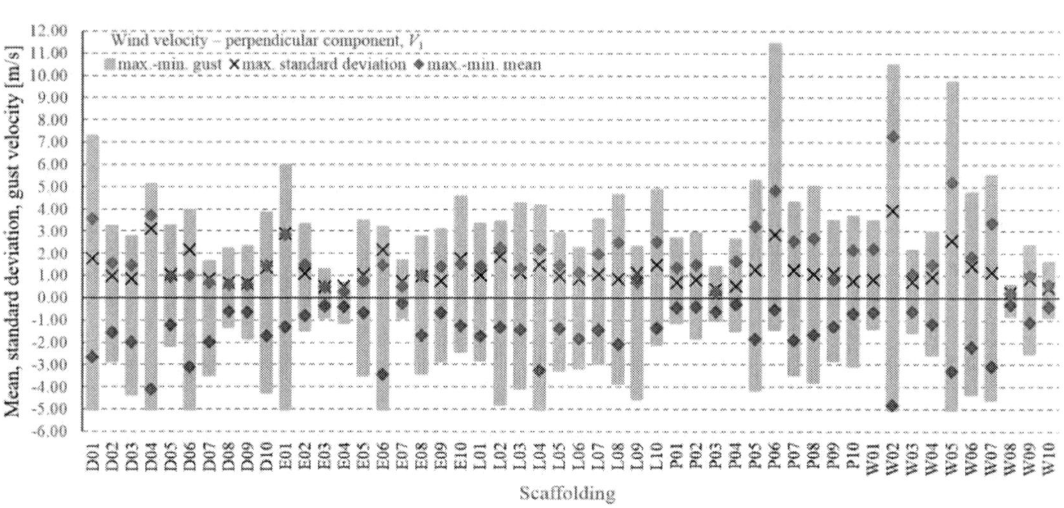

Figure 5. Characteristics of climatic parameters for 50 scaffolding structures.

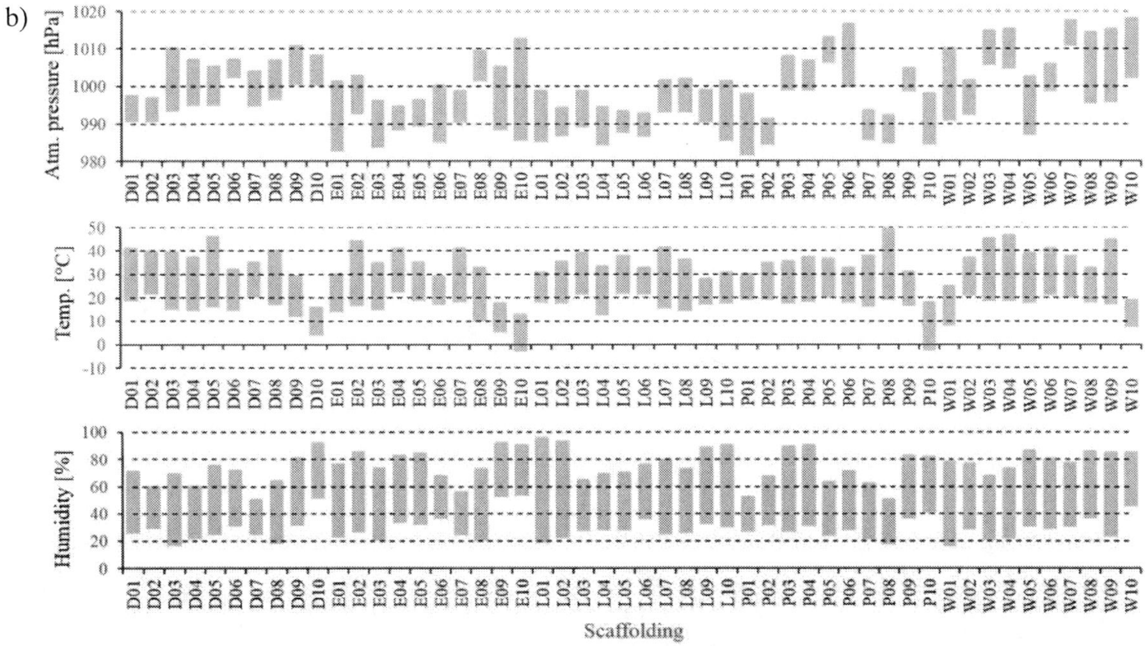

Figure 5. ***(cont.)*** Characteristics of climatic parameters for 50 scaffolding structures.

3.2. Dynamic wind action

The second component of the measuring system eliminates weaknesses of the first one and can be used for determination of the dynamic wind action. The use of anemometers or pressure sensors allows gathering instantaneous data in high resolution, eg every 0.2 sec., which can illustrate the dynamic behavior of the flow or of the structure. The example of its use is illustrated by measurements also performed on scaffolding structures. Five 2D ultrasonic anemometers were hanged on consoles which were fixed to stands of the scaffolding. The 3D ultrasonic anemometer was mounted in a place where the flow is undisturbed by the object – on the roof of the building or fixed to the highest part of stands of the scaffolding. Such location of devices – in undisturbed flow (3D) and on the facade of the scaffolding (2D) – allows to estimate the real instantaneous as well as mean wind action on the structure. The data from 3D anemometer are used as the reference and provide information about the real direction of the flow on the structure and moreover, can be related to the nearest weather station. 2D measurements give the data that take into account the closest neighborhood of the building, presence of other buildings which can produce contraction of the flow or interference effect or edges that induce vortex shedding etc. A few examples of anemometers location on scaffolding structures are presented in Fig. 6.

The example of time history of wind speed and its direction measured in 3 points by 2D anemometers are presented in Fig. 7b-d. In this case anemometers were mounted to the scaffolding structure at the penultimate level of decks. The direction was defined against the facade of the scaffolding, where 0° means the flow perpendicular to the facade, so the value of approximately 100° (as in Fig. 7b-d) means that the wind acted from right. The duration of measurements was 1 hour, and the time step was 0.2 sec. Additionally, time histories of other climatic parameters (h – humidity, T – temperature) are shown in Fig 7a.

Figure 6. Examples of location of 3D and 2D anemometers on scaffoldings.

Another example presents the results of 10-min mean values of the wind speed and direction calculated on two scaffoldings (Fig. 8). The wind speed and its direction were measured on the facade with five 2D anemometers and on the roof of the building – with 3D device. In the above mentioned Fig. 8, the horizontal cross-section of the building and outer scaffolding standards are shown. The vectors are drawn in the appropriate scale and show the values and directions of the wind speed averaged over 10 minutes. Possible directions of the flow indicate complex distribution of the wind load on the scaffolding structure. In the first example, the placement of the scaffolding in the building wake can produce large forces at the edge (right corner) which can further lead to twisting of the structure. It should be underlined that such situation is not considered by any available design standards. The second example shows the situation in which the building façade had openings and the wind acting on the scaffolding could cause its detachment from the building facade. Detailed information about these results can be found in [21].

IOP Conf. Series: Journal of Physics: Conf. Series **1101** (2018) 012019 doi:10.1088/1742-6596/1101/1/012019

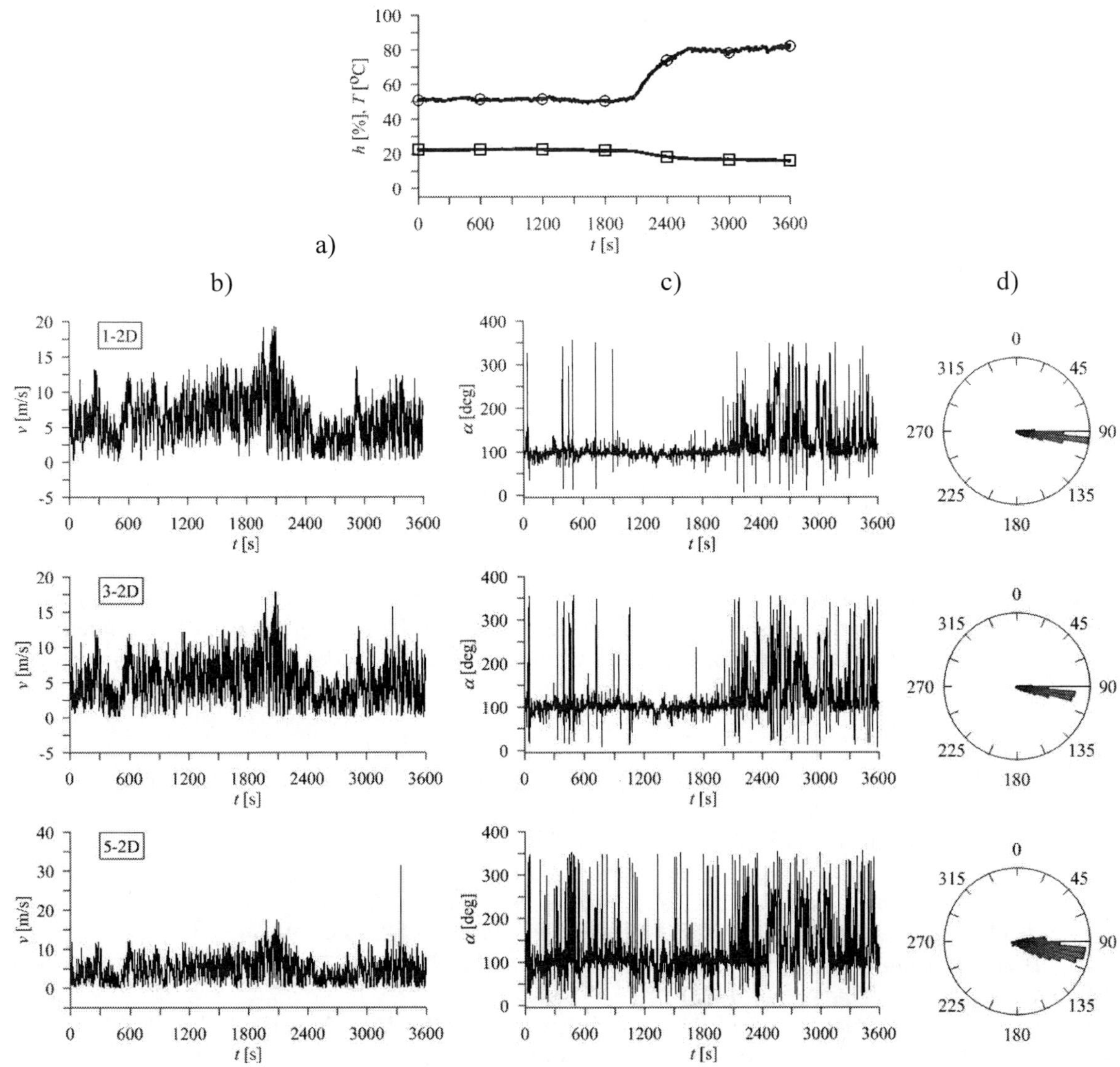

Figure 7. Measured values: a) humidity and temperature, h – O, T – □, b) wind speed, c) wind direction, d) rose plots of wind direction against the facade of scaffolding.

4. Conclusions

The described measuring system is a powerful tool which can help to determine wind conditions and other climatic parameters in the full-scale research. The portable devices, such as multifunctional AMI 310 can definitely be used to assess basic environmental data, but the fact that they can measure only one wind direction component at a time makes it more useful in unidirectional flow condition, rather than in real wind flow. Due to this fact the results can be treated as initial ones giving the general overview on the problem which can be precisely solved with use of the second part of the system. The more advanced part of the system composed of ultrasonic anemometers can allow for measurements of wind flow speed in few points at time, in two or three directions, giving more reliable results. Ultrasonic anemometers provide data which allow consideration of the dynamic wind action on the structure which, especially in urban environment, is usually difficult to predict. Both systems could be applied to determine environmental parameters before design process as well as can provide useful data about environmental actions on already existed structures, as it is presented in the paper.

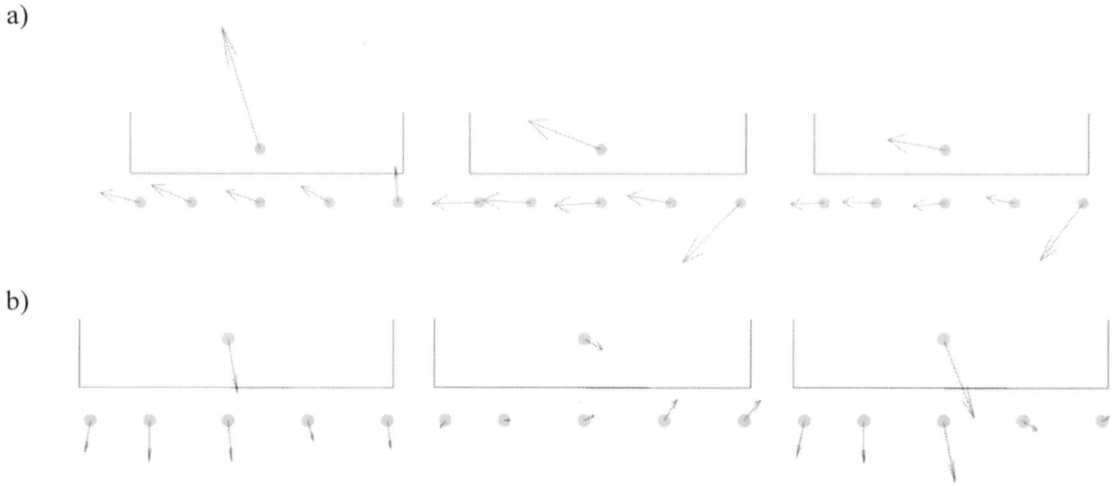

Figure 8. 10-min mean values of wind speed and its direction, a) scaffolding P10, b) scaffolding W07.

Acknowledgements.
The paper has been prepared as a part of the project supported by the National Centre for Research and Development within Applied Research Programme (agreement No. PBS3/A2/19/2015 "*Modelling of Risk Assessment of Construction Disasters, Accidents and Dangerous Incidents at Workplaces Using Scaffoldings*").

References

[1] Gnatowska R, Sosnowski M and Uruba V 2017 *E3S Web of Conference,* 01027
[2] Gnatowska R 2018 *AIP Conference Proceedings*, 110008
[3] Richards P.J and Norris S 2015 *J Wind Eng Ind Aerodyn* **144** 70
[4] Richards P.J and Hoxey R.P 2012 *J Wind Eng Ind Aerodyn* **102** 72
[5] Richards P.J and Hoxey R.P 2012 *J Wind Eng Ind Aerodyn* **102** 87
[6] Snæbjörnsson J.T. and Geurts C.P.W 2006 *J Wind Eng Ind Aerodyn* **94** 845
[7] Breuer P, Chmielewski T, Górski P and Konopka E 2002 *J Wind Eng Ind Aerodyn* **90(3)** 223
[8] Górski P 2009 *Wind Struct.* **12(3)** 259
[9] Li Q.S, Fu J.Y, Xiao Y.Q, Li Z.N, Ni Z.H, Xie Z.N and Gu M 2006 *Eng. Struct.* **28** 1745
[10] Li Q.S, Xiao Y.Q, Wu J.R, Fu J.Y, Li Z.N 2008 J. Sound Vibr. **313** 581
[11] Kent C.W, Grimmond C.S.B, Gatey D, Barlow J.F 2018 *J Wind Eng Ind Aerodyn* **173** 100
[12] Tieleman H.W 2008 *J Wind Eng Ind Aerodyn* **96** 41
[13] Peng Y, Wang S, Li J 2018 *J Wind Eng Ind Aerodyn* **172** 423
[14] Lei L, Wang S and Zhang T.T 2016 *Appl. Therm. Eng.* **95** 70
[15] Tong Y, Wang X, Zhai J, Niu X and Liu L.L 2014 *Build. Environ.* **82** 450
[16] Haw L.C, Saadatian O, Sulaiman M.Y, Mat S and Sopian K 2012 *Energy Build.* **52** 28
[17] de Gracia A, Navarro L, Castell A, Ruiz-Pardo Á, Alvárez S and Cabeza L.F 2013 *Energy Build.* **58** 324
[18] Lipecki T, Jamińska-Gadomska P, Bęc J and Błazik-Borowa E 2017 *Proc. 7th European-African Conference on Wind Engineering,* Liege, Belgium, #180
[19] Szer I, Szer J, Pieńko M, Robak A and Jamińska-Gadomska P 2018 *Int. J. Civ. Env. Eng.* **12(2)** 78
[20] Sumorek A 2015 *IAPGOŚ*. **4** 37
[21] Jamińska-Gadomska P, Lipecki T, Bęc J and Błazik-Borowa E 2017 *Proc. 7th European-African Conference on Wind Engineering,* Liege, Belgium, #179

IOP Conf. Series: Journal of Physics: Conf. Series **1101** (2018) 012020 doi:10.1088/1742-6596/1101/1/012020

Analysis of high-speed photography of cavitating flow in convergent-divergent nozzle

Seweryn Lipiński[1] and Agnieszka Niedźwiedzka[2]

University of Warmia and Mazury in Olsztyn
Faculty of Technical Sciences
[1] Department of Electric and Power Engineering, Electronics and Automation
[2] Department of Mechanics and Basics of Machine Construction
11 Oczapowskiego St., 10-719 Olsztyn, Poland

agnieszka.niedzwiedzka@uwm.edu.pl

Abstract. Cavitation is a phenomenon that is difficult to observe, as well as to analyse. Many methods are being applied in order to study cavitation – high-speed photography is one of them. Analysis of images that were obtained in this way gives the possibility of superficial assessment of the phenomenon, but the use of digital image analysis allows obtaining more information on the studied series of images. In the paper, we propose two methods intended for such analysis - they are destined specifically for obtaining variability maps of cavitating flow based on series of high-speed photography. The work is based on the analysis of three different cavitating flows induced in convergent-divergent nozzles. Obtained results show that described methods can be helpful in analysis of cavitation - variability maps can be used for example as reference images for validation of results of numerical simulations of cavitating flow, which is not possible in case of a direct use of high-speed photographs.

1. Introduction

Cavitation is a phenomenon of liquid evaporation in the areas, where the local fluid pressure drops below the saturated vapour pressure [1]. Cavitation is usually an undesirable phenomenon, as it causes noise, erosion and vibration, hence, among other things, scientific interest in this phenomenon [2]. This first description of cavitation, without using the established term, appeared in 1985 [3]. Such expression as "cavitation" was used not until the work of Thornycroft and Barnaba [4]. One of the most absorbing problems with cavitation is observation of the phenomenon itself, not only its effects. In the first experimental works, mainly photography was applied to investigate this phenomenon. Then the normal photography was replaced with the high-speed photography. Both these tools give a general information about dimensions and intensity of the cavitation area. Many other methods are being applied for the study of cavitating flow [5], but the photography is still used most often.

As the progress in tools for imaging analysis allows obtaining more information on the studied images, the article proposes to implement the methods of digital image processing in order to obtain so-called variability maps that reveal more information on the analysed series of high-speed photographs.

The main aim of this work it to show a new method of high-speed photographs processing, which is intended for analysis of cavitating flows.

The main motivation for the research was to obtain information on the shape and intensity of cavitation cloud for the needs of validation of results of numerical simulations of cavitating flow when using chosen cavitation models applying homogeneous approach [6] but the described method can be successfully utilized for the purpose of analysis of different cavitating flows as well – the only condition is possibility of acquiring high-speed photographs of the flow.

2. Material and methods

2.1. Laboratory stand

The laboratory stand, shown in Figure 1, which was used in the experiment, is intended specifically for investigation of cavitating flows. Its detailed description can be found in [7].

The maximum linear velocity of water in the cavitation chamber is 1 m/s. The length of the cavitation chamber is 900 mm and the internal diameter of the chamber is equal to 50 mm. The test rig is equipped with seven sensors: sensors of inlet and outlet static pressure, sensors of inlet and outlet temperature, sensor of linear fluid velocity, sensors of water level and proximity sensor of the chamber.

Figure 1. Laboratory stand used in the experiment.

2.2. Cavitation inducers

Cavitating flow observed in the cavitation chamber was an effect of the use of two cavitation inducers, each in a form of small-sized nozzle with a small throat ratio. The dimensions of these cavitation inducers are presented in Figure 2. For both nozzles, the following dimensions are common: external diameter – 50 mm, the diameter of the throat – 3 mm and the length of the throat – 6 mm. The nozzles differ in the angles of converging and diverging sections. For the first cavitation inducer (see Figure 2 – left image), the angle of converging section is equal to 30° and the angle of diverging section is equal to 60°. For the second cavitation inducer (see Figure 2 – right image) the both angles are equal to 45°. The throat ratio for each cavitation inducers (i.e. ratio of the throat diameter to the external diameter) is equal to 0.06.

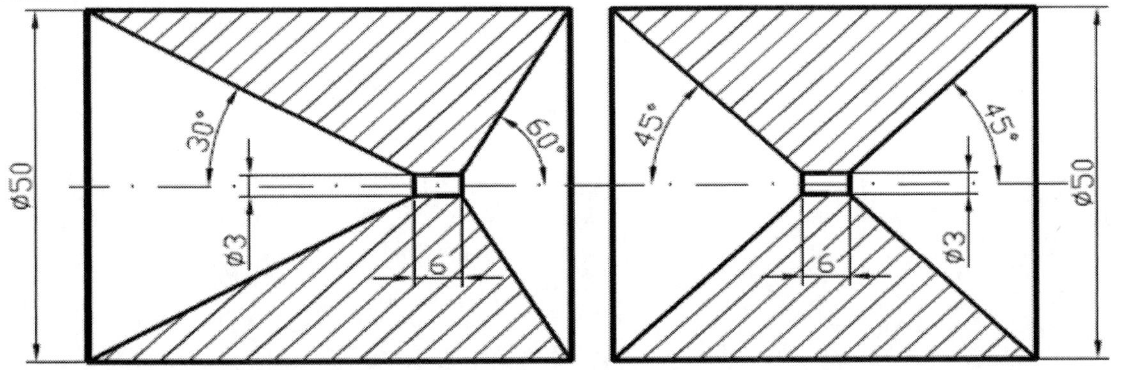

Figure 2. Dimensions of the cavitation inducers used in the experiment [7].

2.3. Experiment

In the whole experiment, three inlet velocities were considered for the first cavitation inducer ($v = 0.1$, 0.2 and 0.3 m/s) and five inlet velocities for the second cavitation inducer ($v = 0.1$, 0.2, 0.3, 0.4 and 0.5 m/s). The medium temperature, i.e. water, was set to 21°C. The outlet total pressure was always equal to the value of the atmospheric pressure. The inlet pressure values, as well as more details on the experiment, are presented in [7].

In the work, three flows were analysed: one for the first inducer ($v = 0.3$ m/s) and two for the second inducer ($v = 0.3$ m/s and 0.5 m/s).

2.4. High-speed photography

To obtain the series of high-speed photographs, the Olympus i-SPEED TR camera with 1000 frames per second was used. Two halogen lamps (with 500 W power) were used to illuminate the scene. Figure 3 shows three exemplary high-speed photographs of cavitating flow from the analysed series, for both used cavitation inducers. In the Figure 3A, a cavitating flow with the inlet velocity v = 0.3 m/s for the first cavitation inducer is shown. In the Figure 3B, a cavitating flow with the inlet velocity v = 0.3 m/s for the second cavitation inducer is shown. In the Figure 3C, a cavitating flow with the inlet velocity v = 0.5 m/s for the second cavitation inducer is shown.

2.5. Computing environment

Obtained images are analysed with the use of Matlab R2013a environment, which is one of the most often used computing environments when it comes to image processing. The connection of basic Matlab functions with Image Processing Toolbox gives an opportunity of simple implementation and testing of created algorithms of digital image processing.

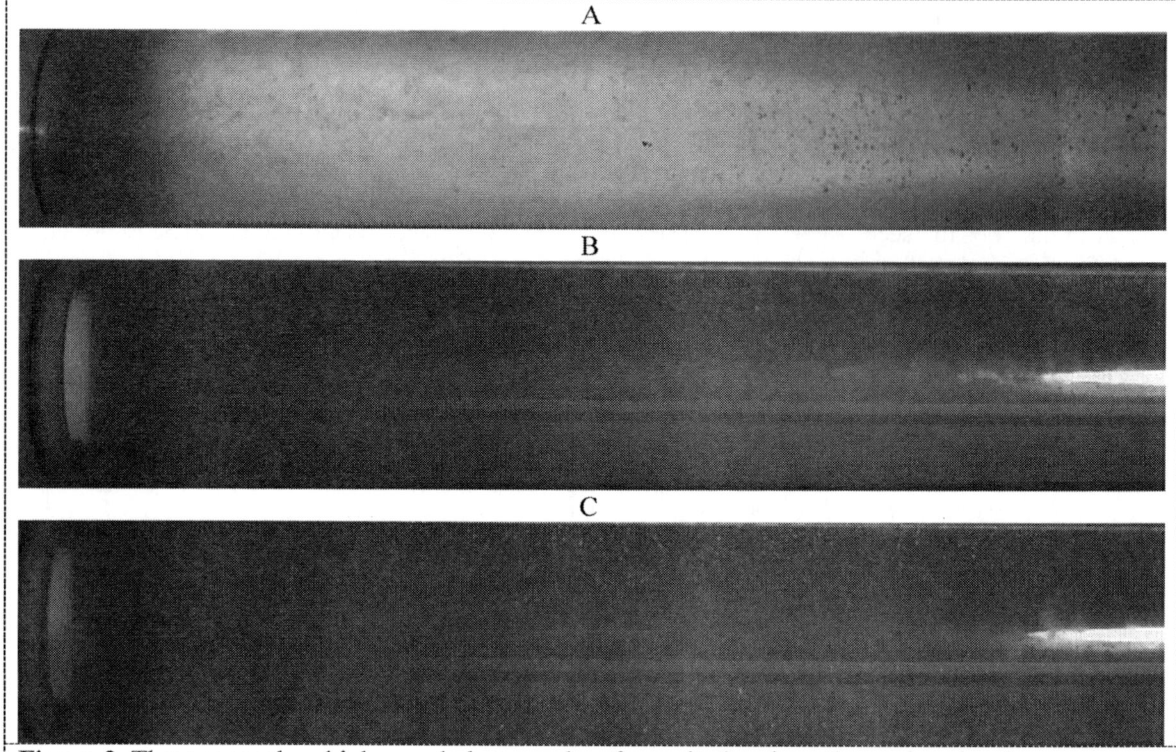

Figure 3. Three exemplary high-speed photographs of a cavitation from the analysed series: A – first inducer (v = 0.3 m/s), B and C – second inducer (v = 0.3 m/s and 0.5 m/s, respectively).

2.6. Used algorithms

Each image from the series is converted to a page of a three-dimensional matrix with number of pages consistent with number of photographs in analysed series. Then, vectors corresponding to each combination of x and y coordinates of created matrix are subjected to variability analysis, performed using two methods.

The first method obtains each pixel of the resulting image based on the minimum, maximum and mean values of each vector corresponding to each combination of x and y coordinates - it is defined as follows:

$$v_2 = \frac{A_{max} - A_{min}}{A_{mean}} \tag{1}$$

The second method obtains the value of each pixel of the resulting image as variance of each vector corresponding to each combination of x and y coordinates and so it is defined as follows:

$$v_1 = \frac{1}{N-1} \sum_{i=1}^{N} |A_i - A_{mean}|^2 \tag{2}$$

As a result, for each single series of high-speed photographs two variability maps are obtained, with dimensions corresponding to dimensions of analysed series of images.

3. Results and discussion

Figure 4 shows obtained variability maps for three different cavitating flows. As it can be seen, performed analysis allows obtaining much more information on the cavitation than the simple observation of series of high-speed photographs (see Figure 3 for comparison). The places where the number of cavitation bubbles increases, have clearly another colour. For the first method of image conversion, places where the number of cavitation bubbles is close to zero, are in shades of blue, while places with a high number of vapour bubbles are in shades of yellow to red. In the second method of image conversion, the obtained images are not as contrast as the images obtained using the first method. The first method of image conversion gives a more detailed view of the intensity of cavitation phenomenon in the analysed region of the cavitation chamber, while the second method gives a general information about the shape of the cavitation cloud. Thanks to achieved variability maps, it is possible to point places, which are putted at risk of cavitation erosion.

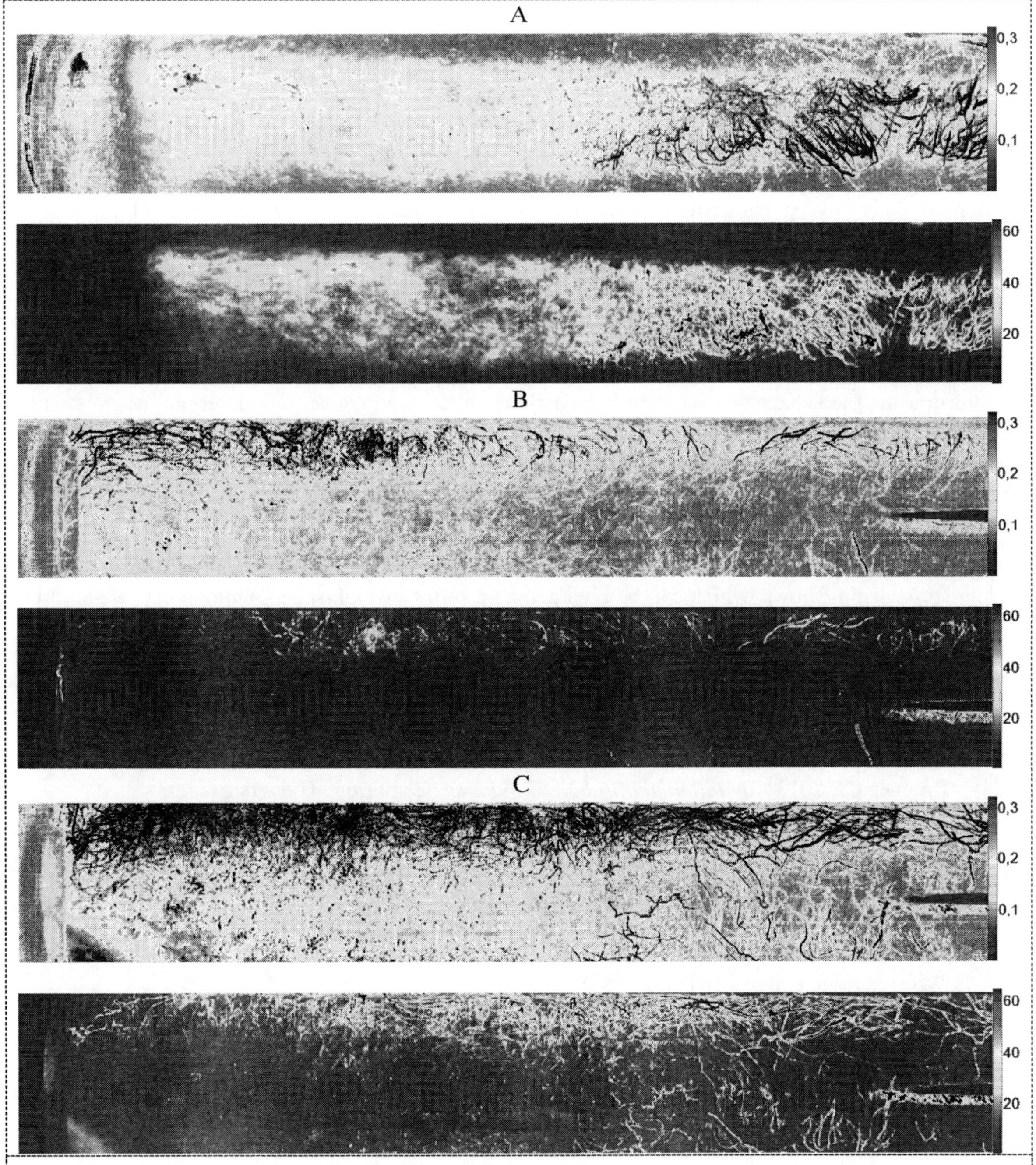

Figure 4. Variability maps of the cavitating flows in a convergent-divergent nozzle:
A – first inducer (v = 0.3 m/s), B and C – second inducer (v = 0.3 m/s and 0.5 m/s, respectively).

For example, variability maps clearly show that the character of cavitation changes with the distance from the inducer. For the first inducer (Figure 4A), bigger bubbles appear further from the nozzle, while for the second and third one (Figure 4B and Figure 4C), they appear closer and higher. Furthermore, areas with greater and lesser intensity of cavitation can be clearly observed, what is clearly visible in Figure 4B and Figure 4C, where for different inducers and the same velocity, we see different character of cavitation (i.e. cloud shape), and of course different intensities. It leads to a conclusion that not only the throat diameter and inlet velocity have influence on the cavitation shape and intensity, but also the angles of converging and diverging sections.

In the experiment it was observed that the changes in the angles of converging and diverging sections could help to limit the maximum inlet velocity, what is not without an influence on the mass flow rate. The performed image analysis shows that the changes in these angles have a big impact on the shape and intensity of the obtained cavitation clouds.

Above discussion leads to the conclusion that while the high-speed photography can be really handy tool, especially for the purpose of the coarse evaluation of the cavitating flow, more accurate differentiation of the cavitating flows is much easier when based on the images shown in Figure 4.

4. Summary and conclusions

The paper shows the methods destined for the purpose of analysis of high-speed photography of cavitating flow, which can be very helpful in analysis of this phenomenon. Obtained variability maps can be used as reference images for validation of results of numerical simulations of cavitating flows, what is not possible in a case of a direct use of high-speed photographs.

For the analysed shape of cavitation inducers, i.e. small-sized converging-diverging nozzles with a small throat ratio (equal to 0.06), the use of described methods allow evaluating the influence of the changes in the angles of converging-diverging sections on the shape and intensity of the cavitation cloud. These results are very important for designers of such devices, as they give valuable information concerning places that are putted in risk of cavitation erosion.

In this paper, presented is a series of images that were obtained for a flow induced in convergent-divergent nozzles, but it should be noted, that the described method could be successfully utilized for different cavitating flows, which can be recorded as a series of high-speed photographs. It should ba also noted, that increased speed of taking photos by the camera would allow observing phenomena with a higher rate of change, but when considering cavitation, there is no need of using faster camera, as the visual assessment of obtained series of images leads to the conclusion that used camera is fast enough.

References

[1] Brennen CE 2013 *Cavitation and Bubble Dynamics* (Cambridge University Press)
Franc JP and Michel JM 2004 *Fundamentals of Cavitation* (Springer Netherlands)
Plesset MS and Prosperetti A 1977 *Ann. Rev. Fluid Mech.* **9** 145
[2] Dular M 2016 *Wear* **346-347** 78
Karimi A and Martin JL 1986 *Int. Metals Rev.* **31(1)** 1
Pędzich Z, Jasionowski R and Ziąbka M 2014 *J. Eur. Ceram. Soc.* **34(14)** 3351
[3] Reynolds O 1894 *Scientific Papers on Mechanical and Physical Subject* tome 2 (London: Cambridge University Press) p 578
[4] Thornycroft J and Barnaby SW 1895 *Minutes Proc. Inst. Civil Eng.* **122** 51
[5] Niedźwiedzka A and Lipiński S 2016 *Meas. Autom. Monit.* **62** 238
Niedźwiedzka A and Lipiński S 2017 *Prz. Mech.* **76(9)** 44
Stutz B and Reboud, J L 1997 *Exp. Fluids* **22** 191
[6] Niedźwiedzka A, Schnerr GH and Sobieski W 2016 *Arch. Thermodyn.* **37** 71
Niedźwiedzka A, Lipiński S and Kornet S 2017 *J. Hydroinform.* **19(5)** 653
[7] Niedźwiedzka A and Sobieski W 2016. *Tech. Sci.* **19** 151

IOP Conf. Series: Journal of Physics: Conf. Series **1101** (2018) 012021 doi:10.1088/1742-6596/1101/1/012021

Comparison of direct flow simulation in a random packed bed of Raschig rings with a macroscale momentum-sink approach

Maciej Marek
Faculty of Mechanical Engineering and Computer Science, Czestochowa University of Technology, Czestochowa, Poland

E-mail: marekm@imc.pcz.czest.pl

Abstract. The purpose of the work is to investigate two numerical models of gas flow in a random packed bed of Raschig rings – one based on direct numerical simulation of flow at the level of individual particles and the other simplified approach in which the bed is treated as a homogeneous domain with prescribed pressure drop calculated from Darcy law. The latter model is usually preferred when only average flow characteristics are important in a given application and is expected to provide reasonably accurate results at the macroscopic level. The results show that when the flow enters the bed in a highly non-uniform manner (here – as a jet), the simplified model is not able to reproduce velocity field even averaged over regions of a considerable size. However, the axial average pressure profile is predicted with a good accuracy.

1. Introduction

Wide applications of random packed beds in many areas of industry (e.g. in chemical engineering as catalytic reactors [1, 2]) sparked large interest in experimental and numerical investigations of flows in such complex structures. However, direct numerical simulations (PRDNS - Particle Resolved Direct Numerical Simulation) of flows in packed beds are very demanding from the computational point of view. They require accurate microscale representation of the bed geometry and high grid resolution for reliable solution of flow between the particles, so usually calculations of this type are limited to beds of relatively small size or small parts of larger structures [3, 4]. For many applications such a detailed study of flow is not necessary and macroscale approach in which only averaged flow characteristics are of interest has become quite popular. In this approach the packed bed is treated, like other porous media, as a homogeneous region of the computational space where the flow equations are supplemented with additional sink terms related to the momentum loss in the inter-particle space (see for instance porous medium model in ANSYS Fluent software [5]):

$$S_i = -\left(\sum_{j=1}^{3} D_{ij}\mu U_j + \sum_{j=1}^{3} C_{ij}\frac{1}{2}\rho|\mathbf{U}|U_j\right) \tag{1}$$

where S_i represents $i-th$ component of the sink term, U_j – $j-th$ component of flow velocity $\mathbf{U}$, μ – dynamic viscosity, ρ – density, and $D_{i,j}$, $C_{i,j}$ are tensors describing potentially anisotropic characteristics of the porous medium. Second sum on the right hand side introduces inertial effects for large velocities and can be neglected in low velocity regime. When the porous medium can be assumed to be isotropic, formula (1) reduces to:

$$S_i = -\left(\frac{\mu}{\alpha}U_i + C_2\frac{1}{2}\rho|\mathbf{U}|U_i\right), \tag{2}$$

where α is the permeability of the medium and C_2 – inertial constant. The first term corresponds to the Darcy law and the formula (2) can be treated as its generalisation (Darcy-Forchheimer law [6]).

Models of this type are used in a considerable amount of literature works – see for example [7, 8, 9] or more complicated versions taking into account variations of local porosity and empirical formulas for pressure drop (e.g. Ergun correlation) [10, 11].

The aim of the present study is to verify how well this simplified momentum-sink approach (here abbreviated as "MSA") reflects the real averaged characteristics of air flow in a random packed bed of Raschig rings. The PRDNS model is based on the methodology previously developed by the author in [4, 12]. The work concentrates on the laminar regime in which viscous stresses are the main component of momentum loss.

2. Methods

2.1. Flow configuration

The considered flow configuration is shown in Fig. 1a. The particles, rings of equal height and diameter, $H_p = D_p = 0.025$ m, with inner diameter $D_i = 0.02$ m, are randomly packed using the algorithm from [4] in a cylindrical container of diameter $D_c = 0.25$ m. Then, a section of the generated bed of height $L = 0.24$ m, sufficiently far from the bottom of bed to eliminate the "bottom effect" [13], is placed in a cylindrical column of height $H_c = 0.36$ m and also of diameter D_c, leaving two empty buffer zones of heights $L_{b,in} = 0.09$ m (above the bed) and $L_{b,out} = 0.03$ m (below the bed). The air flow enters the domain with a uniform profile at the top through a circular orifice of diameter D_{in} which equals: a) D_c, uniform flow case; b) D_p, jet case. The air properties are taken as $\rho = 1.15$ kg/m^3, $\mu = 1.85 \times 10^{-5}$ Pa·s.

On all solid walls (column, particle surface) no-slip boundary condition is imposed, on outflow plane the pressure is assumed to be 0. Flow characteristics are examined on three control planes: $P1$ slightly above the bed inlet ($y = 0.3$ m), $P2$ just below the bed inlet ($y = 0.2$ m) and $P3$ near the bed outlet ($y = 0.1$ m).

2.2. PRDNS model

The model for direct flow simulation consists of two components: bed structure generator [4, 12] and flow solver based on Immersed Boundary Method (IBM) [14, 15]. The random packing of rings is obtained in a seperate simulation by sequential deposition of particles until mechanical equilibrium is attained. Although the mechanics of the process is substantially simplified, the final structure correctly reproduce not only porosity distribution but also distribution of particles' orientation [12, 13]. The flow solver employs structured (Cartesian), staggered computational grid and projection method for solution of the flow equations – Navier-Stokes equations for incompressible fluid of constant density and viscosity – given as follows:

$$\rho\frac{\partial \mathbf{U}}{\partial t} + \rho\mathbf{U}\cdot\nabla\mathbf{U} = -\nabla p + \mu\nabla^2\mathbf{U}\ , \qquad \nabla\cdot\mathbf{U} = 0 \tag{3}$$

where $\mathbf{U} = (U_1, U_2, U_3) = (u, v, w)$ – flow velocity and p – pressure. The complex geometry of the bed influences the flow in the prediction and the correction step of the projection method [16, 17] by damping the flow velocity near the solid boundaries. The details of the method are

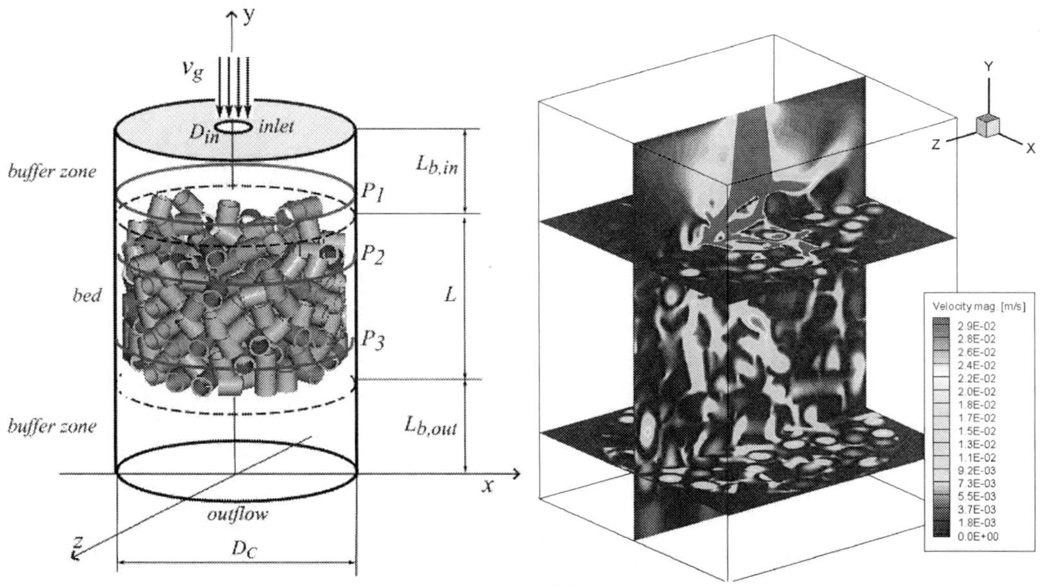

Figure 1. Flow configuration (a) and sample results of PRDNS calculations for flow velocity magnitude in three cross-sections (jet case) (b).

presented in [4] and sample results for the considered flow configuration (jet case) are shown in Fig. 1b.

The grid resolution in all cases was $160 \times 192 \times 160$ cells for the domain of size 0.25 m × 0.36 m × 0.25 m (the second dimension along the column's axis - y), a box in which the physical domain (column) is inscribed. As shown in [4], the selected resolution should be sufficient in the analysed flow regime.

2.3. Momentum-sink model (MSA)

The momentum-sink model differs from the PRNDS model by the treatment of the bed region. Here it assumed that this is an empty space but with additional term **S** on the right hand side of the Navier-Stokes equation (3). As it was verified that Darcy law describes with a good accuracy the relation between flow velocity and pressure gradient in the considered range of velocities and, moreover, the random packed bed of rings (except for near-wall regions) is isotropic at the macroscale [13], the momentum-sink term was taken as: $S_i = -\beta U_i$, with $\beta = \mu/\alpha$. Constant β has been found using uniform flow test case ($D_{in} = D_c$) with different values of inlet velocity v_g. The steady state average pressure gradient along the bed axis turns out to be nearly constant in this case (see [4] for comparison) and the proportionality constant between v_g and dp/dy equals 10.8 [Pa·s/m^2](see Fig. 2a).

The uniform flow case is a perfect configuration from the point of view of momentum-sink model, as the averaged characteristics from the PRDNS simulation almost exactly correspond to the ones obtained with this simplified approach. The purpose of the next section is to examine the jet case i.e. the flow configuration in which the air enters the bed in a highly non-uniform manner.

3. Results

In the main part of this work, simulation has been performed of a jet injected with velocity 0.2 m/s along the column's axis through an orifice of diameter $D_{in} = D_p$ and impinging on the bed

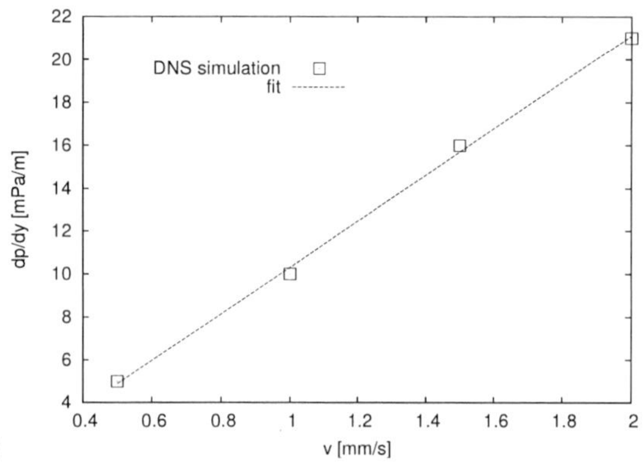

(a)

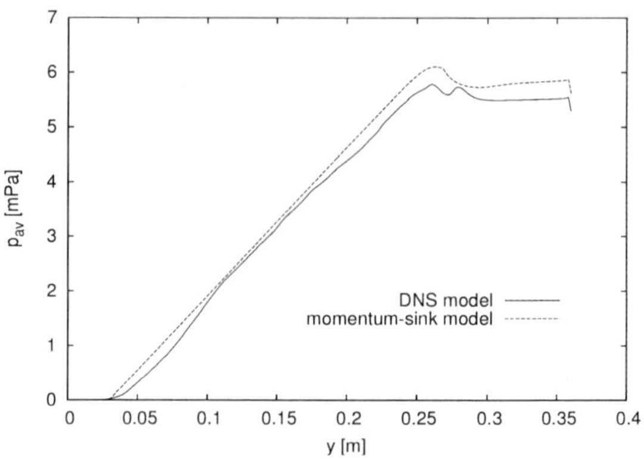

(b)

Figure 2. Relationship between the pressure gradient in the bed and the inlet velocity for uniform flow case (a); the axial average pressure profile for the jet case (b).

($Re = 311$). This test case gave an opportunity to examine developing non-uniform flow in the empty space above the bed and the bed itself. When the steady state was reached, the results obtained with the momentum-sink model could be compared with the spatially averaged results of direct simulations in the real geometry of the packing. The average value of flow property (velocity component or pressure) has been calculated with the PRDNS model at a given point P by finding arithmetic mean of this property in the sphere centred at P and with radius R_{av}. Of course, the averaging is limited to the bed region. Three values of averaging radius have been examined $R_{av} = 10h, 20h, 30h$, where h – size of the computational cell. These values correspond to $R_{av} = 0.625D_p, 1.25D_p, 1.875D_p$, respectively.

The axial average pressure profile has been found by averaging over domain cross-sections normal to the column axis. As it can be seen in Fig. 2b, even for the jet this pressure profile coincides with the solution from the PRDNS model with the differences not larger than a few percent.

Unfortunately, one cannot say the same about flow velocity components. In Fig. 3 three contour plots for streamwise component v (U_2) are shown at the axial cross-section of the column for the PRDNS model (a), averaged solution from the PRDNS model (b) and MSA model (c). For convenience of visual comparison, the colour variations represent the range of negative values of v (mean flow direction) not larger than 10 mm/s in magnitude. What immediately leaps to the eye is the different structure of the flow above the bed and the fact that the flow inside the bed found with the MSA model becomes essentially uniform after quite a short distance ($\simeq 0.5L$).

Fig. 4 gives an opportunity for more quantitative comparisons of the results. Spanwise components u, w are taken from respective control planes ($P1$, $P2$, $P3$) and line segments described by $z = 0$, variable x (u component) or $x = 0$, variable x (w component). Due to symmetry of the domain and the inflow conditions, in the case of MSA model the flow is almost axisymmetric so only one component is shown (u). Of course, in the PRDNS model the geometry of the bed is far from being symmetrical at the particle level (microscale) and significant differences occur between u and w components, even after averaging over a broad region.

The influence of averaging is shown in Fig. 5. It can be noticed that increasing of R_{av} leads to smoothing of irregularities of velocity components (as could be expected) but, at the same time, the non-uniformity of the flow also tends to even out and the very object of the examination dissolves into uniform field (see particularly the evolution of u). Thus, the value of R_{av} cannot

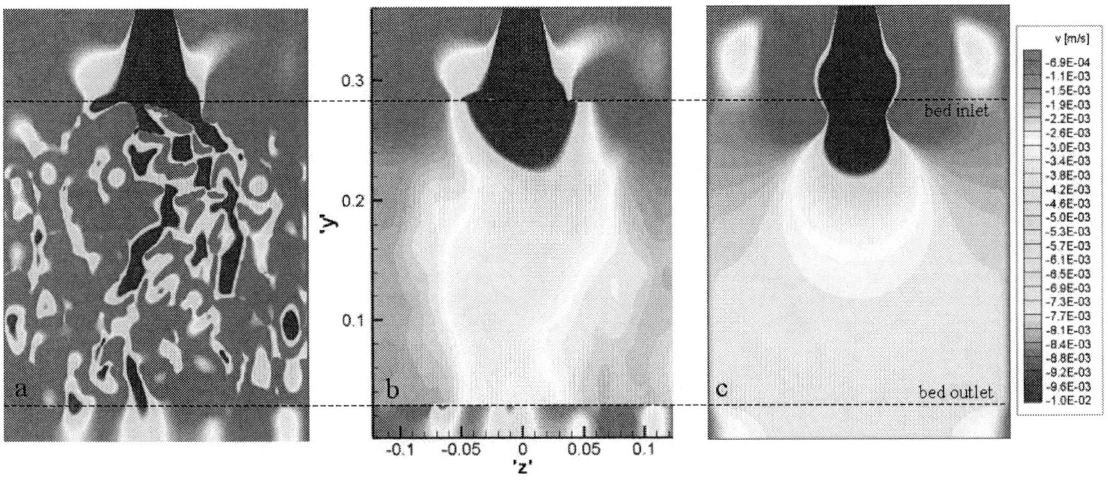

Figure 3. Streamwise velocity component v: the flow in the real geometry of the bed (a), spatially averaged PRDNS solution ($R_{av} = 30h$) (b) and the solution based on the momentum-sink approach (c).

be too large and, as a compromise, $R_{av} = 20h$ was taken for the remaining graphs.

As can be seen from the other graphs in Fig. 4, only streamwise component in plane $P1$ is reproduced well by the MSA model, except for the neighbourhood of the jet entrance into the bed where distinct backflow is wrongly predicted. The difference between magnitudes of velocity components is significant, in terms of relative error particularly large in $P3$ plane where the MSA model predicts considerable uniformisation of the flow, as already noted in Fig. 3.

As for the distribution of average pressure across the column diameter (Fig. 6), in contrast to the axial profile, the MSA model greatly underestimates the maximum near the bed inlet ($P1$) but, on the other hand, reproduces the pressure distribution near the bed outlet ($P3$) with the relative error of order of only 20%.

4. Conclusions

Although the MSA model allows for significant computational advantage comparing to the PRDNS model one should bare in mind the limitations of the simplified approach. At least in the considered test case of an air jet developing in the random packing of Raschig rings, the MSA model is not able to predict the distribution of flow velocity components when averaging is performed at the scales comparable with the diameter of the container. The uniformisation of the flow in the exact model occurs much further beyond the bed inlet and even after distance comparable with the container diameter the streamwise component is dominant near the column axis. On the other hand, the axial average pressure profile and pressure distribution at large distance from the bed inlet are predicted by the MSA model with a accuracy of a few percent. The general observation is that the proper reproduction of flow velocity inside a random packed bed at the macroscale requires much more sophisticated models than a momentum-sink term representing the Darcy law (even in the low velocity regime).

Acknowledgements

The work performed within research project UMO-2014/15/B/ST8/04762 funded by National Centre of Science.

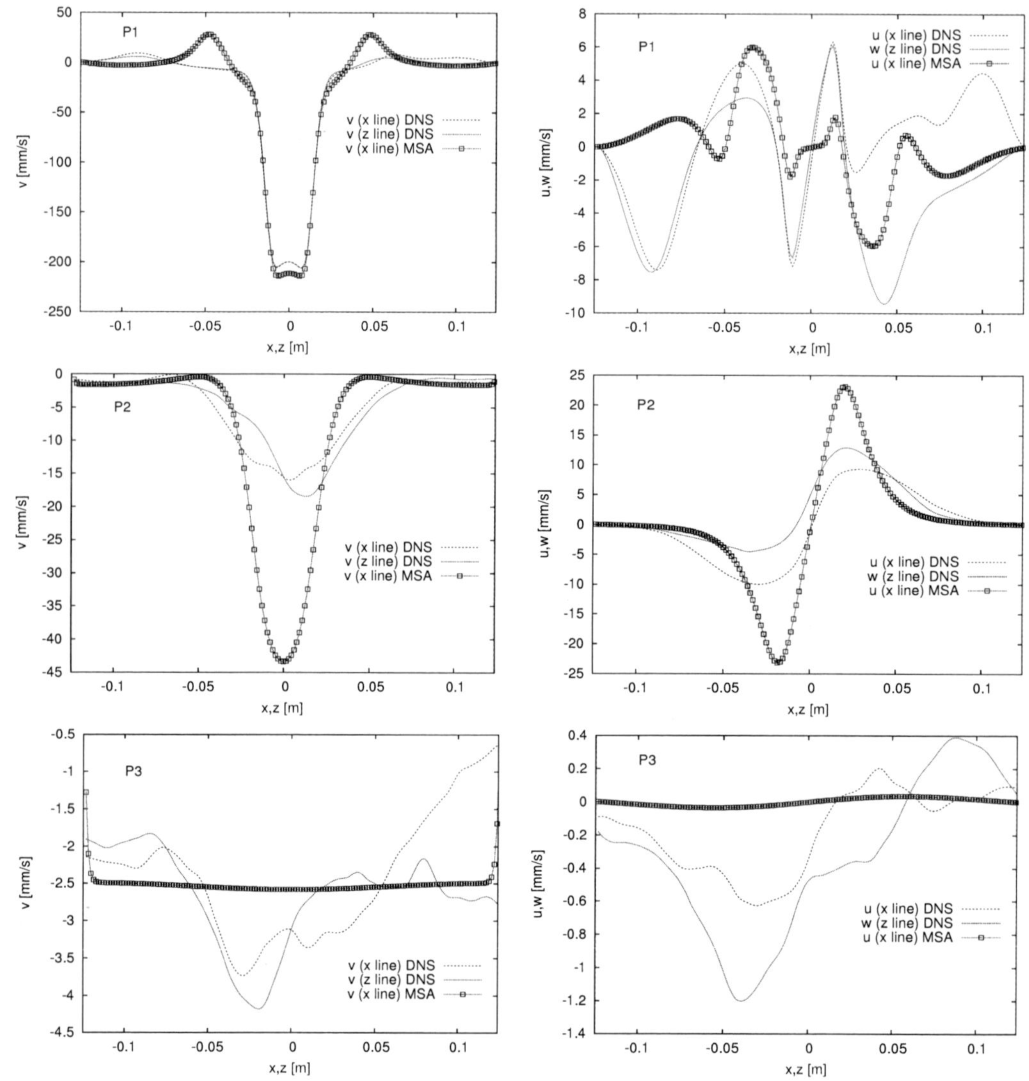

Figure 4. Profiles of streamwise (v) and spanwise velocity components (u,w) in three control planes.

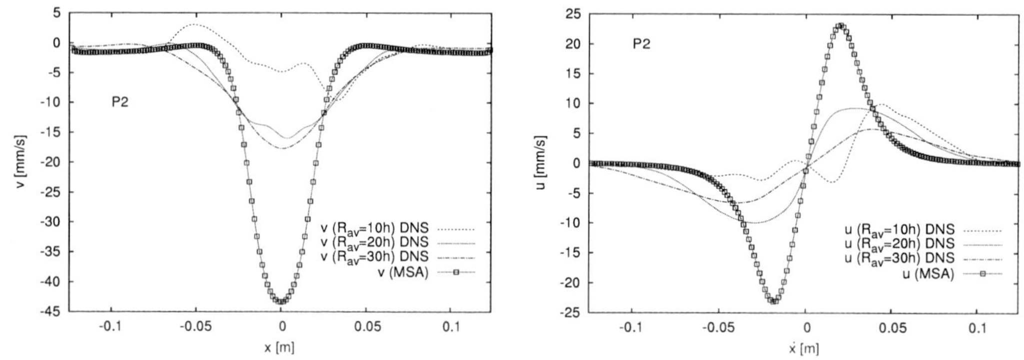

Figure 5. The influence of averaging radius R_{av} on average velocity profiles

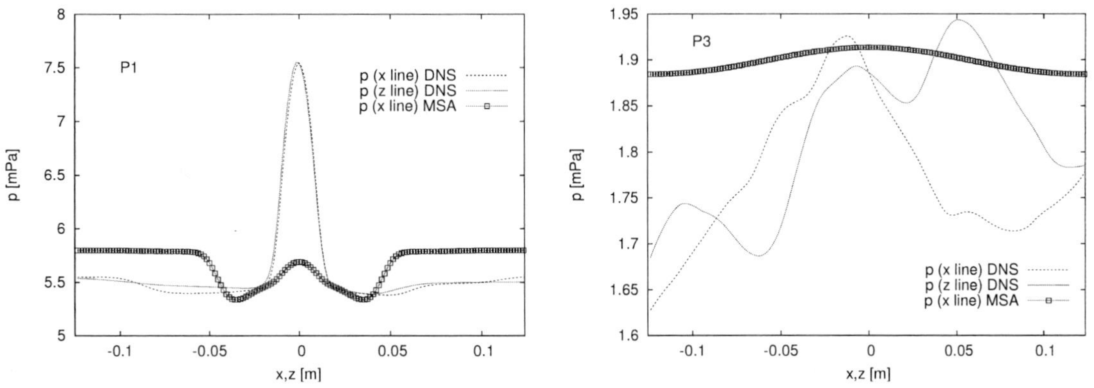

Figure 6. Average pressure profiles in planes $P1$ and $P3$.

References

[1] Billet R 1995 *Packed Towers* (VCH Verlagsgesellschaft mbH, Weinheim)
[2] Maćkowiak J 2010 *Fluid Dynamics Of Packed Columns* (Springer-Verlag, Berlin)
[3] Nijemeisland M, Dixon A G and Stitt E H 2004 *Chem. Eng. Sci.* **59** 5185–5191
[4] Marek M 2017 *Chem. Eng. Sci.* **161** 382–393
[5] ANSYS 2018 *ANSYS Fluent Manual* (ANSYS, Inc)
[6] Andrade J, Costa U, Almeida M, Makse H and Stanley H 1999 *Phys. Rev. Lett.* **82** **(26)** 52495252
[7] Niegodajew P and Asendrych D 2016 *Appl. Math. Model.* **40** **(23-24)** 10222–10237
[8] Asendrych D, Niegodajew P and Drobniak S 2013 *Chem. Proc. Eng.* **34(2)** 269–282
[9] Yang X, Zheng Z C, Winecki S and Eckels S 2013 *Appl. Math. Model.* **37** 9052–9062
[10] Bazmi M, Hashemabadi S and Bayat M 2012 *Int. J. Heat. Mass. Transf.* **39** 736–743
[11] Iliuta I, Hamidipour M, Schweich D and Larachi F 2012 *Chem. Eng. Sci.* **73** 299–313
[12] Marek M 2013 *Chem. Proc. Eng.* **34(3)** 347–359
[13] Niegodajew P and Marek M 2016 *Powder Technol.* **297** 193–201
[14] Iaccarino G and Verzicco R 2003 *Appl. Mech. Rev.* **56(3)** 331–347
[15] Mittal R and Iaccarino G 2005 *Annu. Rev. Fluid. Mech* **37** 239–261
[16] Chorin A J 1968 *Math. Comput.* **22(104)** 745–762
[17] Tryggvason G, Scardovelli R and Zaleski S 2011 *Direct Numerical Simulations Of Gas-liquid Multiphase Flows* (Cambridge University Press, Cambridge)

IOP Conf. Series: Journal of Physics: Conf. Series **1101** (2018) 012022 doi:10.1088/1742-6596/1101/1/012022

Experimental Investigations On The Momentum Pressure Drop During Flow Boiling Of R134a

T Muszynski[1] , R Andrzejczyk[1] and C A Dorao[2]

[1] Gdansk University of Technology, Faculty of Mechanical Engineering, Department of Energy and Industrial Apparatus, Narutowicza 11/12, 80-233 Gdansk, Poland
[2] Norwegian University of Science and Technology, Department of Energy and Process Engineering, N-7491 Trondheim, Norway

Abstract. The article presents experimental investigations of the pressure drop during two-phase flow. Experiments were performed for both adiabatic and heated flow of R134a. Obtained flow patterns were compared with the literature. Obtained data is used to validate momentum pressure drop predictions, a set of graphs showing comparisons, for a representative set of experimental conditions, of the two-phase frictional pressure gradients for the adiabatic and diabatic flow. The model proposed in the article allows to predict both values and peak pressure drop with very good accuracy. Verification of the momentum pressure drop predictions for two-phase adiabatic flow showed that all correlations have good agreement with experimental data.

1. Introduction

Striving for energy efficiency is clearly visible in every part of modern society. Despite the use of alternate energy sources, major gains can be made by increasing the effectiveness of energy utilization, i.e. recovering low-grade waste heat [1], or improving process efficiency [2]. An essential step to assure efficient work of power and process apparatus is their proper design [3]. Modern devices and machines operate with high heat flux densities. Therefore, usage of heat transfer heightening technologies i.e. microjet technology [4,5], flow turbulization [6–8].

A wide range of those devices operates within boiling or condensation of the working fluid to take advantage of high heat transfer coefficients [9]. Two-phase flows are related to large rates of heat transfer because of the latent heat of vaporization and augmentation of the turbulence level between the liquid and the solid surface [10]. Calculating heat transfer coefficient and pressure drop is a challenging task, and has been pursued by numerous research for decades. In the case of diabatic flows, the total pressure drop is due to the conversion of kinetic and potential energy. Also, it is affected by the flow resistance at the wall channel. It has been shown that total pressure drop is the summary of acceleration, static and frictional pressure drop [11]. For two-phase flow acceleration and static pressure drop components are present, a careful selection of void fraction correlation becomes necessary for reliable prediction of these two pressure drops.

The frictional pressure drop of two-phase flow is a function of fluid properties and significantly depends on mass and volume fractions of the mixture components. It was observed that vapor quality and void fraction parameters are unceasingly shifting. This affects overall flow pattern and thus affects pressure drop of the system. Past studies aimed at predicting the frictional pressure drop, includes a large number of studies that rely on either the Homogeneous Equilibrium Model (HEM) or semi-empirical correlations [12]. Sometimes researchers have contradictory views concerning the same issue, such as the application of a given correlation in the calculation of the flow pressure drop or heat transfer coefficient, because of a vast number of existing correlations [13]. This problem is even more pronounced in fluid mixtures flows [14].

The prediction of the two-phase pressure gradient is an essential step in the design of a variety of equipment in the power and process engineering. This study focuses on experimental values of the total, frictional and momentum pressure drop components and their prediction.

2. Experimental test facility

In order to obtain experimental values of momentum pressure drop component, the experimental facility is an R134a loop consisting of a reservoir of refrigerant, a pump, a conditioner, a flow meter, a heated test section, a visualization glass, an adiabatic section and a condenser, see figure 1. The view of the test facility is presented in figure 2.

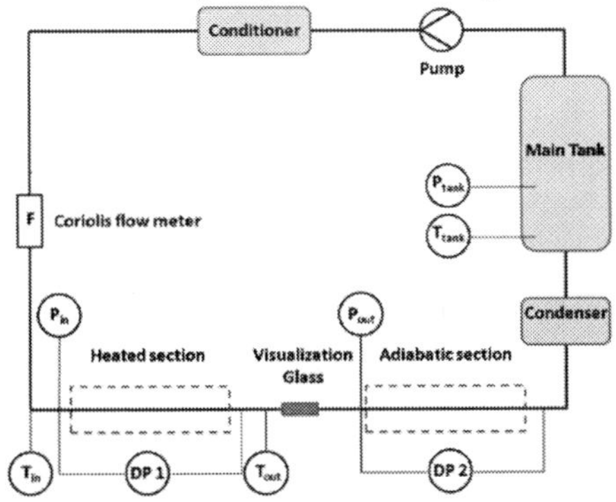

Figure 1. Schematic of the test facility.

Figure 2. The view of the test facility.

The fluid pressure in the loop is set by controlling the temperature in the main tank where the refrigerant is at saturation conditions. The fluid is circulated by a magnetically coupled gear pump. The conditioner is a shell and tube heat exchanger with glycol in the shell side which is used for adjusting the R134a inlet temperature. The fluid flow rate is measured by a Coriolis type mass flow meter. The heated division is made of stainless steel, as a seamless pipe. The tube is electrically heated by Joule effect with the use of a low voltage DC power supply. The section is thermally insulated with a thick layer of mineral wool, thus thermal losses are neglected. The adiabatic test section is a thermally insulated 1m long stainless steel pipe with an inner diameter of 5mm and 8mm outer diameter. The piping system at the supply and outlet of the sections is made with similar pipe dimensions, thus compression and expansions effects are eliminated. Flow visualization allowed for verification of the obtained vapor-liquid mixture patterns, as can be seen in figures 3 and 4

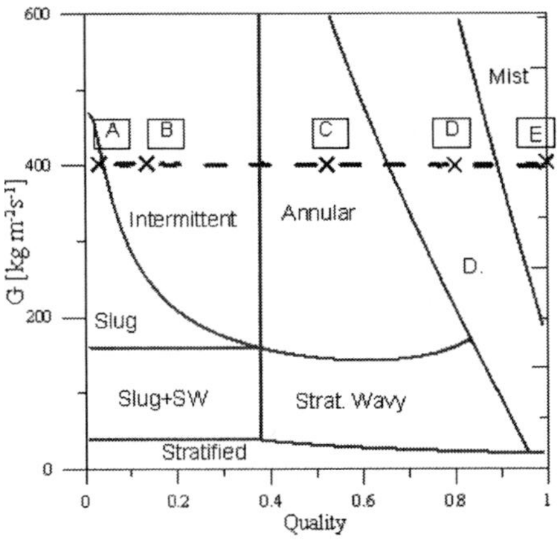

Figure 3. Flow pattern map

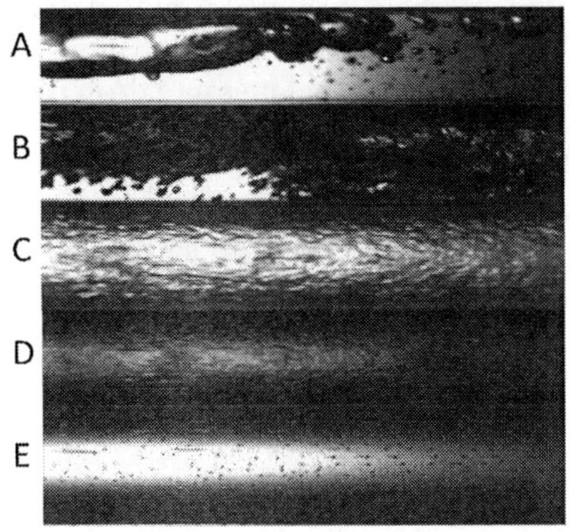

Figure 4. Obtained flow structures.

The measurements data are acquired by National Instruments CompactRIO data acquisition system. The signals from measuring devices were processed with the aid of the LabVIEW application. The temperatures, absolute pressures, pressure differences and mass flow rates were acquired at a frequency of 2 Hz. For every experimental point - about 100 data points were acquired. Additionally, in order to verify the steady-state conditions, data points were doubled after 15 minutes. The range of experimental conditions for the obtained database is presented in table 1.

The local heat flux as the function of generated Joule heat is assumed to be constant during the evaporation process along the length of the tube. Uncertainties were estimated according to the standard procedures described by NIST [15], the measuring device and overall measurement uncertainties are grouped in table 1.

Table 1. Experimental conditions and uncertainty of experiments

Parameters	Range	Uncertainty
x	0 ÷ 1	±5%
D [mm]	5	
L_h [m]	2	
L_{ad} [m]	1	
T_{sat} [°C]	19.4	
G [kg/m^{2}s]	150 ÷ 500	±0.2%
$\dot{q}$ [kW/m^2]	0.1 ÷ 69	±3%

3. Data reduction

The total pressure drop of fluid Δp_{total} is the summation of acceleration, static and frictional pressure drop according to eq. (1) if acceleration and static pressure drop are present.

$$\Delta p_{total} = \Delta p_{mom} + \Delta p_{frict} + \Delta p_{grav} \tag{1}$$

where Δp_{grav} is the elevation head pressure drop and can be omitted in case of a horizontal tube, Δp_{mom} is the momentum pressure drop created by the acceleration of the flow in a heating/cooling process, and Δp_{frict} is the frictional pressure drop of two-phase flow. Most commonly ([16–19] the momentum pressure is expressed by means of eq. (2).

$$\Delta P_{mom} = G^2\left\{\left[\frac{x^2}{\rho_v\varepsilon} + \frac{(1-x)^2}{\rho_l(1-\varepsilon)}\right]_{out} - \left[\frac{x^2}{\rho_v\varepsilon} + \frac{(1-x)^2}{\rho_l(1-\varepsilon)}\right]_{in}\right\} \tag{2}$$

Differences between void fraction correlations are minor, therefore in the further analysis, only three formulas were used. The author's previous experimental studies on pressure drop [20–22] have shown that the predictions given by various methods differ significantly, therefore a modification of Thome [11] approach based on flow pattern map was proposed.

The two-phase frictional pressure drop for annular flow, usually is calculated as:

$$(\Delta p_{fric})_{annular} = 2\cdot(f_i)_{annular}\cdot\frac{\rho_v\cdot u_v^2}{d_i} \tag{3}$$

The modification proposes to calculate the friction pressure drop from Eq.3 at the interface of the liquid-vapor mixture based on modified core flow density ($\rho_v = \rho_c$) due to entrainment. The velocity of the core gaseous flow is affected by the entrainment:

$$u_v = \frac{G\cdot x}{\rho_c\cdot\varepsilon} \tag{4}$$

The average density of gaseous core is calculated based on mass flow rate and the density of the vapor-droplet mixture at the core of the annular flow. It is calculated assuming homogeneous flow for the core flow as follows

$$\rho_c = (1-\varepsilon_C)\rho_L + \varepsilon_C \rho_V \tag{5}$$

where ε_c is the fraction of the core flow with entrained droplets, :

$$\varepsilon_C = \frac{\varepsilon}{\varepsilon + E(1-\varepsilon)} \tag{6}$$

The liquid entrainment fraction correlation by Tibiriçá [23] was used in this study:

$$E = \left(1 + 10^{6.8592}\left(\frac{\rho_V}{\rho_L}\right)^{0.6267} We_{jv}^{-1.1641} Re_L^{-0.3988}\right)^{-1} \tag{7}$$

where the superficial gas velocity Weber number is defined as:

$$We_{jv} = \frac{\rho_v \cdot j_v^2 \cdot D}{\sigma} \tag{8}$$

The superficial gas velocity in [23]is predicted as:

$$j_v = \frac{x \cdot G}{\rho_v} \tag{9}$$

and liquid Reynolds number :

$$Re_L = \frac{G \cdot D}{\mu_L} \; . \tag{10}$$

Therefore friction factor calculated based on homogeneous vapor – droplet mixture flow in the core can be written as:

$$f_i = 0.67 \cdot \left[\frac{\delta}{d_i}\right]^{1.2} \cdot \left[\frac{(\rho_l - \rho_c) \cdot g \cdot \delta^2}{\sigma}\right]^{-0.4} \cdot \left[\frac{\mu_v}{\mu_l}\right]^{0.08} \cdot We_l^{-0.034} \tag{11}$$

where film thickness, Webber number and velocities of phases are calculated as follows:

$$\delta = \frac{d_i}{2} - \left[\left(\frac{d_i}{2}\right)^2 - \frac{(1-\varepsilon_c) \cdot \pi \cdot d_i^2}{2 \cdot (2 \cdot \pi - \Theta_{dry})}\right]^{1/2} \tag{12}$$

$$u_l = \frac{G \cdot (1-x)}{\rho_l \cdot (1-\varepsilon)} \tag{13}$$

$$We_l = \frac{\rho_l \cdot u_l^2 \cdot d_i}{\sigma} \tag{14}$$

Literature observations state that slug and intermittent flow regimes are difficult to separate and have similar behavior. Therefore similar methodology was adopted for calculation pressure drop in these regions:

$$(\Delta p_{fric})_{slug+intermittent} = (\Delta p)_l \cdot \left(1 - \frac{\varepsilon_c}{\varepsilon_{IA}}\right)^{0.25} + (\Delta p)_{annular} \cdot \left(\frac{\varepsilon_c}{\varepsilon_{IA}}\right)^{0.25} \tag{15}$$

The frictional pressure gradient of a mist flow is obtained as:

$$(\Delta p_{fric})_{mist} = \frac{2 \cdot (f_i)_v \cdot G^2}{d_i \cdot \rho_h} \tag{16}$$

When dryout regime does exist, the following linear interpolation was used to capture the variation in frictional pressure gradient across the regime without introducing any jump in the value:

$$(\Delta p_{fric})_{dryout} = (\Delta p)_{x=x_{di}} - \frac{x - x_{di}}{x_{de} - x_{di}} \cdot \left[(\Delta p)_{x=x_{di}} - (\Delta p)_{x=x_{de}}\right] \tag{17}$$

where x_{di} and x_{de} are denoting quality values at the inception and completion of the dryout of the refrigerant respectively:

$$x_{di} = 0.58 \cdot exp\left[0.52 - 0.000021 \cdot We_v^{0.96} \cdot Fr_v^{-0.02} \cdot \left(\frac{\rho_v}{\rho_l}\right)^{-0.08}\right] \quad (18)$$

$$x_{de} = 0.61 \cdot exp\left[0.57 - 0.0000265 \cdot We_v^{0.94} \cdot Fr_v^{-0.02} \cdot \left(\frac{\rho_v}{\rho_l}\right)^{-0.08}\right] \quad (19)$$

$$We_v = \frac{G^2 \cdot d_i}{\rho_v \cdot \sigma} \quad (20)$$

$$Fr_v = \frac{G^2}{\rho_v \cdot (\rho_l - \rho_v) \cdot g \cdot d_i} \quad (21)$$

The obtained experimental pressure drop data were compared with the Darcy-Weisbach correlation, finding the friction coefficient according to the Haaland [24] equation, with pipe internal roughness height given by supplier lower than 0.03mm, having the best agreement. Therefore this correlation was used to predict single phase flow friction factors for vapor and liquid flows.

Electric heating of the channel allows us to assume constant heat flux density along the heated channel length, therefore a linear change in local thermodynamic vapor quality of the flow can be assumed. Therefore the pressure drop along the heated channel can be recreated by using friction pressure drop values at "local" vapor qualities at heated channel length.

Rearranging eq.(1) the momentum pressure drop can be calculated by subtracting frictional pressure drop from measured values of total pressure drop in heated section. This operation for an infinite number of points can be expressed as an integral in form of eq.(22). The accuracy of this method highly depends on polynomial interpolation of the data.

For presented data, the best fit error was below 1%, which corresponds to total momentum pressure drop of 1.15%

$$\Delta P_{ad,decomposition} = \int_0^L \frac{d}{dz} \Delta P_{ad}(x) dz \quad (22)$$

The pressure drop component due to acceleration varies with outlet vapor quality and void fraction. Literature models of a void fraction are created with the assumption that the same flow regimes are obtained at the same local liquid and vapor flow rates. As explained earlier proper selection of void fraction model becomes crucial when evaluating the influence of momentum and elevation pressure drop components. In this study for calculating momentum pressure drop homogeneous:

$$\varepsilon = \frac{1}{1 + \left(\frac{1-x}{x}\right) \cdot \left(\frac{\rho_v}{\rho_l}\right)} \quad (23)$$

and Cioncolini [25] void fractions correlations were used:

$$\varepsilon = \frac{h \cdot x^n}{1 + (h-1)x^n} \quad (24)$$

Where: $h = -2.139 + 3.129(\rho_v \cdot \rho_l^{-1})^{-0.2186}$, $n = 0.3487 + 0.6513(\rho_v \cdot \rho_l^{-1})^{0.5150}$.

4. Experimental results

Adiabatic pressure drops as indicated in the description of the test section can be measured after preparation of vapor-liquid mixture in heated section. Varying the inlet subcooling temperature of working fluid will influence the amount of heat necessary to obtain same vapor quality. Figure 5 shows the experimental two-phase pressure drop of the 5 mm tube as a function of mass flux and heat flux of R134a at different subcooling temperatures, at a saturation temperature of 19.4 °C.

Figs. 6 to 8 shows the experimental adiabatic frictional pressure drops, plotted versus values predicted with the presented model. The pressure drop increases exponentially with increasing mass flux. It can also be seen that the two-phase pressure drop is increasing for higher exit vapor quality with a maximum around vapor qualities around x=0.8-0.9, as reported in the literature. The proposed model modification with entrainment effect allows predicting the peak pressure drop values. As can be clearly seen proposed modification provides very good data representation for selected flow range.

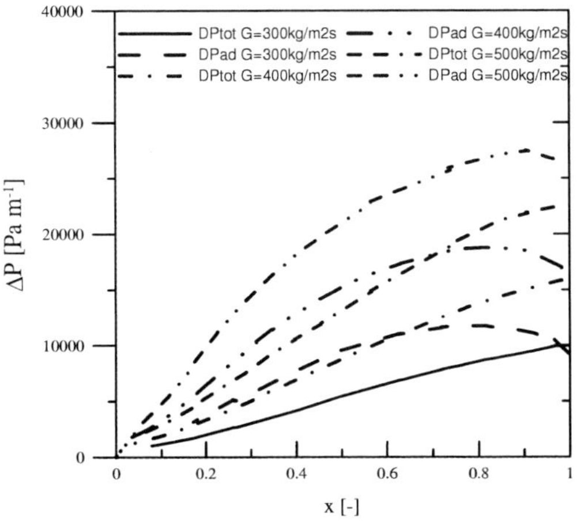

Figure 5. Experimental values of the total and adiabatic pressure drop of R134a versus exit vapor quality of the 5 mm tube at T_{sat}=19.4 °C

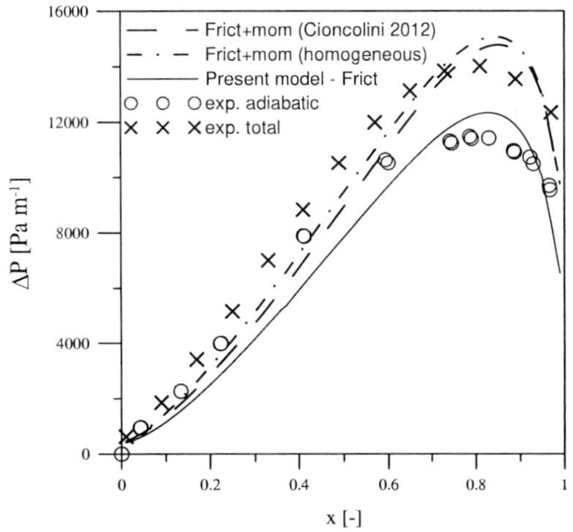

Figure 6. Experimental values of the total and adiabatic pressure drop of R134a versus predictions for G=300 kg/m^{2}s

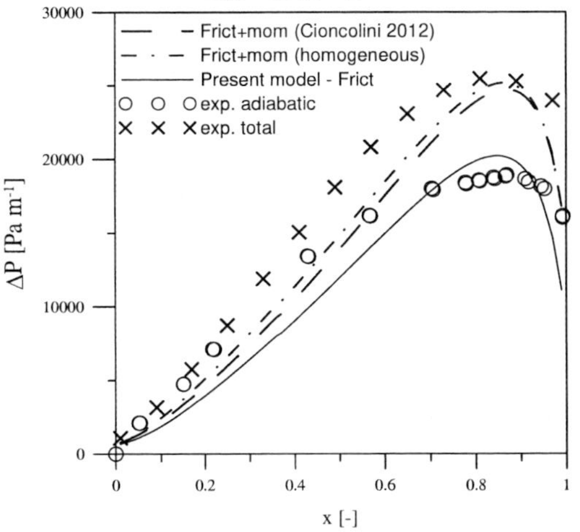

Figure 7. Experimental values of the total and adiabatic pressure drop of R134a versus predictions for G=400 kg/m^{2}s

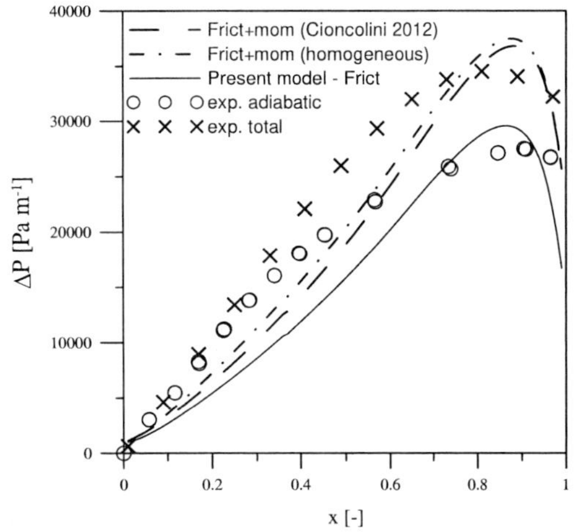

Figure 8. Experimental values of the total and adiabatic pressure drop of R134a versus predictions for G=500 kg/m^{2}s

From the analysis of data presented in figures 6 to 8, a good consistency with the proposed friction pressure drop model and momentum is visible for all of the cases. The influence of various void fraction correlations is more pronounced in higher velocities., as the mass flux is crucial scaling parameter in eq. (2).

5. Conclusions

The experiments in annular flow range were performed to validate void fraction correlations for specific applications of flow boiling momentum pressure drop. An experimental study was undertaken in order to obtain accurate total and frictional pressure drop values R134a in a smooth horizontal 5mm tube, at a saturation temperature of 19.4C. Based on collected experimental data, from a comparison of the adiabatic frictional pressure drop and total pressure drop, momentum pressure drop was calculated. The experimental campaign acquired over 500 experimental data points. Presented data along with data reduction procedure was used to obtain the momentum pressure drop values during flow boiling. Verification of the momentum pressure drop predictions showed that void fraction correlations predict experimental data in the range of ±30%. Proposed modification of the flow pattern based model allows for very good agreement with own experimental data. Presented procedure allowed for prediction of the peak pressure drop in both adiabatic and diabatic flow.

Nomenclature

D	diameter, [m]
E	entrainment fraction [-]
f	friction factor, [-]
Fr	Froude number [-]
Ft	Froude rate, [-]
G	mass flux, [kg/m^{2}s]
g	acceleration due to gravity [m/s^2]
h	enthalpy [kJ/kg]
L	length [m]
ΔP	pressure drop [Pa]
$\dot{Q}$	heat flux [kW]
$\dot{q}$	heat flux density[kW/m^2]
Re	Reynolds number[-]
We	Webber number[-]
x	quality [-]
X_{tt}	Lockhart-Martinelli parameter [-]

Greek symbols

ε	void fraction [-]
μ	viscosity [Pas]
ρ	density [kg/m^3]

Superscripts

ad	adiabatic
A	Annular
c	core
D	dryout
fric	frictional
h	heated
in	inlet
I	Intermittent
l	liquid
mom	momentum
out	outlet
sat	saturation
sub	subcooling
static	gravitational
v	vapor
w	wall

6. References

[1] Muszynski T 2017 Design and experimental investigations of a cylindrical microjet heat exchanger for waste heat recovery systems *Appl. Therm. Eng.* **115** 782–92

[2] Ostrowski P, Pronobis M and Remiorz L 2015 Mine emissions reduction installations *Appl. Therm. Eng.* **84** 390–8

[3] Muszynski T and Koziel S M 2016 Parametric study of fluid flow and heat transfer over louvered fins of air heat pump evaporator *Arch. Thermodyn.* **37** 45–62

[4] Marzec K and Kucaba-Piętal A 2017 Numerical investigation of local heat transfer distribution on surfaces with a non-uniform temperature under an array of impinging jets with various nozzle shapes *J. Theor. Appl. Mech.* **55** 1313–24

[5] Muszynski T and Andrzejczyk R 2016 Applicability of arrays of microjet heat transfer correlations to design compact heat exchangers *Appl. Therm. Eng.* **100** 105–13

[6] Andrzejczyk R and Muszyński T 2016 Performance analyses of helical coil heat exchangers. The effect of external coil surface modification on heat exchanger effectiveness *Arch. Thermodyn.* **37** 137–59

[7] Andrzejczyk R and Muszynski T 2017 Thermodynamic and geometrical characteristics of mixed convection heat transfer in the shell and coil tube heat exchanger with baffles *Appl. Therm. Eng.* **121** 115–25

[8] Muszynski T and Andrzejczyk R 2016 Heat transfer characteristics of hybrid microjet - Microchannel cooling module *Appl. Therm. Eng.* **93** 1360–6

[9] Andrzejczyk R and Muszynski T 2017 The performance of H2O, R134a, SES36, ethanol, and HFE7100 two-phase closed thermosyphons for varying operating parameters and geometry *Arch. Thermodyn.* **38** 3–21

[10] Bohdal T, Charun H and Sikora M 2015 Empirical study of heterogeneous refrigerant condensation in pipe minichannels *Int. J. Refrig.* **59** 210–23

[11] Thome J R 2004 Engineering data book III *Wolver. Tube Inc* **2010**

[12] Xu Y and Fang X 2014 Correlations of void fraction for two-phase refrigerant flow in pipes *Appl. Therm. Eng.* **64** 242–51

[13] Bohdal T, Charun H and Sikora M 2011 Comparative investigations of the condensation of R134a and R404A refrigerants in pipe minichannels *Int. J. Heat Mass Transf.* **54** 1963–74

[14] Targanski W and Cieslinski J T 2007 Evaporation of R407C/oil mixtures inside corrugated and micro-fin tubes *Appl. Therm. Eng.* **27** 2226–32

[15] Taylor B N and Kuyatt C E 1994 Guidelines for Evaluating and Expressing the Uncertainty of NIST Measurement Results *NIST Tech. Note* **1297** 20

[16] Xu Y, Fang X, Li D, Li G, Yuan Y and Xu A 2016 An experimental study of flow boiling frictional pressure drop of R134a and evaluation of existing correlations *Int. J. Heat Mass Transf.* **98** 150–63

[17] Aroonrat K and Wongwises S 2017 Experimental study on two-phase condensation heat transfer and pressure drop of R-134a flowing in a dimpled tube *Int. J. Heat Mass Transf.* **106** 437–48

[18] Revellin R 2005 *Experimental two-phase fluid flow in microchannel. Ph.D. thesis, Ecole Polytechnique Federale De Lausanne, Lyon, France*

[19] Charnay R, Revellin R and Bonjour J 2015 Discussion on the validity of prediction tools for two-phase flow pressure drops from experimental data obtained at high saturation temperatures *Int. J. Refrig.* **54** 98–125

[20] Andrzejczyk R, Muszynski T and Alberto Dorao C 2017 Experimental investigations on adiabatic frictional pressure drops of R134a during flow in 5mm diameter channel *Exp. Therm. Fluid Sci.* **83** 78–87

[21] Muszynski T, Andrzejczyk R and Dorao C A 2017 Investigations on mixture preparation for two phase adiabatic pressure drop of R134a flowing in 5 mm diameter channel *Arch. Thermodyn.* **38** 101–18

[22] Muszynski T, Andrzejczyk R and Dorao C A 2017 Detailed experimental investigations on frictional pressure drop of R134a during flow boiling in 5 mm diameter channel: The influence of acceleration pressure drop component *Int. J. Refrig.* **82** 163–73

[23] Tibiriçá C B, Rocha D M, Sueth I L S, Bochio G, Shimizu G K K, Barbosa M C and Ferreira S dos S 2017 A complete set of simple and optimized correlations for microchannel flow boiling and two-phase flow applications *Appl. Therm. Eng.* **126** 774–95

[24] Haaland S E 1983 Simple and Explicit Formulas for the Friction Factor in Turbulent Pipe Flow *J. Fluids Eng.* **105** 89–90

[25] Cioncolini A and Thome J R 2012 Void fraction prediction in annular two-phase flow *Int. J. Multiph. Flow* **43** 72–84

Acknowledgments
The authors would like to appreciate funding received from the Research Council of Norway under the FRINATEK Project 231529.

IOP Conf. Series: Journal of Physics: Conf. Series **1101** (2018) 012023 doi:10.1088/1742-6596/1101/1/012023

Application of acoustic oscillations in flame extinction in a presence of obstacle

Paweł Niegodajew[1]*, Konrad Gruszka [1], Renata Gnatowska[2], Michal Šofer[3]

[1] Faculty of Production Engineering and Materials Technology, Częstochowa University of Technology Armii Krajowej 19, 42-200 Częstochowa, Poland,
[2] Faculty of Mechanical Engineering and Computer Science, Częstochowa University of Technology Armii Krajowej 21, 42-200 Częstochowa, Poland,
[3] Department of Applied Mechanics, Faculty of Mechanical Engineering, VŠB-Technical University of Ostrava, 17. listopadu 15/2127, 708 33 Ostrava-Poruba, Czech Republic:

* E-mail: niegodajew@wip.pcz.pl

Abstract. Currently, a grooving body of research devoted to the potential use of acoustic waves in suppression of different flame types can be observed. However, all works related to this issue focus on flame quenching in open environment or inside resonator tube. When concerning the use of acoustic waves in a fire/flame extinction, one may expect that the closest environment of a flame would not be free of obstacles. Thus, the present work investigates experimentally how the acoustic screen – being the simplest model of a single obstacle affects the quenching process. To do so, the sound levels required to extinguish a gas burner flame were determined for both different distances between an acoustic screen and a waveguide outlet and different fuel loads. It was found that when decreasing distance between the acoustic screen and the waveguide outlet a higher sound levels are required to suppress the flame – what is quite surprising as the sound level alone also increases when the screen approaches the waveguide outlet. The physical interpretation of this feature formulated based on the present and earlier researches is also included in the present paper.

1. Introduction

Fire is one of the most common hazards and it is often accompanied by a variety of critical situations such as: environmental degradation, material losses or losses in human lives. That is why much efforts have been undertaken worldwide aimed at improving currently available and developing new fire extinguishing methods. In turn, a number of modern and effective technologies have been developed worldwide.

Among a variety of modern fire suppression technologies a few of them are worth mentioning as they have contributed to noticeable progress in the fire-safety discipline. Particularly interesting is the approach utilising a water mist as an extinction agent [1]. It has been proved that the water fog has an excellent extinction properties when droplets are of sufficiently small size (under 60 microns) and move with the proper speed [2]. This method allows to extinguish burning petrol, gas or oil using a water/distilled water or an aqueous solution of inorganic compounds [3,4]. When the fog is produced from the distilled water it does not remain any sediments/precipitates after evaporation. Thus, it can be applied to extinguish fire of artworks, archives, libraries, museum halls and flats. The other interesting solution is the hot aerosol fire extinction technology. The main advantage of this approach over the other available methods is that the hot aerosol fire extinguishing agents do not have to be driven out by pressurised gases and can extinguish class A, B, C, D and K fires at 30 to 200 g/m^3 [5]. This technology is also available in the hand-held portable version (fire extinguishing grenade) suitable for extinguishing small scale fires close to an operator. A successive progress in the field of extinction powders can be observed as a number of new substances characterised by a high adsorption properties are developed [6,7]. Also an extensive research on the so-called clean extinction agents, i.e. extinction gases which are an alternative for already decommissioned halons. According to the standard of NFPA 2001 (National Fire Protection Association) the clean extinction agents include selected halogenated hydrocarbons and inert gases [8].

It should be stressed that all already mentioned approaches are at the high technological level, however, these methods suffer due to the following drawbacks. All of them are (more or less) invasive in nature and usually cannot be applied in hard-to-get areas/places and some of them are toxic due to their chemical compositions. A promising solution for such problems could be a technology utilising the acoustic waves. Such an approach could be used as an alternative or supplement to available methods. In order to do so a fundamental knowledge about extinction properties of acoustic waves must be accessible.

The available literature offers a useful information about experimental studies of a flame behaviour under an action of acoustic waves. A number of works are focused on the response of a gaseous burner flame to acoustic field which can be generally classified as linear or non-linear depending on the excitation frequency [9–15]. The latter one may lead to unpredictable flame behaviour as wrinkling [16,17] or bifurcation [18,19]. Some authors proposed to use an acoustic field in emission reduction of unwanted combustion products as soot [20] or NO_x [21,22] which are a serious problem in a number of industrial applications [23–26]. There are also works related to numerical simulations [27,28] and theoretical consideration [29] of acoustics-flame interaction process.

Much less attention is devoted to the potential use of acoustic excitation in flame extinguishing. McKinney and D. Dunn-Rankin [30] utilised acoustic oscillations in quenching of methanol droplet stream flame. Authors analysed different droplets sizes, frequencies and acoustic pressures to identify flame extinction criteria. It was found that the acoustic pressure required to extinguish the droplet stream flame grows up with increasing both droplets diameters and frequency. In the work of Beisner et al. [31] a Zippo lighter was ignited in microgravity environment and then suppressed using a single tone acoustic wave. The results showed that it is easier to quench the flame in microgravity environment than in regular gravity field and at lower frequencies. In the study of Friedman and Stoliarov [32] a systematic study aimed at elucidating how acoustic waves affect laminar diffusion line-flames was conducted. In particular, authors analysed what conditions have to be meet to suppress such a flame type. The line flames were generated with the use of fuel-laden wicks and different liquid fuels including four alkanes and JP-8. Each experimental trial was performed using the same amount of fuel poured along the centre line of the wick. The acoustic frequency was changed within the range between 30 and 50 Hz, whereas the acoustic pressure ranged from 5 to 50 Pa. It was found that it is easier to extinguish the flame at lower frequencies as the minimum acoustic pressure required to cause an extinction grows with increasing frequency. Authors also observed that with increasing acoustic pressure the average mass loss rate of the fuels also increases. More detailed analysis showed that for the minimum speaker power required to cause extinction the ratio between the modified Nusselt number and the fuel's B number remains constant regardless of the acoustic frequency applied. Bennewitz et al. [33] determined extinction criteria for single droplets of three different fuels located in the vicinity of a velocity antinode of acoustic standing wave produced inside a closed waveguide. Similarly as in the previous works it was found that the extinction pressure increases with increasing excitation frequency. Authors observed that extinction events were associated with temporal increase in the local rate of normal flame strain as the flame moved towards the droplet surface. Niegodajew and co-workers [34] used acoustic waves in suppression process of gas burner flame. The results showed a low impact of the burner power on the acoustic extinction pressure. However, with increasing burner power a substantial growth in critical speaker power (required to suppress a flame) was observed, what was even more pronounced when the distance between a flame and an acoustic source was increased. It was also found that regardless the changes in distance between a burner and a waveguide outlet the acoustic extinction pressure remains the same. The images obtained with Schlieren apparatus showed that the extinction process is caused due to the fuel stream deflection and flame displacement from its original position caused by cumulative effect of oscillatory perturbations and acoustic mean flow. Worthy noticing is also the paper of Hardalupas and Selbach [35]. Authors investigated an impact of acoustic perturbations, of frequency between 200 and 920 Hz, imposed on the coaxial air flow of a Swirl-stabilised burner fuelled with natural gas on the lean-flammability limits. The fuel was injected in twofold way, i.e. axially and radially in the centre of the flow. Also two different burner configurations were investigated, namely with and without diffuser located at the burner outlet. The lean-flammability limits were determined by reducing the gas flow rate until the flame extinction occurred. It was found that the lean flammability limits were noticeably and slightly improved with increased amplitude of acoustic oscillations and frequency, respectively for the case with radial fuel injection and without diffuser. For the other burner configuration a minor impact of acoustic oscillations on the lean flammability limits was observed.

To summarise the literature survey, one may notice that there are only five papers [30–34] strictly related to the study of a flame extinguishing with the use of acoustic waves. In all of these works the different flame types were suppressed in the open environment or inside a resonating tube. When concerning the use of acoustic waves in a fire/flame suppression, one may expect that the closest environment of a flame would not be free of obstacles To the best of authors' knowledge, there is no journal paper investigating the impact of obstacles on acoustically-driven quenching process. Hence, this work is aimed at investigating how the acoustic screen – being the simplest model of a single obstacle affects the quenching process, in particular the sound level measured at the flame location.

2. Experimental setup

Figure 1 illustrates the scheme of experimental setup used in the study of flame extinction phenomenon in a vicinity of acoustic screen. The test rig consist of the following components: generator GW INSTEK SFG-2110 synthesized function generator (1), power amplifier Mosfet MDD. 2108M (2), loudspeaker SONY 1-825-378-11 (3), steel waveguide with the length of 650 mm and inner diameter of 70 mm (4), precise traversing system (5), Bunsen burner (6), sound pressure level (*SPL*) sensor Voltcraft SL-451 (7), acoustic screen (8) and gas flowmeter (9). Presented configuration of the loudspeaker and the waveguide allows to generate the traversing (progressive) longitudinal acoustic wave interacting with the non-premixed flame produced by the burner feed with the natural gas. The burner used is just a simple tube of inner diameter of 16.8 mm. The outlet of the burner was positioned on the axis of the waveguide both horizontally and vertically at a fixed distance of L_B=30 mm from the waveguide end. Three different fuel loads were investigated, namely: q_1=0.02, q_2=0.04 and q_3=0.06 m^3/h. The *SPL* was measured at the distance L_P always located at the axis of the waveguide (see Fig. 1b). The sine signal was used to produce the acoustic wave of 50 Hz frequency. It was assumed that the extinction of the flame occurred when three instantaneous and consecutive quenching events took place.

The experiment was divided into two different parts. In the first one the measurements were aimed at determining *SPL* needed for three consecutive extinction phenomena to occur. Two different independent variables were analysed, i.e. fuel load and distance between the acoustic screen and the waveguide outlet. The second part was devoted to study of the acoustic field between the waveguide outlet and the screen in order to support the findings from the first experiment. To do so the burner was removed from the setup and the *SPL* measurements were performed for various configurations of probe and the acoustic screen. The final part of the experiment was focussed on the verifying whether and if to what extent the acoustic screen influences the *SPL* in its closest vicinity.

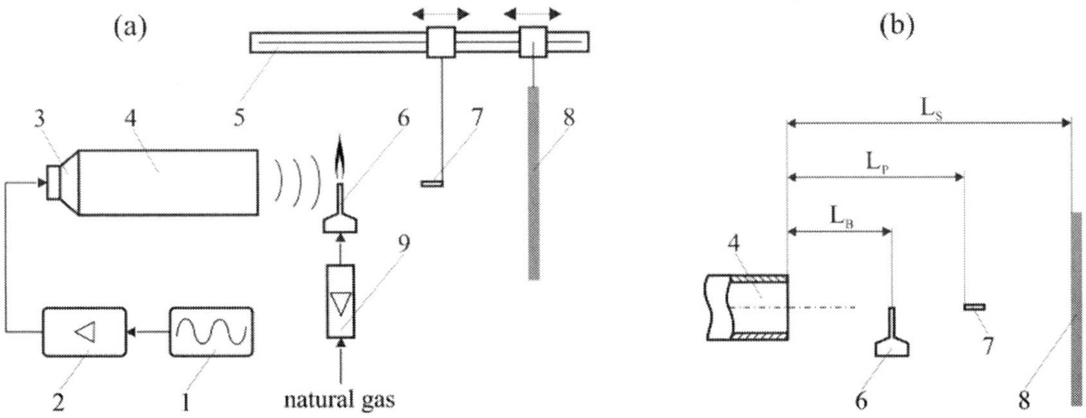

Figure 1. Scheme of the experimental setup (a) and position of the outlet of the burner, the pressure sensor and the screen (b) with respect to the outlet of the waveguide

3. Experimental results

The first stage of the study was aimed at determining the sound pressure level (*ESPL*) required to cause three extinction events for varying wall distance L_S and different fuel loads. Measuring probe was inserted in place of the flame immediately after three extinction events occurred, so that $L_P = L_B$. Results from this part are shown in Fig. 2. As can be clearly seen, the farther the acoustic screen is, the lower *ESPL* is needed for the extinction phenomena to occur. This dependence was measured for three different fuel loads to check its influence on the extinction. It is obvious that with the increase in fuel load, the *ESPL* must be higher also. As demonstrated all three measured curves are similar in character, the differences between highest and lowest *ESPL* are also similar, i.e. 8.4, 8.6 and 8.4 dB for q_1, q_2 and q_3 respectively. One may also observe that when exceeding a certain value of L_S (about 150 mm) the *ESPL* values reach constant levels.

The second part of the study was carried out to support previous findings and clarify the specific effect of acoustic screen on *SPL*. For this stage of experiment, the Bunsen burner was removed and *SPL* distribution was measured for different configurations of acoustic screen and measuring probe.

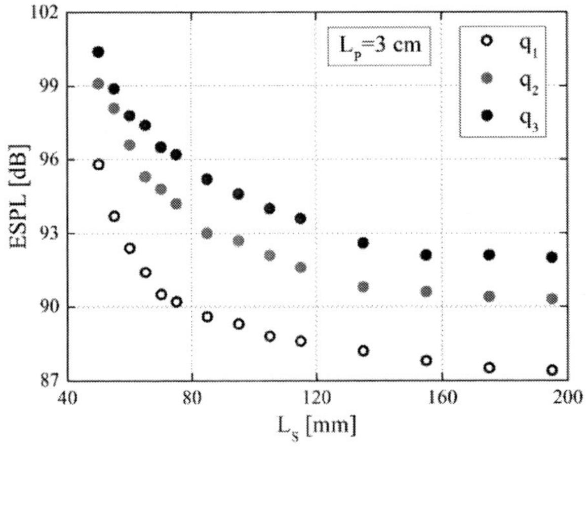

Figure 2. Extinction sound pressure level (ESPL) needed for three consecutive extinction phenomena to occur as a function of acoustic screen distance L_S for different fuel loads.

IOP Conf. Series: Journal of Physics: Conf. Series **1101** (2018) 012023 doi:10.1088/1742-6596/1101/1/012023

First, the dependence between *SPL* near the tube outlet (at the previous location of the burner) and the distance of the acoustic screen (see Fig. 3) was determined. Measurements were performed in two independent runs (denoted as run 1 and run 2 in Fig. 3). For each run the power applied to the loudspeaker was fixed at a constant value and the screen location was changed.

The data clearly show, that the largest range of *SPL* variation lies in the limit of 150 mm for acoustic screen distance L_S. Above this value one may observe strong flattening of the dependence and therefore only small changes in the *SPL* value. Such relationship is observed independently from the power applied to the loudspeaker i.e. both curves presented in Fig. 3 have identical shapes. One may also speculate about the influence of used frequency on the appearance of a standing wave or on changing conditions due to interference effects between the generated and reflected (from the screen) waves. However, due to the used frequency of 50 Hz, the length of acoustic wave in the air is $\lambda \sim 6.8$ m and hence only one kind of interference may occur, which remains irrelevant for studied screen distances. In summary, the result is that above the distance of 150 mm, the screen has an insignificant impact on the *SPL* at the probe location. Moreover, this support the findings from Fig. 2 where a similar dependence was observed.

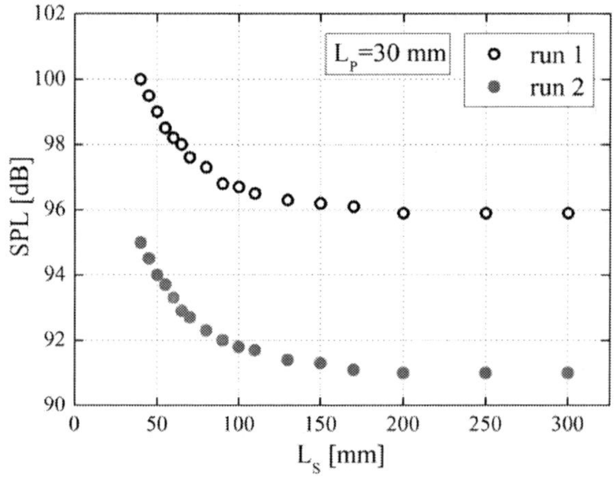

Figure 3. Sound pressure level as a function of screen distance for two different starting *SPL*.

Next, to determine the effect of close proximity of the screen to the probe (and the burner) in isolation from the screen-tube distance, the dependence of *SPL* as a function of probe distance was measured for two cases: when the probe location was varied along with the acoustic screen and without the screen for direct comparison (see Fig. 4).

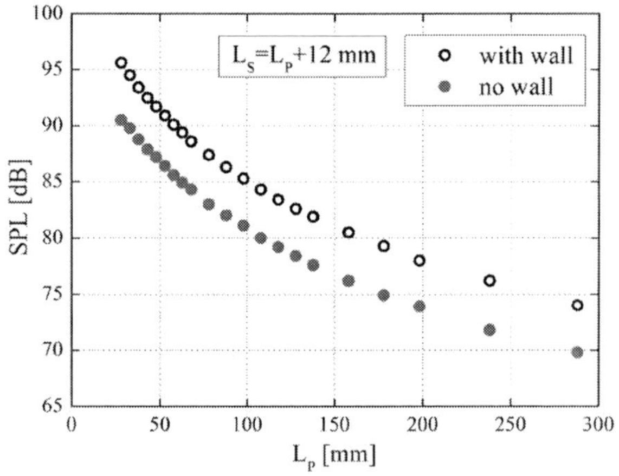

Figure 4. Sound pressure level as a function of probe distance form tube outlet for two cases: when the screen was placed as close as possible to the probe (12 mm), and without screen.

By the analysis of Fig. 4 a clear picture of the impact of close vicinity between the screen and the probe emerges. Only one effect caused by the acoustic screen is observed, namely an increase in the *SPL* in the whole measuring range. Moreover, this increase remains constant independently from the distance between the tube's outlet and the screen, what can be seen by comparing both curves.

Even more information can be acquired when changing the probe location between the waveguide outlet and the screen. Figure 5 illustrates changes in *SPL* with varying L_P for three different screen positions (100, 150 and 200 mm). When bringing the probe closer to the acoustic screen its impact becomes more evident. It is manifested by convergence of all three distributions when reducing probe distance. This means that in the close vicinity to the tube's outlet the impact of the screen (if it is located sufficiently far) on *SPL* is of secondary importance. When analysing *SPL* distribution for case when L_S=100 mm and for L_P=20 mm a slight increase (about 0.2 dB) in *SPL* with respect to the other distributions. This suggests that the smallest analysed L_S lies very close to the critical distance above which the influence of the screen is negligible.

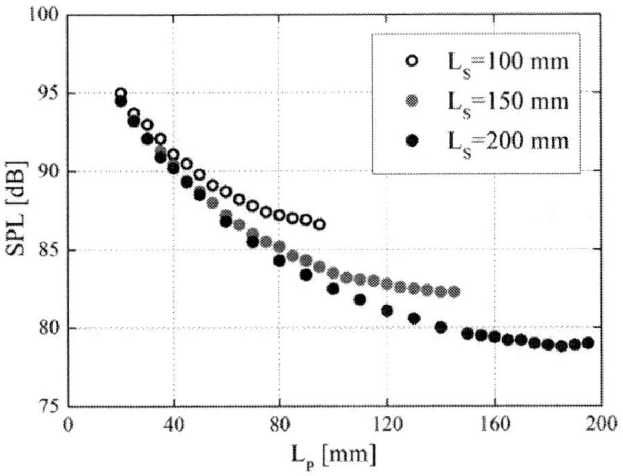

Figure 5. Sound pressure level as a function of probe distance form tube outlet when the screen was fixed at L_s=100 mm, L_s=150 mm and L_s=200 mm.

It should be stressed that the quenching process depends not only on the acoustic oscillations but also on the mean flow effect as it was elucidated in [34]. As it is also known, the normal velocity component decreases when the flow approaches the wall [36]. Therefore one may expect that when placing the wall just behind the flame will reduce the impact of the mean flow due to change in the momentum transport direction. Hence, the conclusion is that the mean flow for the given configuration and analysed parameters ranges has a dominative character on the quenching process. It is also manifested by increase in measured *SPL* at the flame position (due to the presence of the screen), which is still insufficient to suppress the flame.

4. Conclusions

The main conclusion resulting from the present study is that immediate surroundings has substantial impact on the ability to quench fire when utilising acoustic oscillations. If near a fire source (i.e. directly behind it) there is an object, flame extinguishing will be much more difficult, although the sound pressure in this area will increase. Moreover, the closer the object to the flame is, the higher acoustic pressure is required to cause an extinction event. This means that the increase in *SPL* itself is not sufficient to suppress the flame.

Based on present results supported by previous observations it was concluded that mean flow effect has significant impact on the extinction process next to the acoustic oscillations.

Obtained results allowed to determine the distance at which the impact of the screen on the flame quenching diminishes. This finding allowed to conclude that this distance is the same regardless the adopted fuel loads.

5. References

[1] Yu H Z, Zhou X and Carpenter J 2017 Physical scaling of water mist fire extinguishment in industrial machinery enclosures *Fire Saf. J.* **91** 596–605

[2] Yoshida A, Kashiwa K, Hashizume S and Naito H 2015 Inhibition of counterflow methane/air diffusion flame by water mist with varying mist diameter *Fire Saf. J.* **71** 217–25

[3] White J P, Verma S, Keller E, Hao A, Trouvé A and Marshall A W 2017 Water mist suppression of a turbulent line fire *Fire Saf. J.* **91** 1–9

[4] Tianwei Z, Hao L, Zhiyue H, Zhiming D and Yong W 2017 Active substances study in fire extinguishing by water mist with potassium salt additives based on thermoanalysis and thermodynamics *Appl. Therm. Eng.* **122** 429–38

[5] Zhang X, Ismail M H S, Ahmadun F R B, Abdullah N B H and Hee C 2015 Hot aerosol fire extinguishing agents and the associated technologies: A review *Brazilian J. Chem. Eng.* **32** 707–24

[6] Zhang X, Shen Z, Cai C, Yu X, Du J, Xing Y and Ma S 2012 Modification of ADP extinguishing powder by siliconization in spray drying *Particuology* **10** 480–6

[7] Koshiba Y, Iida K and Ohtani H 2015 Fire extinguishing properties of novel ferrocene/surfynol 465 dispersions *Fire Saf. J.* **72** 1–6

[8] Anon 2015 *NFPA 2001: Standard on Clean Agent Fire Extinguishing Systems* (NFPA Publishing)

[9] Kornilov V N, Schreel K R A M and De Goey L P H 2007 Experimental assessment of the acoustic response of laminar premixed Bunsen flames *Proc. Combust. Inst.* **31 I** 1239–46

[10] Wangher A, Searby G and Quinard J 2008 Experimental investigation of the unsteady response of premixed flame fronts to acoustic pressure waves *Combust. Flame* **154** 310–8

[11] Chen L W and Zhang Y 2015 Experimental observation of the nonlinear coupling of flame flow and acoustic wave *Flow Meas. Instrum.* **46** 12–7

[12] Saurabh A and Paschereit C O 2017 Dynamics of premixed swirl flames under the influence of transverse acoustic fluctuations *Combust. Flame* **182** 298–312

[13] Worth N A and Dawson J R 2012 Cinematographic OH-PLIF measurements of two interacting turbulent premixed flames with and without acoustic forcing *Combust. Flame* **159** 1109–26

[14] Wang Q, Huang H W, Tang H J, Zhu M and Zhang Y 2013 Nonlinear response of buoyant diffusion flame under acoustic excitation *Fuel* **103** 364–72

[15] Baillot F and Lespinasse F 2014 Response of a laminar premixed V-flame to a high-frequency transverse acoustic field *Combust. Flame* **161** 1247–67

[16] Karimi N 2014 Response of a conical, laminar premixed flame to low amplitude acoustic forcing - A comparison between experiment and kinematic theories *Energy* **78** 490–500

[17] Magina N, Steele W, Emerson B and Lieuwen T 2017 Spatio-temporal evolution of harmonic disturbances on laminar, non-premixed flames: Measurements and analysis *Combust. Flame* **180** 262–75

[18] Kashinath K, Waugh I C and Juniper M P 2014 Nonlinear self-excited thermoacoustic oscillations of a ducted premixed flame: bifurcations and routes to chaos *J. Fluid Mech.* **761** 399–430

[19] Kozlov V V., Grek G R, Korobeinichev O P, Litvinenko Y A and Shmakov A G 2016 Combustion of hydrogen in round and plane microjets in transverse acoustic field at small Reynolds numbers as compared to propane combustion in the same conditions (Part I) *Int. J. Hydrogen Energy* **41** 20231–9

[20] Foo K K, Sun Z, Medwell P R, Alwahabi Z T, Dally B B and Nathan G J 2017 Experimental investigation of acoustic forcing on temperature, soot volume fraction and primary particle diameter in non-premixed laminar flames *Combust. Flame* **181** 270–82

[21] Kim M, Choi Y, Oh J and Yoon Y 2009 Flame-vortex interaction and mixing behaviors of turbulent non-premixed jet flames under acoustic forcing *Combust. Flame* **156** 2252–63

[22] Oh J, Khan Q S and Yoon Y 2009 Acoustic excitation effect on NOx reduction and flame stability in a lifted non-premixed turbulent hydrogen jet with coaxial air *Int. J. Hydrogen Energy* **34** 7851–7861

[23] Poskart A, Radomiak H, Niegodajew P, Zajemska M and Musiał D 2016 The analysis of nitrogen oxides formation during oxygen - enriched combustion of natural gas *Arch. Metall. Mater.* **61** 1925–30

[24] Jamrozik A, Tutak W, Pyrc M and Sobiepański M 2017 Experimental Investigations on Combustion, Performance, and Emission Characteristics of Stationary CI Engine Fueled with Diesel–Methanol and Biodiesel–Methanol Blends *Environ. Prog. Sustain. Energy* **36** 1151–63

[25] Jamrozik A, Tutak W, Gnatowski A, Gnatowska R, Winczek J and Sosnowski M 2017 Modeling of Thermal Cycle Ci Engine With Multi-Stage Fuel Injection *Adv. Sci. Technol. Res. J.* **11** 179–86

[26] Tutak W, Jamrozik A and Gnatowska R 2017 Combustion of different reactivity fuel mixture in a dual fuel engine *Therm. Sci., https://doi.org/10.2298/TSCI170606299T*

[27] Pera C and Reveillon J 2007 Direct numerical simulation of spray flame/acoustic interactions *Proc. Combust. Inst.* **31 II** 2283–90

[28] Laverdant A, Gouarin L and Thévenin D 2007 Interaction of a Gaussian acoustic wave with a turbulent non-premixed flame *Combust. Theory Model.* **11** 585–602

[29] Strobio Chen L, Bomberg S and Polifke W 2016 Propagation and generation of acoustic and entropy waves across a moving flame front *Combust. Flame* **166** 170–80

[30] McKinney D J and Dunn-Rankin D 2000 Acoustically driven extinction in a droplet stream flame *Combust. Sci. Technol.* **161** 27–48

[31] Beisner E, Wiggins N D, Yue K-B, Rosales M, Penny J, Lockridge J, Page R, Smith A and Guerrero L 2015 Acoustic Flame Suppression Mechanics in a Microgravity Environment *Microgravity Sci. Technol.* **27** 141–4

[32] Friedman A N and Stoliarov S I 2017 Acoustic extinction of laminar line-flames *Fire Saf. J.* **93** 102–13

[33] Bennewitz J W, Valentini D, Plascencia M A, Vargas A, Sim H S, Lopez B, Smith O I and Karagozian A R 2018 Periodic partial extinction in acoustically coupled fuel droplet combustion *Combust. Flame* **189** 46–61

[34] Niegodajew P, Łukasiak K, Radomiak H, Musiał D, Zajemska M, Poskart A and Gruszka K 2018 Application of acoustic oscillations in quenching of gas burner flame *Combust. Flame* **194** 245–249

[35] Hardalupas Y and Selbach A 2002 Imposed oscillations and non-premixed flames *Prog. Energy Combust. Sci.* **28** 75–104

[36] van Hout R, Rinsky V and Grobman Y G 2018 Experimental study of a round jet impinging on a flat surface: Flow field and vortex characteristics in the wall jet *Int. J. Heat Fluid Flow* **70** 41–58

IOP Conf. Series: Journal of Physics: Conf. Series **1101** (2018) 012024 doi:10.1088/1742-6596/1101/1/012024

Wetting phenomena and droplet-wall interaction modelling with Smoothed Particle Hydrodynamics approach

M Olejnik and J Pozorski

Institute of Fluid-Flow Machinery, Polish Academy of Sciences, Fiszera 14, 80-231, Gdańsk

E-mail: {michal.olejnik, jp}@imp.gda.pl

Abstract. Smoothed Particle Hydrodynamics (SPH) is a meshless CFD method, remaining a promising alternative for multiphase flow calculations due to straightforward interface treatment. In the present work we particularly focus on the accurate modelling of wetting phenomena and contact angles. For that purpose we modify the surface tension model in the neighbourhood of the triple line. The necessity of proper boundary conditions in SPH is discussed along the preliminary validation of proposed approach. Results show a great potential of the chosen method.

1. Introduction

Wetting phenomena and interaction of droplets with different surfaces are fundamental processes governed by inertial, capillary and viscous forces. Accurate theoretical description and numerical modelling of such problems remain a difficult task, despite a superficial simplicity of the phenomena. The present work presents an attempt in tackling this challenge by means of the meshless Smoothed Particle Hydrodynamics (SPH) approach. SPH is the particle based method for fluid-flow modelling. Its Lagrangian nature and lack of numerical grid make it exceptionally suitable for cases where large deformations and topology changes occur, like interfacial flows. There is no need to track the interface or reconstruct it like in Eulerian approaches, since its position and shape can be explicitly obtained form positions of particles representing different phases. Furthermore SPH can handle large density ratios of liquid-gas systems fairly well. Examples of application of SPH to the problem of air bubbles rising in the liquid [1] or break-up of the liquid ligaments [2] show its usefulness as a multiphase flow solver. The present work is focused on the proper boundary conditions and contact angle treatment. To enable modelling of interactions of droplet with various type of surfaces we modify the interface normals used for surface tension forces calculations. The proposed method is tested for a range of static cases, with the preliminary study of a droplet in an air flow as a cherry on top.

2. SPH method for multiphase flows

SPH has been used for fluid dynamics problems for over twenty years, and many articles can introduce the Reader into the method, e.g. the review paper by Price [3] or comprehensive handbook by Violeau [4]. The general idea behind SPH lies in interpolation theory. If we consider any field A defined on the space Ω, its value at point $\mathbf{r}$ can be interpolated from the integral

$$A(\mathbf{r}) \approx \int_{\Omega} A(\mathbf{r}')W(\mathbf{r}-\mathbf{r}', h)d\Omega, \tag{1}$$

where $W(\mathbf{r}, h)$ is the weighting interpolation kernel and h is the so called smoothing length, defining the range of interpolation (i.e. $W(\mathbf{r}, h) \equiv 0$ for $\mathbf{r} \geq h$). The discretisation of Ω into a set of particles of volumes Ω_b changes the integral in Eq. (1) into a summation, yielding basic SPH interpolation given as

$$\langle A \rangle_a = \sum_b A_b W_{ab} \Omega_b, \tag{2}$$

where $W_{ab} = W(\mathbf{r}_a - \mathbf{r}_b, h)$, $A_b = A(\mathbf{r}_b)$ and a, b are indices of the SPH particles. Thanks to the properties of the kernel function the derivatives can be obtained by shifting differential operators from the fields to the function $W(\mathbf{r}, h)$. Approximation of gradient is simply taken as

$$\langle \nabla A \rangle_a = \sum_b A_b \nabla W_{ab} \Omega_b, \tag{3}$$

with higher derivatives obtained in analogous manner. Using this methodology differential equations can be translated into the SPH summation formulae and solved by calculating interactions between particles.

2.1. Governing equations

The set of the governing equations for viscous flow consists of the Navier-Stokes (momentum) equation

$$\frac{d\mathbf{u}}{dt} = -\frac{1}{\varrho}\nabla p + \frac{\mu}{\varrho}\Delta u + \mathbf{a} \tag{4}$$

and the continuity equation

$$\frac{d\varrho}{dt} = -\varrho \nabla \cdot \mathbf{u}, \tag{5}$$

where $\mathbf{u}$ denotes the velocity vector, ϱ is the fluid density, p is the pressure and μ is the dynamic viscosity. In Eq. (4), $\mathbf{a}$ stands for the interfacial term, detailed in Sec. 2.2. Due to the Lagrangian nature of SPH we also include the advection equation for positions and velocities.

Depending on the purpose and assumptions, different SPH formalisms for the fluid flow can be obtained by using Eqs. (2) and (3). In the present study we use a formulation proposed by Hu and Adams [5]. To the best of our experience, their approach is best suited for modelling multiphase flows with large density ratios [6]. Due to limited space we will not present the SPH forms of governing equations, as they can be found in recent (and also decent) paper by Olejnik and Szewc [2]. The key feature of used formulation is replacing Eq. (5) with direct density calculation, i.e.

$$\varrho_a = m_a \sum_b W_{ab}. \tag{6}$$

This allows the density field to be represented only by the spatial distribution of particles and not by their masses. Thanks to this, densities of particles near the interface are not affected by the other fluid, which is important in multiphase flow modelling. In our work we use weakly compressible form of SPH, called WCSPH, hence the set of equations is closed with the equation of state

$$p = \frac{s^2 \varrho_0}{\gamma}\left[\left(\frac{\varrho}{\varrho_0}\right)^{\gamma} - 1\right], \tag{7}$$

where s is the artificial speed of sound, ϱ_0 is the reference (initial) density and γ is a numerical parameter. Values of s and γ are chosen to ensure density fluctuations at level of 1% or below.

2.2. Surface tension model

The influence of surface tension is modelled with Continuum Surface Force method (CSF), originally proposed by [7], with SPH implementation due to [8]. In this approach surface tension forces are converted into a force per unit volume

$$\mathbf{F}_s = \mathbf{f}_s \delta_s = (\sigma \kappa \hat{\mathbf{n}})\, \delta_s, \tag{8}$$

where δ_s is a suitably chosen surface delta function, $\mathbf{f}_s$ is the force per unit area, σ is the surface tension coefficient, κ is the local curvature of the interface and $\hat{\mathbf{n}}$ is the unit vector normal to the interface. Using the so-called color function c (say, $c = 0$ for the first phase and $c = 1$ for the second one) $\hat{\mathbf{n}}$ can be calculated using the formula

$$\hat{\mathbf{n}} = \frac{\mathbf{n}}{|\mathbf{n}|} = \frac{\nabla c}{|\nabla c|}. \tag{9}$$

The local curvature is obtained from

$$\kappa = -\nabla \cdot \hat{\mathbf{n}}. \tag{10}$$

Assuming that $\delta_s = |\mathbf{n}|$, the influence of surface tension can be included in Eq. (4) by adding the term

$$\mathbf{a}_a = \frac{\sigma}{\varrho_a} \kappa_a \mathbf{n}_a. \tag{11}$$

3. Boundary conditions and contact angle modelling

To accurately model wetting phenomena properly defined boundary conditions are of the essence. Lagrangian nature of SPH and use of moving particles as computational nodes makes modelling of solid wall a bit more challenging than in classic Eulerian approaches. The problem was intensively investigated for single phase flows, resulting in many schemes for boundary conditions to choose from. The most basic method is the so called dummy-particles approach, in which solid wall is built from immovable particles. The governing equations are solved for those “dummies”, but their positions and velocities are not evolved in time. This method works well despite its simplicity. However, for multiphase systems a simple question immediately arises – from which phase should the wall be built? The answer is by no means trivial. In the work of Yeganehdoust *et al.* [9] this approach was used for modelling the droplet-wall interaction. The colour function value for the dummies was derived from the temporary position of droplet, to allow for accurate calculation of the interface normals in the vicinity of the triple line. The proposed approach allowed for satisfactory results, but explicit tracking of the relative position of the droplet and the wall is problematic from the algorithmic point of view and limits versatility of the method.

In our work we use the so-called ghost particle approach for solid boundaries. In this method the wall acts as a mirror – each time step the particles close to the wall (at distance less or equal to the interpolation range of the kernel function used) are mirrored onto its opposite side. Their physical properties can be adjusted so that the desired boundary conditions, i.e. no-slip or free-slip, the Neumann b.c. for pressure, are satisfied. Detailed description of this approach can be found in [6] and [10]. In the case of multiphase flows the material properties of mirrored particles – density, viscosity and colour function, are exactly the same as those of parent ones. This is in fact very convenient, since (i) the boundary conditions are being constructed on the fly and there is no need for tracking the phase distribution in the flow; (ii) all summation formulae are calculated accurately without risk of under/overestimation of density or similar error. The main drawback of the ghost particles approach is its applicability only to cuboid domains, however, for the purpose of the basic research this is more than enough.

As for the contact angle calculation we decided to follow the idea used in the Volume of Fluid approach by Šikalo *et al.* [11]. We modify the unit vector normals to the interface near the

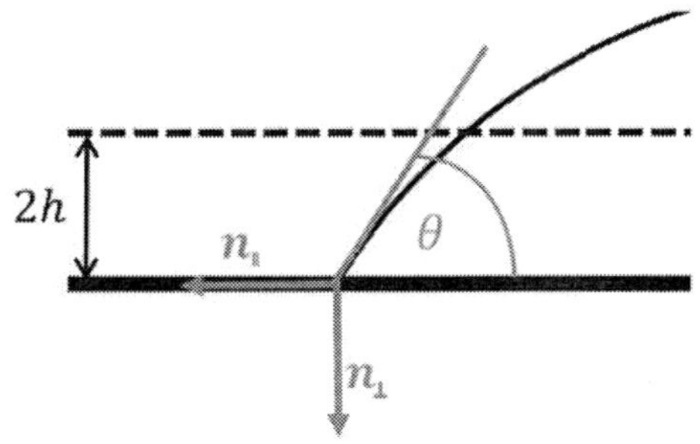

Figure 1. Schematic illustration explaining the way in which interface normals are modified.

triple line as

$$\left[\frac{\nabla c}{|\nabla c|}\right]_{\mathrm{mod}} = \mathbf{n}_{\perp}\cos\theta_D + \mathbf{n}_{\parallel}\sin\theta_D, \tag{12}$$

where θ_D is the desired contact angle whereas $\mathbf{n}_{\perp}$ and $\mathbf{n}_{\parallel}$ are the unit vectors perpendicular and parallel to the wall, see Fig. 1. By "near the triple line" we mean particles that satisfy the following conditions: (i) they are closer than $2h$ from the wall; (ii) they have a non-zero ∇c. The choice of $2h$ is simply the range of the kernel function employed in our calculations. Equation (12) is used for curvature calculation only, hence Eq. (11) becomes for these particles

$$\mathbf{a}_a = \frac{\sigma}{\varrho_a}\left(-\nabla\cdot\left[\frac{\nabla c}{|\nabla c|}\right]_{\mathrm{mod}}\right)\mathbf{n}_a. \tag{13}$$

To avoid problems with a non-smooth transition from the modified to primal vectors normal to the interface, we employ the correction proposed by Breinlinger *et al.* [12]

$$\left[\frac{\nabla c}{|\nabla c|}\right]_{\mathrm{mod}} = \alpha\left(\mathbf{n}_{\perp}\cos\theta_D + \mathbf{n}_{\parallel}\sin\theta_D\right) + (1-\alpha)\frac{\mathbf{n}}{|\mathbf{n}|}, \tag{14}$$

where α is the linear blending function equal 0 on the wall and 1 at the distance $2h$ apart. The described model is very robust, and, as proved in Sec. 4., can yield accurate results.

We also need to stress that within the ghost particles approach framework, the solid wall is not physically present in the calculations. Due to the formulation, the particles are not penetrating it or getting exactly to its position. This means that there are no issues related to the no-slip boundary condition and velocity following from the acceleration computed in Eq. (13).

4. Results

4.1. Static cases

The model was first applied to the 2D static case. The initially hemispherical droplet of liquid was placed on the wall as in the first panel of Fig. 2. The density and viscosity ratios of gas G and liquid L were set to $\varrho_L/\varrho_G = 1000$ and $\mu_L/\mu_G = 100$, respectively. The resolution was set to $h/R = 1/16$ and $h/\Delta r = 2$, where R is the initial radius of the droplet and Δr is the mean spacing between particles. Only surface tension forces were inducing the motion (no gravity or another forcing). The domain was periodic in the left-right direction and high enough to ignore the influence of the top wall and with no-slip b.c. on the bottom. The model was tested for values of θ_D ranging from 30° to 150°. Due to the modification of the interface normals near the triple line and resulting acceleration, the droplet should spread on the wall or minimise its contact area, depending on the value of θ_D. As shown in Figs. 2 and Fig. 3 this is exactly what happens. In both cases the droplet will finally reach the desired shape and the modified vectors

IOP Conf. Series: Journal of Physics: Conf. Series **1101** (2018) 012024 doi:10.1088/1742-6596/1101/1/012024

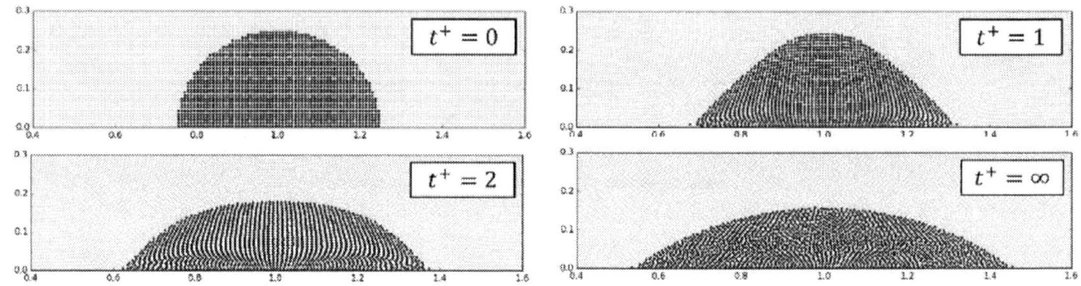

Figure 2. Evolution of the droplet placed on a hydrophilic wall ($\theta_D = 30°$).

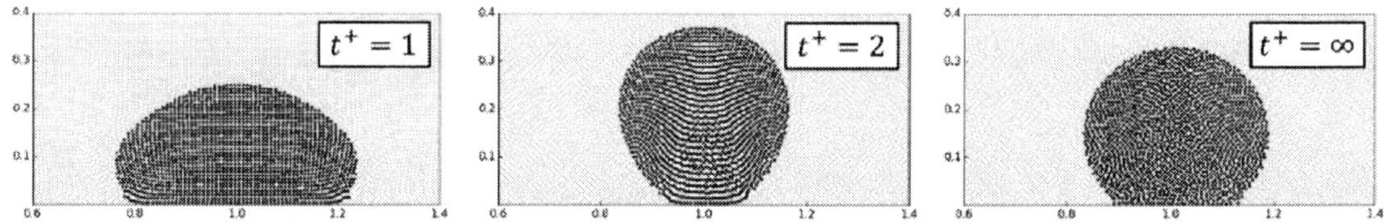

Figure 3. Evolution of the droplet placed on a hydrophobic wall ($\theta_D = 150°$).

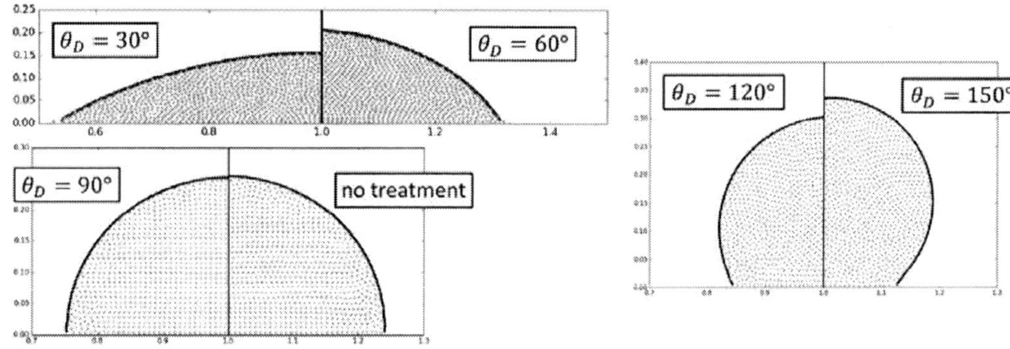

Figure 4. Comparison of the steady-state solution for different values of θ_D. The black line marks isoline of the colour function 0.5.

will be equivalent to those of the CSF model, and the motion will cease. Figure 4 shows a result obtained for the whole range of the θ_D tested. What is curious, but also expected, applying no special treatment yields the same result as imposing $\theta_D = 90°$.

4.2. Dynamic cases

As was already mentioned, the model does not pose any problems related to the contradiction between no-slip b.c. and triple line velocity, hence we applied it to the dynamic case. We modelled exactly the same set-up as in previous Sec. 4.1, but with increased resolution ($h/R = 1/32$ and $h/\Delta r = 2$). Furthermore, the mass force was imposed along the channel's length to induce the motion. The computations were performed for two values of θ_D, namely 30° and 150°, to model the interaction of droplet with entirely different walls. In both cases the creeping air flow forced movement of the droplet. The dimensionless numbers defined with the mean velocity at the channel centreline were $Re < 1$ and $Ca \approx 1$. As expected, for the hydrophilic wall we observed smearing of the droplet, as shown in the upper panel of Fig 5. In

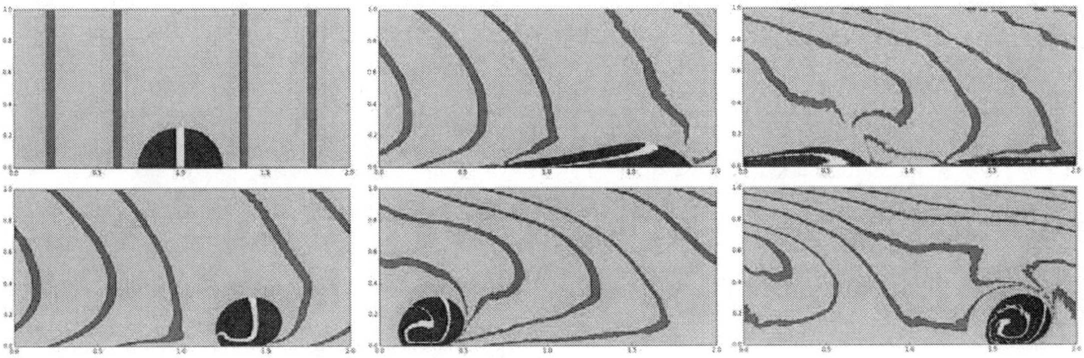

Figure 5. Evolution of the liquid droplet (royal blue and green) in the creeping gas flow (light blue and revolutionary red). Upper panel - hydrophilic wall, bottom - hydrophobic. The gas flow direction is from left to right.

the case of the hydrophobic wall, droplet started to steadily roll, which is also in agreement with physical intuition.

5. Conclusions

In the present study we have presented a robust, yet effective method of the wetting phenomena modelling within the framework of ghost particles boundary conditions for SPH. The model was able to capture wide range ($30° - 150°$) of the desired static contact angle. Simulation of the droplet placed on the wall, subjected to the air flow, yielded satisfactory results. The outlook for future development is promisingly bright. The next step of the research should focus on moving into the 3D set-up (which will be straightforward) and thorough, quantitative validation. Based on the problems and inaccuracies encountered we will further improve the model.

Acknowledgements

This research was partly supported by the National Science Center (2013/11/B/ST8/03818).

References

[1] Szewc K, Pozorski J, Minier J-P 2013 *Int. J. Multiphase Flow* **50** 98
[2] Olejnik M, Szewc K 2018 *J. Theor. App. Mech. (Pol.)* **56** in press
[3] Price DJ 2012 *J. Comp. Phys.* **231** 759
[4] Violeau D 2012 *Fluid Mechanics and the SPH Method* (Oxford: Oxford University Press)
[5] Hu XY, Adams NA 2006 *J. Comp. Phys.* **213** 844
[6] Szewc K, Pozorski J, Minier J-P 2012 *Int. J. Num. Meth. Eng.* **91** 343
[7] Brackbill JU, Kothe DB, Zemach C 1992 *J. Comp. Phys.* **100** 335
[8] Morris J P 2000 *Int. J. Num. Meth. Fluids* **33** 333
[9] Yeganehdoust F, Yaghoubi M, Emdad H, Ordoubadi M 2016 *App. Math. Model.* **40** 8493
[10] Cummins SJ, Rudman M 1999 *J. Comp. Phys.* **152** 584
[11] Šikalo Š, Wilhelm HD, Roisman IV, Jakirlić S, Tropea C 2005 *Phys. Fluids* **17** 062103
[12] Breinlinger T, Polfer P, Hashibon A, Kraft T 2013 *J. Comput. Phys.* **243** 14

IOP Conf. Series: Journal of Physics: Conf. Series **1101** (2018) 012025 doi:10.1088/1742-6596/1101/1/012025

Effect of Hot Spot Location on Flow Structure in Nozzle Guide Vane

Michal Piotrowicz, Pawel Flaszynski, Piotr Doerffer

Institute of Fluid-Flow Machinery, Polish Academy of Sciences,
Fiszera 14, 80-231 Gdansk, Poland

E-mail: michal.piotrowicz@imp.gda.pl

Abstract. Interaction of a combustor and turbine in aircraft engines generates still a lot of open questions. Investigation of combustor and turbine separately have been done for a years by research institutes and aircraft engine companies, but there is lack of knowledge about the interaction effect. In the paper, numerical simulations results for combustor simulator and nozzle guide vane (NGV) of the first turbine stage are presented. No combustion is modelled, but the hot flow at the swirler inlet is applied. Geometry and flow conditions are defined according to the turbine stage designed for experimental investigations in FACTOR (Full Aerothermal Combustor-Turbine Interactions Research) project within a framework of European Commission 7th Framework Programme. The main objective of the presented investigations is an analysis of the location of the high temperature zone upstream of the nozzle guide vane and its effect on the wall temperature. Numerical simulations are carried out by means of Fine/Turbo Numeca with two turbulence models: Spalart-Allmaras and Explicit Algebraic Reynolds Stress Model. As a model validation, a comparison of total pressure, total temperature and flow angles downstream of combustor simulator with experimental data from University of Florence is also presented.

1. Introduction

In order to reduce fuel consumption and improve efficiency, modern gas turbines operate at high speeds and high temperatures. The high uncertainty in predicting the interaction between the combustor and the turbine leads to the application of increased safety margins in the design process. Therefore, a better understanding of interaction of the combustor – turbine is required to maintain service life while optimizing high-pressure turbines (HPT) together with the lean burn combustion chamber.

Currently, quite often the design system is divided into two steps. As the first step, combustion chamber is designed [6,8], resulting in a temperature distribution at the outlet. Then, the turbine is designed according to boundary conditions defined at the combustor – turbine interface [7,10]. A disadvantage of such approach is very weak interaction between the combustor and the turbine, what can be highly important for a durability and performance of the first stage of gas turbine.

There is great experience developed over the years by the aircraft engines companies and different test facilities are available to investigate combustor and turbine separately, but there was no test rig allowing for the interaction analysis. Such gap in the knowledge drove the idea to build a test stand giving opportunity for experimental and numerical investigations of combustor-turbine interaction. The two test sections were designed and built within a framework of the FACTOR (Full Aerothermal

Combustor-Turbine Interactions Research) project: at the German Aerospace Center (Deutsches Zentrum für Luft- und Raumfahrt; DLR) in Goettingen and at the University of Florence. DLR test rig is large facility including combustor simulator and turbine stage, while at the University of Florence a tri sector rig (3 swirlers and 6 nozzle guide vanes downstream) is available.

Last years a few investigations focused on a similar problem are carried out [5,11]. Results presented by authors are related to the interaction between the combustor and nozzle guide vane (NGV), but no effect of hot spot circumferential location was investigated. Below, numerical simulations results for combustor simulator and NGV for the two circumferential locations and the effect of hot spot (high temperarure zone) location on NGV wall temperature is presented. As the first step, a comparison of numerical results with experimental data from University of Florence for combustor without NGV is shown.

2. Geometry

Computational geometry (figure 1) is defined according to the design calculations [2] and final test section configuration.

Validation of the calculation model shown in figure 1 (left) is based on a comparison with measurements [3] carried out by University of Florence (UNIFI).

Further analysis of the influence of the nozzle guide vane (NGV) relative position to the swirler on the flow structure is investigated by means of model shown in figure 1 right.

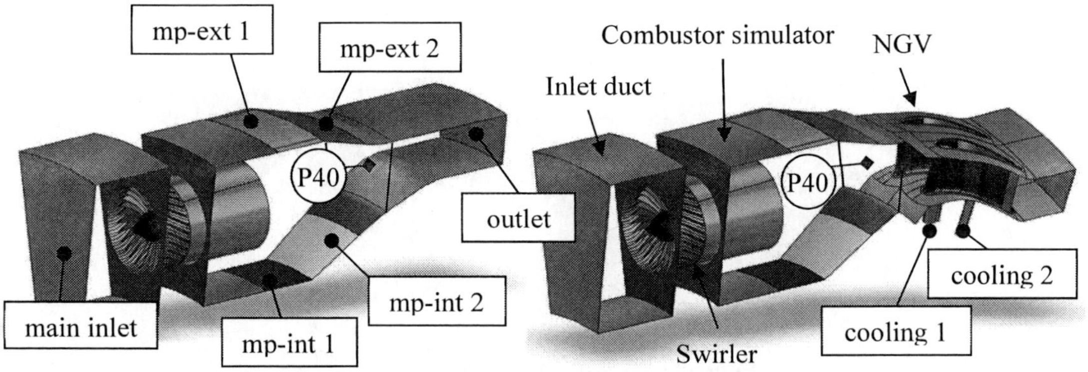

Figure 1. Computational domain left side only combustor, right combustor simulator + NGV

Figure 2 shows two configurations that differ from each other in the position of the NGV relative to the swirler. In the LEC case, the leading edge of NGV 1 is fixed right downstream of the swirler. In the second configuration (PAC) the blade passage is located downstream of the swirler.

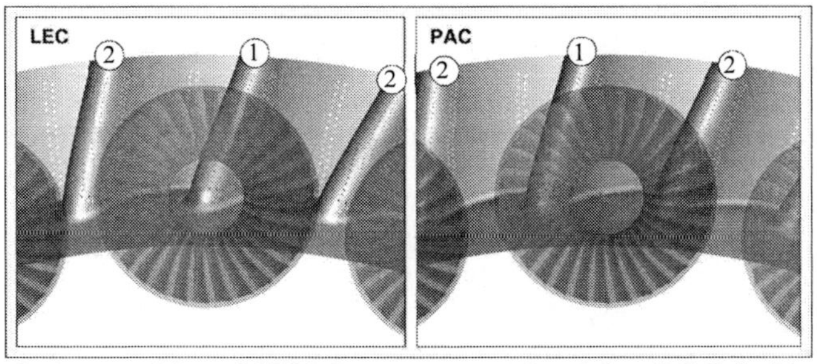

Figure 2. Two relative positions NGV downstream of swirler: LEC (left) and PAC (right).

IOP Conf. Series: Journal of Physics: Conf. Series **1101** (2018) 012025 doi:10.1088/1742-6596/1101/1/012025

3. Numerical model description

Numerical simulations were carried out for the flow domain shown in figure 1 by means of Fine^{TM}/Turbo NUMECA. The flow solver is a structured, three-dimensional density-based, multi-block Reynolds Averaged Navier Stokes code based on a finite volume method. Presented results are obtained for Spalart-Allmaras (SA) and Explicit Algebraic Reynolds Stress Model (EARSM) turbulence models [12, 13]). For spatial discretization (central, 2nd order) with Jameson type artificial dissipation is applied, while for the temporal discretization a four-stage explicit Runge-Kutta scheme is employed. In order to increase the convergence rate the local time-stepping, implicit residual averaging and full multi-grid techniques are used in the solver.

The mesh has been generated with IGG^{TM} and AutoGrid5^{TM}. The mesh size varies from $7.75 \cdot 10^6$ grid cells for the combustor chamber simulator only, up to $33.7 \cdot 10^6$ grid cells for the model including cooled nozzle guide vanes. Full non matching boundary connection (FNMB) is applied at interfaces: inlet duct-swirler, swirler-combustor simulator, combustor simulator – NGV. The mesh is refined close to the solid walls in order to keep y^+ below 1. An example of a mesh for a swirler, combustor and NGV is shown in figure 3.

The unsteady simulations were performed with a time step 5e-05s. The simulation time corresponding to 10 full flow-through the domain is used for the time statistics.

At the main inlet the, mass flow rate conditions is set to ~0.155 kg/s, total temperature is 512K, the turbulence intensity 1% with turbulent/laminar viscosity ratio 10. Cooling of the combustor simulator has been simplified from the flow through the perforated plate to the surface on which the parameters are distributed uniformly. Detailed information are shown in table 1.

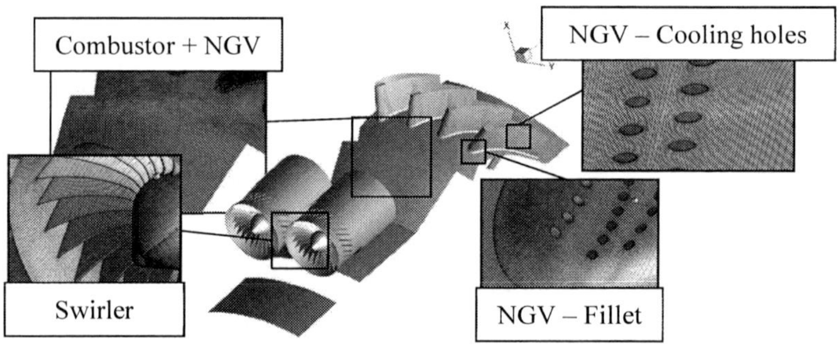

Figure 3. Details of the computational grids

Table 1. Operating conditions at combustion chamber

mass flow main inlet (%)	static temperature [K]	angle/wall [deg]	turbulence intensity (%)	viscosity ratio(-)
15	297.65	60	1	10
6		30		
19		60		
12		30		

For configurations with NGV, additional cooling needs to be considered where mass flow rate for each plenum channel is: cooling 1 = 0.009 kg / s and cooling 2 = 0.0045 kg / s. The temperature, turbulence intensity and viscosity ratio are the same as shown in table 1.

4. Model validation - comparison with measurements

The validation of the numerical model was carried out by means of comparison of experimental data in the plane P40. At this location, the interface between combustor simulator (CS) and nozzle guide vane (NGV) is defined.

IOP Conf. Series: Journal of Physics: Conf. Series **1101** (2018) 012025 doi:10.1088/1742-6596/1101/1/012025

The normalized distribution of total pressure normalised by the mean total pressure is shown in plane 40 (figure 4). Low pressure area can be observed in the middle of the control plane what arises from the wake generated downstream of the swirler. Moving away from the core of the vortex generated by swirler, pressure is increasing reaching maximum closer to the endwalls and then it decreases. The distribution is highly non-uniform and it is driven by the interaction of the swirling flow in combustor middle zone and cooling flow ejected at the upper/lower endwalls.

The EARSM turbulence model better predicts the level of losses both in the vortex generated by swirler and outside. Additionally, it correctly predicts the minimum value as well as a circumferentially averaged results as presented in figure 4 (right). In the case of SA turbulence model, the distribution of these parameters is more uniform which may result from less accurate prediction of swirling flow.

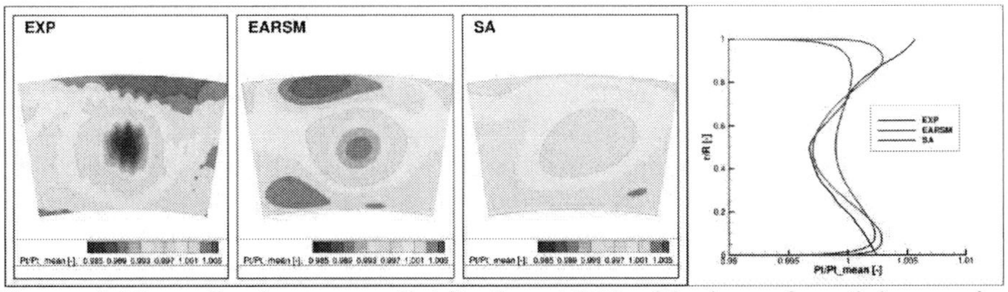

Figure 4. Normalized pressure distribution at P40 with circumferential averaging

The temperature distribution shown in figure 5 has been defined according to the following equation:

$$T_{nd} = \frac{T-\bar{T}}{\bar{T}-T_{cool}} \tag{1}$$

where T - total temperature, $\bar{T}$- averaged total temperature and T_{cool} – total temperature of coolant

The EXP results indicate that this geometry allows to transport "hot-spot" directly to the analyzed plane. It is important effect which needs to be modelled as no combustion is considered and temperature is much lower than in the real combustor. There are two zones where the maximum can be seen (figure 5). The high temperature in the center is related to the hot flow downstream of the swirler, while the zones at the sides/outline result from the presence of neighbouring swirler. It should be remembered that the results for three sectors limited by side walls are presented for EXP, as mentioned in the article describing the experiment [3].

Likewise the EARSM model seems to be closer to measurements and again it is noticeable in the centre of the plane. Additionally this model correctly predicts the areas with local minimum. This may suggest that intensity of mixing inside the combustor simulator is closer to the existing in the test section. Much larger discrepancies are shown by the SA model. It can be observed that the hot air is transported further than in experimental data and its maximum is closer to the hub. It can also be clearly seen that the distribution of the temperature in close proximity to the hub and shroud is much lower than measured. This is confirmed by the circumferential average shown in plot of figure 5.

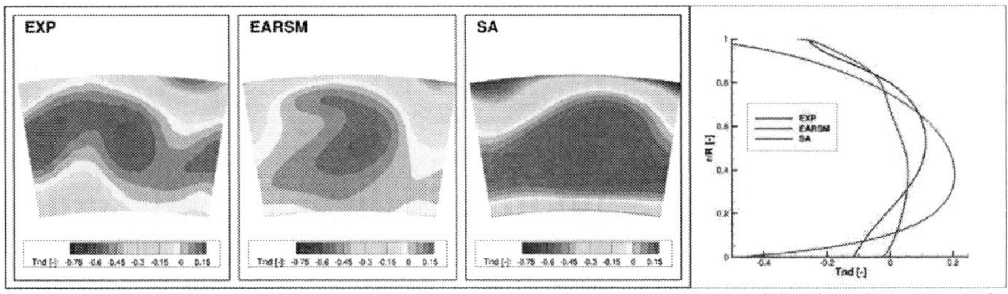

Figure 5. Non dimensional temperature distribution at P40 with circular averaging

IOP Conf. Series: Journal of Physics: Conf. Series **1101** (2018) 012025 doi:10.1088/1742-6596/1101/1/012025

Swirl angle distribution (figure 6) indicates that both turbulence models correctly predicts distribution in the middle of the channel, the differences starts appearing in the zones close hub and shroud. These differences may result from the simplifications of boundary conditions at perforated plates in CS, what is mentioned in section 3.

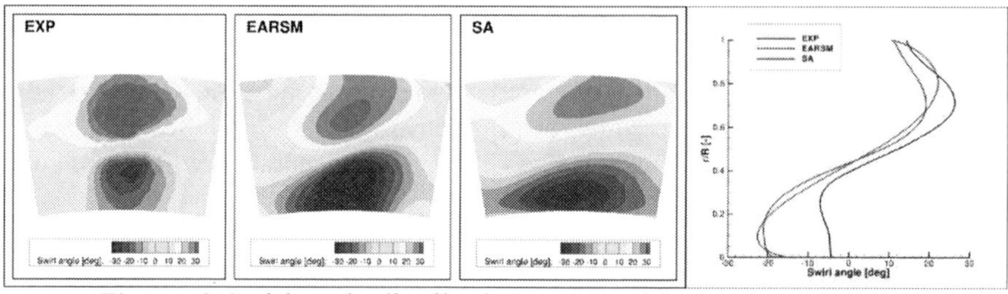

Figure 6. Swirl angle distribution at P40 with circular averaging

Finally, the pitch angle distribution confirms earlier observations that the EARSM model seems to give results closer to experiment. This is indicated by the deviation of the flow figure 7. The areas with the positive (red) and negative (blue) components are compared in size and value in relation to EXP. These differences are not so noticeable when they are averaged circumferentially (plot in figure 7).

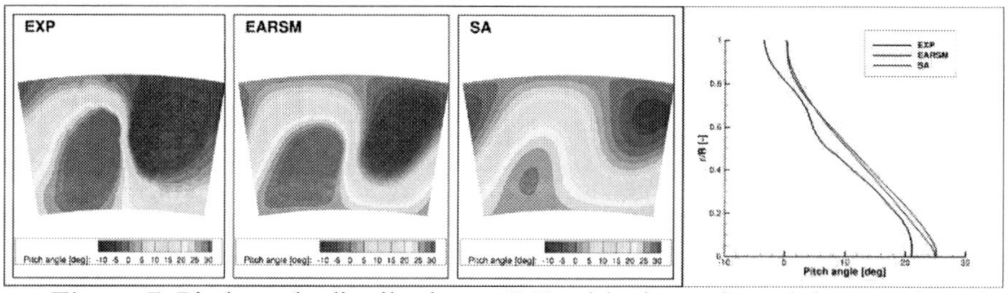

Figure 7. Pitch angle distribution at P40 with circumferential-averaging

The complex geometry of combustor simulator creates a strongly three-dimensional flow structure. Applied simplifications of the existing test section to only one sector (periodic conditions) and replacing perforated plates for CS cooling flow with uniform mass flow distribution does not cause significant discrepancies, as proven above. Satisfactory compliance of EARSM model with the experimental data allows the use this model for analysis of NGV position influence on flow structure in cascade.

5. Hot spot location effect on Nozzle Guide Vane

Results presented above are obtained for combustor simulator without NGV. In such situation a question arises if there is potential effect of NGV located downstream of plane 40 and if this affects flow conditions in plane 40. As shown in figure 8, there is insignificant effect. It should be emphasized that the control plane is very close to NGV, approximately 50% of NGV chord upstream.

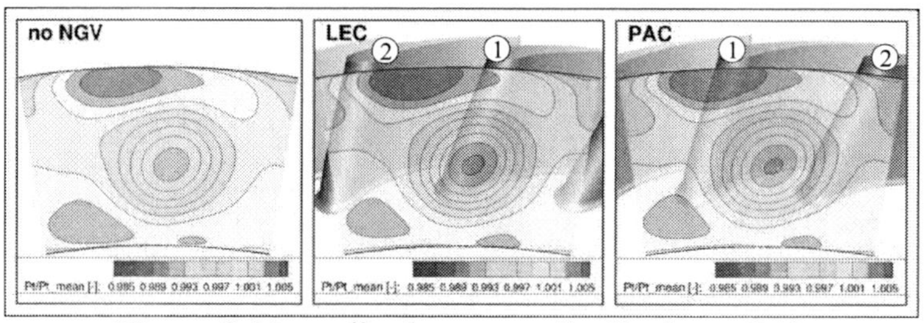

Figure 8. Normalized pressure distribution at P40

IOP Conf. Series: Journal of Physics: Conf. Series **1101** (2018) 012025 doi:10.1088/1742-6596/1101/1/012025

Comparing the non-dimensional temperature distribution in the plane P40 (figure 9), more differences can be seen, nevertheless the main flow features are the same. The similar shape of the high temperature areas is the result of swirl generated by the swirler. The vortex direction is marked with a dashed line in figure 9.

The presence of the blades causes the vortex to be stretched and transported more intensively towards the hub.

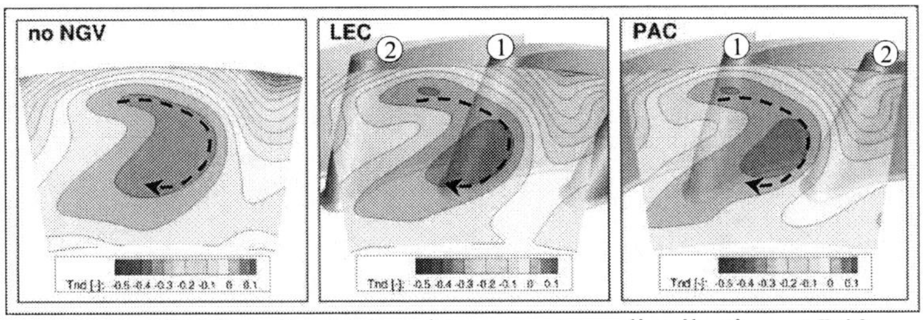

Figure 9. Non-dimensional temperature distribution at P40

The evolution of hot air zone, visible in the plane at leading edge location (figure 10), indicates that the flow core in the case of LEC is separated by the leading edge of NGV 1 into two structures transported at both sides of one NGV. A significant part of the hot stream migrates towards the suction side of the NGV 1. In the PAC configuration the hot stream is in the middle of the blade passage and also affects both NGVs.

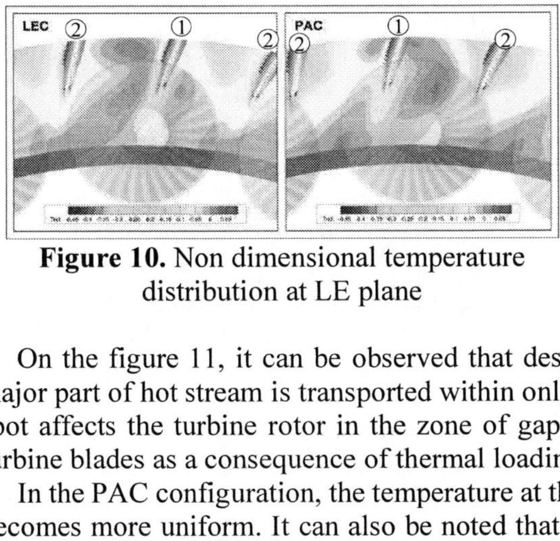

Figure 10. Non dimensional temperature distribution at LE plane

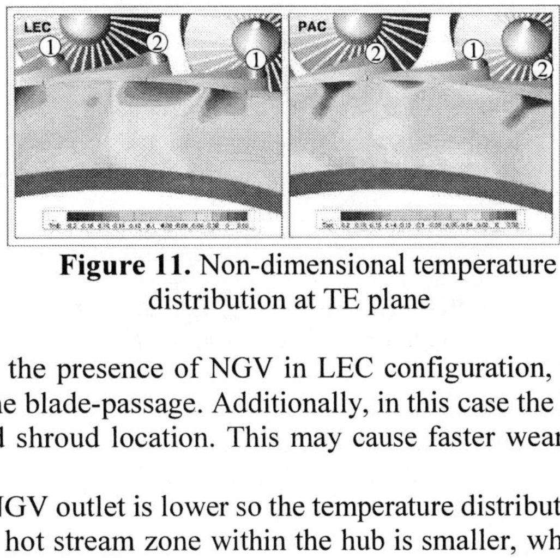

Figure 11. Non-dimensional temperature distribution at TE plane

On the figure 11, it can be observed that despite the presence of NGV in LEC configuration, the major part of hot stream is transported within only one blade-passage. Additionally, in this case the hot spot affects the turbine rotor in the zone of gap and shroud location. This may cause faster wear of turbine blades as a consequence of thermal loading.

In the PAC configuration, the temperature at the NGV outlet is lower so the temperature distribution becomes more uniform. It can also be noted that the hot stream zone within the hub is smaller, which may be important for the turbine durability.

The iso-surface Tnd=0.01 is shown below (figure 12), where one can observed the hot zone penetration in the blade-passage.

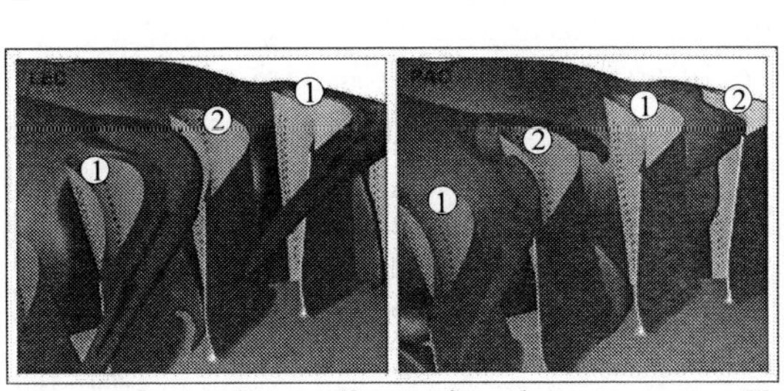

Figure 12. Iso surface of non-dimensional temperature Tnd=0.01 [-]

IOP Conf. Series: Journal of Physics: Conf. Series **1101** (2018) 012025 doi:10.1088/1742-6596/1101/1/012025

The change in the flow structure, forced by the NGV location, directly affects the wall temperature on NGV. In the LEC case, the temperature distribution on the pressure side of both NGVs is very similar (figure 13 and figure 14). Larger differences appear when the area near the NGV 2 hub is compared (figure 13). In case of PAC position, the wall temperature on NGV2 is much higher than on NGV 1. As in the previous configuration (LEC), it can also be noticed that higher temperature zone is located close to NGV 2 (PAC) hub region.

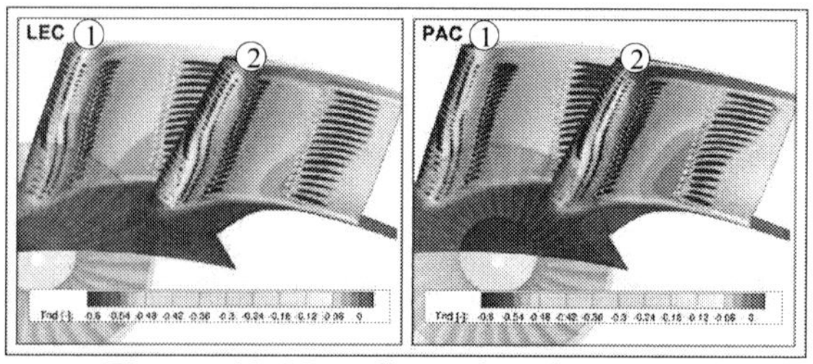

Figure 13. Non dimensional temperature distribution at NGV 1 and NGV 2

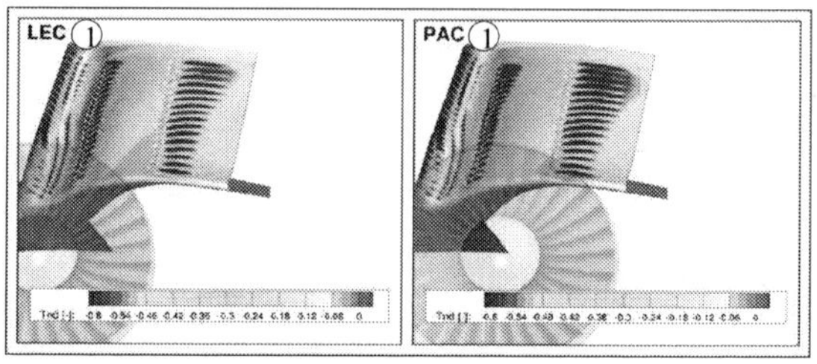

Figure 14. Non dimensional temperature distribution at NGV 1

Comparison of the temperature distribution on the suction side (figure 15) indicates that on NGV 1 for the LEC configuration, located downstream of the swirler, temperature is higher than for the PAC configuration. On the suction side of NGV 2 differences are lower, but slightly higher temperature is for PAC case (figure 16).

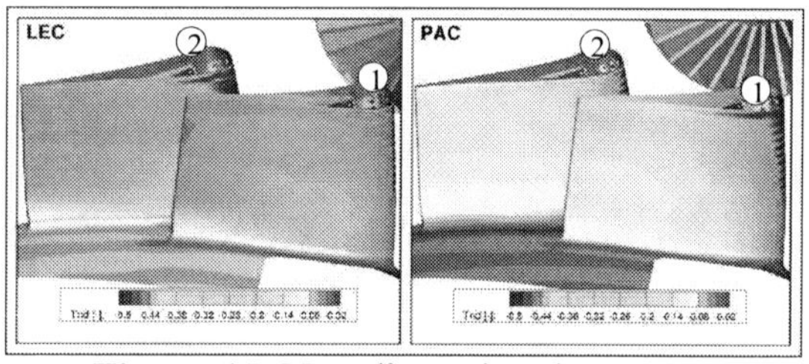

Figure 15. Non dimensional temperature distribution at NGV 1 and NGV 2

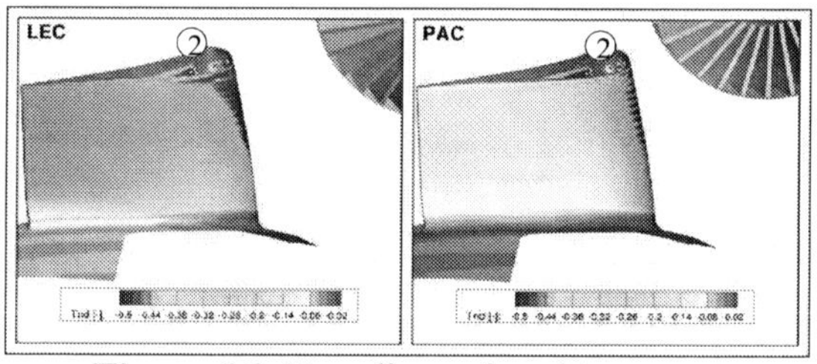

Figure 16. Non dimensional temperature distribution at NGV 2

6. Conclusions

Comparison of numerical simulations results against experimental data in plane upstream of nozzle guide vane allows to validate the model applied to investigation of flow structure for selected configurations. Both turbulence models (Spalart-Allmaras and Explicit Algebraic Reynolds Stress Model) show good prediction of the total pressure and flow angles. Some differences in the predicted and measured temperature are shown, but in both cases the hot spot existence is clearly visible. Nevertheless, better prediction of the total temperature upstream of the nozzle guide vane is obtained by EARSM model, so this one was selected for flow structure investigations and analysis of wall temperature in turbine cascade.

Circumferential position of guide vanes downstream of the swirlers strongly affects the wall temperature and flow conditions upstream of rotor cascade. Effect of the hot spot location upstream the leading edge or upstream the passage is presented on locally higher temperature on both sides of the neighbouring vanes. It influences on thermal loading of turbine stator, but has also strong impact on temperature level and flow non-uniformity at turbine rotor inlet.

Future investigations will be focused on unsteady simulations for the investigated configurations including rotor cascade.

Acknowledgements.
The presented analysis are carried within a framework of EU FP7 FACTOR project and authors are particularly grateful to Tommaso Bacci from University of Florence for the sharing of experimental results. Numerical simulations were carried out in Computational Centre of TASK (Trojmiejska Akademicka Siec Komputerowa) and supported by PL-Grid Infrastructure.

References

[1] Barringer M, Thole K and Polanka, M 2004 Developing a Combustor Simulator for Investigating High Pressure Turbine Aerodynamics and Heat Transfer, *Proceedings of ASME Turbo Expo 2004*: Power for Land, Sea, and Air, June 14-17, 2004 Vienna, Austria, **GT2004-53613**
[2] Flaszynski P, Piotrowicz M 2017 An investigation of hot spot location at turbine inlet – numerical simulations for factor project test rig, *Transactions of the Institute of Fluid-Flow Machinery*, No. **135**, 2017, 101–115
[3] Andreini A, Bacci T, Insinna M, Mazzei L and Salvadori S 2016 Hybrid RANS-LES Modeling of the Aero-Thermal Field in an Annular Hot Streak Generator for the Study of Combustor-Turbine Interaction, *Proceedings of ASME Turbo Expo 2016*: Turbine Technical Conference And Exposition, June 13-17, Seoul, Soutch Korea, **GT2016-56583**
[4] Tsao J, and Lin C 1999 Reynolds stress modelling of jet and swirl interaction inside a gas turbine combustor, *International Journal for Numerical Methods in Fluids*, vol. **29**, 1999, pp. 451–464.
[5] Klapdor E 2010 Simulation of Combustor - Turbine Interaction in a Jet Engine, *PhD thesis*, TU Darmstadt
[6] Qureshi I and Povey T 2011 A combustor-representative swirl simulator for a transonic turbine research facility, Proceedings of the Institution of Mechanical Engineers, Part G: *Journal of Aerospace Engineering*, vol. **225**, 2011, pp. 737–748.
[7] Rahim A, Khanal B, Romero E and He L 2015 Effect of NGV Lean Under Influence of Inlet Temperature Traverse, *Journal of Turbomachinery*, vol. **136**, 2013, pp. 1–12.
[8] Cresci I, Ireland P, Bacic M, Tibbott I and Rawlinson A 2015 Realistic Velocity and Turbulence Intensity Profiles At the Combustor-Turbine Interaction (Cti) Plane in a Nozzle Guide Vane Test Facility, European Turbomachinery Conference, 2015, pp. 1–11.
[9] Koupper C, Bonneau G, Gicquel L and Duchaine F 2016 Large eddy simulations of the combustor turbine interface: study of the potential and clocking effects, *Proceedings of ASME Turbo Expo 2016*: Turbine Technical Conference And Exposition, June 13-17, 2016, Seoul, Soutch Korea, **GT2016-56443**
[10] Barigozzi G, Mosconi S, Perdichizzi A and Ravelli S 2017 The Effect of Hot Streaks on a High Pressure Turbine Vane Cascade with Showerhead Film Cooling, *International Journal of Turbomachinery, Propulsion and Power*, vol. **2**, 2017, p. 15.
[11] Legrenzi P A 2017 Coupled CFD Approach for Combustor-Turbine Interaction, *PhD thesis*, Loughborough University
[12] Spalart P and Allmaras S 1992 A One-Equation Turbulence Model for Aerodynamic Flows, Proc. of 30th Aerospace Sciences Meeting and Exhibit
[13] Menter F, Garbaruk A and Egorov Y 2012 Explicit algebraic reynolds stress models for anisotropic wall-bounded flows, *Progress in Flight Physics* **3** (2012) 89-104

IOP Conf. Series: Journal of Physics: Conf. Series **1101** (2018) 012026 doi:10.1088/1742-6596/1101/1/012026

On the 3D structure of the flow-field in the vicinity of inclined plate

P Prochazka[1] and V Uruba[1,2] and V Skala[1]

[1] Institute of Thermomechanics, Czech Academy of Sciences, Dolejškova 5, Praha 8, Czech Republic
[2] University of West Bohemia, Faculty of Mechanical Engineering, Department of Power System Engineering, Universitní 8, Plzeň, Czech Republic

prochap@it.cas.cz

Abstract. The motivation of the presented study is supporting new ideas about principle of flight by Hoffman and Johnson, see [1]. The new hypothesis of physical mechanism of flight relies on existence of streamwise vortical structures on the suction side of the airfoil and within its wake. The vortices origin is supposed to be the instability of the boundary layer subjected to adverse pressure gradient. The vortices are of highly dynamical nature, changing theirs position, size and other parameters in time very rapidly. For this experiment the simplest airfoil possible was chosen represented by a flat plate in uniform flow and moderate angle of attack. In the suggested paper detailed measurement of the zone of interest will be carried out using stereo PIV method. System of measuring planes perpendicular to the flow will be explored. POD analysis is to be utilized.

1. Introduction

The generation of the lift seems to be resolved issue in the community of people dealing with the aerodynamics. Nevertheless, the traditional Newton mechanics did not give an answer how the lift (and also drag) is created. For example, potential flow for incompressible flow gives zero drag and also lift (D'Alembert's Paradox). Kutta-Zukovskij equation uses an auxiliary variable (circulation) to simulate the lift. The circulation is however unphysical and it was not observed in the nature. To uncover the secret of flight D'Alembert's Paradox should be resolved. The true physical principal of lift and flight has been investigated (experimentally and numerically) by authors Hoffman and Johnsson [1, 2]. They formulated some new ideas - the downwash effect is a necessary condition to generate the lift and the downwash is coming from a low-pressure rotational slip separation. So the main source of the lift generation should be the existence of the streamwise vortical structures located above the wing and in the wake behind it. The streamwise vorticity is presented as a consequence of the natural instabilities of the boundary layer developing under adverse pressure gradient on the airfoil suction side. Some experiments of flow instabilities and their effects were presented recently [3, 4].

This experimental study is dealing with streamwise-oriented vortical structures investigated in parallel planes around and behind the simplest case of the airfoil – flat plate with semi-cylindrical edges. Previous study within this research were oriented both on streamwise vorticity behind the plate [5] and on the investigation of the boundary layer in suction side [6]. Presented study focuses on detailed investigation in several parallel planes perpendicular to the incoming flow using stereo PIV technique (2D3C approach). The flow field is studied not only in the wake and suction side but also the pressure side is examined. The downwash effect is fully detectable as well as streamwise vortices which are characterized by highly dynamical nature, rapid changes of their positions, size and other parameters. The dynamic flow is decomposed using POD (Proper Orthogonal Decomposition) into most dominant flow structures taking effect on this phenomenon.

This article contains preliminary results from just one type of measuring planes (fig. 1). Another articles will be presented concurrently and in the future for parallel planes with the wing surface and also for 2D3C vector field in the perpendicular planes to currently one used.

2. Experiment description

2.1 Model geometry

The test model of simple flat plate with semi-cylindrical leading and trailing edges was placed in a uniform low turbulence stream. The stream was produced using blow-down wind tunnel situated in the Laboratory of turbulence and shear layers. The distribution of velocity was perfect uniform, intensity of turbulence was under 0,2% and the flow was perfect steady in time for the incoming velocity 5 m/s. The plate was inclined at the angle of attack 7° with respect to the channel flow. (The pressure side of the profile was measured for AoA -7° due to technical reasons). The span dimension of the airfoil is 450 mm, the chord length is c = 100 mm and the thickness is 2 mm. Corresponding Reynolds number is about 33 000.

The schematic view of the experimental layout is plotted in figure 1. There is a rotatably mounted wing just behind the wind tunnel nozzle with cross-section dimension 250 x 250 mm. The Cartesian coordinate system was introduced so that the origin was located in the middle of the trailing edge. The x- and y-axis are in the plane of measurement and z-axis is oriented perpendicular to the FoV (field of view) with the direction of stream. There are three green parallel planes in the airfoil wake at dimensionless positions z/c =0.5, 1 and 1.5. Finally, twenty distinct measuring planes (7 upstream the trailing edge and 13 downstream) were investigated with the distance between two adjacent planes 10 mm. This gave us quite good image about flow field topology of flow domain. The trailing edge position is characterized by the rate z/c = 0.

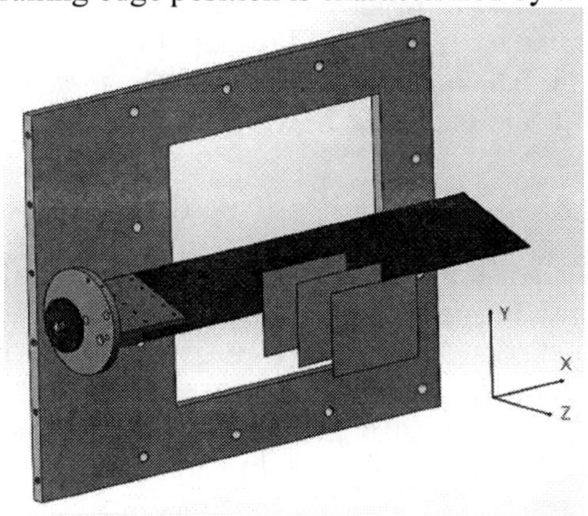

Figure 1. The setup of inclined plate and wind tunnel orifice, z/c = 0.5, 1 and 1.5.

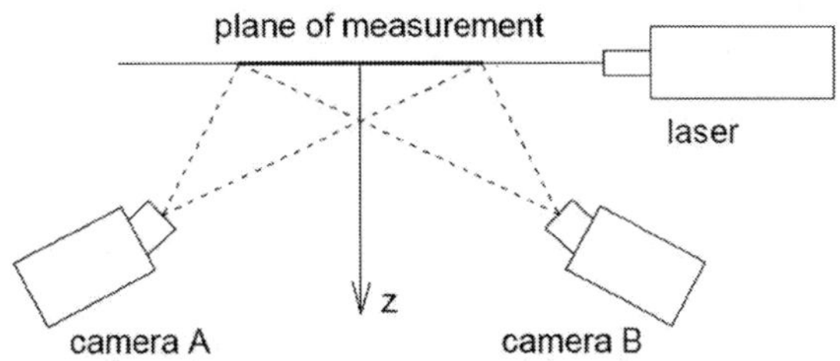

Figure 2. Stereo PIV layout.

2.2 Measurement technique

All three velocity component of each individual measuring planes were detected using stereoscopic time-resolved PIV technique. The measuring apparatus is from Dantec company. The laser New Wave Pegasus, Nd:YLF double-head, was used to illuminate the safex particles. It emits the coherent light with wavelength of 527 nm, maximal frequency is 10 kHz, shot energy is 10 mJ (for 1 kHz) and corresponding power is 10 W per head. Two CMOS cameras NanoSense MKIII with Scheinpflug mounting of lenses were used to capture the tracing particles distributed in the flow. Each camera has resolution of 1280 x 512 pixels, maximal corresponding acquisition frequency is 512 double-snaps per second. The dimensions of FoV was 100 x 60 mm. The data acquisition was conducted in Dynamic Studio software ver. 3.4. Preceding calibration was performed using standard calibration target 100 x 100 mm. The shifting in z-direction was precisely performed by stepper motor in five positions (-2, -1, 0, 1, 2 mm). The calibration (Pinhole camera model) error was set to approx. 0,2 pixels. The 3-component vel. vector field was consequently evaluated and each plane was created by 99x61 vectors. Further analysis was performed in newer version of this software 6.2. The schematic view of stereo PIV layout is in figure 2.

2.3 POD analysis

The typical way, how to apply decomposition methods POD is on two-dimensional dataset. In the presented article, the POD will be applied for three-component vector fields gained from stereo PIV anemometry. POD analysis is used here to extract the most energetic dominant coherent structures presented in the random turbulent flow. This analysis is coming from the vel. fluctuations which means that the mean value is subtracted from all snapshots. The number of resulting orthogonal topological POD modes is equal to number of all PIV snaps. The detailed information about POD calculation can be seen e.g. in [7]. The result is a set of POD mode ordered according to the relative kinetic energy content. Low order modes (high-energy) contain the most dominant flow structures. These ones will be investigated above, below and behind the airfoil as well as their evolution in longitudinal direction. This research was dealing with statistical quantities evaluation as well as POD modes. For this purpose, the acquisition frequency of 100 Hz was set as desirable.

3. Results

The presented data were acquired by the TR-PIV system and will be mainly presented in the form of vectors and scalar distribution in the measuring plane xy perpendicular to the channel flow with direction z. Previously, some preliminary investigation of isolated planes behind the plate were acquired. The area above and below the plate surface (suction and pressure side) are emphasized in this article. Generally in this paper, black vectors represent the velocity components projected into the measuring plane xy. The color represents scalar value, e.g. streamwise vel. component W or turbulent kinetic energy (TKE). The black line represents the position of trailing edge (or its projection in the incoming flow direction) and dashed line is used to emphasize the projection of the trailing edge in the direction of the inclined plate (7°). Black streamlines were added to make the instantaneous vortices visible and green lines were used to plot the z-vorticity (ω_z) in the case of POD flow topology.

The results are to be presented in dimensionless form. All coordinates are divided by the chord length c = 100 mm. The velocity magnitude is normalized using incoming flow velocity U_e = 5 m/s. Turbulent kinetic energy is normalized by U_e^2.

3.1 Instantaneous velocity

There are presented three distinct measuring planes z/c = -0.5, 0 and 0.5 (figure 3) to see how the instantaneous streamwise vortices are evolving along streamwise coordinate z. While the pressure side more or less does not show any strong perturbations, the suction side is full of vortices already in the middle of the plate chord (z/c = -0.5). These PIV snapshots show that these structures are irregular, they are changing very rapid in time and in space (they are not regular along x-axis unlike mean image). The vortices are extending in their size in downstream direction (z/c = 0) and merging with each other which leads to decreasing of their number and strength(z/c = 0.5). The streamlines clearly show downwash effect, the wake position (minimal velocities – blue and green color) is located a little above the dashed projected line. There is impossible to follow the structures snapshot per snapshot with rather low acquisition frequency 100 Hz. That is why, further investigation of these vortices will be provided by POD decomposition.

3.2 Time-mean velocity

Twenty measuring planes were investigated in total (seven of them above and below the profile) which gave us very detailed information about mean flow field topology around the inclined flat plate with AoA 7°. The statistical values were calculated from approx. 1600 PIV snaps measured in 16 seconds. There are plotted isolines of velocity magnitude (70% U_e, 80% U_e and 90% U_e) in figure 4. The flat plate is displayed in its whole span. The origin of coordinate system is located in the middle of trailing edge. A different representation of acquired results provides figure 5. There one can see all measuring planes and three different distribution of scalar value (velocity modulus, TKE and V-component). Notice, that the distribution is not dependent on x-axis. The flow seems to be 2D from these results. However inst. images have revealed fully three-dimensional behaviour of the flow. Maximal velocities (higher than 100% U_e) are located just behind the leading edge (very first plane) - a small separation bubble was detected further upstream during other experiments. This clear velocity overshoot is locally oriented, further downstream the maximal velocities does not exceed U_e. The

IOP Conf. Series: Journal of Physics: Conf. Series **1101** (2018) 012026 doi:10.1088/1742-6596/1101/1/012026

formation of wake region is clearly seen as well as downwash angle. Maximal TKE values, which characterize locally dynamic activity, can be observed above the plate. The fluctuation activity takes a half value in the wake region. Notice, the TKE is almost negligible in the region of pressure side.

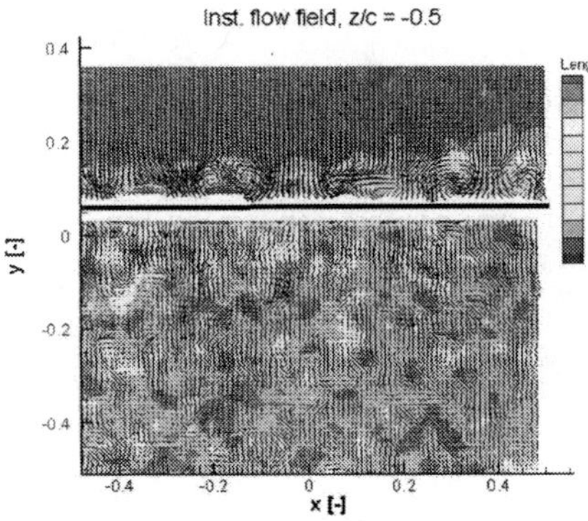

Figure 3a. Instantaneous vel. vector map for plane z/c = -0.5, distribution of velocity length.

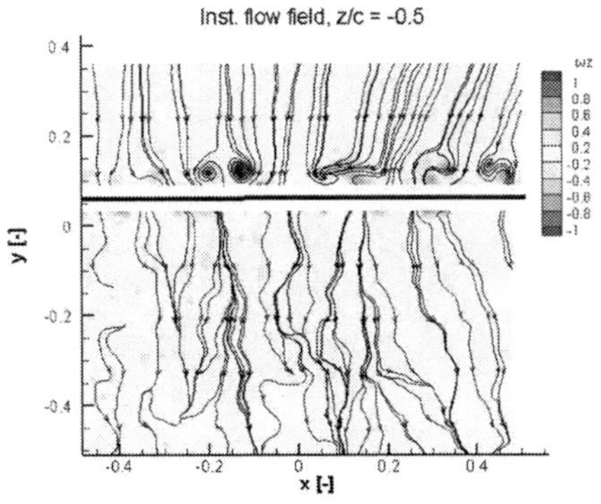

Figure 3b. Instantaneous vel. vector lines for plane z/c = -0.5, distribution of vorticity.

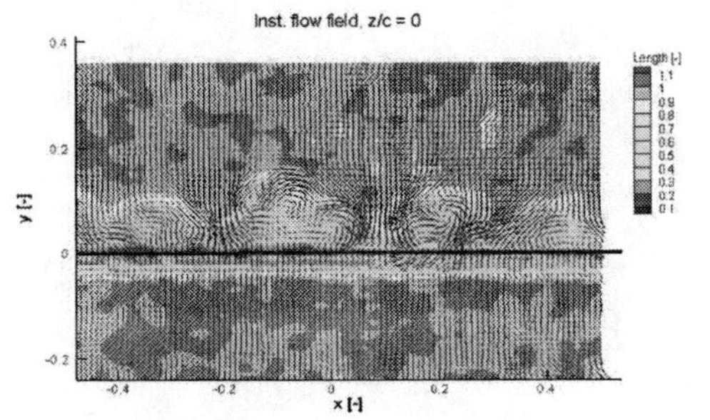

Figure 3c. Instantaneous vel. vector map for plane z/c = 0, distribution of velocity length.

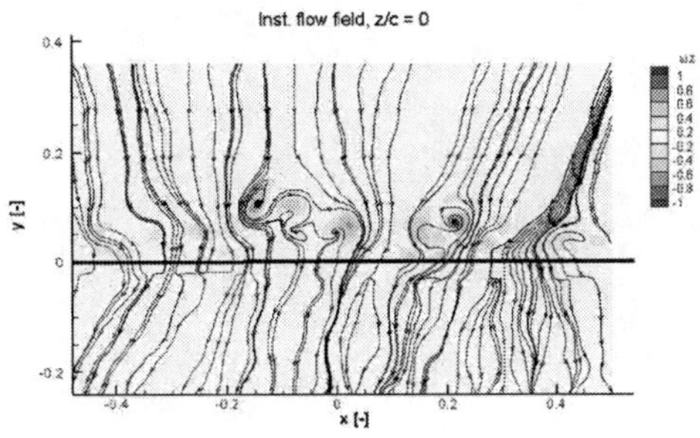

Figure 3d. Instantaneous vel. vector lines for plane z/c = 0, distribution of vorticity.

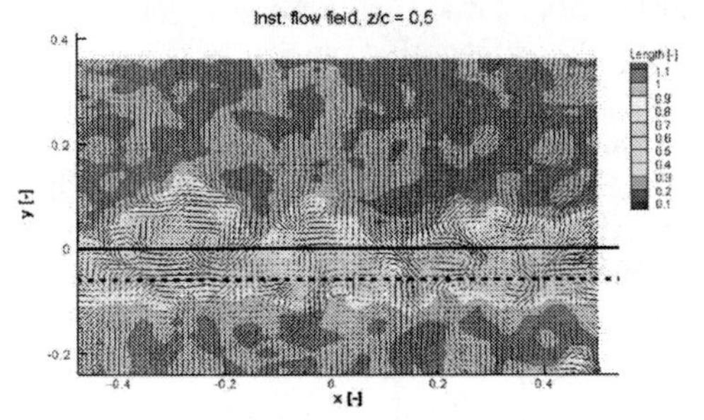

Figure 3e. Instantaneous vel. vector map for plane z/c = 0.5, distribution of velocity length.

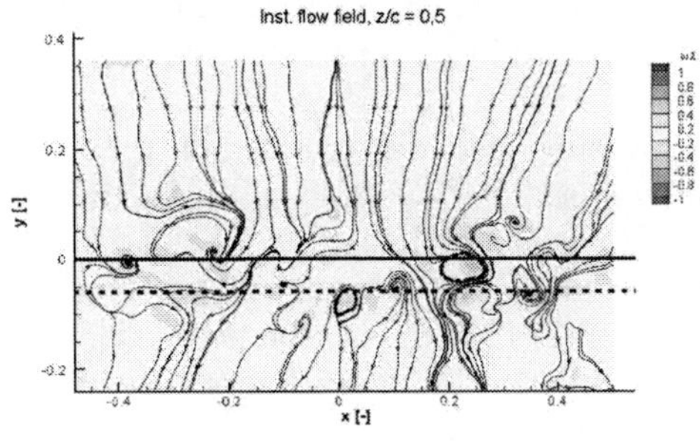

Figure 3f. Instantaneous vel. vector lines for plane z/c = 0.5, distribution of vorticity.

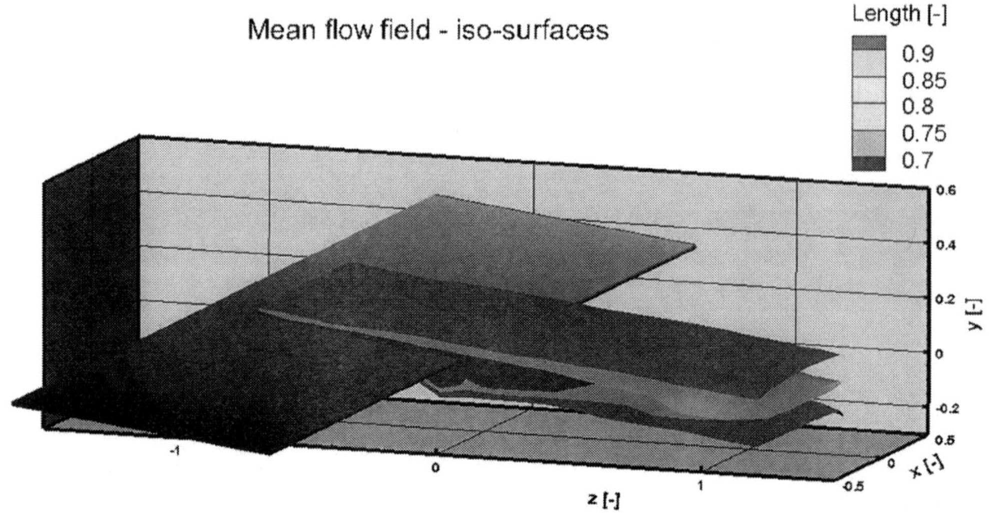

Figure 4. Iso-surfaces of velocity length, time-mean flow field behind inclined plate.

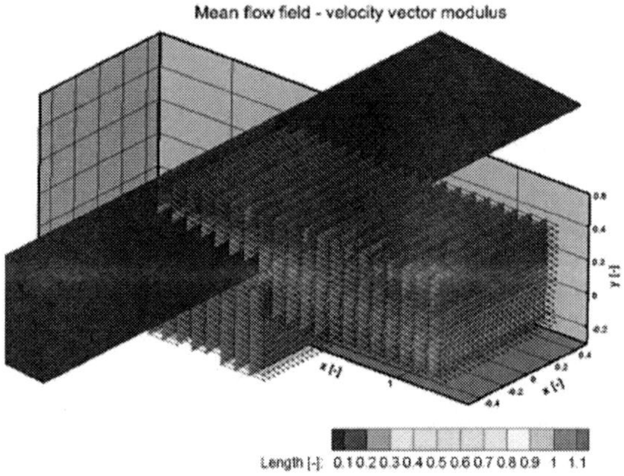

Figure 5a. Time-mean flow, distribution of velocity length in 20 consecutive planes.

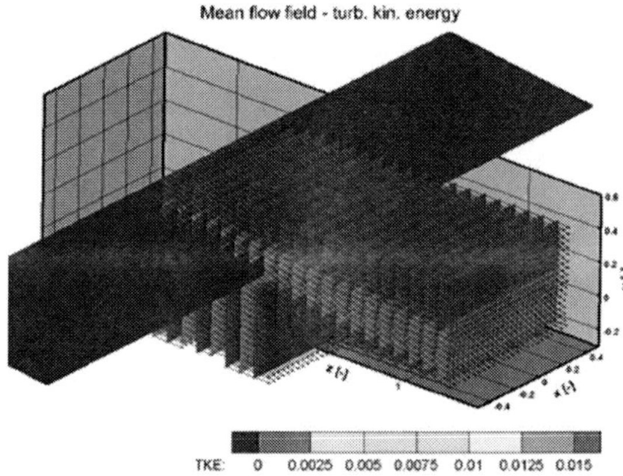

Figure 5b. Time-mean flow, distribution of TKE in 20 consecutive planes.

The pressure side did not demonstrate any coherent structures. On the other hand, the velocity evolution is slightly dependent on z-axis (figures 6a,b). The flow is decelerated a little to 90% U_e (z/c = -0.7) to be re-accelerated more close to the trailing edge. An obvious increase of the boundary layer thickness along z-coordinate can be seen in suction side. It was not possible to measure so close to the lower surface from some technical reasons nevertheless the boundary layer of pressure side does not reach such extension in thickness. The maximal fluctuation activity at 50% of chord length is situated close to the upper surface (y = 0.12). No velocity perturbations were revealed for pressure side. Figure 6d depicts another mean flow field evolution further downstream in the wake (z/c = 0). The maximal TKE values are locates still slightly above the surface at the plane of the leading edge. The position of the wake center is shifted below with increasing distance from trailing edge.

The minimum is located near the dashed line (this is valid for all investigated planes). This implies that the downwash angle should be close to the AoA. This is valid but only for narrow region inside the wake. Figure 7 shows the distribution of v-velocity component in all domain. The green color corresponds to approx. 10% of U_e and blue is even higher - 15%. Biggest value (more than 7°) is above the plate and near the leading edge. On the contrary the smallest value was detected in the pressure region where the value is 5°. In the wake in substantial distance from the trailing edge, the downwash effect becomes weaker. The downwash effect is nicely seen in figure 8 where the zy plane is reconstructed.

IOP Conf. Series: Journal of Physics: Conf. Series **1101** (2018) 012026 doi:10.1088/1742-6596/1101/1/012026

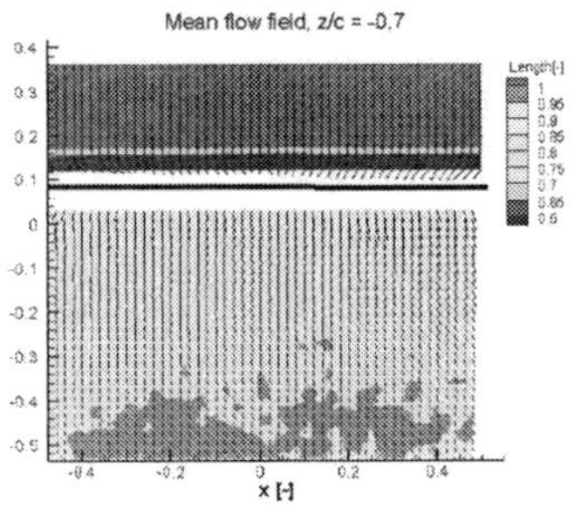

Figure 6a. Velocity distribution for suction and pressure side in z/c = -0.7.

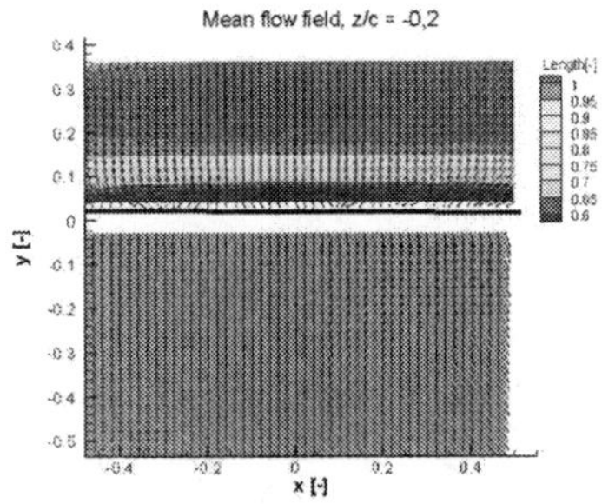

Figure 6b. Velocity distribution for suction and pressure side in z/c = -0.2.

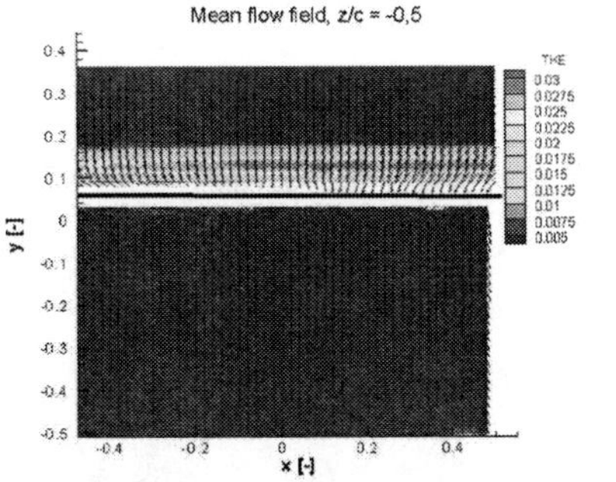

Figure 6c. Distribution of TKE for both side in z/c = -0.5.

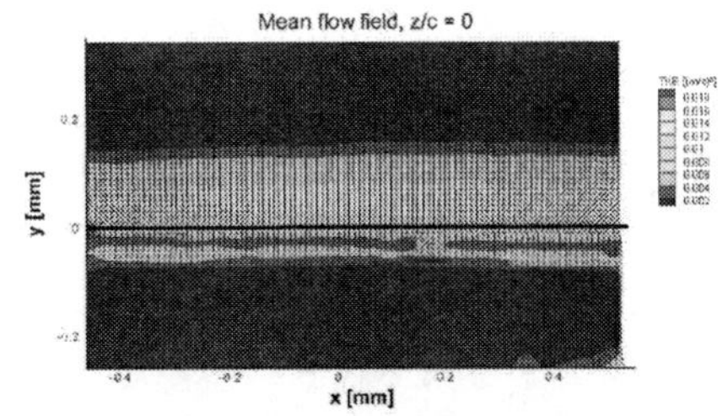

Figure 6d. Distribution of TKE for both side in z/c = 0.

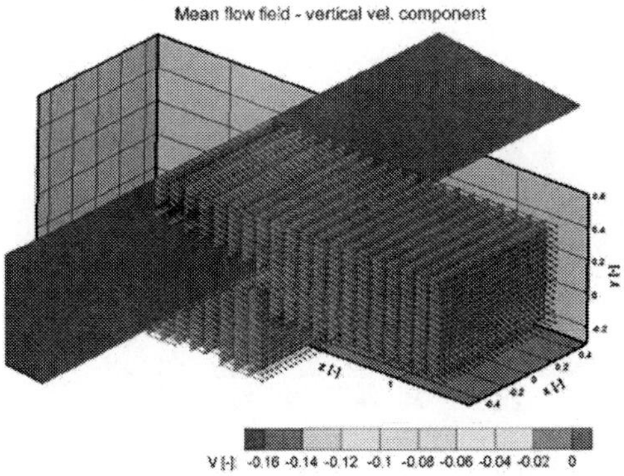

Figure 7. Time-mean flow, distribution of vertical velocity in 20 consecutive planes.

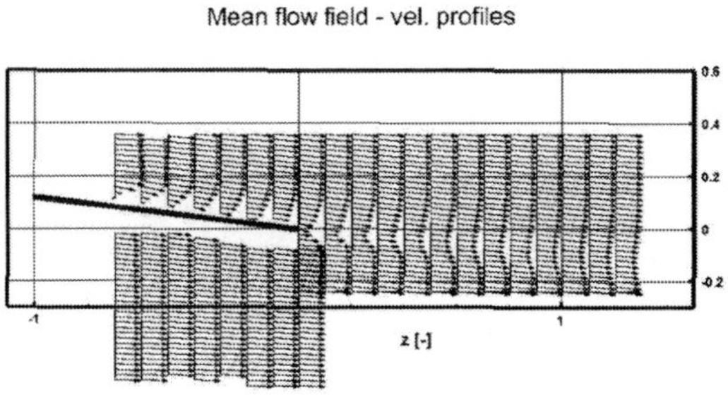

Figure 8. The effect of downwash to vector field.

3.3 POD modes

POD is a great tool to identify most relevant vortex structures from turbulent flow field containing significant coherent structures. Although the highest-order modes represent very often only noise, all 1600 modes were valuated. It is possible to reconstruct the whole flow field around the airfoil using modes recombination. The second POD mode, which is characterized by the existence of one pair of contra-rotating vortices, was selected to reconstruct one of the possible forms of the flow field. The figure 9 shows two vortex cores (detected by negative and positive vorticity). The inclination due to the flate plate is about 1° smaller. This effect can also be seen from other types of modes, e.g. at the plane z/c = 0.8. Figure 10 is plotted for other two types – 7. and 16. modes and they are characterized by the existence of three pairs and several pairs of streamwise vortices. The location of them is situated slightly above dashed line (the projection of trailing edge in the incoming flow direction). Moreover the second plotted mode contents the structures that are also placed below. The green lines

IOP Conf. Series: Journal of Physics: Conf. Series **1101** (2018) 012026 doi:10.1088/1742-6596/1101/1/012026

are isolines of vorticity (positive or negative) and the space between two adjacent vortices are filled by the accelerating or decelerating fluid (streaks). The vortices are located in the centers of isolines.

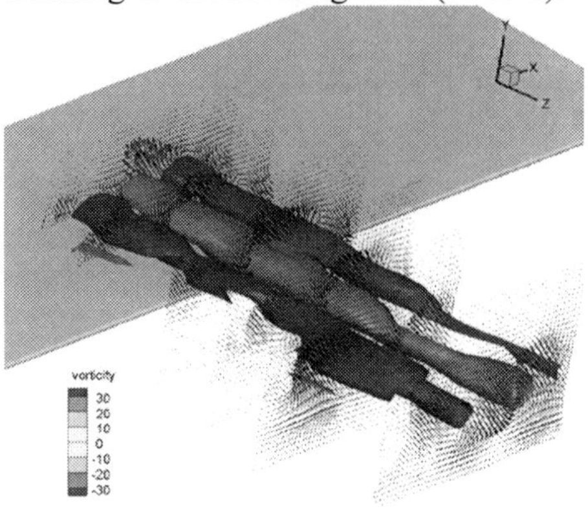

Figure 9. The reconstruction of two counter-rotating streamwise-oriented vortices in the airfoil wake. One of the conceivable form of the flow field.

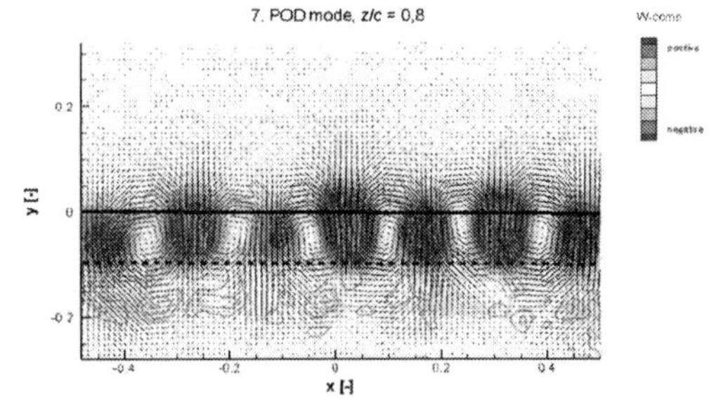

Figure 10a. 7th POD mode in z/c = 0.8 in the form of three pairs of counter-rotating vortices.

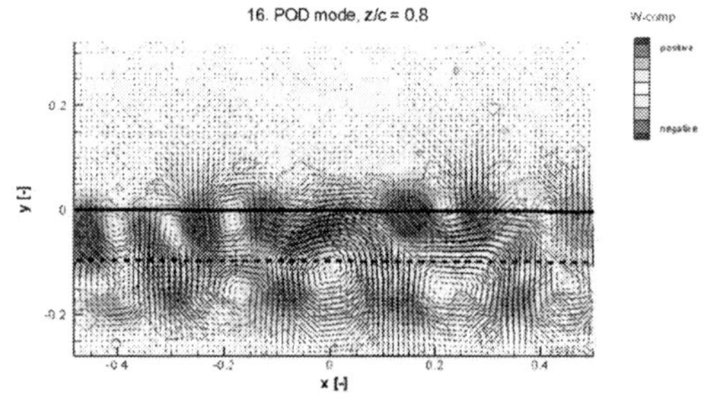

Figure 10b. 16th POD mode in z/c = 0.8 in much complex form, vortices also below dashed line.

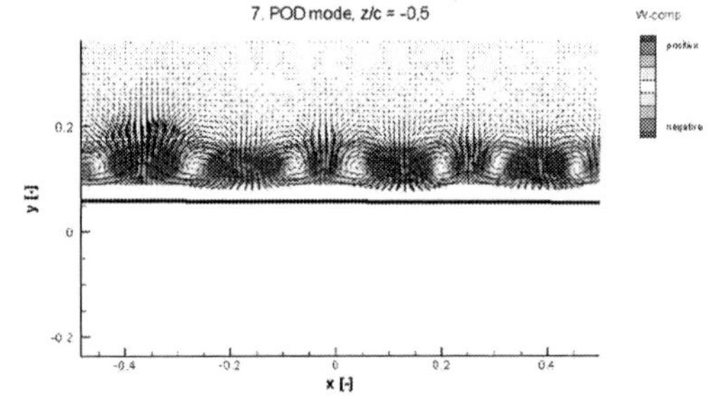

Figure 11a. 7th POD mode in z/c = -0.5 for suction side.

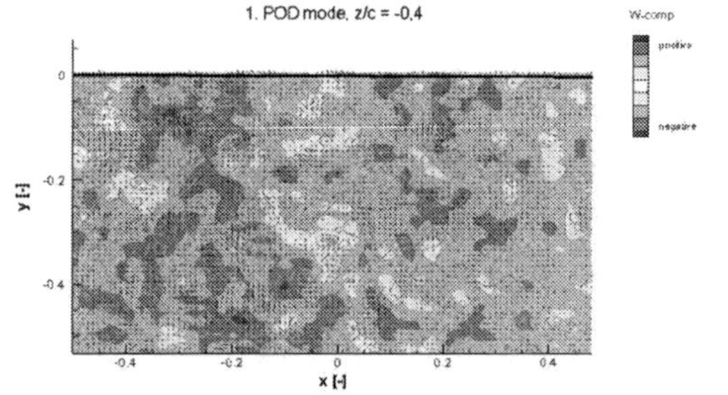

Figure 11b. First POD mode in z/c = -0.4 for pressure side, orderly smaller amplitudes.

The suction side is the place where one can find the origin of these structures. Figure 11a is plotted for seventh POD mode in z/c = -0.5 which is characterized by the existence of vortex pairs located very close to the airfoil surface. Their distribution is perfect regular. We can decompose similar modes revealing streamwise vortices in downstream direction, but there were not detected such structures closer to the leading edge. From TKE distribution, it is obvious that there is no strong dynamics in pressure side. This corresponds with the first mode evaluated at z/c = -0.4 below the airfoil. One cannot see any velocity streaks and even this is the high-energy mode, there are no dominant coherent structures (very low amplitude). Little importance of POD mode in pressure side considering to suction side is visible at the table 1. Sum of ten first mode contain only 5% of energy for pressure side unlike 30% for suction side at z/c = -0,5. Fifty first modes contain more than 2/3 of total energy for suction side unlike 16% of energy for pressure side.

The relative energy content of first twenty modes for both side are compared in graph 12. Especially first five suction modes are dominating over pressure modes. Figure 12b illustrates the development along z-coordinate. POD modes evaluated in measuring planes corresponds to the trailing edge position contain slightly more kinetic energy than modes located more downstream.

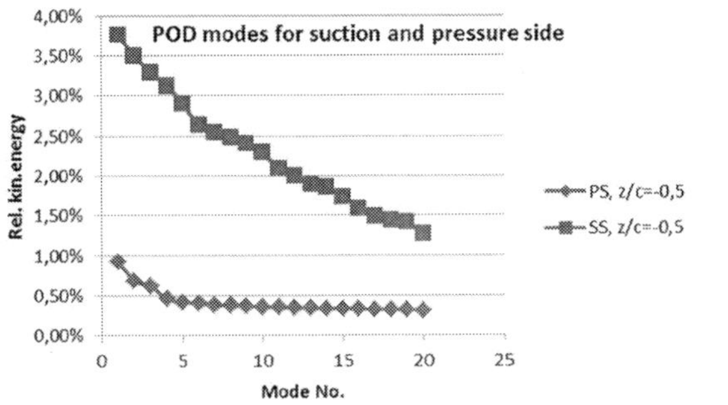

Figure 12a. The importance of 20 POD modes for pressure (blue) and suction sides.

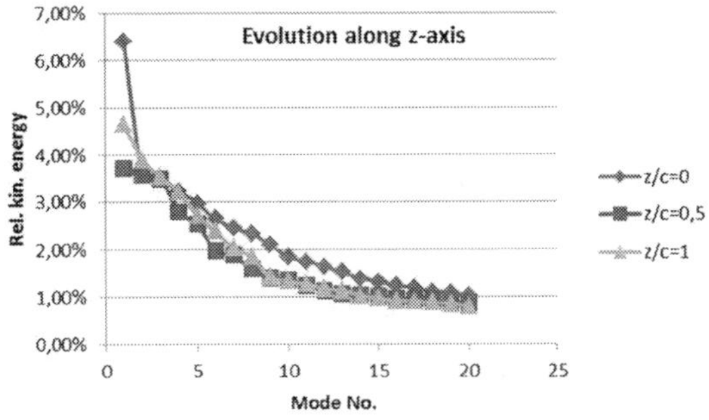

Figure 12b. Rel. kin. energy content of 20 POD modes along streamwise coordinate.

Table 1. Kinetic energy fraction, the comparison of pressure and suction sides.

	Pressure side			Suction side		
	z/c = -0.5	z/c = -0.3	z/c = -0.1	z/c = -0.5	z/c = -0.3	z/c = -0.1
10 modes	4,96%	4,40%	5,35%	28,96%	31,47%	32,52%
20 modes	8,24%	7,73%	8,61%	45,74%	47,88%	48,25%
50 modes	16,50%	16,21%	16,89%	67,25%	68,37%	68,19%

4. Conclusion

The experimental evaluation of streamwise vortices was done. The downwash effect was depicted in detail and downwash angle was set for various flow domain location. Rotational slip separation was studied in the trailing edge location as well as the vortices origin was investigated close to the leading edge. No relevant velocity disturbances were found at pressure side. Low-order POD modes imply that the flow is based on the existence of stremwise vortices above the plate and in the wake and that these structures are subjected to the downwash effect. Both finding support a new theory of flight and lift generation [1].

References

[1] Hoffman J and Johnsson C 2009 The Mathematical Secret of Flight *Normat 57*, Vol.4
[2] Hoffman J and Johnsson C 2010 Resolution of d'Alambert's paradox *J. Math. Fluid Mech.*, 12(3)
[3] Uruba V 2016 Force Interaction of an Airfoil in Fluid-Flow *ES 2016*, 7375
[4] Uruba V 2016 On Aerodynamics Forces Physical Mechanism *SKMTaT 2016*
[5] Uruba V, Pavlík D, Procházka P, Skála V and Kopecký V 2017 On 3D flow structures behind an inclined plate *EPJ Web of Conferences*, vol 143 Article number 02137
[6] Uruba V, Procházka P and Skála V 2017 On 3D Flow Structures of the Boundary Layer on the Suction Side of a Plate *EPJ Web of Conferences*, vol 180 Article number 02112
[7] Uruba V 2012 Decomposition methods in turbulent research *EPF Web of Conferences*, vol 25 Article number 01095

Acknowledgement:
The authors gratefully acknowledge financial support of the Grant Agency of the Czech Republic, project No. 17-01088S.

IOP Conf. Series: Journal of Physics: Conf. Series **1101** (2018) 012027 doi:10.1088/1742-6596/1101/1/012027

Developing micro-scale heterogeneous numerical simulation of a solid oxide fuel cell anode

Tomasz A Prokop, Janusz S Szmyd, Grzegorz Brus
AGH University of Science and Technology, Krakow, 30-059, Poland
E-mail: prokopt@agh.edu.pl

Abstract. In this article a fully three dimensional, multiphase, meso-scale Solid Oxide Fuel Cell (SOFC) anode transport phenomena model is described. The Butler-Volmer model is combined with empirical relations for conductivity and diffusivity - notably the Fuller-Shetler-Giddings equation - for transport of gas reagents. Numerical simulation is used to obtain partial pressure and electric potential fields for each phase, accounting for activation and concentration overpotential as well as ohmic losses. Mesh generator and a solver have been developed to provide an in-house code for the computations. The findings are presented alongside a parametric study of a specific YSZ electrolyte-based SOFC anode microstructure. Despite high dependence of SOFC anode performance on the geometry of its anisotropic, three-phase microstructure, meso-scale numerical models simulating transport phenomena within these electrodes are not very common.

1. Introduction

Fuel cells are devices capable of direct conversion of chemical energy to electrical energy. Lack of combustion chambers and heat engines facilitates elimination of irreversibility sources from the system which in turn results in a highly improved efficiency. This is accomplished by separating oxidation and reduction half-reaction, which occur on the anode and the cathode respectively. The conduction of ions between the separate electrodes takes place in the electrolyte. Solid Oxide Fuel Cell is a type of fuel cell where the electrolyte is composed of a ceramic material and its electrodes are made of ceramic-metal composites. Each phase of the composite performs a peculiar transport function. In an Yttrium-Stabilized Zirconia SOFC anode, the ion conduction takes place in the oxide phase, the electron transport is carried out in the metallic phase and gaseous reagents are delivered via a network of pores. This results in a highly complex microstructure which impacts the performance of SOFC devices (see Fig. 1). Thus, SOFC technologies have become a prominent subject of research in the field of energy engineering [1–8]. Attempts at modelling three dimensional transport phenomena in the depth direction of the cell have been relatively rare [9–11]. Yet, as we have discussed in our previous publication, this may be the most comprehensive method of investigating the impact of heterogeneity in a complex SOFC microstructure [12]. At the same time, the impact of the microstructure on the SOFC device performance still remains a subject of the contemporary research [13]. Such a state of affairs justifies attempts at increasing insight into SOFC anode transport phenomena, including gas diffusion within the porous medium. In this research, a fully three dimensional, multiphase, micro-scale (pore-scale), heterogeneous (non-continuous) SOFC anode model is proposed and verified using a semi-analytical approach. The Butler-Volmer model is combined with recently

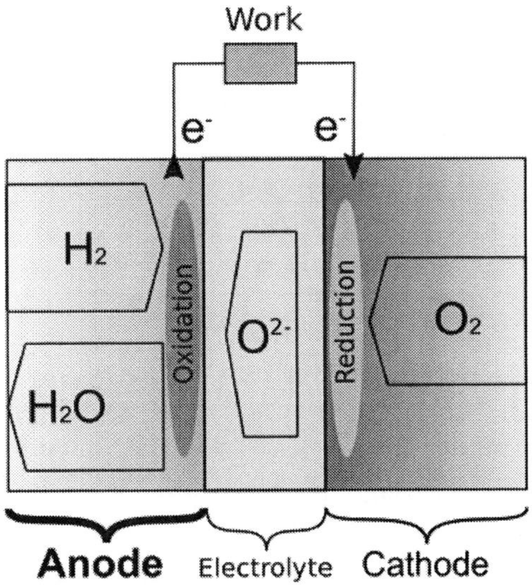

Figure 1. Principle of SOFC operation

developed empirical relations for conductivity and aspects of Maxwell-Boltzmann kinetic theory to describe the transport of mass within the porous medium. The numerical mesh construction is based on the data acquired using FIB-SEM tomography. Methods for localized quantification of electrode morphology parameters (such as Triple Phase Boundary length) are implemented and tested. Exchange current distribution in both electrodes, partial pressure and electric potential fields for each phase are computed numerically. The method of verification relies on a comparison of a three-dimensional, numerical solution to an analytical solution of a one-dimensional case within the range of its applicability. A predictive, parametric study involving localized second-law analysis is performed to predict activation, concentration and ohmic losses, as well as overpotentials at different working conditions of an SOFC stack. Furthermore, the paper contains a parametric study in which different kinds of heterogeneity (varying in placement and the material type of the introduced disturbance) have been introduced.

2. Mathematical model

The mathematical model is based on a set of Poisson equations, describing the transfer of charge and mass in the system. Transport of each of the substances - oxygen ions, electrons, gaseous species - is considered in the following set of equations:

$$\begin{cases} 0 = -i + \dfrac{\partial}{\partial x}\left(\sigma_{\mathrm{el}}\dfrac{\partial \phi_{\mathrm{el}}}{\partial x}\right) + \dfrac{\partial}{\partial y}\left(\sigma_{\mathrm{el}}\dfrac{\partial \phi_{\mathrm{el}}}{\partial y}\right) + \dfrac{\partial}{\partial z}\left(\sigma_{\mathrm{el}}\dfrac{\partial \phi_{\mathrm{el}}}{\partial z}\right) \\ 0 = i + \dfrac{\partial}{\partial x}\left(\sigma_{\mathrm{ion}}\dfrac{\partial \phi_{\mathrm{ion}}}{\partial x}\right) + \dfrac{\partial}{\partial y}\left(\sigma_{\mathrm{ion}}\dfrac{\partial \phi_{\mathrm{ion}}}{\partial y}\right) + \dfrac{\partial}{\partial z}\left(\sigma_{\mathrm{ion}}\dfrac{\partial \phi_{\mathrm{ion}}}{\partial z}\right) \\ 0 = -i + \dfrac{\partial}{\partial x}\left(\dfrac{2FD_{\mathrm{H_2}}}{RT}\dfrac{\partial p_{\mathrm{H_2}}}{\partial x}\right) + \dfrac{\partial}{\partial y}\left(\dfrac{2FD_{\mathrm{H_2}}}{RT}\dfrac{\partial p_{\mathrm{H_2}}}{\partial y}\right) + \dfrac{\partial}{\partial z}\left(\dfrac{2FD_{\mathrm{H_2}}}{RT}\dfrac{\partial p_{\mathrm{H_2}}}{\partial z}\right), \\ 0 = i + \dfrac{\partial}{\partial x}\left(\dfrac{2FD_{\mathrm{H_2O}}}{RT}\dfrac{\partial p_{\mathrm{H_2O}}}{\partial x}\right) + \dfrac{\partial}{\partial y}\left(\dfrac{2FD_{\mathrm{H_2O}}}{RT}\dfrac{\partial p_{\mathrm{H_2O}}}{\partial y}\right) + \dfrac{\partial}{\partial z}\left(\dfrac{2FD_{\mathrm{H_2O}}}{RT}\dfrac{\partial p_{\mathrm{H_2O}}}{\partial z}\right), \end{cases} \tag{1}$$

where i is charge transfer rate, ϕ_i is electric potential at ion-conducting or electron-conducting phase, p_i is pressure of species i in the gas mixture. T is temperature in Kelvins, F is the

Table 1. Scheme of boundary conditions

$z = 0$	$z = z_b$	Vertical boundaries
$p_{H_2} = p_{H_2,b}$	$\frac{\partial p_{H_2}}{\partial z} = 0$	No transfer
$p_{H_2O} = p_{H_2O,b}$	$\frac{\partial p_{H_2O}}{\partial z} = 0$	
$\phi_{el} = \phi_{el,b}$	$\frac{\partial \phi_{el}}{\partial z} = 0$	
$\frac{\partial \phi_{ion}}{\partial z} = 0$	$\phi_{ion} = \phi_{ion,b}$	

Faraday constant R is the universal gas constant, while σ_i and D_i signify conductivities and diffusivities respectively. The values of conductivities are obtained with the following semi-empirical relationships by Anselmi-Tamburini et al. [14] and Bessette et al. [15], while D_i is obtained from Fuller-Schettler-Giddings correlation [16]:

$$\begin{cases} \sigma_{el} & = 3.27 \cdot 10^6 - 1065.3T \\ \sigma_{ion} & = 3.4 \cdot 10^4 \cdot \exp\left(-\frac{10350}{T}\right) \\ D & = 10^2 \frac{T^{1.75}\left(\frac{1}{M_{H_2}} + \frac{1}{M_{H_2O}}\right)}{p\left(V_{H2}^{1/3} + V_{H_2O}^{1/3}\right)} \end{cases}, \tag{2}$$

The charge transfer rate is derived from the Butler-Volmer model. Additionally, empirical relationship for equilibrium charge transfer rate i_0 has been employed.

$$i = i_0 l_{tpb}\left[\exp\left(\frac{\alpha F \eta_{act}}{RT}\right) - \exp\left(-\frac{\alpha F \eta_{act}}{RT}\right)\right] \tag{3}$$

$$i_0 = 31.4 \cdot p_{H_2}^{-0.03} \cdot p_{H_2O}^{0.4} \cdot \exp\left(-\frac{152155}{RT}\right) \tag{4}$$

where i_0 (A m^{-1}) is exchange current density derived from the data by de Boer [17], l_{tpb} (m m^{-3}) is local TPB length density and η_{act} is activation overpotential defined as follows:

$$\eta_{act} = \phi_{el} - \phi_{ion} - \eta_{conc}\ ,\ \eta_{conc} = \frac{RT}{2F}\ln\left(\frac{p_{H_2}^{bulk}}{p_{H_2}}\frac{p_{H_2O}}{p_{H_2O}^{bulk}}\right), \tag{5}$$

where η_{conc} (V) is concentration overpotential.

The model is discretized using Finite Volume Method and the resulting set of equations is solved using Successive Over-Relaxation method with local linearization of the source term.

2.1. Analytical approach

The performance of an anode can be described by a discussion of potential fields present in its interior, as well as electronic losses occurring during its operation. These values may, under some assumptions and approximations be obtained by means of analytical modelling. If the transport phenomena for a given case may be considered as one-dimensional, the problem can be described by the Set of Equations 7. The general scheme of case boundary conditions is presented in Table 1.

$$\begin{cases} \frac{\partial j}{\partial x} = i_0 l_{tpb}\left[e^{\frac{\alpha F \eta}{RT}} - e^{-\frac{\alpha F \eta_{act}}{RT}}\right] \\ j = \frac{V_{ion}}{\tau_{ion}}\sigma_{ion}\frac{\partial \eta}{\partial z} \end{cases}, \tag{6}$$

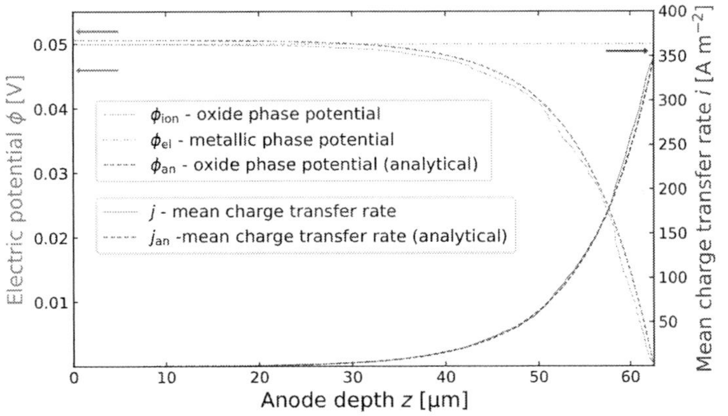

Figure 2. Verification. Ionic potential and exchange current density fields. Boundary conditions: $\phi_{\mathrm{el,b}} = 0.05$ V, p=100000 Pa, $p_{\mathrm{H_2O}} = 3000$ Pa, $p_{\mathrm{H_2}} = p - p_{\mathrm{H_2O}}$, thickness: 62.6 μm, $T = 1073$ K

where V_{ion} is ion-conducting phase volume fraction, and τ_{ion} is ion-conducting phase tortuosity. For small effective conductivity and large anode thickness, the following solution holds true:

$$\begin{cases} \phi_{\mathrm{ion}} = \dfrac{RT}{\alpha F} \cdot \sinh \hat{\eta}_{\mathrm{b}} \cdot \left(1 - \dfrac{\cosh \frac{\hat{z}}{\hat{\varepsilon}}}{\cosh \frac{1}{\hat{\varepsilon}}}\right) \\ j = \dfrac{V_{\mathrm{ion}}}{\tau_{\mathrm{ion}}} \dfrac{\sigma_{\mathrm{ion}} RT}{L \alpha F} \dfrac{\sinh \hat{\eta}_{\mathrm{b}}}{\hat{\varepsilon}} \cdot \dfrac{\sinh \frac{\hat{z}}{\hat{\varepsilon}}}{\cosh \frac{1}{\hat{\varepsilon}}} \end{cases}, \tag{7}$$

where $\hat{\eta} = \frac{\alpha F \eta}{RT}$, $\hat{\varepsilon} = \sqrt{\frac{V_{\mathrm{ion}} \sigma_{\mathrm{ion}} RT}{2 \tau_{\mathrm{ion}} i_{\mathrm{an}} \alpha F z_{\mathrm{b}}^2}}$, and $\hat{z} = \frac{z}{z_b}$. j (Am^{-2}) signifies current density at the ion-conducting phase. The ionic potential and mean exchange current density obtained from the numerical simulation have been compared to the analytic solution showing good agreement within the common range of applicability. The analytical solution does not account for concentration losses in thick anodes and is accurate when large currents are considered. Furthermore, it appears that at lesser depths, convergence occurs at a much slower rate - this is caused by small values of ϕ_{ion} gradient and barely significant difference between ϕ_{el} and ϕ_{ion}. Since mean values of macroscopic overpotential and ohmic losses arise from the product of the local potential differences and the local exchange current density, resulting error should be very small. Verification has also been performed for other sets of boundary conditions. The numerical model appears to match the analytical solution in other applicable cases not presented in this paper (similar consistency is observed).

3. Results and discussion

The simulation has allowed to determine active layer thickness as well as current density for a number of boundary condition sets, allowing for a parametric analysis (see Fig. 3). In the presented example, it can be noted that when the electrode's thickness is lower, the distribution of the potential fields is more uniform. This is related to the fact that, as the thickness increases, charge transfer rate is concentrated closer to the electrolyte, while a significant part of the anode experiences very small reaction rate. This is consistent with the experimental data on active thickness reported in the literature [18]. The uneven distribution of charge transfer rate is also related to variations in the local TPB density.

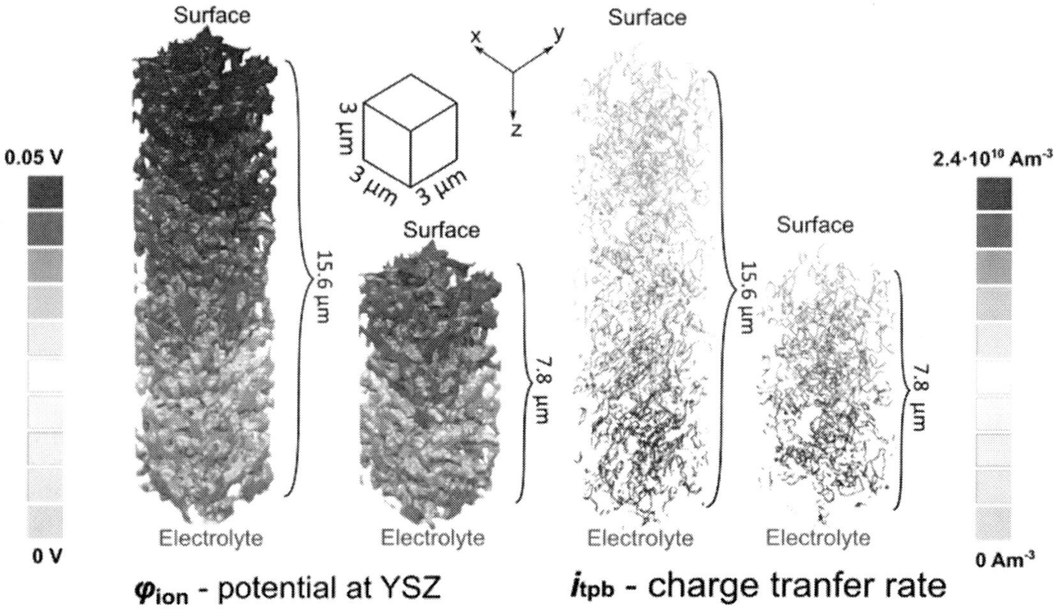

Figure 3. Electric potential field and exchange current distribution in oxide phase of a Solid Oxide Fuel Cell anode for electrodes of different thickness. Boundary conditions: $\phi_{\mathrm{el,b}} = 0.05$ V, p=100000 Pa, $p_{\mathrm{H_2O}} = 3000$ Pa, $p_{\mathrm{H_2}} = p - p_{\mathrm{H_2O}}$, , $T = 1073$ K

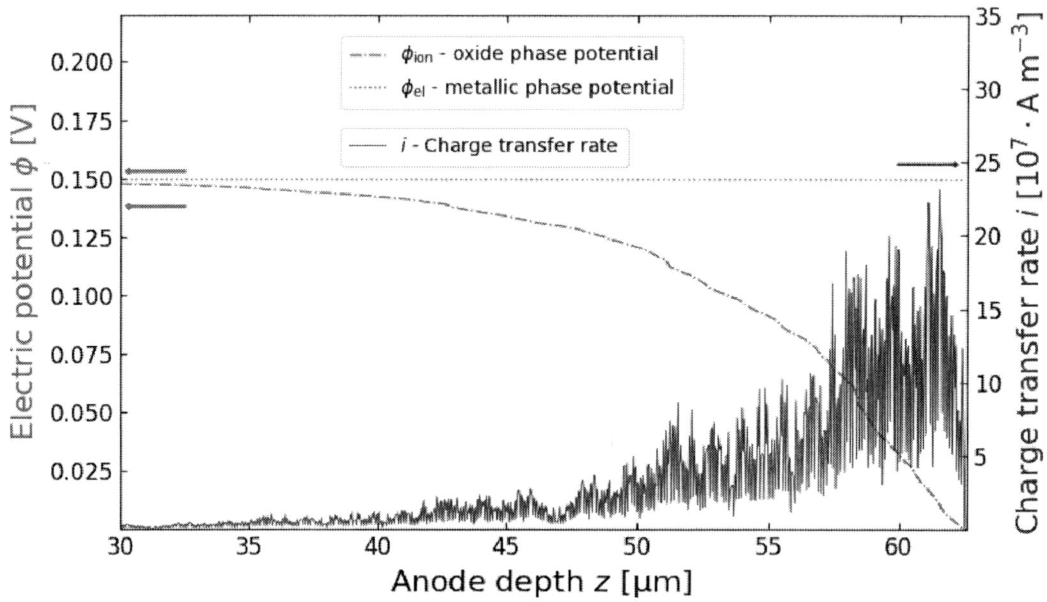

Figure 4. Electric potential field and exchange current distribution in oxide phase of a Solid Oxide Fuel Cell anode at different depths of the electrode. Boundary conditions: $\phi_{\mathrm{el,b}} = 0.05$ V, p=100000 Pa, $p_{\mathrm{H_2O}} = 3000$ Pa, $p_{\mathrm{H_2}} = p - p_{\mathrm{H_2O}}$, Thickness: 62.6 µm, $T = 1073$ K

4. Conclusions

It has been concluded that, in the case of a homogeneous electrode, the problem may be treated as one-dimensional, however only a three-dimensional, heterogeneous model offers an opportunity to address local phenomena within the electrode. The results support the claim that, in a non-homogeneous electrode, the local, microscopic morphology (visible in Fig. 3) may impact the macroscopic parameters of the microstructure. This is evident from irregularities in the potential and current distribution (see Fig. 4). The model provides a useful tool for analysis of cell degradation over its operation period, as well as other phenomena dependent on electrode heterogeneity.

Aknowledgments

The presented research is the part of the "Easy-to-Assemble Stack Type (EAST): Development of a solid oxide fuel cell stack for the innovation in the Polish energy sector" project, carried out within the FIRST TEAM programme (project number First TEAM/2016-1/3) of the Foundation for Polish Science, co-financed by the European Union under the European Regional Development Fund. The research is partially supported by the Polish Ministry of Science and Higher Education (AGH grant No. 11.11.210.312). The authors are grateful for their support. This research made use of computational power provided by the PL-Grid Infrastructure. Matplotlib [19] library for Python programming language and Blender open source 3D graphics software were used for the result visualization.

References

[1] Molenda J *et al.* 2017 *International Journal of Hydrogen Energy* **42** 4366–4403

[2] Iwai H *et al.* 2010 *Journal of Power Sources* **195** 955–961

[3] Wilson J R, Kobsiriphat W, Mendoza R, Chen H Y, Hiller J M, Miller D J, Thornton K, Voorhees P W, Adler S B and Barnett S A 2006 *Nature Materials* **5** 541–544

[4] Jiao Z and Shikazono N 2016 *Science Bulletin* **61** 1317 – 1323 ISSN 2095-9273

[5] Shearing P, Golbert J, Chater R and Brandon N 2009 *Chemical Engineering Science* **64** 3928 – 3933

[6] Kishimoto M, Iwai H, Saito M and Yoshida H 2011 *Journal of Power Sources* **196** 4555–4563

[7] Brus G A, Iwai H, Otani Y, Sciazko A, Saito M, Yoshida H and Szmyd J S 2015 *ECS Transactions* **68** 1237–1244

[8] Brus G A, Iwai H, Otani Y, Saito M, Yoshida H and Szmyd J S 2015 *Fuel Cells* **15** 545–548

[9] Shikazono N, Kanno D, Matsuzaki K, Teshima H, Sumino S and Kasagi N 2010 *Journal of The Electrochemical Society* **157** B665–B672

[10] Kanno D, Shikazono N, Takagi N, Matsuzaki K and Kasagi N 2011 *Electrochimica Acta* **56** 4015–4021

[11] Rhazaoui K, Cai Q, Kishimoto M, Tariq F, Somalu M, Adjiman C and Brandon N 2015 *Electrochimica Acta* **168** 139–147

[12] Prokop T A, Berent K, Iwai H, Szmyd J S and Brus G 2018 *International Journal of Hydrogen Energy* **43** 10016 – 10030

[13] Brus G, Iwai H, Mozdzierz M, Komatsu Y, Saito M, Yoshida H and Szmyd J S 2017 *Journal of Applied Electrochemistry* **47** 979–989

[14] Anselmi-Tamburini U, Chiodelli G, Arimondi M, Maglia F, Spinolo G and Munir Z A 1998 *Solid State Ionics* **110** 35–43

[15] Bessette II N F, Wepfer W J and Winnick J 1995 *Journal of the Electrochemical Society* **142** 3792–3800

[16] Fuller E N, Schettler P D and Giddings J C 1966 *Industrial and Engineering Chemistry* **58** 18–27

[17] de Boer B 1998 *Hydrogen oxidation at porous nickel and nickel/yttria stabilised zirconia cermet electrodes* Ph.D. thesis Universiteit Twente

[18] Miyawaki K, Kishimoto M, Iwai H, Saito M and Yoshida H 2014 *Journal of Power Sources* **267** 503–514

[19] Hunter J D 2007 *Computing In Science & Engineering* **9** 90–95

The actuator cell model for the Darrieus wind turbine

K Rogowski[1]

[1]The Institute of Aeronautics and Applied Mechanics, Nowowiejska 24, 00-665 Warsaw, Warsaw University of Technology, Poland

Abstract. Development of existing and innovative aerodynamic models for the Darrieus wind turbine has become very popular in recent years. Since research in the field of aerodynamics of the Darrieus concept is very limited, the development of simplified aerodynamic methods is very difficult. Therefore, the major objective of the present study is to present the concept of a new aerodynamic model for the Darrieus wind turbine – the actuator cell model (ACM). Aerodynamic loads are added to the unsteady incompressible Navier-Stokes equations as momentum source terms. The source terms are computed basing on instantaneous aerodynamic forces taken from the literature. The numerical results of wake structure computed by ACM are compared with the experimental data. Agreement between the numerical results of velocity profiles and the experimental data is reasonably good.

1. Introduction

Development of simplified aerodynamic models giving reasonable results of aerodynamic loads and wake structure for the Darrieus wind turbine is a challenge. The most popular aerodynamic model for the Darrieus wind turbine is the double multiple streamtube model (DMS) developed by Paraschivoiu [1]. Computations of aerodynamic blade loads basing on this model are very fast, however, its use is limited [2]. The DMS model fails for a large rotor solidity and for heavily loaded blades because the flow past the rotor is assumed to be quasi steady. Vortex models, based on vorticity equations, are another group of simplified aerodynamic models for the Darrieus concept. These models are computationally expensive and more accurate in comparison with momentum-based methods [3, 4]. Nowadays, Computational Fluid Dynamics (CFD) is becoming an important tool for calculating the complex unsteady flow around the rotor of a vertical axis wind turbine. CFD simulations of VAWTs are, however, very expensive and they are also limited [5-7].

A new trend in modelling of large onshore and offshore wind farms are CFD models that are combined with simplified aerodynamic models. CFD models allow modelling of complex terrain or complex forest environment whereas simplified models, such as, for example, blade element momentum method (BEM), can efficiently calculate aerodynamic blade loads. Such simplified CFD models allow a significant reduction of the computation time. The combination of the BEM code with the incompressible Navier-Stokes equations for aerodynamic analysis of wind turbine with a horizontal axis of rotation was performed by Mikkelsen [8]. The 3D Navier-Stokes equations with the large eddy simulation (LES) model combining with the actuator line technique were applied by Troldborg [9] for the analysis of wake behind horizontal axis wind turbine operating at various flow conditions. Rajagopalan and Fanucci [10] were among the first who performed the computations of a two-dimensional vertical axis wind turbine using a finite difference procedure where turbine blades were replaced by a porous cylindrical shell having a thickness of one volume control. A similar approach for the Darrieus concept was applied by Fortunato et al. [11]. The 2D actuator surface technique for a two-dimensional two-bladed vertical axis wind turbine has been used by Shen et al. [12]. In this approach, the two-dimensional Navier–Stokes equations are used with the k-ω shear stress transport (SST) turbulence model and the turbine blades are modeled as body forces distributed on two rotating actuator surfaces.

In this paper, the concept of the new simplified aerodynamic model for the Darrieus wind turbine has been presented. The motivations for this topic are still growing demand for electricity causing that makes it necessary to look for alternative energy sources as well as high computing costs of typical full CFD calculations.

2. Darrieus-type vertical-axis wind turbine

2.1. Experimental model of a wind turbine

Numerical model of a one-bladed Darrieus-type vertical-axis wind turbine (VAWT) is developed based on the experiments conducted at Texas Tech University [3, 13]. The experiments were carried out in a water tow tank with a depth of 1.25 m, a width of 5 m and a length of 10 m. The choice of water as an operating fluid has some advantages, they are as follows: easier visualization of the flow field; a lower rotational speed of the rotor for a given Reynolds number and easier to measure aerodynamic forces because of lower blade inertial forces. During the experiment, the one bladed rotor made of aluminium with a NACA 0012 airfoil and a chord length of 9.14 cm was investigated. The diameter of the turbine rotor was 1.22 m. The towing velocity was 9.1 cm/s giving the tip to speed ratio of 5.0. The tip speed ratio is defined as the ratio of the tangential velocity of the blade and the wind velocity. During the experiments, the angular velocity of the rotor was kept constant and equal to 0.75 rad/s. Operating conditions of the rotor and its dimensions were chosen to give the blade Reynolds number of 40,000. The rotor solidity for the one-bladed rotor, defined as the ratio of the chord length to the rotor radius, was 0.15. Because of the limited space, more details of the experiments are not presented in this paper, but they are available in the research reports [3, 13].

2.2. Aerodynamic characteristics of wind turbine blades and wake structures

The flow past the Darrieus wind turbine is very complex because the angle of attack varies periodically with the azimuth from positive to negative values. A continuous change in the angle of attack causes unsteady induced velocities. Furthermore, depending on the operating conditions of the rotor and its geometry, blade-wake interaction and dynamic stall can occur. The operating conditions of the rotor are determined by the tip speed ratio *TSR* and the blade Reynolds number. The tip speed ratio is defined as $TSR=\omega R/V_0$, where: ω is the angular velocity of the rotor; R is the rotor radius and V_0 is the wind speed. The geometry of the rotor depends on such factors as: the number of blades, the chord length, the rotor swept area, the rotor radius and the airfoil shape of the blade. Figure 1a presents the silhouette of the one-bladed rotor investigated in this paper.

Aerodynamic characteristics of the wind turbine blades are aerodynamic forces exerted on a moving blade by the fluid. Figure 1b shows the cross-section of the Darrieus-type wind turbine rotating with an angular velocity ω. The relative velocity W is the sum of the local wind velocity V and the tangential velocity of the blade V_t, definded as $V_t=R\omega$. An aerodynamic force having lift and drag components, L and D respectively, is projected to the chordwise and the radial directions, giving respectively the tangential and normal blade loads, F_T and F_N. The local angle of attack α is visible between the tangential velocity of the blade and the relative velocity W. The normal and tangential loads, given by the dimensionless coefficients *CFN* and *CFT*, are defined as:

$$CFN, CFT = \frac{F_N, F_T}{0.5\rho V_0^2 c} \tag{1}$$

where: ρ is the medium density; V_0 is the medium velocity and c is the chord length.

Velocity profiles presented in this article are taken within one rotor diameter downstream of the rotor axis. The actuator cell model makes it possible to obtain two velocity components in the wake, streamwise V_x and lateral V_y (see Fig. 1c). The obtained results of velocities are compared directly with the experiment [13]. The velocity profiles presented in this paper are defined as:

$$u = \frac{V_x}{V_0} - 1 \tag{2}$$

$$w = \frac{V_y}{V_0} \tag{3}$$

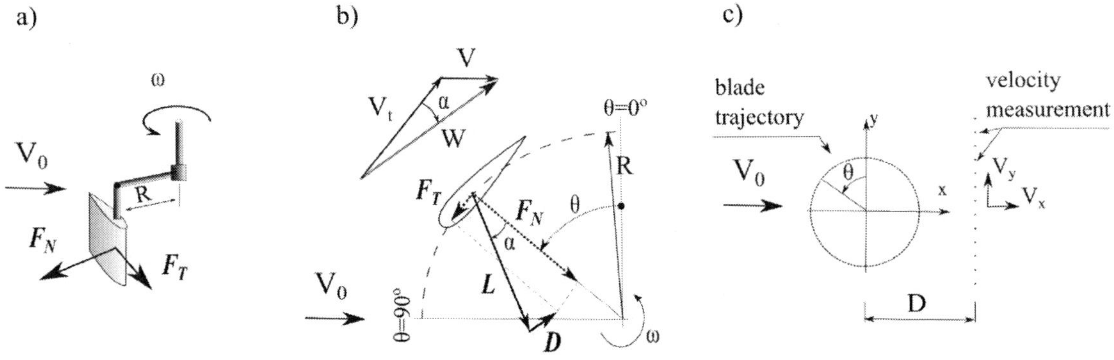

Figure 1. Model of a one-bladed vertical-axis wind turbine (a); aerodynamic blade loads, velocity vectors and angles (b); wake velocity measurement (c).

3. Method

3.1. Actuator Cell Model

Since the flow past a Darrieus-type wind turbine is inherently unsteady, a dime-dependent approach is required. In this paper, unsteady flow past a Darrieus wind turbine is analyzed by using the actuator cell concept proposed by the author of this paper. The concept utilises the sliding mesh model (SSM), available in ANSYS Fluent, to solve the unsteady Navier-Stokes equations. Aerodynamic blade loads are introduced into the model using momentum source terms.

In the ACM concept, the equations of mass and momentum conservation are solved. No turbulence model is used during simulations, therefore additional transport equations are not solved. The equations for mass and momentum conservation in an inertial reference frame for laminar flow as are follow [14]:

$$\frac{\partial \rho}{\partial t} + \nabla \cdot (\rho \vec{v}) = 0 \tag{4}$$

$$\frac{\partial}{\partial t}(\rho \vec{v}) + \nabla(\rho \vec{v} \vec{v}) = -\nabla p + \nabla \cdot \tau + \vec{S} \tag{5}$$

where: p is the static pressure; τ is the stress tensor and $\vec{S}$ are source terms (body forces).

The sliding mesh model is one of the most advanced CFD model available in ANSYS Fluent. In this concept, transient flow around the rotor can be analyzed using the rotating sub-domain mesh. The author of this paper created three cell zones A, B and C that are bounded by interface zones (Fig. 2). During the calculation, the cell zones B and C rotate relative to the cell zone A in discrete steps. The cell zone C does not move relative to the zone B.

In order to simulate aerodynamic blade loads, the momentum source terms $\vec{S}$ (see Eq. 5) are applied to the smallest cell zone C (Fig. 2). The components of the source terms $\vec{S}$ in a two-dimensional coordinate system are given by:

$$S_x, S_y = -\frac{F_x, F_y}{V_C} \tag{6}$$

where: V_C is the volume of the cell zone C consisting of one mesh cell; F_x and F_y are components of the aerodynamic blade load in the coordinate system presented in Fig. 1c. These components can be written:

$$F_x = F_N \sin\theta - F_T \cos\theta \tag{7}$$

$$F_y = -F_N \cos\theta - F_T \sin\theta \tag{8}$$

In this version of ACM model, the instantaneous aerodynamic forces F_N and F_T are taken directly from the experiment [13]. In this work, validation of ACM model is performed by comparing velocity profiles in the aerodynamic wake downstream behind the Darrieus-type rotor. The momentum source terms S_x and S_y are applied to the cell zone C using the user defined functions (UDF). Values of F_x and F_y have been tabulated and stored in a separate file. Using the UDF code, the information about an actual position of the cell zone C are taken from the ANSYS Fluent solver and then the appropriate values of F_x and F_y forces are interpolated from table data. Figure 3 presents non-dimensional components of the instantaneous aerodynamic force, *CFN* and *CFT*, as a function of azimuth θ.

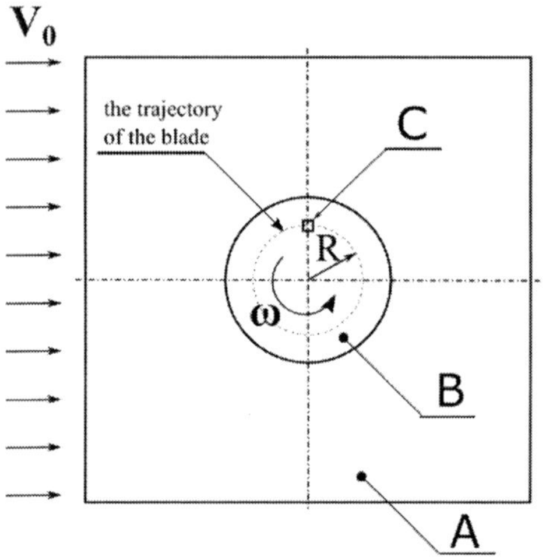

Figure 2. CFD model of the Darrieus wind turbine.

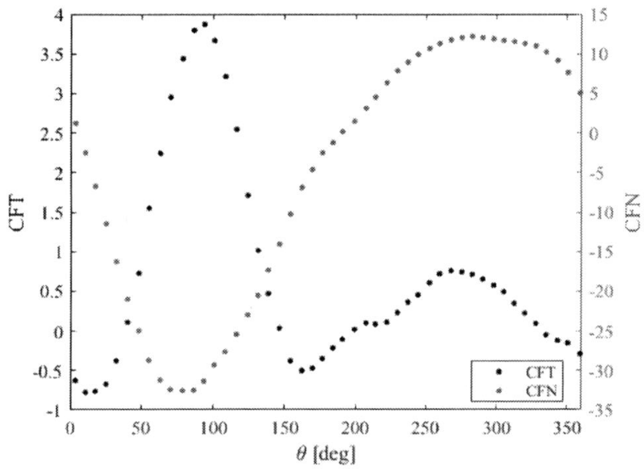

Figure 3. Non-dimensional components of the instantaneous the aerodynamic force, *CFN* and *CFT*.

3.2. Computational Mesh

A two-dimensional computational domain consists of three cell zones A, B and C (see previous Section) which are meshed independently using quadrilateral elements. Figure 4 shows the computational mesh for the actuator cell model used for all results presented here. The cell zone A contains only one mesh cell. The mesh density in the area behind the rotor is higher in order to obtain accurate results of velocity profiles. Computations are also performed using coarser mesh presented in Fig. 5. The results of velocity profiles at an example azimuth of 70 degrees obtained for the coarser and refined meshes are presented in Fig. 6. As it can be seen, the quality of the mesh does not affect the streamwise velocity component and only slightly affects the lateral velocity. Therefore, a distribution of mesh elements in a spanwise direction is an important factor. However, since the streamwise velocity is more important in comparison with the lateral velocity component, more mesh refinements have not been taken into account in these investigations. The number of mesh cells presented in Fig. 4 is 43,707.

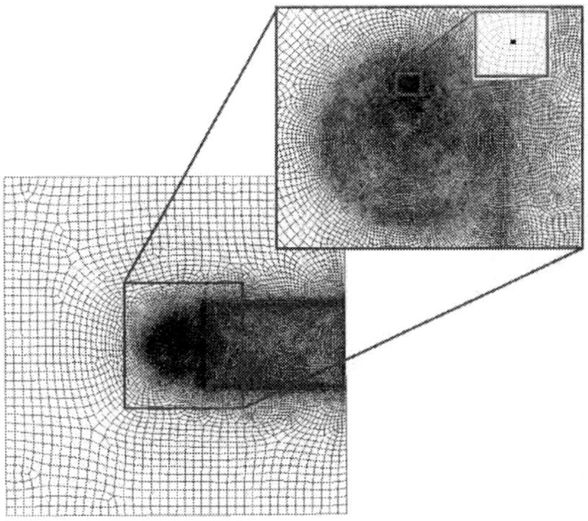

Figure 4. Finer mesh.

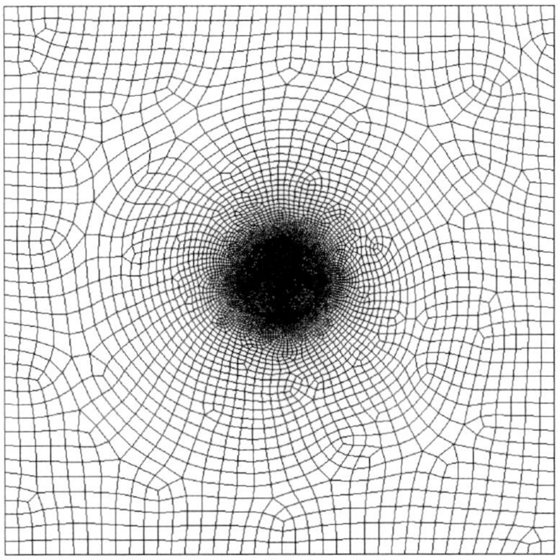

Figure 5. Coarser mesh.

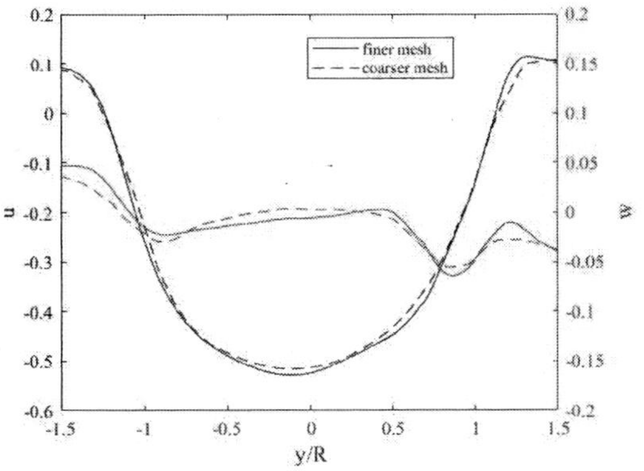

Figure 6. Wake velocity profiles for the finer and the coarser grids.

3.3. Numerical scheme and boundary conditions

All simulations presented in this paper were obtained using the CFD code ANSYS Fluent 15. The pressure based solver with the SIMPLE pressure-velocity coupling scheme and a second-order spatial discretization of momentum and pressure was employed to solve the resulting set of equations. A physical time step size Δt was established to be 0.00023 s. This value of the time step size corresponds to the increment of the azimuthal angle $\Delta\theta$ of 0.02°. In order to match the experimental set-up the following boundary conditions were chosen: velocity inlet at the domain inlet; pressure outlet and symmetry at the upper and lower sides of the computational domain. The symmetry boundary condition can be used in order to model zero-shear slip walls in viscous flows.

4. Results and discussion

Numerical results of velocity profiles for two-dimensional rotor are compared with experimental results of Strickland et al. [13] and with the analytical results of these authors using a vortex model. Figure 7 shows the results of instantaneous streamwise and lateral velocity components for six different azimuth positions of the rotor. The dimensionless streamwise and lateral velocities, u and v respectively, given by equations 2 and 3 are shown as a function of the location y normalized by the rotor radius R (see Fig. 1). The calculated velocity profiles are also compared with numerical results of Rogowski et al. [6]. These authors carried out a full CFD analysis of the one-bladed wind turbine employing, among others, the RNG k-ε turbulence model. The level of agreement between the numerical results of velocity profiles obtained by the actuator cell model and the experimental results by Strickland et al. [13] is reasonable. Velocity profiles given by ACM are also very similar to the full CFD results [6]. Figure 7 shows that wake characteristics downstream behind the rotor given by the vortex model seem to be underestimated.

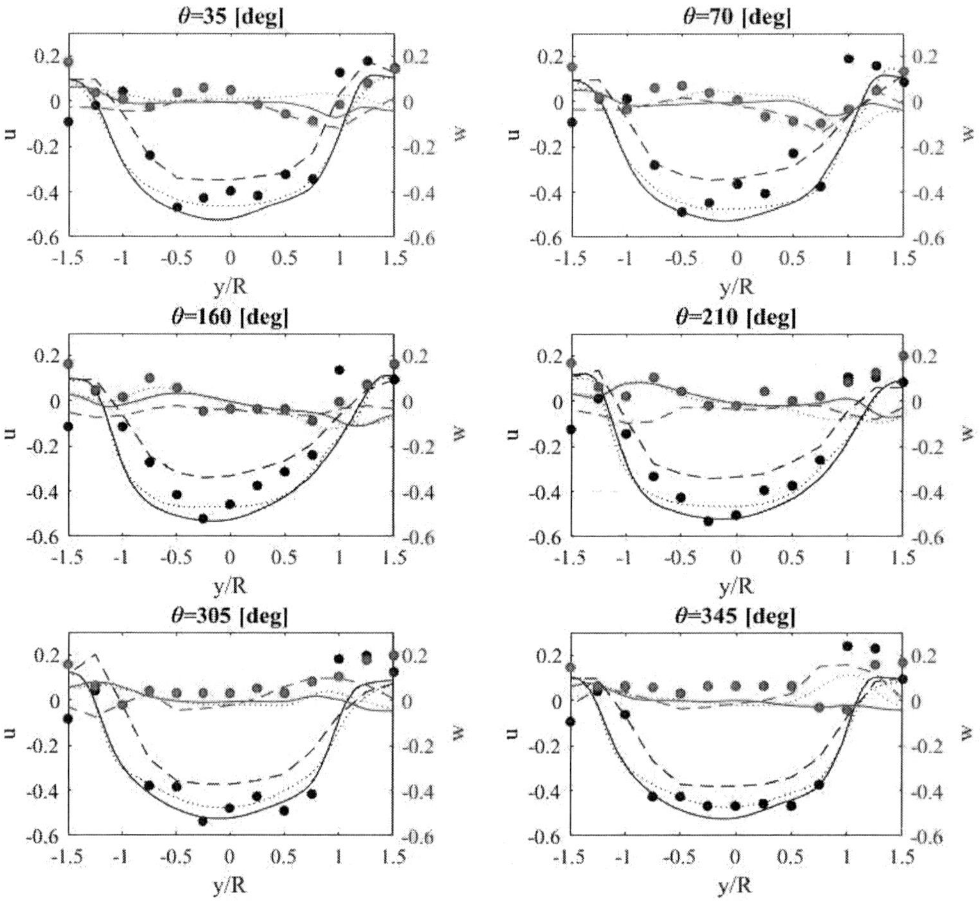

Figure 7. Profiles of streamwise and lateral velocity components. The results obtained by the virtual cell model – black and red solid lines; experiment [13] – circles; vortex model [13] – dashed lines; RNG k-ε results – dotted lines [6].

Wake structure downstream behind the rotor is also analysed qualitatively by using streaklines. Streaklines were obtained by Strickland at al. [3] by means particles which flowed over the trailing edge of the rotor blade. The patterns of streaklines were recorded by using a camera. Numerical results of streaklines are obtained employing the discrete phase model (DPM) available in ANSYS Fluent. In the Euler-Lagrange approach the water phase (fluid phase) is considered as a continuum, while the particle trajectories are solved by tracking the particles through the computed flow field. In these investigations masless particles were injected to the main flow from the surface C (Fig. 2). The experimental results are compared with streaklines calculated by using the ACM model developed by the author of this paper. Figures 8a and 8b present respectively streaklines calculated by CFD and photographed by Strickland et al. [3]. The comparison between ACM and the experimental data is given in Fig. 8c. The comparison (Fig. 8c) shows good agreement in the area in which the dye pattern is not too diffusive. The streakline presented in Fig. 8 indicates the vortex sheet produced by the airfoil.

IOP Conf. Series: Journal of Physics: Conf. Series **1101** (2018) 012028 doi:10.1088/1742-6596/1101/1/012028

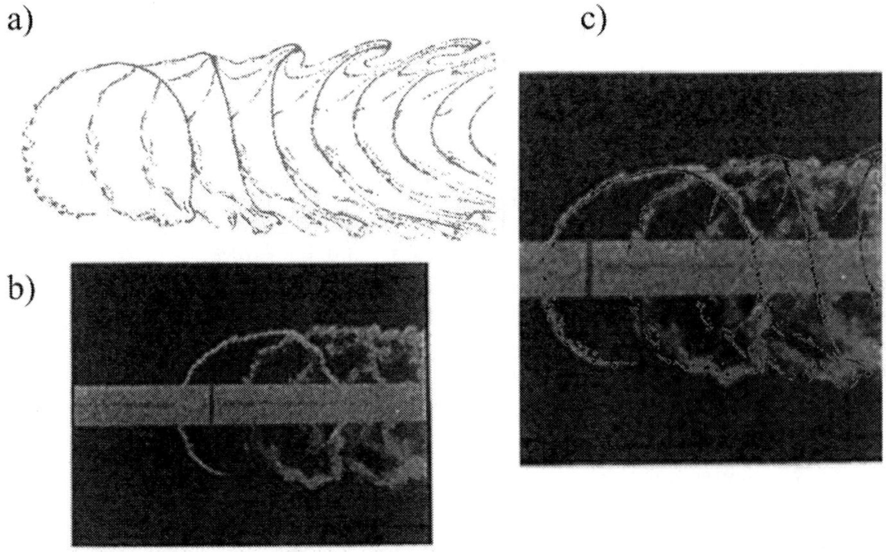

Figure 8. Streaklines computed by ACM (a); streaklines from the experiment [3] (b); streaklines – the comparison between ACM and the experiment [3] (c).

Figure 9 presents instantaneous static pressures computed by the actuator cell model. The obtained results are also compared with full CFD analysis presented by Rogowski at al. [6]. In the upwind part of the rotor (for the azimuthal angle range between 0° and 180°) overpressure areas are seen on the outer part of the airfoil. For the downwind part of the rotor (for azimuthal angles from 180° to 360°) the angle of attack changes the sign and overpressure areas are seen along the inner part of the airfoil. With increasing of the azimuthal angle up to $\theta \approx 100°$ the static pressure increases because of the angle of attack increases and then decreases. In the downwind part of the rotor pressures are lower because flow velocity at the rotor is decreased. Pressure distributions obtained by ACM are similar to these given by the RNG k-ε turbulence model [6].

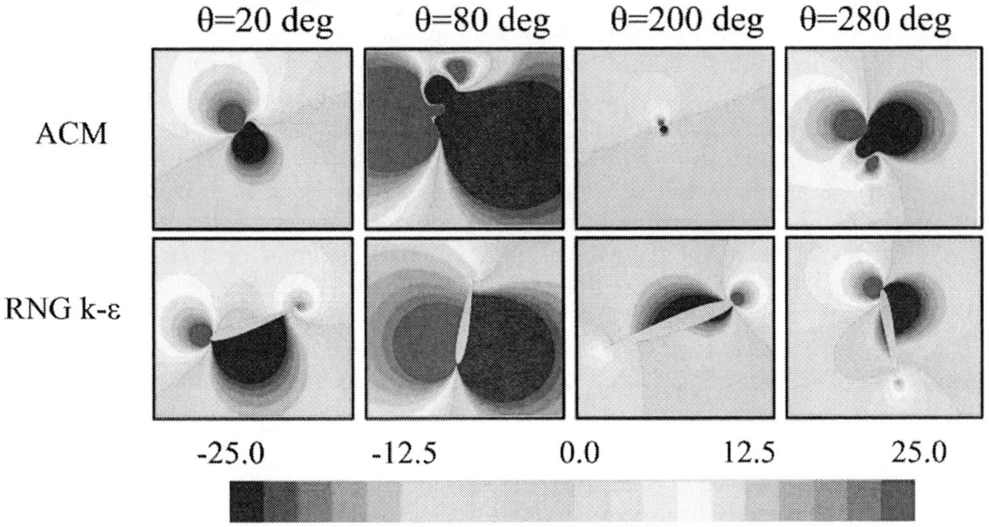

Figure 9. Instantaneous static pressure at three azimuthal angles – the comparison between ACM and full CFD simulation with the RNG k-ε turbulence model [6].

Acknowledgments
The presented numerical computations were performed in the Interdisciplinary Centre for Mathematical and Computational Modelling of the Warsaw University.

5. Conclusions

The main purpose of the author is validation and verification of the simplified CFD model for the Darrieus wind turbine – the actuator cell model.

The presented results both instantaneous streamwise and lateral velocities obtained by the ACM model seem to be reasonable even for a coarse mesh.

The computational effort for the ACM is very low. In this work, only one processor (Intel Xeon x86_64) and 4GB of RAM were used. Simulations of 10 full revolutions of the rotor with very low time step size (corresponding to the increment of the azimuthal angle $\Delta\theta$ of 0.02°) took about one week. Full CFD analysis of the Darrieus-type wind turbine would take much longer.

A small number of mesh cells give results much quicker than the classical full CFD modelling. Therefore the ACM model can be coupled with momentum-based model in order to determine aerodynamic blade loads. In addition, the advantage of the presented model is the possibility of using it with a turbulence model to simulate a flow past a Darrieus-type rotor with shaft, struts or other rotor elements.

The presented studies are as a part of more extensive numerical research on a simplified numerical model for a Darrieus wind turbine.

References

[1] Paraschivoiu I 2002 *Wind Turbine Design With Emphasis on Darrieus Concept* International Press, Canada
[2] Beri H and Yao Y 2011 Double Multiple Stream Tube Model and Numerical Analysis of Vertical Axis Wind Turbine *Energy and Power Engineering* **3** 262-270
[3] Strickland J H, Webster B T and Nguyen T 1979 *A Vortex Model of the Darrieus Turbine: An Analytical and Experimental Study* Technical Report SAND79-7058, Sandia Laboratories
[4] Scheurich F, Fletcher T M and Brown R E 2011 Simulating the Aerodynamic Performance and Wake Dynamics of a Vertical-Axis Wind Turbine *Wind Energy* **14**(2) 159-177
[5] Allet A, Hallé S and Paraschivoiu I 1999 Numerical Simulation of Dynamic Stall Around an Airfoil in Darrieus Motion *Journal of Solar Energy Engineering* **121**(1) 69-76
[6] Rogowski K, Hansen M O L, Maroński R and Lichota P 2016 Scale Adaptive Simulation Model for the Darrieus Wind Turbine *Journal of Physics: Conference Series* **753** 022050
[7] Rogowski K, Maroński R and Piechna J 2017 Numerical analysis of a small-size vertical-axis wind turbine performance and averaged flow parameters around the rotor *Archive of Mechanical Engineering* **64**(2) 205-218
[8] Mikkelsen R F and Sørensen J N 2003 *Actuator Disc Methods Applied to Wind Turbines* PhD Thesis, Technical University of Denmark, Denmark
[9] Troldborg N, Sørensen J N and Mikkelsen R F 2008 *Actuator Line Modeling of Wind Turbine Wakes* Ph.D. thesis, Technical University of Denmark, Denmark
[10] Rajagoplan R G and Fanucci J B 1985 Finite Difference Model for Vertical Axis Wind Turbines *Journal of Propulsion and Power* **1**(6) 432-436
[11] Fortunato B, Dadone A and Trifoni V 1995 A two-dimensional methodology to predict vertical axis wind turbine performance *Journal of Solar Energy Engineering* **117**(3) 187–193
[12] Shen W, Zhang J and Sørensen J 2009 The Actuator Surface Model: A New Navier–Stokes Based Model for Rotor Computations *Journal of Solar Energy Engineering* **131**(1) pp. 011002-1-9
[13] Strickland J H, Smith T and Sun K 1981 *A Vortex Model of the Darrieus Turbine: An Analytical and Experimental Study* Technical Report SAND81-7017
[14] Batchelor G K 1967 *An Introduction to Fluid Dynamics* Cambridge Univ. Press, Cambridge, England

IOP Conf. Series: Journal of Physics: Conf. Series **1101** (2018) 012029 doi:10.1088/1742-6596/1101/1/012029

Experimental and numerical investigations of Tesla turbine

K Rusin[1], W Wróblewski[1] and M Strozik[1]

[1]Silesian University of Technology, Konarskiego 18, 44-100 Gliwice, Poland

E-mail: krzysztof.rusin@polsl.pl

Abstract. Tesla turbine is a bladeless turbine and its principle of operation is based on shear stress arising from fluid viscosity and turbulence. A prototype of Tesla turbine is an object of numerical and experimental investigations presented in the paper. Experimental investigations were carried out for different inlet pressures and compared to results from numerical investigations. The roughness of the disc surface was measured and taken into account in the numerical analysis. Numerical investigations were performed in CFX 17.1 commercial software for steady state and transient. Comparison of power and efficiency vs rotational velocity characteristics shows relatively good agreement between experiment and CFD.

1. Introduction

Tesla turbine, known also as a bladeless turbine or boundary layer turbine, was invented in 1913 by Nikola Tesla [1]. It is a radial turbine, whose rotor consists of multiple discs, mounted in parallel on the shaft. Its principle of operation is based on shear stresses arising from fluid viscosity and turbulence. The supply system consists of one or multiple nozzles. Their aim is to accelerate working medium and to deliver it at an optimal angle [2]. Fluid flows tangentially to the surface of the disc from the outer to the inner radius and due to adhesive forces, sticks to it. Tangential stresses occur on the disc surface due to momentum diffusion, which is an effect of fluid viscosity and makes disc rotating.

Advantages [3,4] like resistance to erosion, flexibility in the choice of working medium and reversibility of flow make bladeless turbine well suited for systems powered by renewable energy sources with the low boiling medium [5]. Structure of the turbine is unsophisticated, which means low costs of manufacturing. Consequently, the economic viability of an installation with Tesla turbine is good.

Tesla turbine has been often an object of investigations. Kim and Yoon [6] investigated the possibility of using a bladeless expander in a cogeneration installation supplied with waste heat. Song et al. [7] tested the influence of low boiling medium on Tesla turbine performance. Carey [8] analysed a 4kW solar cogeneration system and developed a mathematical model in order to determine its optimal working conditions. Schosser et al [9] conducted numerical investigation of flow in the rotor and compared the results with data obtained from analytical solution.

Preliminary experimental investigation of Tesla turbine and comparison of performance with the prediction of numerical analysis is presented in the paper. The object of interest was Tesla turbine model MK3 manufactured by Gyroscope company. The experiment involves the determination of rotor disc roughness, efficiency characteristic of the generator and finally power and efficiency vs rotational velocity characteristics. Numerical investigations are performed with the use ANSYS 17.1 package: Meshing and CFX.

2. Experimental investigations set up

Experimental investigations were performed in Institute of Power Engineering and Turbomachinery in the Silesian University of Technology. Test stand, which is presented in Figure 1, was divided into two parts: (a) supply system and (b) measurement system.

a) b)

Figure 1. Test stand: (a) supply system and (b) measurement system

The supply system consisted of air tank (1) with 0.5 m^3 volume, reciprocating compressor (2) and two valves (3, 4). The aim of this system was to produce air flow with demanded parameters. The compressor was filling the tank with the air up to pressure 8 bar. The outflow from the tank was regulated by means of two valves and was carried out through the duct of 13 mm in inner diameter. The main valve (3) served as a safety valve and could quickly cut off the flow. The membrane control valve (4) was supposed to stabilize the pressure level and keep it constant. Steady state, which is required to perform series of measurements, occurs as long as pressure upstream of the valve is higher than the set value, owing to which time for measurement is relatively short (e.g. 1 minute). The second part was measurement system. A pressure transducer (5) was placed right after the control valve, which allowed control of inlet pressure to the Tesla turbine (6). No loss was assumed in the short duct between pressure transducer and turbine, so pressure value obtained from transducer was considered as an inlet pressure to the turbine. The turbine was connected to the generator (7). The generator was brushless motor model HC 2816-1220, which is often used in flying radio controlled models. It was connected to the tachometer (8), which was supposed to measure rotational velocity of generator and turbine. Its principle of operation was based on counting impulses produced in the generator, where every 3 impulses corresponded to 1 revolution. Electric load (9) was responsible for loading generator with demanded value. It was possible to set either resistance or exact power value. The electric load measured also power produced in the generator.

In order to make a valid comparison between CFD analysis and experimental investigations, it was necessary to obtain the same quantity that is internal power. Internal power is an effect of thermodynamic processes taking place in turbine nozzles and in a rotor and does not take into account losses in bearings or in the generator. This quantity can be obtained in CFD calculations, but in case of experimental investigations only electrical power is measured directly. Therefore it is necessary to estimate the efficiency of generator and power lost in bearings due to friction. It was assumed that efficiency of bearings is equal to 99%, but to obtain efficiency characteristic of the generator, the test rig was designed.

The test rig is presented in Figure 2. A power supply (1) produced electric current with proper voltage and amperage, which propelled electric motor (2). The rotational velocity of the motor was changing with the use of the potentiometer (3). The electric motor was connected to the investigated generator (4). Efficiency characteristic of the motor was given by the manufacturer, so the shaft power propelling the generator was known. The generator was loaded by the electric load (5), which forced the exact value of produced power. As in the main test stand, the rotational velocity was measured by the tachometer (6). The electric current produced in generator was rectified in (7). The measurement procedure was based on efficiency estimation with respect to rotational velocity and applied electric load. It was possible to estimate efficiency by measuring power propelling the motor and power produced in the generator.

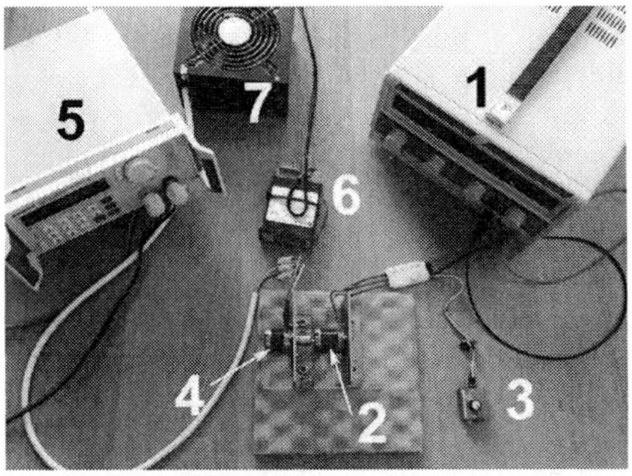

Figure 2. Generator test rig

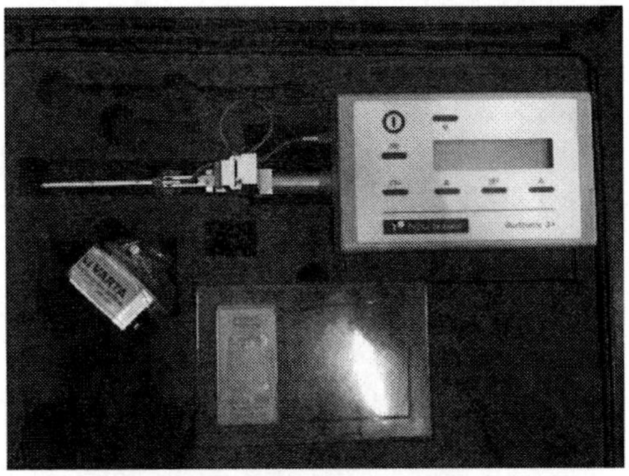

Figure 3. Taylor-Hobson Surtronic 3+ Roughness Gage

The roughness of the disc surface is of significant importance for the turbine performance. It may cause an increase of turbulence in the boundary layer and therefore increase of shear stresses occurring on disc walls. In order to make CFD analysis more precise, it is recommended to determine this parameter. The roughness of disc surface was measured with the use of Taylor-Hobson Surtronic 3+ Roughness Gage, which is presented in Figure 3. Two sides of the one rotor disc were surveyed. The measurement was carried out along 5 radii crossing outlet sections of the rotor and in two directions: parallel and normal with respect to the radius. The length of the measuring section totalled 0.8 mm. As the result, roughness expressed as arithmetical mean deviation R_a was obtained. R_a was calculated from the equation:

$$R_a = \frac{1}{a}\sum_{i=1}^{a}|h_i| \tag{1}$$

Where:
h – roughness height
a – number of measurements

Results obtained from experiments are presented in section "Results".

3. Numerical analysis

The numerical analysis was performed in ANSYS 17.1 commercial software. The computational mesh was generated using ANSYS Meshing tool and the setup of numerical model was done in CFX. This software uses an implicit finite volume formulation to resolve discretized unsteady Reynolds-averaged Navier-Stokes equations. Computations were performed both in steady state and in transient.

3.1. Model of geometry

Geometry model, presented in Figure 4, was reflecting dimensions of Tesla turbine model, which was investigated experimentally. For simplicity reasons, only half of the turbine was modelled that is two inter-disc gaps and the gap between the last disc and the turbine casing. Such configuration takes into consideration two discs and one side of the third disc and does not influences the results of numerical simulations, but decreases computational time. The inter discs gaps were 73mm in diameter and 1.5 mm thick and the discs were 1.3 mm thick. The spacers of spiral shape were also taken into account. The rotor domain was supplied by the inlet system, which consisted of cylindrical chamber (1) with two inlet orifices with 1.8mm in diameter (2) providing the working medium directly in between the discs. The chamber was 6mm in diameter, 13mm in length and was positioned at an angle 45° with respect to the vertical symmetry axis of the rotor. Tip clearance (3) between the rotor and the casing was 0.5 mm in height. The outlet from the rotor consisted of 5 orifices with 7.5 mm in diameter, whose centres were located 10mm away from the rotor axis. The working medium flowed out from the rotor to the converging collecting chamber (4) which was linked by four cylindrical ducts (5) with second converging collecting chamber (6). Outflow into the ambient took place by means of 20mm length duct (7) with 13 mm in diameter.

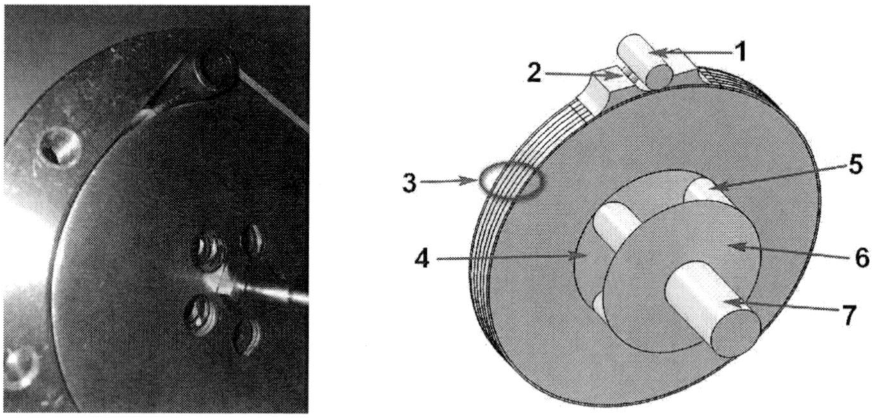

Figure 4. The geometry of the numerical model

3.2. Numerical model and boundary conditions

Numerical simulations were performed for air treated as an ideal gas. Changes in laminar fluid viscosity due to temperature changes were taken into consideration, as it may influence shear stresses occurring on disc walls. It was modelled with the use of Sutherland formula [10]:

$$\mu(T) = \mu_{ref}\left(\frac{T}{T_{ref}}\right)^{3/2}\frac{T_{ref}+T_S}{T+T_S} \tag{2}$$

where: Sutherland temperature T_S=110.4 K, reference viscosity $\mu_{ref}=1.719\cdot10^{-5}$ Pa·s, reference temperature T_{ref}=273.15 K. Turbulence model used for simulations was k-ω SST [11], which is a combination of k-ω turbulence model near the wall and k-ε turbulence model in the far field. The roughness of the surface was modelled using sand grain roughness ε. It can be derived using arithmetical mean deviation of roughness height R_a [12]:

$$\varepsilon = \frac{2R_a}{\left(\frac{\pi}{2} - cos^{-1}\left(1-\frac{\pi^2}{16}\right)^{\frac{1}{2}} - \frac{\pi}{4}\left(1-\frac{\pi^2}{16}\right)^{\frac{1}{2}}\right)} \tag{3}$$

Sand grain roughness was calculated according to results obtained from measurement of roughness. Details of the numerical model and boundary conditions are presented in Figure 5. Total pressure and total temperature were assumed at the inlet sections of the supplying chamber. Inlet pressure was equal

IOP Conf. Series: Journal of Physics: Conf. Series **1101** (2018) 012029 doi:10.1088/1742-6596/1101/1/012029

to 3 or 4 bar and the total temperature was equal to 303 K. Two inlet sections ensured equal conditions for both inlet orifices, which is the case for whole turbine geometry. The supply system and the tip clearance domains were stationary, but bottom walls of the tip clearance, which were in contact with the discs, were rotating. Tip clearance was connected to the inlet system and to the rotor with General Grid Interface. Symmetry boundary condition was applied on one lateral wall of the tip clearance. The rotor domain was modelled as stationary, but lateral walls were rotating. Additionally, sand grain roughness and no-slip wall were applied on these walls. Spacer domains were rotating and connected to the rotor with frozen rotor interface. The outlet system was modelled as stationary, thus it was connected to spacer domain with frozen rotor interface. Static pressure equal to 1 bar was assumed at the outlet section. All walls of the model were adiabatic.

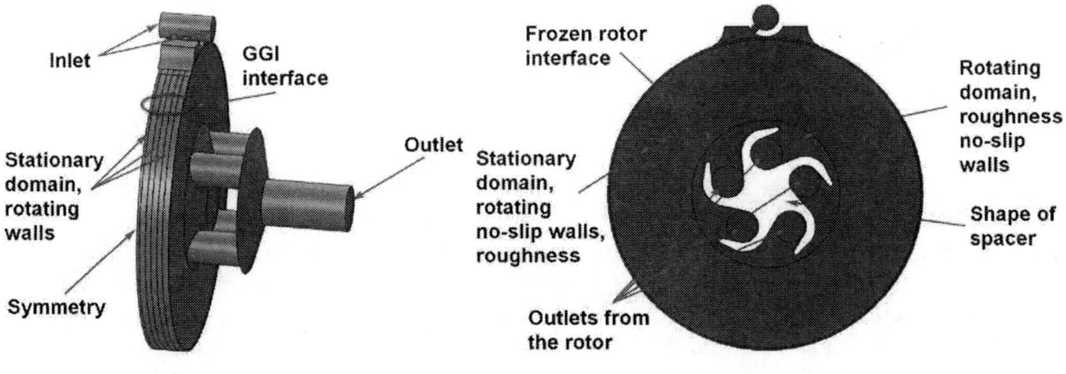

Figure 5. Details of the numerical model

Conservation equations for mass, momentum and energy were solved:

$$\frac{\partial \rho}{\partial t} + \nabla \cdot (\rho \mathbf{v}) = 0 \tag{4}$$

$$\frac{\partial (\rho \mathbf{v})}{\partial t} + \nabla \cdot (\rho \mathbf{v} \otimes \mathbf{v}) = -\nabla p + \nabla \cdot \boldsymbol{\tau} \tag{5}$$

$$\frac{\partial (\rho h_{tot})}{\partial t} - \frac{\partial p}{\partial t} + \nabla \cdot (\rho \mathbf{v} h_{tot}) = \nabla \cdot (\lambda \nabla T) + \nabla \cdot (\mathbf{v} \cdot \boldsymbol{\tau}) \tag{6}$$

These equations were discretized in space with the use of the high-resolution scheme. Integration with respect to time was done using implicit second-order Euler scheme.

Sensitivity study of mesh density influence on results was performed. 5 numerical meshes with 2M, 2.7M, 4.1M, 5.2 M and 7.7 M elements were created and value of power was treated as an indication of mesh quality. Sensitivity study of time step for transient simulations was performed as well.

Results of mesh sensitivity study and numerical simulations are presented in paragraph "Results".

4. Results

Results of roughness measurement are presented in Figure 6. Roughness in all directions is small and the differences with the change of direction are insignificant. Average roughness totals R_a=0.28μm, which proves that surface was improved with polishing. Such a small value roughness has minor contribution to increase of turbulence viscosity and thereby generated power.

Characteristic of generator efficiency with respect to produced power and rotational velocity is presented in Figure 7. Characteristic is of paraboloid shape with maximum efficiency equal to 84% achieved for 18000 min^{-1} and 65 W. Region of efficiency higher than 70% is limited by 25-80W range and 10000-25000 min^{-1} range. The rapid drop in efficiency occurs for higher rotational velocities i.e. above 27500 min^{-1} and for small power values.

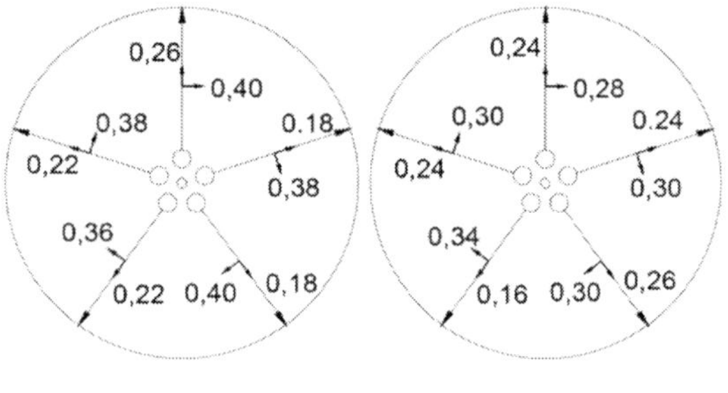

Figure 6. Results of disc roughness measurement

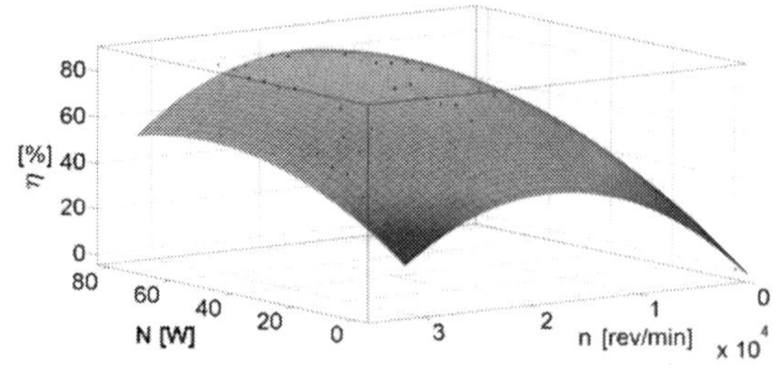

Figure 7. Characteristic of generator efficiency with respect to power and rotational velocity

The efficiency of the generator can be estimated using the equation:

$$\eta = 104.2 - 0.003869n + 0.8053N + 2.052 \cdot 10^{-5} n \cdot N + 3.9 \cdot 10^{-8} n^2 - 0.00978N^2 \tag{7}$$

Results of mesh independence study are presented in Figure 8. Reference power N_{ref} was obtained from the highest density mesh, to which other results were compared. Analysis proved that mesh with 4.1M does not influence results. Fragment of discretization is presented in Figure 9. Areas with highest velocity gradient, i.e. the supply system and the upper part of the inter-disc gap have the finest mesh. Boundary layer was created on disc walls, cylindrical chamber, two inlet orifices, the upper wall of tip clearance, two collecting chambers and walls of 4 cylindrical linkers. It provided distribution of non-dimensional distance $y^+ \leq 1$. Time step equal to 10^{-5}s was determined for transient computations.

Comparisons of power and efficiency characteristics for experiment and numerical investigations are presented respectively in Figure 10 and Figure 11. Power in numerical investigations was calculated basing on averaging power values from each time step of the transient simulation. All characteristics of power are of a parabolic shape and relatively flat, which means that in a wide range of rotational velocity power changes insignificantly, e.g. for 3 bar CFD power varies only 25% of maximal value. Values obtained from numerical investigations are higher than values obtained from the experiment.

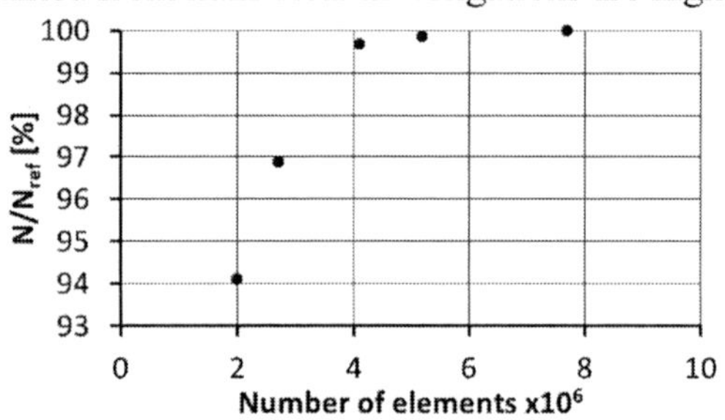

Figure 8. Analysis of mesh independence study

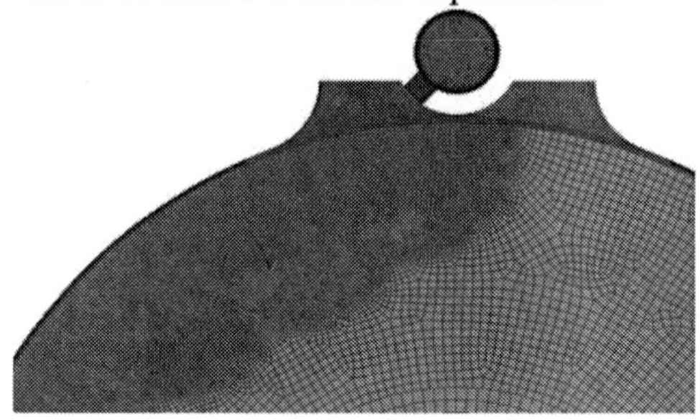

Fig. 9 Model discretization

Maximal power values are: 55.6 W (25000 min^{-1}) for 3 bar CFD, 42.5 W (20500 min^{-1}) for 3 bar experiment, 98.3 W (30000 min^{-1}) for 4 bar CFD and 71.5 W (20500 min^{-1}) for 4 bar experiment. Efficiency was calculated using equation:

$$\eta = \frac{N}{\dot{m} \cdot T_{in} \cdot c_p \left(1 - \left(\frac{p_{out}}{p_{in}}\right)^{\left(\frac{\kappa-1}{\kappa}\right)}\right)} \tag{8}$$

where: $\dot{m}$ – mass flow, T_{in} – inlet temperature, c_p – specific heat capacity at constant pressure, $p_{in,out}$ – pressure at inlet and outlet, κ – heat capacity ratio. Mass flow for the experiment was estimated assuming choked flow in the inlet orifices. Maximal efficiency values correspond to maximal power and are equal to: 11.2 % for 3 bar CFD, 8.4% for 3 bar experiment, 11.8% for 4 bar CFD and 8.9% for 4 bar experiment. An increase of inlet pressure causes a shift of maximal efficiency toward higher rotational velocities.

Several factors influenced disparities between experimental investigations and CFD analysis. The most important one is precision of turbine manufacturing. Discs are not ideally parallel to each other after mounting on a shaft, therefore the width of the inter-disc gap is not constant. Lack of appropriate sealing of inlet system might cause volumetric loses. It should be also noted that efficiency characteristic of the generator was estimated only for 500-30000 min^{-1} range due to overheating of the electric motor, therefore it is not recommended to use it beyond this range of rotational velocity. However, values obtained from experiment partially exceeded this limit, thus estimation of power can be misleading. It can be seen that for case 4 bar, the higher rotational velocity, the bigger differences between results obtained from CFD and experiment.

Numerical model has its own flawns as well. The biggest challenge comes from flow rotation. Additional body force occurs, which interact with the turbulence [13]. This may lead to false prediction of eddy viscosity and thereby power.

Figure 12 and Figure 13 present Mach number distribution for inlet pressure 4 bar and rotational velocity 30000 min^{-1} obtained from steady-state numerical investigations.

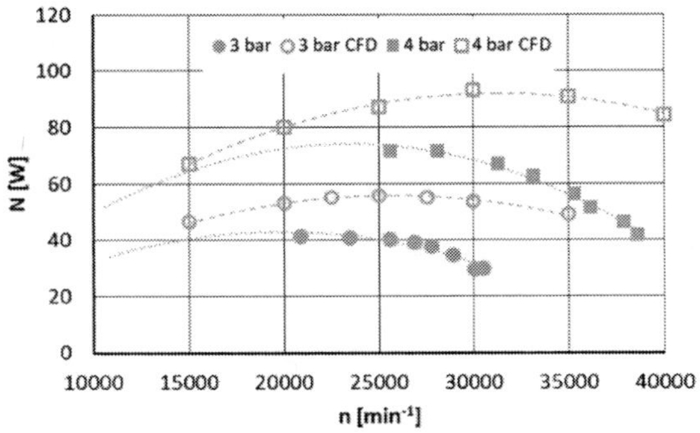

Figure 10. Characteristics of power vs rotational velocity

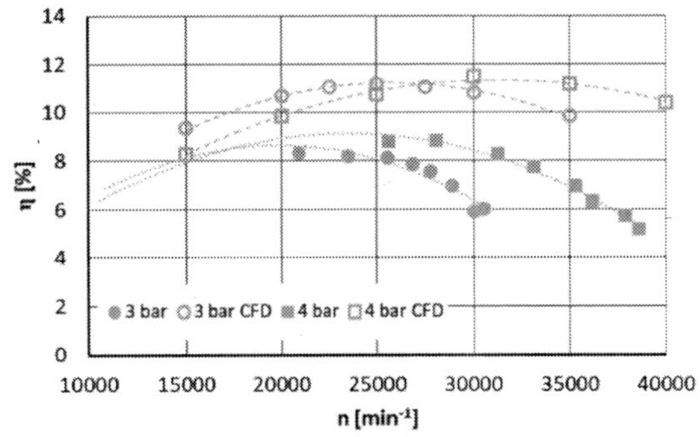

Figure 11. Characteristics of efficiency vs rotational velocity

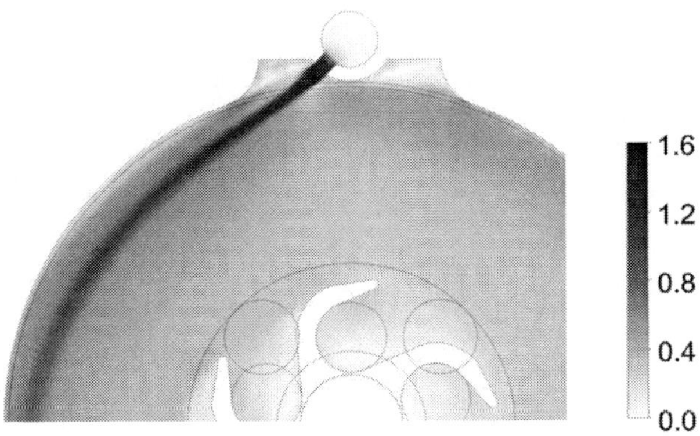

Figure 12. Mach number distribution in the inter-disc gap for 4 bar inlet pressure, n=30000 min^{-1}

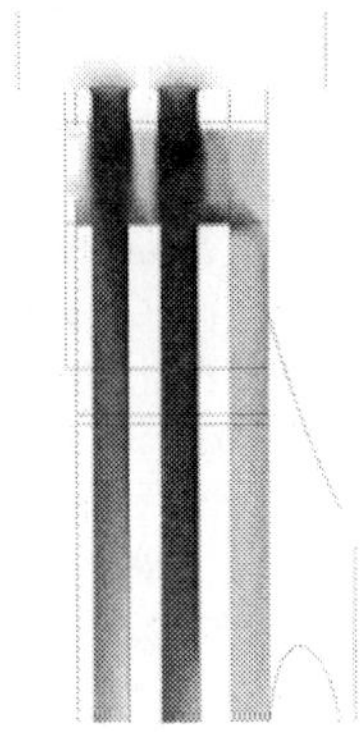

Figure 13. Mach number distribution in the cross-section of an inter-disc gap for 4 bar inlet pressure, n=30000 min^{-1}

It can be seen in Figure 13 that the working medium in the inlet chamber has to perpendicularly change direction, therefore vena contracta occurs in the inlet orifices. Air mainly flows into the inter-disc gaps, but there is also an interaction with disc tips. The working medium forms a jet flowing out from the inlet, which is visible in Figure 12. The highest Mach number is equal to 1.6 and it occurs at the outlet section from the inlet orifices. The jet flowing into the inter-disc gap is bending due to the occurrence of centrifugal forces coming from rotational velocity. An inlet orifice angle equal to 45° does not provide satisfactory flow path, as the jet flowing to the rotor should be moved as remotely as possible from the outlet. This would ensure higher torque and generated power. Fluid velocity decreases approaching the outlet system, and average Mach number in the outlet orifices is equal to 0.16. The higher fluid velocity at the outlet, the bigger kinematic loss and lower overall turbine efficiency.

5. Conclusions

The paper presents experimental and numerical investigations of Tesla turbine. Three experimental stands were designed. Their aim was to measure generator efficiency, rotor disc roughness and Tesla turbine performance characteristics. Rotor disc roughness equal to R_a=0.28 indicated polished surface, which did not contribute in a significant way to shear stress generation. Maximal power values obtained from experimental investigations performed for inlet pressure 3 and 4 bar were respectively equal to 42.5 W and 71.5 W, which corresponded with efficiency 8.4% and 8.9%. Steady-state and transient numerical investigations were performed in CFX 17.1 commercial software. The numerical model took into account most of the important phenomena, including change of the laminar viscosity with the temperature and disc roughness. Mesh independence study was also performed. Power and efficiency characteristics based on averaged transient results for each time step of transient computations were obtained. The highest power and efficiency values were: 55.6 W, 11.2% for inlet pressure 3 bar and 98.3 W, 11.8% for 4 bar. The highest Mach number was equal to 1.6. An inlet orifice angle should be increased in order to improve turbine performance.

Investigations indicated on challenges associated with measurement of Tesla turbine performance parameters but also revealed turbine potential for low power applications. However, in order to achieve higher efficiency, retooling of the inlet nozzles, as well as the outlet system geometry, has to be done.

Acknowledgements

The presented work was performed within the Silesian University of Technology statutory research funds for young scientists. This research was partially supported by the PL-Grid Infrastructure.

References

[1] Tesla N, *Turbine, Patent No: 1,061,206*, United States Patent Office of New York, 1913
[2] Neckel A L and Godinho M 2015 *Experimental Thermal and Fluid Science* **62** 131
[3] Lampart P and Jędrzejewski ł 2011 *Journal of Theoretical and Applied Mechanics* **49** 447
[4] Sengupta S and Guha A 2012 *Journal of Power and Energy* **226** 650
[5] Lampart P, Kosowski K, Piwowarski M and Jędrzejewski Ł 2009 *Polish Maritime Res.* **63** 28
[6] Kim Ch K and Yoon J Y 2016 *Energy* **101** 411
[7] Song J, Gu Ch and Li X 2018 *Applied Thermal Engineering* **110** 28
[8] Carey V P 2010 *Journal of Engineering for Gas Turbines and Power* **132** 122301-1
[9] Schosser C, Lecheler S and Pfitzner M 2016 *Periodica Polytechnica Mech Eng* **61** 12
[10] Sutherland W 1893 *Philosophical Magazine* **5** 507
[11] Menter F 1993 *24th Fluid Dynamics Conference,* Orlando, USA
[12] Adams T, Grant Ch and Watson H 2012 *Int. J. of Mech. Engineering and Mechatronics* **1** 66
[13] Tucker P G 2013 *Progress in Aerospace Sciences* **63** 1

IOP Conf. Series: Journal of Physics: Conf. Series **1101** (2018) 012030 doi:10.1088/1742-6596/1101/1/012030

Application of Background Oriented Schlieren for quantitative measurement of transonic flows

L Santos [1] and W Stryczniewicz [2]

[1] Aerospace Engineering Department, Universidade de Brasilia, Brasilia, Brazil
[2] Aerodynamics Department, Institute of Aviation, Warsaw, Poland

E-mail: [1] `lsantos.moraiss@gmail.com`, [2] `wit.stryczniewicz@ilot.edu.pl`

Abstract. Flow visualization techniques have progressed from methods based on qualitative description of flow-field into techniques providing both qualitative and quantitative results. In this paper we present implementation of Background Oriented Schlieren (BOS) technique for measurements of density values at transonic speeds. The density values obtained by BOS were then compared with results from CFD analysis of the same flow characteristics. In both qualitative and quantitative terms, the BOS and CFD results agreed, which proves the applicability of BOS for transonic flows.

1. Introduction

Flow visualization techniques have been more frequently applied to wind tunnel tests, replacing the usual setup of pressure taps and other intrusive instrumentation. These techniques are an improvement in terms of whole flow-field visualization and the possibility of processing complex types of flow.

One of these methods, the *Background Oriented Schlieren* (BOS) technique [3,7] is based on imaging of background distortion caused by changes of the refraction index n of medium in the region between camera and background. The changes of refraction index are proportional to the density change as stated in equation 1 and therefore determination of apparent movement of background pattern can be used for quantitative determination of the density gradients in the investigated flow.

$$\frac{n-1}{\rho} = G(\lambda) \tag{1}$$

The constant $G(\lambda)$, in equation 1, is called the Gladstone-Dale constant. It depends on characteristics of the gaseous media and also on the frequency of light used.

The technique is also based on image displacements obtained by comparing *wind-off* and *wind-on* pictures[1]. The refraction index n is related to the obtained image displacements [1], (Δx and Δy), by the partial differential equation :

$$\frac{\partial^2 n}{\partial x^2} + \frac{\partial^2 n}{\partial y^2} = \frac{n_0}{MZ_d h}\left(\frac{\partial \Delta x}{\partial x} + \frac{\partial \Delta y}{\partial y}\right) \tag{2}$$

[1] Wind-off and Wind-on pictures stand for pictures taken of the test section of a wind tunnel with flow at rest and during wind tunnel operation respectively.

The constants n_0, M, Z_d and h in equation 2 represent refractive index of undisturbed flow, optical magnification, distance between background and schilieren object and object thickness respectively. Once this equation is integrated, equation 1 can be used to obtain density.

This visualization technique is therefore not only capable of extracting qualitative information about the flow-field but also quantitatively extract density values. This makes BOS an attractive method for analyzing complex flow types.

This shows that visualization techniques have progressed from methods based on qualitative description of fluid flow into techniques providing both qualitative and quantitative results.

Transonic flows are a good example of such complex phenomenon where significant gradients occur. These occur due to the mixed subsonic and supersonic regions which are both present in the same flow-field. There is very little analytical theory available on the behavior of transonic flow, which makes the *Background Oriented Schilieren* technique suitable for this type of application.

In this paper we present the implementation of *Background Oriented Schlieren* technique for measurements of the density gradients at transonic speeds. We perform our analysis on a conical body submitted to transonic flow of Mach 0.9 inside a wind tunnel test section. In the past decade, the BOS technique has been applied for measurements of compressible flows in wind tunnels, supersonic jets [4] and full-scale measurements, such as visualization of shock waves surrounding jet fighter [10] and path of rotor blade vortices in helicopter flight [4,10]. Although, the possibility of using density measurements with BOS technique at supersonic speeds [4,9,10] has been confirmed, the feasibility of BOS for density field reconstruction at transonic flow regime has not been explored. The aim of the presented research was to perform BOS flow visualization and develop a procedure for determination of density field at transonic speeds. The results obtained by this procedure are then validated by *Computational Fluid Dynamics* (CFD) simulations that numerically determine the density field around the conical body. In the paper, development of BOS measurement setup as well as data reduction procedures are presented. A dedicated procedure was developed for retrieving the density field from the experimental data. The presented methodology can be used for investigations of compressible flows at transonic speeds as well as can be extended for supersonic flows.

2. Experimental Setup

The flow over a cone placed in the test section of a transonic wind tunnel was investigated at Mach number 0.9 by the BOS technique.

The N-3 wind tunnel is a closed circuit blow down type wind tunnel with partial recirculation of the flow. The cross-section of the test chamber is a square of side H = 0.6 m. The Mach number can be set with accuracy 0.01.

The Background Oriented Schlieren is based on digital analysis of image displacements. The background pattern is illuminated by source of light providing constant level intensity over the whole test time and imaged through a fluid containing spatial density gradients (figure 1). Since coherent beam of light is not required, the BOS setup is much simpler than classical Toepler schlieren or shadowgraph schlieren apparatus.

The images were captured with Nikon d800 digital SLR camera with 36 MP CMOS sensor and Nikon Nikkor AF-S 70-200 mm f2.8 lenses. The focal length was set to 200 mm. The background was created by spraying black paint on white paper. A series of images was created by varying the pressure in the airbrush. The series varied from very small (1-2 mm in diameter) and dense spots to big (5-8 mm) and sparse dots. The optimal density, providing images of dots size from 1 to 3 pixels on wind-off images, was chosen from the series of background. In order to provide stable illumination of the background, a HardSoft IL-106X pulsed LED illuminator [5] with a green LED was utilized. During the test, the illuminator was in continuous work mode and the images were acquired by remote shutter release.

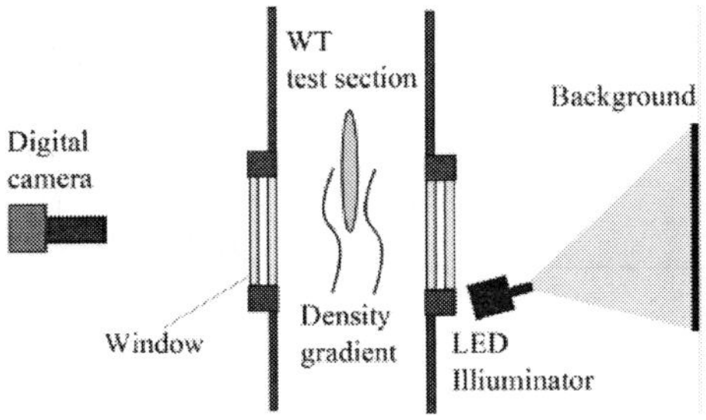

Figure 1. Experimental setup.

3. BOS Algorithm

The Background Oriented Schilieren technique can be summarized in the following steps:

(i) Acquisition of reference image (*wind-off*) of background without flow and images during the wind tunnel run (*wind-on*).

(ii) Determination of the apparent image displacements between the wind-off and wind-on images by digital image processing.

(iii) Integrate Poisson equation (equation 2) in order to obtain the density field.

The images acquired for both *wind-off* and *wind-on* conditions are displayed in figure 2.

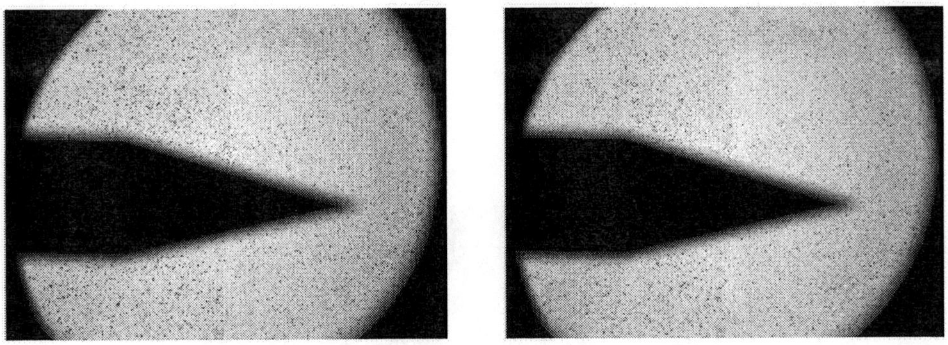

(a) (b)

Figure 2. Images acquired of the wind tunnel test section: (a) Wind-off. (b) Wind-on

There is no apparent difference between wind-off and wind-on images as one can easily notice in figure 2. However, when both images are represented by matrices of pixel values, the differences become clear. Based on that, the *PIV Cross-Correlation Analysis* [6] computes the difference of both images pixel by pixel. The maximum difference corresponds to the most probable displacement, and generates a corresponding vector in each interrogation area of the image.

Adaptive PIV algorithm [8] was applied to both images, where the areas with no displacement (body area) was removed. The resultant displacement field is shown in figure 3. One can easily notice that the greater displacements are concentrated in the regions where expansion and shock waves appear (greater density gradients).

Once the displacement field is obtained, we have all the information required to integrate equation 2. In order to restrict our analysis to the flow area of interest, we crop the domain as demonstrated in figure 4.

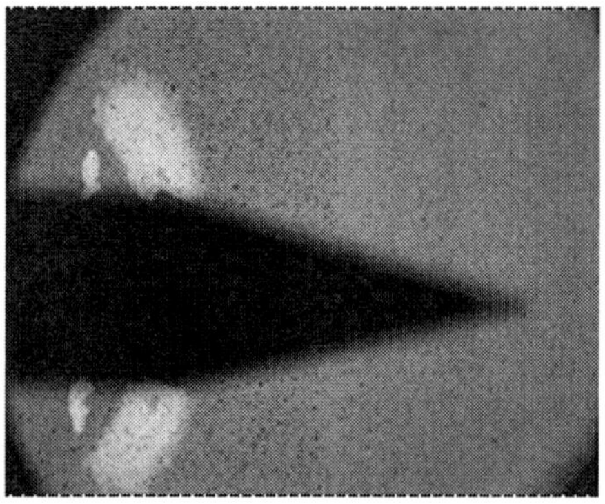

Figure 3. Displacement field obtained by PIV Correlation Algorithm.

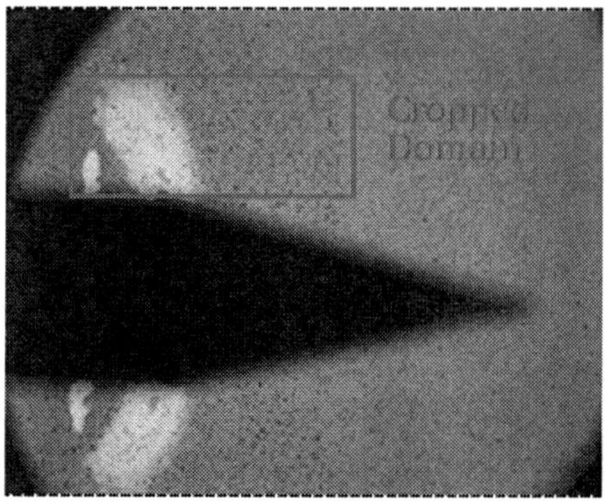

Figure 4. Area of interest.

Equation 2 is then integrated over the cropped domain using the five point finite difference method with successive over relaxation [2]. Dirichilet and Neumann boundary conditions are applied to the edges parallel and perpendicular to the flow respectively [1] as shown in figure 5. The Dirichilet boundaries were set to be equal the undisturbed flow refractive index ($n = n_0$) and the Neumann boundaries were set to zero. Once the refractive index field was obtained, density was calculated using the *Gladstone-Dale* equation (1).

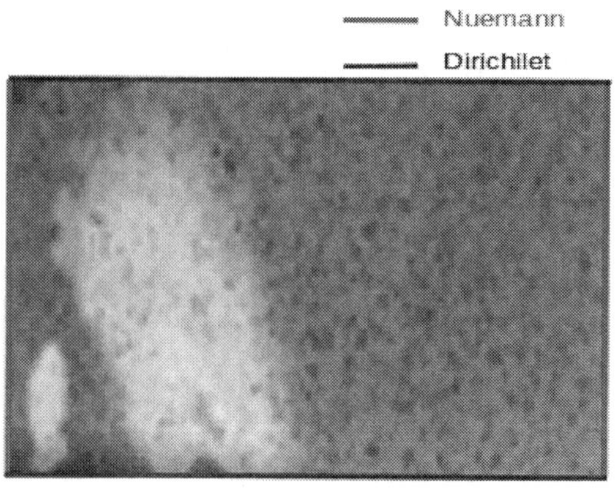

Figure 5. Boundary conditions applied to the cropped domain.

A *Computational Fluid Dynamics* (CFD) simulation was also performed using the commercial code *Ansys Fluent* in order to check density values feasibility. *Fluent* is a *Pressure based* solver, where the governing equations are solved sequentially (segregated from one another).

The simulation domain consisted in a 2-D wedge that represents the cross section of the conical shaped body under analysis. The simulation domain and the mesh are represented in figure 6 and 7. A total of 26400 elements were used. The boundary conditions were set as follows:

- Farfield of Mach 0.9 on the right, top and left edges.
- Physical walls on the diagonal edge and the adjacent to it.
- Symmetry line ahead of the conical body.

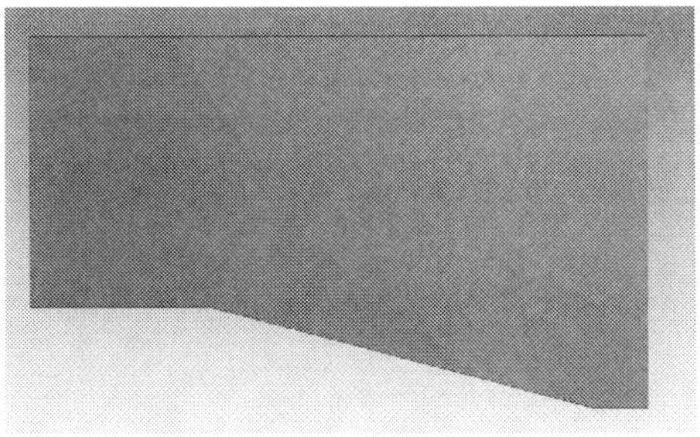

Figure 6. Simulation domain.

Figure 7. Simulation Mesh.

4. Results and Discussion

In this section, the outputs obtained from the BOS algorithm are discussed and compared to CFD results in both qualitative and quantitative terms. The density contours obtained by CFD are displayed in figure 8. In this figure it is possible to notice the presence of characteristic elements of transonic flows such as shock and expansion waves.

Figure 8. Density field - CFD

However, a comparative analysis is only possible if the two flow domains of BOS and CFD are the same. In order to do so, the CFD is cropped identically to the BOS displacement field as shown in figures 9 and 10.

Once the domains are equal, they can be compared as shown in figure 11. Although the CFD contours are smoother, the similarity between CFD and BOS is clear in terms of location of flow gradients. The location of density decrease is similar in both cases as show in blue in figure 11. Re-compression of the flow is also evident in both plots on the left hand side of the blue part. Therefore, both plots demonstrate the qualitative similarity between BOS and CFD.

In order to compare both techniques in quantitative terms, lines were traced on both domains at different locations in the y direction. Density was acquired across those lines for both domains, and the values were plotted in figure 12. The density lines show good similarity between BOS

IOP Conf. Series: Journal of Physics: Conf. Series **1101** (2018) 012030 doi:10.1088/1742-6596/1101/1/012030

and CFD results, as well as almost the same behavior in terms of increase and decrease of density.

Table 1 compares the maximum, minimum and average values extracted from both domains, as well as the difference between them. The average and maximum values are within a small error range when compared to the minimum values. This happens because of wall effects that considerably affect CFD results. Comparing the numerical values shows that the BOS outputs match the CFD results within an acceptable error range. This technique is therefore capable of not only displaying qualitative characteristics of the flow, but also quantitatively calculating density values. BOS is indeed a proof that flow visualization techniques are powerful tools that can measure flow properties.

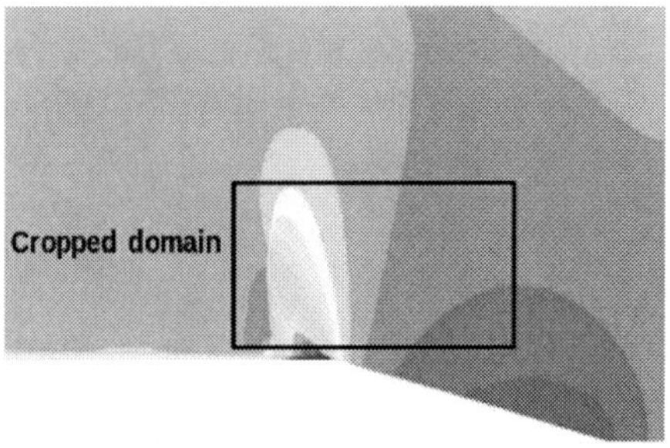

Figure 9. CFD domain cropped.

Figure 10. BOS region of interest.

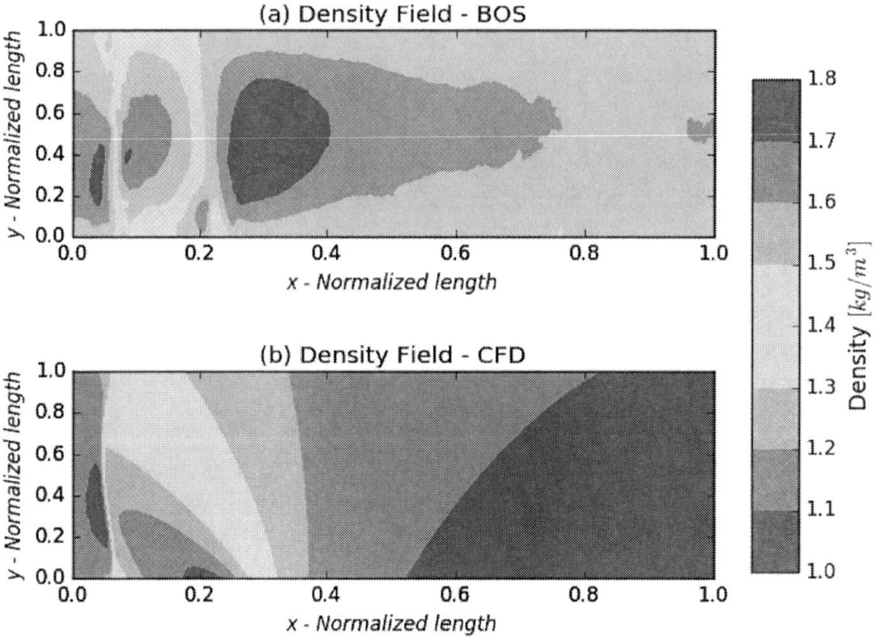

Figure 11. Density field comparison: (a) Density field - BOS. (b) Density field - CFD

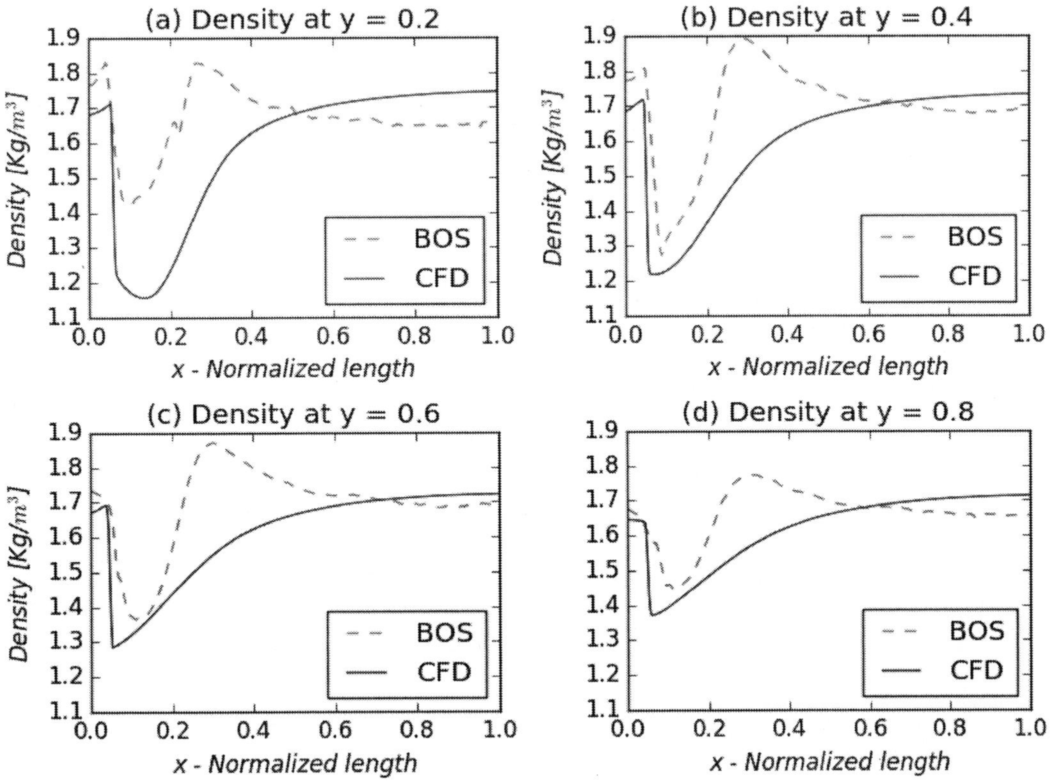

Figure 12. Density lines at different y locations: (a) y = 0.2. (b) y = 0.4. (c) y = 0.6. (d) y = 0.8.

Table 1. Density values comparison.

-	Density (BOS)	Density (CFD)	Difference
Maximum Value [Kg/m^3]	1.89	1.76	7.60 %
Minimum Value [Kg/m^3]	1.26	1.06	18.4 %
Average Value [Kg/m^3]	1.66	1.60	3.73 %

5. Conclusions

Flow visualization techniques have greatly progressed from methods of qualitative description to methods capable of measuring flow properties.

The results demonstrated by figures 11 and 12 show that the BOS technique was able to match CFD results within an acceptable error range in both qualitative and quantitative terms. *Background Oriented Schilieren* together with the *PIV Correlation* technique are capable of not only providing qualitative information about the flow, but also compute the density flow-field within a small error range. This technique also proved to be very convenient for the analysis of transonic flows, and can be further applied to more complex types of flow. The simplicity of the BOS setup makes it attractive to use for comparison of the CFD results with Wind Tunnel experimental results for verification of the numerical simulations results.

IOP Conf. Series: Journal of Physics: Conf. Series **1101** (2018) 012030 doi:10.1088/1742-6596/1101/1/012030

References

[1] Aguayo Ortiz A., Cardoso Sakamoto H., Echeverría Arjonilla C., Porta Zepeda D., Stern Forgach C., and Monsivais Galindo G. Calibration of a background oriented schlieren (bos). In *Recent Advances in Fluid Dynamics with Environmental Applications*, pages 103–114, Cham, 2016. Springer International Publishing.

[2] John H Mathews and Kurtis D Fink. *Numerical Methods Using MATLAB.* Third edition, 1999.

[3] Markus Raffel. Background-oriented schlieren (bos) techniques. *Experiments in Fluids*, 56(3):60, Mar 2015.

[4] H. Richard, M. Raffel, M. Rein, J. Kompenhans, and G. E. A. Meier. Demonstration of the applicability of a background oriented schlieren (bos) method. In *Laser Techniques for Fluid Mechanics*, pages 145–156, Berlin, Heidelberg, 2002. Springer Berlin Heidelberg.

[5] Boleslaw Stasicki, Juergen Kompenhans, Christian Willert, and Krzysztof Ludwikowski. Pulsed led illuminator for visualization, recording and measurements of high-speed events in mechanics. *In Proceedings of The Int. Conf. On Experimental Mechanics 2010*, pages 1–6, 01 2010.

[6] W Stryczniewicz. Algorithm for determining the vector field of velocity by the method of image anemometry. *Principles of Mechatronics*, 9, 2012.

[7] W Stryczniewicz. Quantitative visualization of compressible flows. *Transactions of the Institute of Aviation*, 2018.

[8] R Theunissen, F Scarano, and M L Riethmuller. An adaptive sampling and windowing interrogation method in piv. *Measurement Science and Technology*, 18(1):275, 2007.

[9] L. Venkatakrishnan and G. E. A. Meier. Density measurements using the background oriented schlieren technique. *Experiments in Fluids*, 37(2):237–247, Aug 2004.

[10] Shota Yamamoto, Yoshiyuki Tagawa, and Masaharu Kameda. Application of background-oriented schlieren (bos) technique to a laser-induced underwater shock wave. *Experiments in Fluids*, 56(5):93, Apr 2015.

IOP Conf. Series: Journal of Physics: Conf. Series **1101** (2018) 012031 doi:10.1088/1742-6596/1101/1/012031

Hybrid MPI/Open-MP acceleration approach for high-order schemes for CFD

Michal Saczek, Karol Wawrzak, Artur Tyliszczak, Andrzej Boguslawski

Faculty of Mechanical Engineering and Computer Science, Czestochowa University of Technology, Czestochowa, Poland

E-mail: michal.saczek@icis.pcz.pl

Abstract. The paper presents hybrid MPI+OpenMP (message Passing Interface/Open Multi-Processor) used for paralleled programs including high-order compact method. The main tools used to implement parallelism in computations are OpenMP and MPI which differ in terms of memory they are based on. OpenMP works on shared-memory and MPI on distributed-memory whereas hybrid model is based on combination of those methods. Tests performed and described in this paper present significant advantages provided by hybrid MPI/OpenMP. Testing computations needed for verifying possibilities of MPI, Open-MP and Hybrid of both are carried out using an academic high-order SAILOR solver. Obtained results seem to be very promising to accelerate simulations for fluid flows as well as for application using high order methods. The proposed approach was tested up to 96 cores with up to 4 nodes.

1. Introduction

For ages researchers have struggled to understand precisely a complex structure of fluid flows. Impressive improvements and achievements in computer technology and increasingly common use of numerical simulations have enabled more elaborate experiments thanks to which there can be observed constant progress in the field of Computational Fluid Dynamics (CFD). Issue that has received lots of attention in the CFD community are high-order discretisation methods that are irreplaceable in DNS (Direct Numerical Simulation) and LES (Large Eddy Simulation) studies focused on very deep and detailed analysis of the fluid flow problems. High-order methods are expected to open the door to high solution accuracy with lower cost, their present status and ways of applications are described in the review paper [1]. In terms of accuracy there are spectral and pseudo-spectral methods that are considered superior to other high-order discretisations ones, however, they can be used mainly in simple computational domains.

One of the most attractive high-order method is compact difference one [2], which combines global approximation (spectral method) and flexibility of the computational domain (finite volume method). However, in this method the approximations of derivatives are obtained by solving linear system of equations (tridiagonal or pentadiagonal) that creates huge difficulties in parallelization as it makes it a necessity to execute algorithm step by step in a serial fashion. In order to divide linear system of equations into independent computational groups special methods have to be applied. There are many methods, such as cyclic reduction method[3], partitioning methods [4, 5, 6, 7] and various implementations of the Gauss elimination method [8, 9, 10], which allow the linear system to be solved in a parallel way.

As shown in the exhaustive review paper [11] the dominant programming tools which are used to implement parallelism in CFD codes are OpenMP, MPI, hybrid OpenMP+MPI and CUDA. However, the CUDA tools is not included in the paper. The main difference between OpenMP and MPI is different access to memory, OpenMP works on shared-memory[12] whereas MPI is based on distributed-memory that means the exchange of data relies on communications between nodes[13]. Hybrid programming is based on the idea of using OpenMP threads to employ the multiple cores in a certain socket/node and MPI to communicate among the sockets/nodes. OpenMP, MPI methods and their combination are presented in Fig. 1. All of the tools mentioned above were widely used by many researches in CFD investigation: pure OpenMP[14, 15], pure MPI[16, 17], as well as hybrid approach MPI+OpenMP[18, 19].

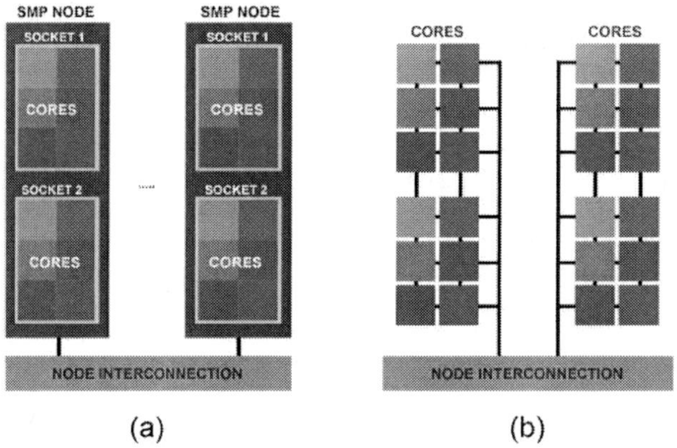

Figure 1. SMP(Symmetric multiprocessing) cluster: a - typical multi-socket multi-core cluster, b - mapped MPI processes to each core[20].

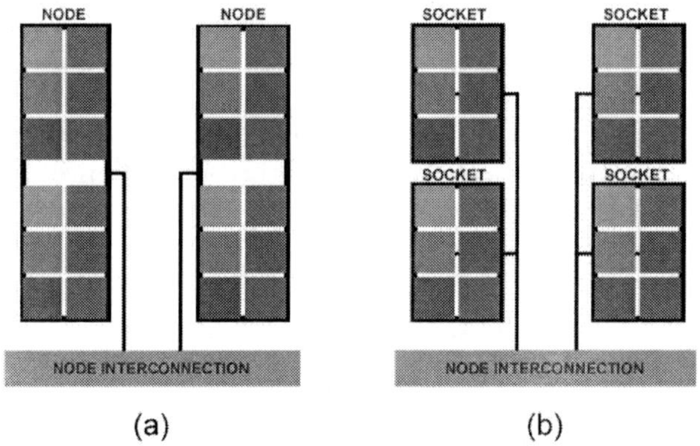

Figure 2. SMP cluster: a - mapped MPI to each node and OpenMP threads to each core on that node, b - mapped MPI processes to each socket and OMP threads to each core on that socket[20].

Modern multi-socket multi-core SMP cluster, showed in Fig. 1a, allows a program to be run on various configuration[20]. Initial research, which is not presented in the paper, proved that the best configuration for hybrid approach is to assign MPI to each socket and then to use every core available on that socket as OpenMP threads, as is shown in Fig. 2b. Another option is shown on Fig. 2a, every MPI process is assigned to nodes and then every core available on that node works as threads. The last option is showed in Fig. 1a where MPI process is assigned to every core available in cluster then threads can be pined to the same core. This method is especially useful for pure MPI programs.

The aim of this work is to present hybrid MPI+OpenMP parallelized algorithm for high-order compact method in three spatial dimensions. The test program which was created for this research is included in SAILOR solver used in various LES studies including free jet flows [21], multi-phase flows [22] and flames[23].

2. Numerical algorithm

The SAILOR code is based on a projection method for the pressure-velocity coupling. The time integration is performed by a predictor-corrector (Adams-Bashforth/Adams-Moulton) approach and the spatial discretization is based on the 6th order compact difference method for half-staggered meshes. The detailed description of the SAILOR code was presented in [24].

2.1. Compact Difference Scheme

The 6th order compact finite difference formula for the first order derivative on a collocated mesh is given as [2]:

$$af'_{i-1} + f' + af'_{i-1} = \frac{b_1(f_{i+1} - f_{i-1})}{h} + \frac{b_2(f_{i+2} - f_{i-2})}{2h} \tag{1}$$

where the coefficients are equal to $a = 1/3$, $b_1 = 7/9$, $b_2 = 1/36$. The equation (1) written for every node leads to the system of equations:

$$\mathbf{Af}' = \mathbf{Bf} \tag{2}$$

where the matrix $\mathbf{A}$ is tridiagonal.

2.2. Basic tridiagonal matrix algorithm

The tridiagonal matrix is a special type of matrix that the nonzero elements only on the main diagonal $\mathbf{b}$ and two neighbors before ($\mathbf{a}$) and after ($\mathbf{c}$) main diagonal. A general system of

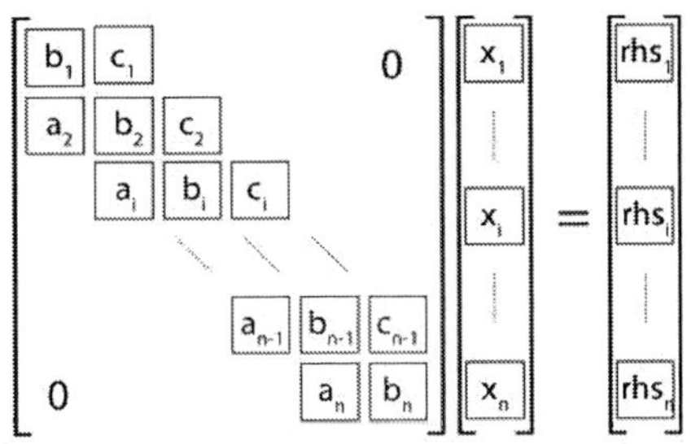

Figure 3. Tridiagonal matrix.

equations showed in Fig. 3 is:

$$\mathbf{Ax} = \mathbf{RHS} \tag{3}$$

and it could be solved by tridiagonal matrix algorithm (TDMA) also known as the Thomas algorithm. It could also be solved by the product of inverse tridiagonal matrix and **RHS** ($\mathbf{x} = \mathbf{A}^{-1}\mathbf{RHS}$). Nevertheless, the later method is more time-consuming than TDMA. The TDMA algorithm is carried out in two steps:

- **forward** loops from 1 to n and preparing d and γ coefficients

$$\begin{cases} d_i = c_i/b_i & \text{where } i = 1 \\ d_i = (b_i - a_i * d_{i-1})/c_i & \text{where } i = 2, 3, ..., n \end{cases} \tag{4}$$

$$\begin{cases} \gamma_i = rhs_i/b_i & \text{where } i = 1 \\ \gamma_i = (rhs_i - a_i * \gamma_{i-1})/(b_i - a_i * d_{i-1}) & \text{where } i = 2, 3, ..., n \end{cases} \tag{5}$$

- **backward** loops form n to 1 and getting a result x

$$\begin{cases} x_i = \gamma_i & \text{where } i = n \\ x_i = \gamma_i - d_i * x_{i+1} & \text{where } i = n-1, n-2, ..., 1 \end{cases} \tag{6}$$

and requires to be done step by step in a serial fashion.

2.3. Parallel TDMA "arrowhead" method

The arrowhead method [7] is one of partitioning methods which allows solving equation (3) in a parallel way. The method relies on reducing initial matrix to some new independent diagonal blocks and sparse auxiliary ones along the right and bottom side and a coupling supplementary block (Fig. 4). Reduced matrix is presented in Fig. 5.

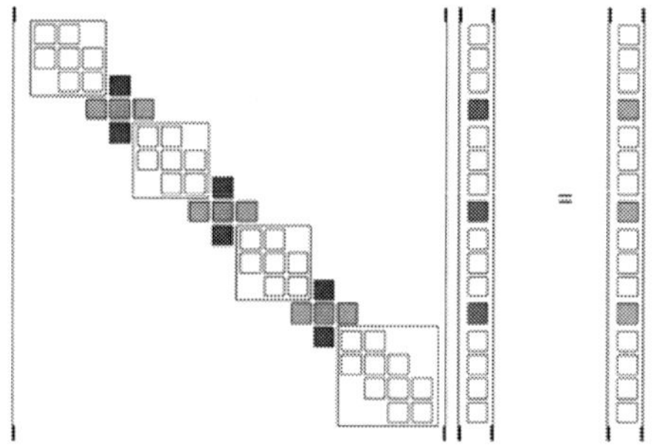

Figure 4. Initial tridiagonal block divide by number of processes[7].

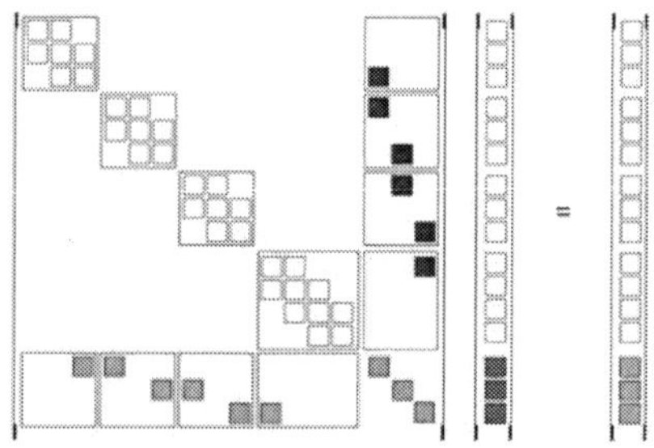

Figure 5. Matrix after rearranged to arrowhead form[7].

The matrix rearrangement is performed in the following way:

- define number of subsystems M which can rely on number of MPI tasks. Every subsystem is shown in Fig. 4 marked by thin black square,
- mark $M-1$ separations block-rows (orange, green, red and grey colors) for every subsystem without last one and corresponding block-column (blue) what is shown in Fig. 4
- marked separation block-rows are shifted to the bottom
- marked separation block-columns are shifted to the right

Global formula of this system is written as

$$\begin{pmatrix} S & W_r \\ W_l & H \end{pmatrix} \begin{pmatrix} x_s \\ x_h \end{pmatrix} = \begin{pmatrix} rhs_s \\ rhs_h \end{pmatrix} \tag{7}$$

where S is subsystem, H - head (orange squares), W_{lr} - wings (green and blue squares).

The solution of the system is given by the relations

$$\begin{cases} x_s = S^{-1}rhs_s - S^{-1}W_r x_h \\ x_h = (H - W_l S^{-1} W_r)^{-1}(rhs_h - W_l S^{-1} rhs_s) \end{cases} \tag{8}$$

The system is solved in the following steps:

- preparing $Z_k = (S^k)^{-1}W_l^k$ on each process, which leads to TDMA solution for $S_k Z^K = W_l^k$,
- preparing $H - \sum_{k-1}^{M} W_l^k Z^k$ on one process,
- preparing $z_k = (S^k)^{-1}rhs_s^k$ on each process, which leads to TDMA solution for $S_k z^K = rhs_s^k$,
- preparing $rhs_h - \sum_{k-1}^{M} W_l^k z^k$ on one process.

Synchronizing data to one process and preparing coefficients for result:

- preparing x_h which is a result for $M-1$ subsystems $x_h = (H - W_l S^{-1} W_r)^{-1}(rhs_h - W_l S^{-1} rhs_s)$ which leads to TDMA solution for $(H - W_l S^{-1} W_r)x_h = (rhs_h - W_l S^{-1} rhs_s)$,
- Sending x_h by scattering and correction of result on every process $x_s^k = z^k - Z^k x_h$.

This method is very simple to implement and provide speed-up, especially compared to basic TDMA method. It could be applied for other diagonals like five-diagonal matrix or seven-diagonal matrix.

2.4. Test program

The SAILOR code is parallelized using MPI library by dividing the computational domain in the one direction (y) as presented in the Fig. 6a. Within this approach derivatives in the others directions (x, z) are obtained in parallel way (the communication between MPI processes is not required). Each subdomain whereas the y-derivatives are computed using serial TDMA algorithm or the arrowhead method (the communication between MPI processes is required). As it was mentioned in the Introduction in order to present hybrid MPI+OpenMP algorithm the test program was created. The test program prepares analitic function and performs calculations of derivatives in all spatial directions (x; y; z), hovewer, it does not carry out whole CFD simulation. To allow OpenMP threads cooperate with MPI the domain could be additionally divided. For y-derivatives (communication between MPI processes is required) division should be done in second direction as presented in the Fig. 6b. For another derivatives (communication between MPI processes in not required) division should be done in the same direction as MPI processes, as shown in the Fig. 6c.

3. Results

Tests were run on PLGRID platform, which allows about 2399 TFlops computational power. Running and compiling application was done by using Intel MPI library version 2018 Update 1 and OpenMP in version 5.0. Specification of cluster is shown in table 1. Tested domain had to be fitted to cluster specification and was performed for two mesh sizes, for seven different numbers of computational units and for four different numbers of nodes. The meshes were

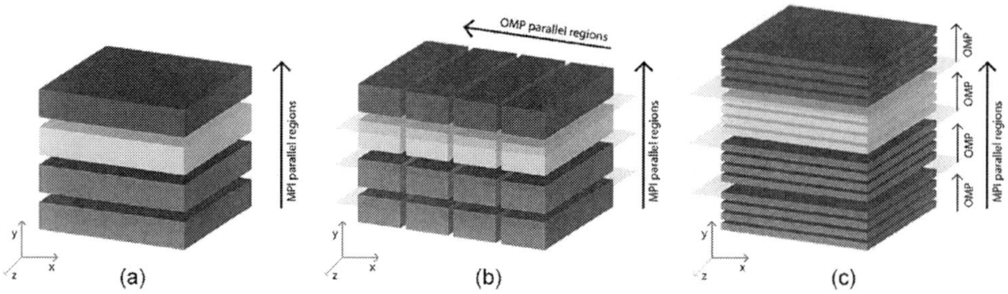

Figure 6. Domain dividing: a - to MPI processes, b - to MPI processes and OpenMP in second direction, c - to MPI processes and OpenMP in the same direction.

Table 1. Specyfication of the test cluster.

Processors	Intel E5-2680v3
Socket per node	2
Cores per socket	12
Clock rate	2,50 GHz
Memory size	128GB
Nodes	2160
Network	Infiniband FDR 56Gb/s

Table 2. Number of nodes for computational units and meshes.

computational units	1	6	12	24	48	60	96
240x240x240	1	1	1	1	2	3	-
480x480x480	1	1	1	1	2	3	4

structured and the distance between mesh points was constant. Details are shown in table 2. The tests were performed for three different cases: pure MPI tool and serial TDMA, pure MPI and parallel TDMA (arrowhead), hybrid (MPI/OpenMP) and parallel TDMA (arrowhead).

Speed-up is measured by function: $P = T^1/T'$, where T^1 is time for particular configuration, T' is time for basic TDMA solver on one computation unit.

As it can be observed in Figs 7 and 8 the hybrid solution offers great performance especially for many CPUs and it is less efficient for smaller number of CPUs. For minimum number of CPUs it offers similar efficiency that serial TDMA with pure MPI. The maximum performance what was achieved for hybrid solution is:

- for 240x240x240 mesh size:
  - about 7 times faster than serial TDMA and pure MPI
  - about 4.7 times faster than parallel TDMA method and pure MPI
- for 480x480x480 mesh size:
  - about 10 times faster than serial TDMA and pure MPI

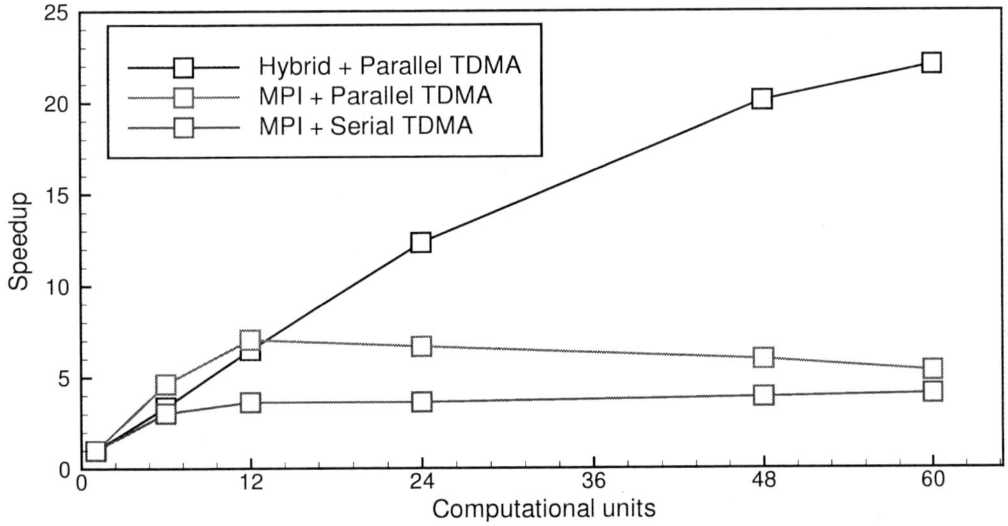

Figure 7. Results for mesh size 240x240x240 points.

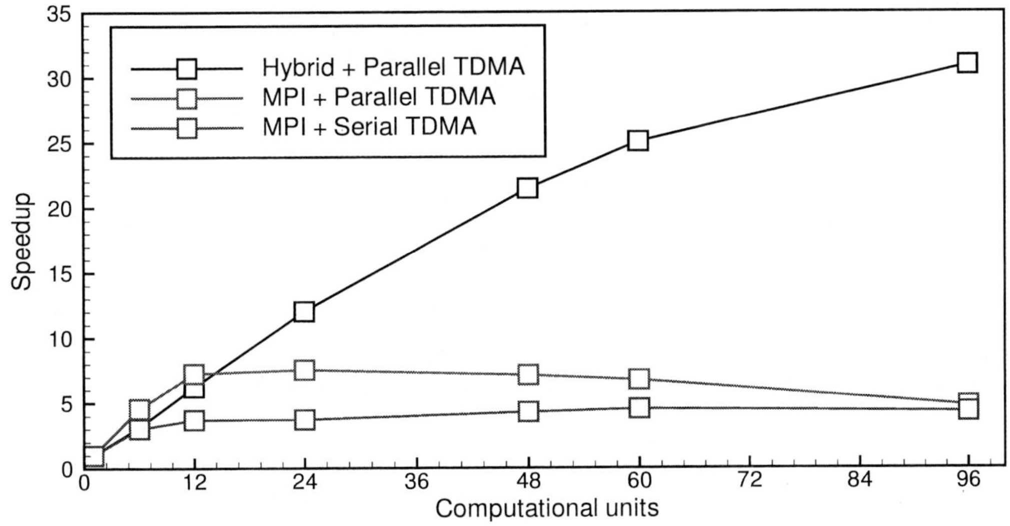

Figure 8. Results for mesh size 480x480x480 points.

– about 6.7 times faster than parallel TDMA method and pure MPI

It can also be observed that pure MPI with serial TDMA accelerates with increasing computational units. That is because measured speed-up includes preparing RHS, creating f and performing calculations in the second and third spatial directions. It is also important that both approaches serial TDMA and parallel TDMA method with pure MPI slow down after some "sweet point" depending on mesh size.

4. Conclusions

There is a great demand for fast and efficient systems in today's word. Hybrid MPI/OpenMP approach is supposed to enhance the speed of calculations for high order methods, what has been proved by the investigations performed. Comparison of all three sorts of approaches shows the important advantages of the hybrid model over the pure MPI with serial TDMA and pure MPI with parallel TDMA. It is worth mentioning that the size of computational domain directly influences the performance of calculations. The tests indicate that the larger domain is used the larger acceleration is achieved. In the case of computational units below 12 it is sufficient to use pure MPI with parallel TDMA, however, to gain best performance in the case of computational units over 12 taking advantage of the hybrid MPI/OpenMP approach is preferable. As a part of future work this approach will be implemented to whole CFD simulation. Moreover other approaches of performing calculations for high-order methods on heterogeneous clusters with different numbers of socket and core per socket may be compared.

Acknowledgements.
The research was supported by Polish National Science Centre, project no. 2016/21/B/ST8/00414. The computations were carried out using PL-Grid Infrastructure.

References

[1] Wang Z, Fidkowski K, Abgrall R,Bassi F, Caraeni D, Cary A, Deconinck H, Hartmann R, Hillewaert K, Huynh H, Kroll N, May G, Persson P, Leer B and Visbal M 2013 *Int. J. Numer. Methods Fluids* **72** 811-845
[2] Lele S 1992 *J. Comput. Phys.* **103** 16-42
[3] Hockney R 1965 *J. ACM* **12** 16-42
[4] Wang H 1981 *ACM Trans. Math. Softw.* **7** 170-183
[5] Mattor N, Williams T, and Hewett D 1995 *Parallel Comput.* **21** 1769-1782
[6] Sun H, Zhang H, and Ni L 1992 *IEEE on Trans.Comput.* **41** 286-296
[7] Belov P, Nugumanov E and Yakovlev S 2015 *Preprint* arXiv 1505.06864
[8] Stone H 1975 *ACM Trans.Math. Softw.* **1** 289-307
[9] Rao S C S 2008 *Parallel Comput.* **34** 177-197
[10] Qin J and Nguyen D 1998 *ADV ENG SOFTW* **29** 395-397
[11] Afzal A, Ansari Z, Faizabadi A, and Ramis M 2017 *ARCH COMPUT METHOD E* **24** 337-363
[12] Dagum Land Menon R 1998 *IEEE Comput. Sci. Eng.* **5** 46-55
[13] William D, Lusk E and Skjellum A 1999 *Using MPI: Portable Parallel Programming With The Message-Passing Interface* **1**
[14] Amritkar A, Deb S and Tafti D 2014 *J.Comput.Phys*
[15] Mavriplis D 2002 *Int. J. High Perform. Comput. Appl.* **16** 395-407
[16] Maknickas A, Kačeniauskas A, Kačianauskas R, Balevičius R, and Algis Džiugys A 2006 *Informatica* **17** 207-224
[17] Jia R and Sundén B 2004 *Comput. Fluids* **33** 57-80
[18] Mininni P, Rosenberg D, Reddy R, and Pouquet A 2011 *Parallel Comput.* **37** 316-326
[19] Gropp W, Kaushik D, Keyes D, and Smith B 2001 *Parallel Comput.* **27** 337-362
[20] Rabenseifner R, Hager G, and Jost G 2009 In *Proc Euromicro Int Conf Parallel Distrib Netw Based Process* **17** 427-436
[21] Wawrzak K, Boguslawski A, and Tyliszczak A 2015 *Flow Turbul. Combust.* **95** 437-459
[22] Aniszewski W, Boguslawski A, Marek M, and Tyliszczak A 2012 *J. Comput. Phys.* **231** 7368-7397
[23] Wawrzak A and Tyliszczak A 2017 *ARCH MECH* **69** 157-175
[24] Tyliszczak A 2014 *J. Comput. Phys.* **276** 438-467

IOP Conf. Series: Journal of Physics: Conf. Series **1101** (2018) 012032 doi:10.1088/1742-6596/1101/1/012032

A Computational Study of the Fluid Particles Distribution in an Helicopter Wake

T P Seredyn[1]

[1] Polish Air Force Academy, Department of Applied Science, Dywizjonu 303 nr 35, 08-521 Dęblin, Poland

E-mail: t.seredyn@wsosp.pl

Abstract. This paper presents a Computational Fluid Dynamics (CFD) study of droplet spray dispersal in the wake of a Mi-2 helicopter up to 600 m downstream. The Reynolds-averaged Navier-Stokes (RANS) equation use a Lagrangian (droplet phase) and Eulerian (fluid phase) procedure to predict the droplet trajectories trough the turbulent aircraft wake. The methods described have the potential to improve current models for aerial spraying and help in the development of new spraying procedure. In this study, models are developed for the sprays released from atomizers mounted on a helicopter. A parametric study of the aircraft model examines the effects of crosswinds on the helicopter's vortex structures and the resulting droplet trajectories. The study shows, that such influence is underestimated in the current models. A comparison of our results with those obtained with AGricultural DISPersal (AGDISP) is also presented.

1. Introduction

The spread of various substances by aircraft takes place in many areas of human life, from military applications and protection of forests to agricultural aviation. It should be mentioned that a very large and very substantial amount of aircraft are employed in crop protection treatments, as well as in the management of forests. Chemical agents are most commonly used, due to their high degree of effectiveness in killing pests, diseases and weeds, at relatively low costs. However, one of the fundamental limitations in the application of this technique is the chance of chemical agent drift [12]. By drift we mean, the unintentional movement and settlement of agents, outside the area that is being treated.

For the past fifty years from the first study by Reed [11], many theoretical models and a large data-collection effort has been undertaken [16]. From the 1980s until now, the development of Lagrangian droplet trajectory models has been suggested. Experimental research and the revolution in computer technology have enabled better modeling of aircraft vortex wakes and the tracking of droplets within them. This has ultimately led to the development of the Lagrangian AGricultural DISPersal (AGDISP) model, which has become the industry standard. AGDISP was developed by the USDA Forest Service and is based on equations of motion governing the behavior of droplets in a fluid. However, the most important limitations of AGDISP are connected to a very simple representation of the wakes produced by an aircraft. Even so, the Lagrangian models are more capable of capturing the near-wake flow field effects of aircraft than the Gaussian approach.

Subsequent authors, such as Деревянко [5] developed this theory, considering a bi-plane system. They assumed two pairs of horseshoe vortices, which took into account the influence of the propeller slipstream and cross winds. Moreover, Деревянко included the evaporation of droplets. These issues were also addressed in further scientific research and a Lagrangian trajectory model was adopted [2]. An extensive field study and model validation effort [3] confirmed the predictive capability of the Lagrangian model [17] to approximately 500 m downwind. In Poland, an elaborate model of spraying from airplane was suggested by Pietruszka [10], who developed the most appropriate, vortex-swirling theory. During the same period, other researchers independently developed their own spray-drift models or contributed essential parts of the modeling process. These models are shown in [13,14,18].

2. Aim and theoretical background

The aim of this study is to present mathematical models that describe the physics of the spread and distribution of atomized liquid droplets, in the air, by a helicopter. Based on analysis of the models, the quality and usefulness of these theories, in fieldwork and any possible indications of future directions of research will be considered.

The flight of any aircraft results in a disturbance of the velocity field, in a large area of its flight path. This is due to the turbulence field flowing from the wings, tail-planes or rotors and propellers. In further distances behind the aircraft, vorticity is concentrated into two vortices, at a distance of about 0.8 of a wingspan or rotor to one another [8,9]. The energy of these disturbances makes it necessary to take into account their effect on spray droplet trajectory. This trajectory depends mainly on droplet size and the position of the nozzle along the wing or the rotor [13].

The AGDISP model is based on the following assumptions for a helicopter: when an aircraft flies at constant altitude and speed, the aerodynamic lift generated by the aircraft equals the aircraft's weight. The majority of lift is carried by the wings or rotors and generates one pair of swirling masses of air (vortices) downstream of the aircraft. The rollup of this trailing vorticity is approximated as occurring immediately downstream of a helicopter and the local swirl velocity, around each of the two vortices, is given by:

$$v = \frac{\Gamma}{2\pi} \frac{r}{max(r,r_c)^2} , \tag{1}$$

where, Γ is vortex circulation strength, given by:

$$\Gamma = \frac{(1-F)W}{2\rho_a R U_\infty} , \tag{2}$$

r is the distance from the vortex center to the droplet position, r_c is vortex core radius, W is aircraft weight, R is the main rotor radius and U_∞ is a helicopter speed. Vortex strength decays with time, because of atmospheric turbulence, following a simple decay model:

$$F = \exp\left(-\frac{x}{R}\right) . \tag{3}$$

More explanations one can find in the original work by Bilanin et all. [2] and later papers [16,17]. The influence of area coverage on the flow velocity profile is explained more precisely in [7].

3. Motion of droplets

Sprayed droplets, with a high, initial velocity, have a significant distribution of diameters, within the range of few to several hundred μm (depending on the type of sprayers). This corresponds to Reynold's numbers, with respect to their sedimentation velocity, of $0.1 \leq Re \leq 100$. According to the work [4], drag force acting on the droplet versus the Reynolds number $Re = \frac{vd}{\nu}$ can be presented in the form:

$$C_D = \begin{cases} 0.424 & Re \geq 1000 \\ \frac{24}{Re}(1 + 0{,}16Re^{0.667}) & 0.5 < Re < 1000 \\ \frac{24}{Re} + 4.5 & Re \leq 0.5 \end{cases} . \tag{4}$$

For this range of Reynold's numbers, an error in the drag coefficient does not exceed 6%. The equation of droplet motion, in its general form, can be written as:

$$\frac{d}{dt}\left(m\dot{\overrightarrow{R_d}}\right) = m\vec{g} + \overrightarrow{F_D} + \overrightarrow{F_A} + \overrightarrow{F_M} + \overrightarrow{F_B} , \tag{5}$$

where, $m = \frac{\pi}{6}\rho_d d^3$ is the mass of a droplet, while $\overrightarrow{F_D}, \overrightarrow{F_A}, \overrightarrow{F_M}, \overrightarrow{F_B}$ are aerodynamic drag force, buoyancy force and Magnus and Basset forces, respectively. In the present form of motion, the last two forces are omitted. This is due to the fact, that such assumptions were adopted in previous works (so in this work to compare result), without significantly affecting the results of the calculations. This equation is solved

IOP Conf. Series: Journal of Physics: Conf. Series **1101** (2018) 012032 doi:10.1088/1742-6596/1101/1/012032

having given the velocity field of the air around droplets. The trajectory equation $\frac{dR_d}{dt} = v_d$ can be solved as well using the implicit and the trapezoidal schemes, where the new location of a droplet is given in the form $R_d^{n+1} = R_d^n + \frac{1}{2}\Delta t(v_d^n + v_d^{n+1})$. More details can be found in [4]. The evaporation of droplets is given in the formula:

$$m_d(t + \Delta t) = m_d(t) - N_i A_d M_{w,i} \Delta t, \tag{6}$$

where A_d is the surface and $M_{w,i}$ is the molecular weight of the droplet and N_i - molar flux of vapour given in the form:

$$N_i = k_c(c_{i,s} - c_{i,\infty}). \tag{7}$$

Here $c_{i,s}$ and $c_{i,\infty}$ is the vapour concentration at the droplet surface and the vapour concentration in the bulk gas, respectively. k_c can be received from the equation:

$$\frac{k_c d_d}{D_{i,m}} = 2 + 0.6Re^{1/2}Sc^{1/3}, \tag{8}$$

where the Schmidt number has the form $Sc = \mu/(\rho D_{i,m})$ for the diffusion coefficient of vapour in the bulk $D_{i,m}$. More detailed discussion of these issues are presented in the paper [4].

4. Object of research

The object of the study is the PZL Świdnik Mi-2 helicopter, equipped with apparatus for ULV spraying in the form of rotating sprayers (atomizers) AR-86.00.00 produced by the WSK PZL Świdnik, used in chemicalization treatment. For the study it has been assumed an altitude of the aircraft equal to $h = 5\ m$ and a speed corresponding to the most commonly used in treatments by means of helicopters $v_r = 22\ m/s$. The helicopter had the equipment that allows registration of: flight speed, pump rotations, atomizers rotation and the pressure of the pump.

Figure 1. Mi-2 helicopter geometry.

Field experiments were carried out in the years 1987-1992 on the Academy of Agriculture and Technology in Olsztyn research training ground, constituted at the former airfield in Gryźliny near Olsztyn at the area of 150 hectares covered with grass. In this area, parallel to the direction of the wind, a 800 m long measurement line was delineated with the samplers enabling to determine not only the size and density of the spray, but also, which is very important, changes in the diameter of the droplets for the given drift distance. The computational model of the plane is a three-dimensional full-scale model of the Mi-2 helicopter designed in by means of ANSYS 16.2 Design Modeller. The model includes the fuselage and two rotors treated as pressure drop. In order to simplify the geometry, certain components of lesser significance were not included [4]. The fluid domain is given in the **Figure 3**. Dimensions of the domain: 600 m long, 160 m wide and 33 m height. Its total volume is 3168000m³.

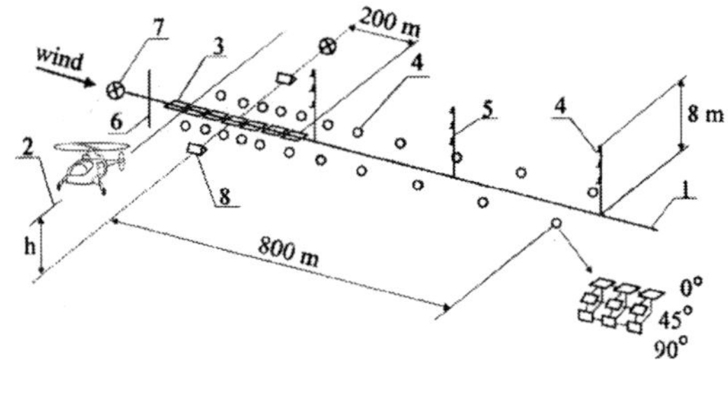

Figure 2. The schematic drawing of the research experiment site and the measurement line: 1- measure line, 2 – flight path, 3 – mass samplers, 4 – droplet samplers, 5 – masts, 6 – measurements of meteorological parameters, 7 – cameras, 8 – markers.

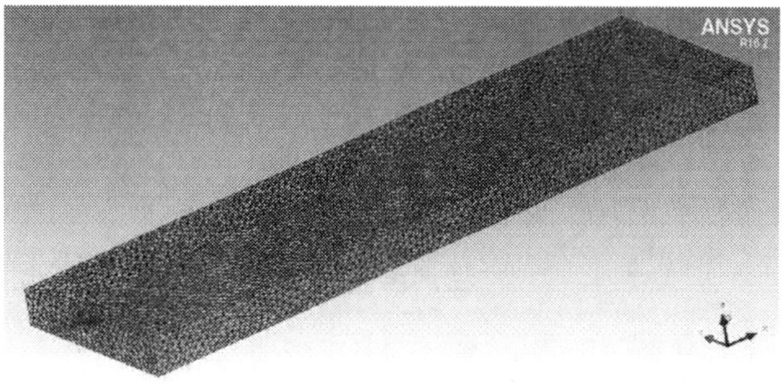

Figure 3. The geometry and mesh of the domain.

The domain was meshed as shown in the **Figure 3.** It contains over $7.5*10^6$ elements with the unstructured mesh containing tetra-shaped elements. The mesh is refined around the helicopter to ensure accurate resolution in these regions. During the calculations, the mesh was refined again in the regions behind the helicopter according to the appearance of swirls and droplets, until the results were independent of the grid structures.

The domain boundaries allow the flow to enter or exit the domain, depending on the local flow direction. Four sides act as a far field with specified velocity conditions that impose the wind, one acts as an outlet and one as a ground. The flow simulations have been computed using Reynolds-averaged Navier-Stokes (RANS) equations solver based on the finite volume method.

Figure 4. The refined mesh around the helicopter.

A commercial code, ANSYS Fluent, has been chosen for this purpose [4]. The basic set of gas phase governing equations is given in (9). This is a general equation with symbols given in the **Table 1**. The equations have been closed with the Menter K-ω SST 4-equations turbulence model, more adequate in terms of the flow turbulization behind the aircraft [4]. Since the pressure farfield condition was used on fluid boundary of the domain, the ideal gas model of density, as demanded by this condition, was also assumed.

$$\frac{\partial(\rho\phi)}{\partial t}+\frac{\partial(\rho u\phi)}{\partial x}+\frac{\partial(\rho v\phi)}{\partial y}+\frac{\partial(\rho w\phi)}{\partial z}=\frac{\partial}{\partial x}\left(\Gamma\frac{\partial\phi}{\partial x}\right)+\frac{\partial}{\partial y}\left(\Gamma\frac{\partial\phi}{\partial y}\right)+\frac{\partial}{\partial z}\left(\Gamma\frac{\partial\phi}{\partial z}\right)+S_\phi. \tag{9}$$

Discrete phase was chosen as interacting with continuous phase and droplets diameter distribution was represented by Rosin-Ramler distribution with total flow rate 1.378 kg/s.

Table 1. Gas phase governing equations.

Main equation	$\frac{\partial(v_j\rho\emptyset)}{\partial x_j}=\frac{\partial(\Gamma\emptyset)}{\partial x_j\partial x_j}+S_\emptyset$		
	$\emptyset$	Γ	$S_\emptyset$
Mass	1	0	0
Momentum	v_i	μ	$-\frac{\partial p}{\partial x_i}-\frac{\partial(\rho\overline{v_i'v_j'})}{\partial x_j}$

5. Results.

The theoretical models should be useful for predicting the behavior of spray released from aircraft and be able to calculate the distribution of a mass aimed at the target field. The AGDISP model was shown to overpredict deposition rates, relative to experimental field data [6], although recent improvements have allowed the latest version of AGDISP to match field data. However, the study shows that the mass distribution generated by AGDISP software does not exactly match the distribution received during the field experiments.

Figures below present a comparison of the theoretical calculations and data collected, during field experiments carried out at the research training ground of the Academy of Agriculture and Technology, which the experimental data was presented in the study by Seredyn et al. [15]. Analysis was done using a 30% water solution of urea, with an additional 2% of nigrosine, with settings for a 20.23 dm^3/ha dose. Atmospheric conditions were measured as: wing velocity – 5.1 m/s, temperature – 290.2 K and humidity

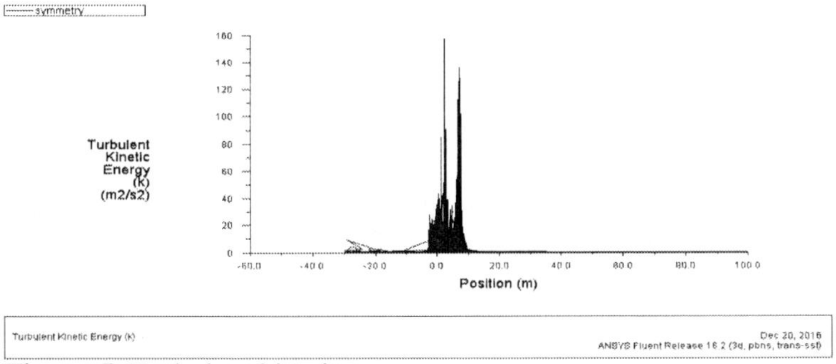

Figure 5. Turbulent kinetic energy.

– 65%. For the helicopter, the following conditions occurred: flight velocity – 22.8 m/s, height – 5 m and two sprayers on the boom.

First of all, after calculating the solution, it is quite different from that one received by means of the AGDISP model. In the **Figure 5** it is clearly visible that the turbulence around the helicopter is very high and disappears very quickly according to CFD predictions. In AGDISP model the turbulence is constant.

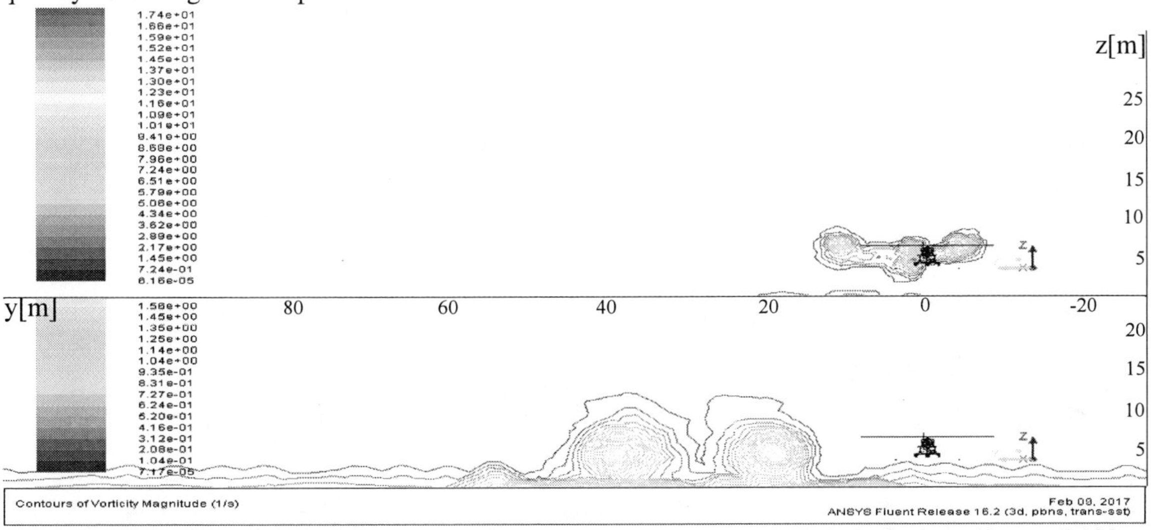

Figure 6. The contours of the vorticity magnitude (1/s) according to CFD in 2 planes – 10 m and 100 m behind the helicopter (x axis).

Now, in the **Figure 6** the vorticity generated by the helicopter in flight is presented. One can conclude very simply, that the assumption about two counter rotating masses of the air is not justified. The conclusions form the CFD calculations are close to the Joulain or Misaka predictions [8,9]. It is stated, that the vorticity roll-up is a quite long lasting process and quite complicated, especially after the helicopter such as Mi-2. From the figure we can see that 10 m after the helicopter the vorticity is more complicated, that those after 100 and more meters. On the other hand, according to the AGDISP model predictions the paths of the droplets exhibit the symmetry with respect to the plane passing through the axis of the helicopter. Whereas the paths of droplets, which were calculated in the CFD analysis, are quite complicated what is obvious from the observations. These paths are presented in the **Figure 7**.

Now the main problem will be considered. Of course, the main aim of the study is to receive the mass distribution on the ground, which can be received, using the Mi-2 aircraft with the agricultural equipment set mounted on it. First of all, it is very convenient to present the contours of the droplets in their spatial distribution. The droplets diameter distribution was set according to studies presented it the paper [15] and ranges from 10.77 to 320.6 micrometres.

Figures 8 and **9** give the overview of such contours. From the first picture one is able to read the droplets distribution for spraying in the air with respect to their mass distribution in planes perpendicular to the direction of the flight. At the beginning, droplets are gathered in very small volumes, but the situation is changing very soon, what is obvious, having given information about the velocity field behind the aircraft with the wind. The symmetrical structures vanish and the droplets are moving in the wind direction as well as maintaining the rotational movement caused by the swirling air.

IOP Conf. Series: Journal of Physics: Conf. Series **1101** (2018) 012032 doi:10.1088/1742-6596/1101/1/012032

The **Figure 9** gives the other information, mainly about the distribution of the droplets with respect to their diameters. At the beginning, the released droplets are very quickly intercepted by the air stream. The droplets with greater diameters are more cumulated at the beginning. After a very short time small droplets are caught by the vortices generated by the helicopter, whereas the bigger ones are falling down due to gravity. Of course, from the pictures it is readable the general behaviour of the droplets. Almost all droplets are moving outwardly from the vertical plane oriented with the flight velocity and including aircrafts center of gravity. It means, that more mass is moving in this direction comparing to models with the simple modelling of vortex lines, assuming its fixed position and not including the influence of the ground.

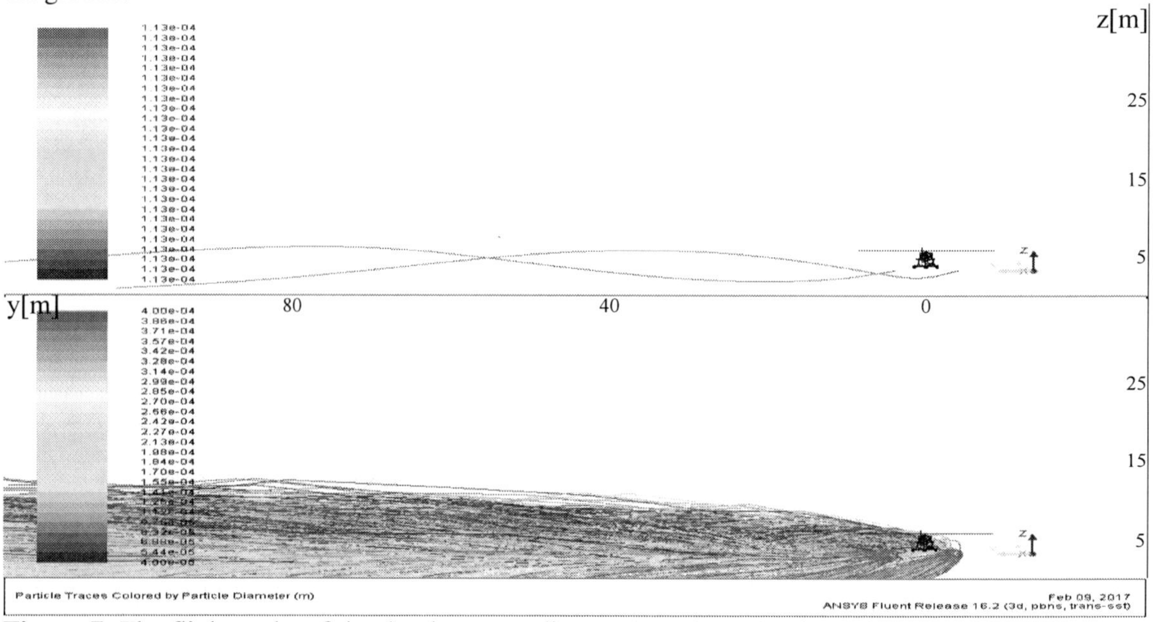

Figure 7. The flight paths of the droplets according to CFD (two droplets of 113 μm diameter – top, many droplets – bottom).

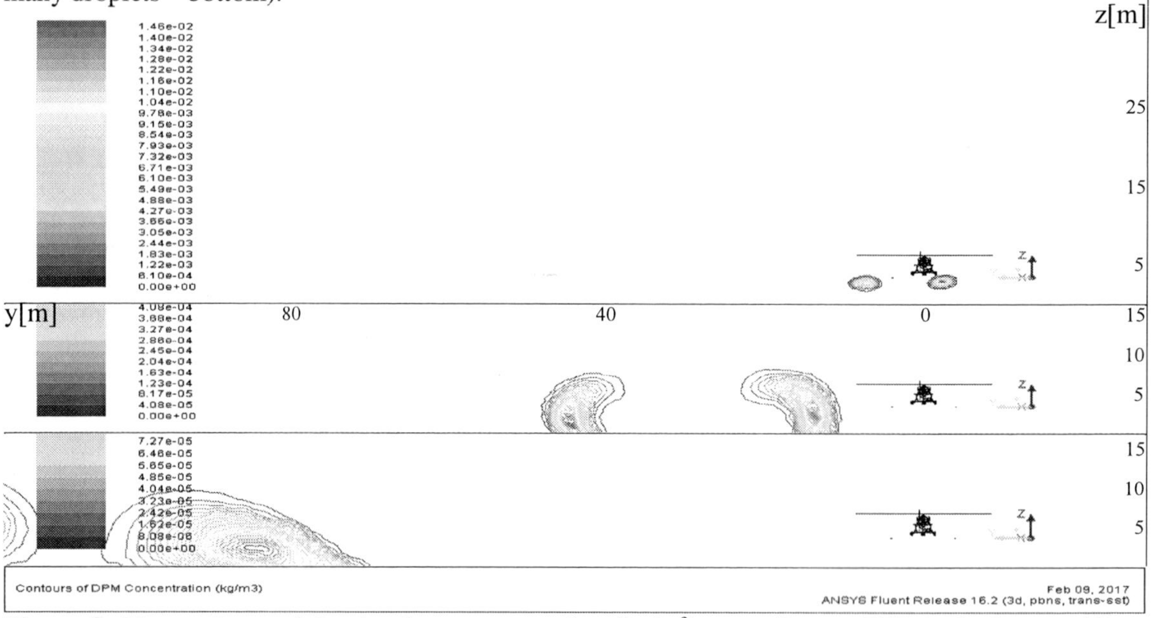

Figure 8. The contours of the droplets concentration (kg/m^3) according to CFD in 3 planes - 10 m, 100 m and 400 m behind the helicopter (x axis).

Finally, the experimental volume fraction of the droplets in the direction of the wind compering to the theoretical models predictions is presented. From the **Figure 10** it is seen, that the mass distribution on the ground is more complicated in the experimental case, compared to the AGDISP model predictions. The results given by the AGDISP model is very similar to the 'free' Bache and Sayer model predictions [1]. The 'free' model, although is closer to the experiment in the stripe from 0 to 10 m. The AGDISP model fits better in the stripe from 25 to 150 m.

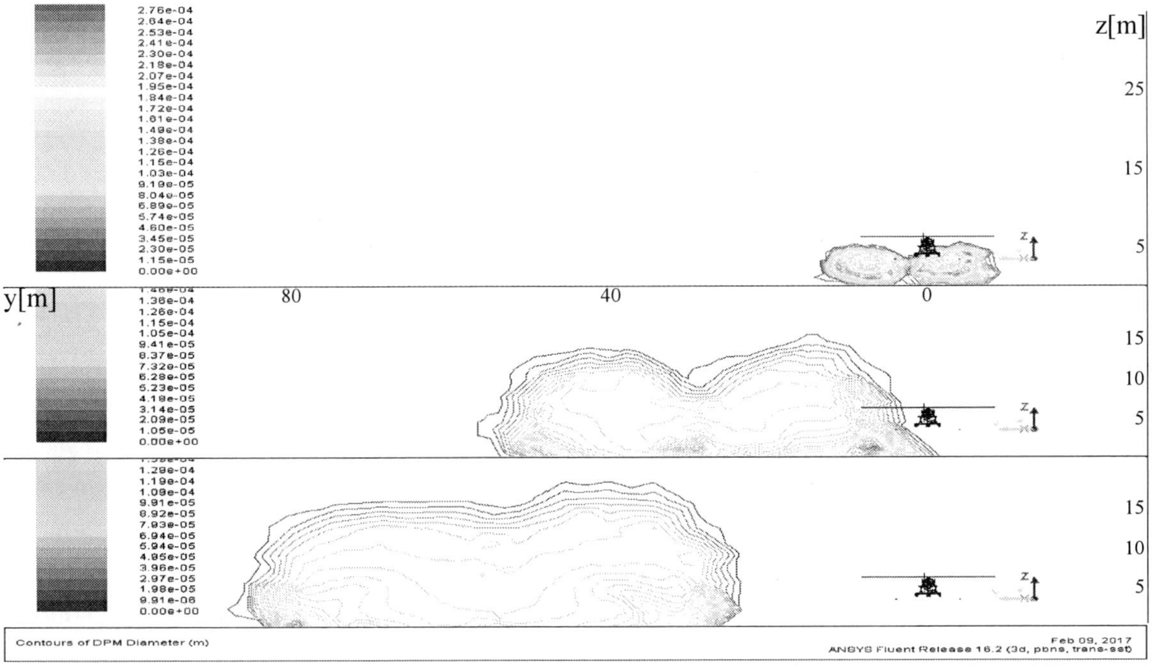

Figure 9. The contours of the droplets diameters according to CFD in 3 planes – 10 m, 100 m and 200 m behind the helicopter (x axis).

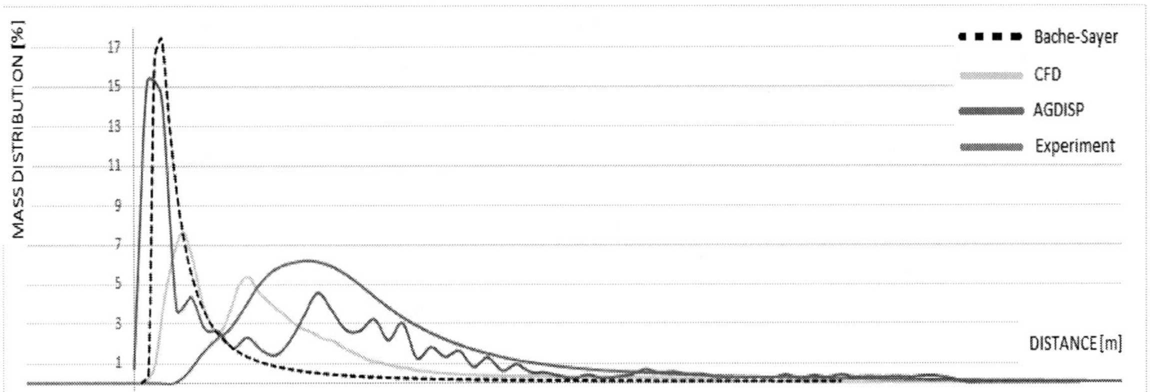

Figure 10. Mass distribution for the Mi-2 helicopter: the blue line – field test, the orange line – AGDISP, the yellow line – CFD, the black dotted line – Bache and Sayer ('free'- without the wake) model.

6. Major conclusions

In this paper the behavior of droplets released from aircraft is approached by a CFD model based on Reynolds-averaged Navier-Stokes (RANS) models, a Lagrangian model for the droplet phase and Eulerian model for fluid phase. The authors analyzed the spray of droplets from the Mi-2 agricultural helicopter, by means of CFD simulations, using the Discrete Phase Model (DPM) method. The three-dimensional full-scale CFD model of the helicopter and surrounding flow field has been created. The CFD models demonstrated very important influence of vortices on droplets. The CFD results were compared to field data as well as to AGDISP and other theoretical models predictions.

In summary, however, the need for further study is highly noticeable. Firstly, the modeled results, which were obtained from the CFD calculations, are more consistent with experimental data showing two maxima of mass concentration. However, there is still a compulsion to implement the physics of flow around an aircraft to a more advanced degree trough the refinement of the geometry of the helicopter. Secondly, such theoretical analysis and their improvements provide a low-cost alternative to field experiments aiming to optimize the lateral distribution of sprayed droplets. Finally, further research, both theoretical and experimental, is needed to produce a higher degree of conformity between results received by means of mathematical models and field tests. In the authors' opinion the final step in the development of such research is the use of Computational Flow Dynamics calculations.

7. References

[1] Bache D H and Sayer W J D 1975 *Transport of Aerial Spray. A Model of Aerial Dispersion* (Agricultural Meteorology 15) p 257-271
[2] Bilanin A J, Teske M E and Barry J W 1989 *AgDisp: the Aircraft Spray Dispersion Model, Code Development and Experimental Validation* (Trans. ASAE 32) p 327–334
[3] Bird S L, Perry S G, Ray S L and Teske M E 2002 *Evaluation of the AgDISP Aerial Spray Algorithms in the AgDRIFT Model* (Environ. Toxicol. Chem. 21) p 672–681
[4] SAS IP 2013 *ANSYS Fluent Theory Guide* (ANSYS)
[5] Деревянко В С 1974 *Влияние аэродинамических возмущений на процессы авиационного опыливания и опрыскивания* (Москва: Транспорт)
[6] Duan B, Yenodol W G and Mierzejewski K 1992 *Statistical Comparison of the AGDISP Model with Deposit Data* (Atmos. Environ. A 26(9)) p 1635-1642
[7] ESDU No. 72026 1972 *Characteristics of the Wind Speed in the Lower of the Atmosphere Near the Ground: Strong Winds* (London: Neutral Atmosphere)
[8] Joulain A, Desvigne D and Alfano D 2017 *Numerical investigation of the vortex roll-up from a helicopter blade tip using a novel fixed-wing adaptation method* (CEAS Aeronautical Journal 8(2)) p 245-260
[9] Misaka T, Holzapfel F and Gerz T, 2012, *Wake Evolution of Wing-Body Configuration from Roll-Up to Vortex Decay* (AIAA Paper 2012-0428, 50th Aerospace Sciences Meeting, USA) p 1-15
[10] Pietruszka J and Rowiński R S 2004 *Computer Simulation of Aerial Spraying* (Annual Review of Agricultural Engineering 3(1)) p 125-140
[11] Reed W H 1954 *An Analytical Study of the Effect of Airplane Wake on the Lateral Dispersion of Aerial Sprays* (NACA Report 1196)
[12] Rowiński R S and Ferenc M 2000 *Some Problems Concerned with the Theory of Drift* (Annual Review of Agricultural Engineering 2(1)) p 148-156
[13] Ryan S D, Gerber A D and Holloway A G L 2013 *A computational Study on Spray Dispersal in the Wake of an Aircraft* (Trans. ASABE 56(3)) p 847-868
[14] Saputro S and Smith D B 1990 *Expert System for Aerial Spray Drift* (ASAE Paper No. 901018. St. Joseph, Mich.: ASAE)
[15] Seredyn T P 2014 *Experimental Investigations of a Drifting Cloud of Droplets Dispersed from Aircraft* (Archive of Mechanical Engineering 3) p 393-407
[16] Teske M E, Bird S L, Esterly D M, Curbishley T B, Ray S L and Perry S G 2002 *AgDRIFT: a Model for Estimating Near-field Spray Drift from Aerial Applications* (Environ. Toxicol. and Chem. 21(3)) p 659-671
[17] Teske M E, Thistle H W and Ice G G 2003 *Technical Advances in Modeling Aerially Applied Sprays* (Trans. ASAE 46(4)) p 985-996
[18] Wallace D J, Picot J J C and Chapman T J 1995 *A Numerical Model for Forestry Aerial Spraying* (Agric. Forestry Meteorology 76) p 19–40

XXIII Fluid Mechanics Conference (KKMP 2018) IOP Publishing
IOP Conf. Series: Journal of Physics: Conf. Series **1101** (2018) 012033 doi:10.1088/1742-6596/1101/1/012033

Aeroelastic analysis of a helicopter rotor in hover

A Sieradzki

Center of New Technologies, Institute of Aviation, Warsaw, Poland

E-mail: adam.sieradzki@ilot.edu.pl

Abstract. The paper deals with the numerical analysis of a helicopter rotor in hover conditions, including aeroelastic effects. In existing literature the effects of an aircraft structural deformation on airloads distribution and performance are fairly well described but mainly for airplanes. The purpose of the presented work is to indicate the possible influence of blades elasticity on helicopter rotor loads and performance in hover. Additionally, it aims to validate a new computational method based on a set of experimental results for a UH-60A helicopter rotor. The method used combines a high fidelity Navier-Stokes aerodynamic model coupled with a low-order beam structural model. This approach made it possible to take into consideration blade deformations in simulations and revealed the high influence of aeroelastic effects on rotor loads and performance at hover conditions. To authenticate the results, the presented method was validated with experimental data from two different flight test programs, three small-scale wind tunnel tests and one full-scale wind tunnel test. They were conducted for the UH-60A helicopter rotor by different organizations. The comparison of results revealed that the created computational method has high accuracy.

1. Introduction

Determining the loads and performance of a helicopter rotor is a very complex research problem. This is the result of a large variety of phenomena connected with the rotating rotor and the highly multidisciplinary nature of the issue. In existing literature the effects of an aircraft structural deformation on airloads distribution are fairly well described but mainly for airplanes. It is difficult to find information about the possible impact of blades deformability on rotor performance. Some conclusions can be drawn from the articles presenting the results of a rigid blade simulation compared with experimental data. The example is the simulation of W-3A Sokół helicopter rotor presented in [1]. In this article the analysis is carried out with a very accurate - from the rotor aerodynamics point of view - computational method, but is omitting blade deformability and rotor trimming. As a result, an error of 20% was obtained for global rotor performance parameters. However, in the aforementioned work the rotor was simulated in a forward flight, not hover. The purpose of this article is to indicate the possible influence of blade elasticity on helicopter rotor loads and performance in hover (on the example of UH-60A helicopter), and to validate created computational method based on a set of experimental data.

2. Computational method

In the presented work, a simplified structural model is used, based on the Euler-Bernoulli beam theory (also known as engineer's beam theory or classical beam theory). [2] It assumes that there are no shearing effects (cross-section perpendicular to the neutral axis of the beam remains perpendicular after deflection) and can be used for analysis of beams with high elongation. Due to the fact that the elongation of helicopter blades is relatively large, it is possible to use the classical theory of Euler-Bernoulli beams for the analysis of this type of construction. Its advantage in comparison to the e.g. Timoshenko theory is simplicity, and thus the ease of numerical implementation. Then, the blade deformations were obtained from the following relationships:

$$\frac{\partial^2 z}{\partial r^2} = \frac{M_g}{EI_x} \tag{1}$$

$$\frac{\partial \theta}{\partial r} = \frac{M_s}{GI_s} \tag{2}$$

where r is the radius, z is the deflection, M_g is the bending moment, EI_x is the bending stiffness, Θ is the torsional deformation angle, M_s is the twisting moment, and GI_s is the torsional stiffness.

The described beam theory was used to formulate partial differential equations for blade bending/flapping and blade torsion/pitch steering. Both equations were coupled together. The lead-lag motion of a blade and bending in the plane of the highest stiffness was neglected in the presented model. The derived system of equations was similar to that described in [3, section 9-4.3]. The beam model equations were numerically implemented using a finite difference method to create Blade Deformation Solver (BDS). In BDS, the unknown values of the individual quantities are obtained in the consecutive time steps using an explicit time stepping method.

The high fidelity aerodynamic model is based on the Navier-Stokes equations solver and the method of Moving Deforming Mesh (MDM), that determines the type of analysis as transient. This is one of the most accurate and time-consuming rotor aerodynamics models available. A similar approach is presented in [4][5]. This method uses the sliding mesh interface separating the part of the domain rotating with the rotor and the remaining stationary part. Additionally, the changes of the blade pitch angle and the articulated motion and deformations of the blades require the use of algorithms that deform and rebuild the computational grid. In order to get the final convergent solution, it is necessary to perform calculations for at least a few rotations of the rotor to get the required periodicity or constant values of monitored parameters. The MDM method is able to capture high variety of aerodynamic phenomena (interference, flow separations, shock waves, etc.).

The tight coupling between structural and aerodynamic model was implemented in the User Defined Functions (UDF) environment of the well-known CFD commercial code - ANSYS Fluent. This type of coupling makes it possible to get a time accurate solution. At each time step, this requires gathering information about the blade loads, mapping the loads to the beam model, and determining the beam deformations and corresponding nodal displacements in the CFD mesh. The obtained blade aerodynamic loads were applied gradually to the BDS beam model to avoid rapid blade response and undesirable vibrations. All of these processes and algorithms were implemented in the developed UDF.

In some cases concerning rigid blades, which were analyzed for comparison purposes, Moving Reference Frame (MRF) method and the steady state solution type were used. This approach was also utilized to get the initial solution for other calculations in a reasonably short time. It is worth mentioning that aeroelastic analysis of a helicopter rotor in hover could be performed with similar accuracy using the coupling between BDS and MRF model to get final steady state solution. However, it was decided to use the unsteady MDM model just to allow the developed method to be used later in a forward flight cases, which cannot be treated as steady state problems.

3. Test case and computational process

Simulations were performed for the full-scale, isolated UH-60A helicopter rotor in hover conditions. Three different cases regarding blade elasticity and articulation (relative motions) were analyzed: the no-relative-motion rigid blade (MRF simulation), the rigid blade with the possibility of flapping motion and the elastic blade with the possibility of flapping motion. The calculations included a number of collective pitch angle (Θ) values and were compared with a set of experimental results [6-13] to validate developed computational method.

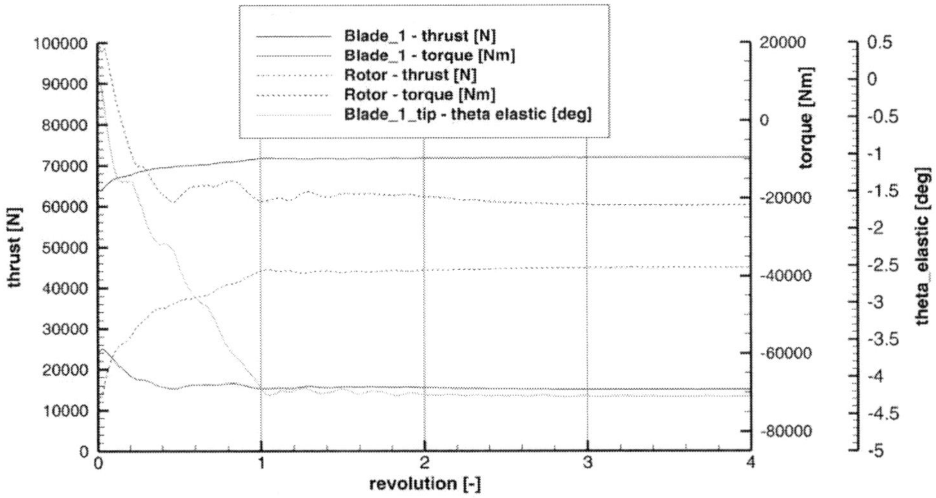

Figure 1. History (convergence) of thrust and torque acting on the blade and the rotor, and the angle of torsional deformation (*theta_elastic*) at the blade tip.

In order to speed up the calculations, the MRF model was initially used with the intention of generating approximate flow field around the rotor, which would require a lot of time in the case of transient analysis. MRF simulations were performed for collective pitch angles equal to 8, 10 and 12 degrees. Unsteady analysis using the MDM-BDS module was additionally performed for 14 degrees and 16 degrees (as start-up cases the results of MDM-BDS simulations for 12 and 14 degrees were used respectively). The convergence of the thrust and torque acting on the blade and the entire rotor obtained with the MDM-BDS module is shown in figure 1. The initial values visible in the graph correspond to the results of the MRF analysis (no-relative-motion rigid blades). Through the first rotation of the rotor, the forces determined in CFD were gradually applied to the blade beam model, which resulted in its deformation and the largest change in the monitored forces and moments. Their values converged after about 4 revolutions of the rotor - without using the MRF initial solution it would take about 15 revolutions.

4. Simulation results discussion

The obtained values of thrust (C_T) and power (C_P) coefficients divided by rotor solidity (σ) are shown in figure 2. They were calculated based on the following relationships:

$$C_T = T/(\rho A(\Omega R)^2) \tag{3}$$

$$C_P = P/(\rho A(\Omega R)^3) \tag{4}$$

where T is the thrust, P is the power, ρ is the fluid density, A is the area of the rotor, Ω is the angular speed and R is the radius of the rotor.

The results showed that in the presented case, the elasticity has significant impact on global performance characteristics of the rotor and blades airloads distribution. For example, at collective pitch angle of 12°, the values of C_T/σ and C_P/σ were reduced respectively by 37% and 51% (figure 2). It can be easily seen that when the flexibility of the blades is taken into account, the required collective pitch angle should be increased by about 4-5 degree to ensure a similar level of thrust. This is the angle close to the average torsional deformation angle at the blade tip, obtained as a result of the simulations and shown in figure 3. Neglecting of the blade deformability causes excessive load occurring at the tip of the blade. The level of aerodynamic loads in this region, obtained for a collective pitch angle of 10° and rigid blades, is comparable to that obtained for a pitch angle of 16° with flexible blades. One of the factors conducive to such differences can be the swept blade geometry at the tip, which increases the torsional moment and deformation angle of the blade.

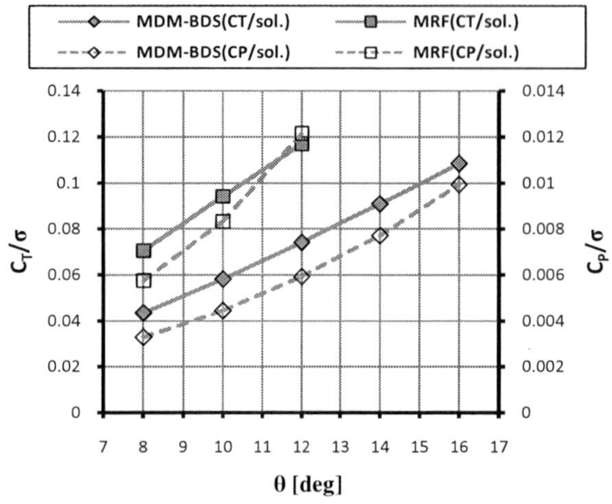

Figure 2. Comparison of thrust and power coefficients for elastic blade (MDM-BDS) and no-relative-motion blade (MRF).

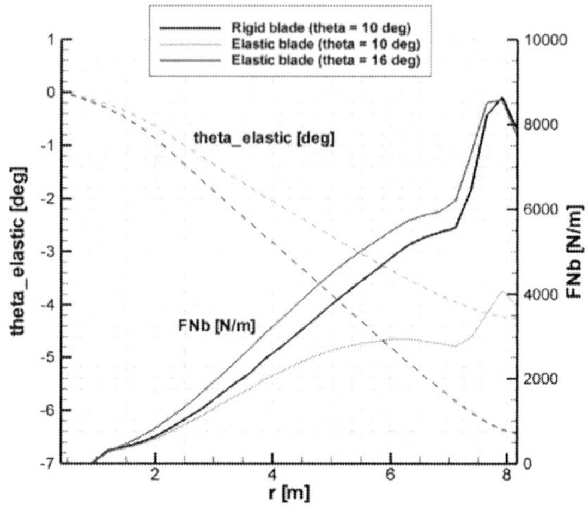

Figure 3. The comparison of the torsional deformation angle (*theta_elastic*) and blade normal load (*FNb*) for Θ=10° and Θ=16°.

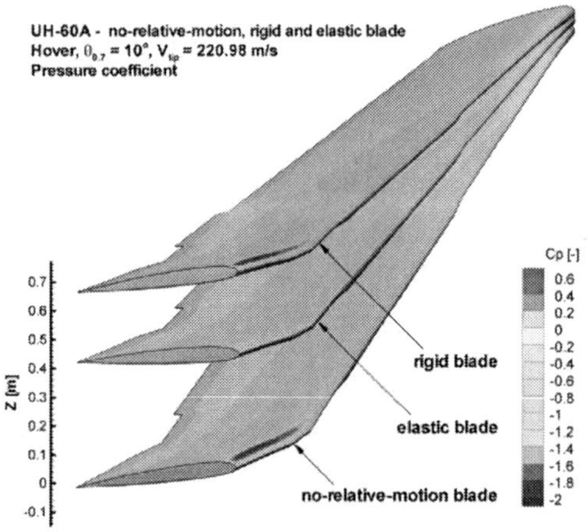

Figure 4. Graphic representation of the flap angles with pressure coefficient distributions for a no-relative-motion blade, a rigid blade and an elastic blade.

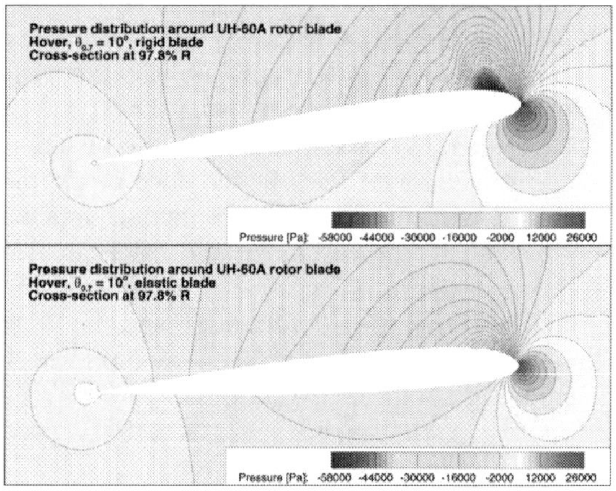

Figure 5. Comparison of pressure distributions around the blade in the section of 97.8% radius for a rigid and elastic blade.

The changes in the blade pitch angles are of key importance when analyzing the performance of a helicopter rotor, especially if the helicopter is equipped with modern blades with an advanced tip geometry. The impact of blade deformation on blade surface pressure distribution and pressure distribution in the plane perpendicular to the blade axis (in the region of the tip), is shown in figures 4 and 5 respectively. When comparing the variants presented, it can be noticed that the rigid blades have very similar pressure distributions. This is in line with expectations, because in the case of hover, the flapping angle of the isolated rotor does not depend on the blade's azimuth (it is constant). Therefore, the effective angle of attack in a given blade section is similar to that obtained for a blade that does not have the ability to make a flapping motion. However, the pressure distribution looks different in the case of the elastic flapping blade. Also the much smaller aerodynamic forces acting in the plane normal to the rotor disk cause the blade to have a smaller deflection angle than the rigid blade, as shown in figure 4. For the presented case (Θ=10°), the value of deflection at the tip of rigid flapping blade is 0.66 m, compared to 0.42 m for the elastic blade. The maximum torsional deformation angle of the blade is -4.26°, which significantly affects the flow around the tip. A decreased angle of attack results in a significantly lower loading in the region of the airfoil nose (figure 5).

5. Simulation results validation

The obtained results of calculations were compared with the following experimental data for the UH-60A helicopter rotor in hover (two flight test programs, three small-scale and one full-scale wind tunnel program - more information about them could be found at given references):

- First set of flight test programs was conducted by U.S. Army Aviation Engineering Flight Activity (AEFA) and consisted of a series of flight test on different UH-60A aircraft. The data presented in this paper is the hover data from the 1st Year [6] and 12th Year [7] production aircraft - the primary difference between them is the installation of Extended Stores Support System on the second aircraft.
- The second flight test program was the UH-60A Airloads Program conducted by Army/NASA. [8] The noticeable scatter of data is due in part to variable wind conditions during testing. [9]
- The first two small-scale programs were hover tests conducted at the Sikorsky Model Hover Test Facility. The first test (Balch [10]) used a 1/5.97-scale model of the UH-60 helicopter, the second (Lorber [11]) used a 1/5.73-scale model. The hover performance data presented in this paper is concerned with the isolated rotor measurements. The third small-scale program was conducted at the Duits-Nederlandse Windtunnel (DNW) and used the same model as the second test. [12]
- The full-scale wind tunnel program was conducted by Army/NASA at the NASA Ames 80- by 120-Foot Wind Tunnel. The presented results were obtained at low to medium thrust conditions, with different values of the shaft angle α_s to indicate the effect of wind tunnel walls on the hover performance measurements. The test model included NASA's Large Rotor Test Apparatus (LRTA) - it was not the isolated rotor. [13]

The values of C_P/σ as a function of C_T/σ are shown in figure 6. Very good agreement of the results was achieved. In order to more accurately present the differences between individual experimental tests and the results of numerical simulations, the Figure of Merit (F_M) parameter was determined according to the following formula:

$$F_M = {C_T}^{3/2}/(C_P\sqrt{2}) \tag{5}$$

The F_M parameter is the equivalent of the aerodynamic energy function determined for fixed wing aircraft and is characterized by greater sensitivity to any changes than thrust and power coefficients. A comparison of the obtained F_M values with experimental data is shown in figure 7. It reveals the advantage of the aeroelastic analysis (MDM-BDS) over the simplified simulation (MRF). The advantage is especially visible for a large values of thrust, where an incorrect distribution of angles of attack (and therefore loads) for the rigid blade causes an underestimation of the F_M value. It should also be noted that the visible difference would be much larger if the F_M values were presented not as a function of thrust coefficient but collective pitch angle of the blade.

The discrepancies in rotor performance calculated by MRF (rigid blades) and MDM-BDS (elastic blades) models are caused by, among others, the flow separation at the blade tip that appears in the case of rigid blade computations. It is visualized by the chordwise wall shear contours in figure 8. Negative values of the wall shear correspond to the reversed flow over the blade surface and indicate flow separation. The flow separation also begins to appear at high collective pitch values for elastic blade cases (figure 9). This effect cannot be fully captured with Reynolds-averaged Navier-Stokes (RANS) simulation, so the flow separation could also be an explanation of the differences between experiment and simulation results, especially at high thrust conditions.

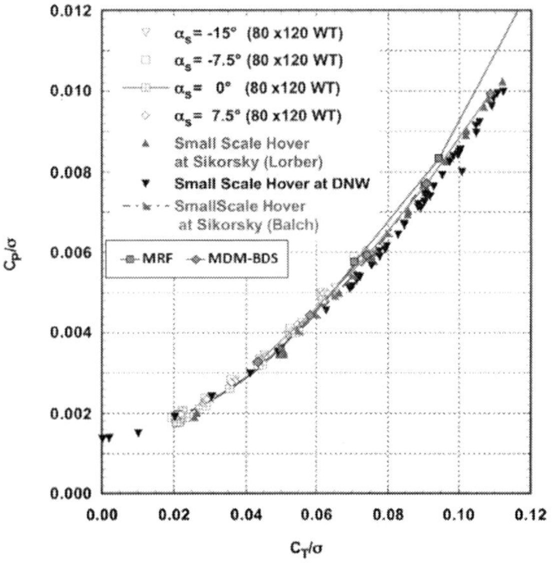

Figure 6. Rotor power comparison - results of MDM-BDS and MRF cases compared with one full-scale and three model-scale experiments. [13]

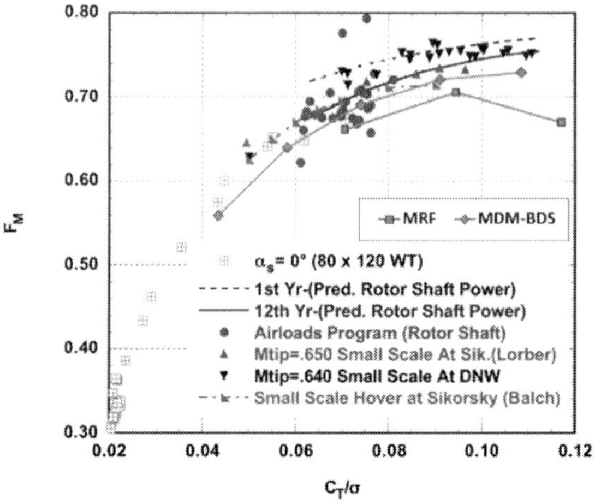

Figure 7. Comparison of F_M values - results of MDM-BDS and MRF cases compared with all available experimental data. [13]

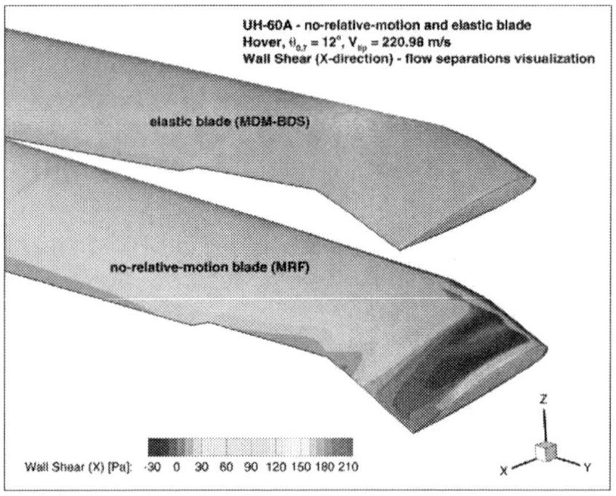

Figure 8. Wall shear (X-direction) contours on the surface of a no-relative-motion and elastic blade for Θ=12° (max Θ for MRF simulation). Negative values mean reversed flow (separation).

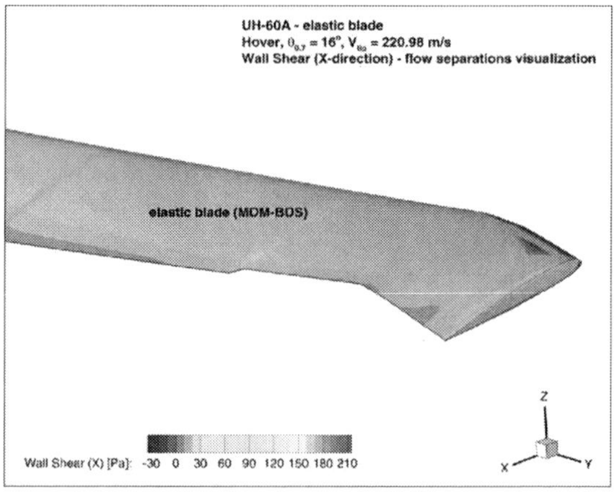

Figure 9. Wall shear (X-direction) contours on the surface of an elastic blade for Θ=16° (max Θ for MDM-BDS simulation). Negative values mean reversed flow (separation).

It can be noticed that the MDM-BDS model also underestimates the value of the F_M parameter in relation to the average value from experiments. It is approximately a constant error, independent of the C_T/σ value. Known factors that may have contributed to this error are as follows:

- Spalart-Allmaras turbulence model used in simulations (it is expected that using finer mesh and one of transitional models should increase the F_M values slighlty, due to possible decrease of aerodynamic drag and power reduction caused by laminar flow regions),
- inaccurately reproduced tip geometry of the blade which could happen due to some uncertainties in geometry definition (based on report [14], no access to CAD file).

Nevertheless, the consistency of the F_M results should be considered good, because, as already mentioned, the F_M parameter could be very sensitive to geometric changes and many other factors. It is also worth noting the discrepancies between different experimental data themselves.

6. Conclusions

Aeroelastic effects can have significant influence on helicopter rotor loads and performance in hover conditions. Changes in the angle of attack due to elastic blade twisting are the main cause of airloads reduction. Deformability of the blades influences in particular the flow conditions around the blade tip and at given collective pitch angle could reduce the rotor thrust by up to 37% and the rotor power by up to 51%. These values were obtained for the UH-60A helicopter rotor at high thrust hover conditions. The comparison of simulation results with a set of experimental data for the UH-60A helicopter revealed good accuracy of the developed computational method in terms of global performance parameters prediction.

Simulation of a helicopter rotor is a multidisciplinary problem. Of which, a strong fluid structure interaction should not be omitted. Otherwise, this can result in the outcomes that differ considerably from the actual performance of the helicopter.

Acknowledgements

Computational support was obtained from University of Warsaw Interdisciplinary Centre for Mathematical and Computational Modelling, in the Computational Grant No. G66-11.

References

[1] Szulc O, Doerffer P, Żółtak J and Małecki J 2013 Time-accurate simulation of flow past PZL W-3A "Sokół" (Falcon) helicopter main rotor in forward flight *Task Quart.* **17** pp 43-61
[2] Timoshenko S 1953 *History of strength of materials* (New York: McGraw-Hill)
[3] Johnson W 1980 *Helicopter theory* (New York: Dover Publications Inc.)
[4] Stalewski W and Sznajder K 2014 Computational investigations of active flow control on helicopter-rotor blades *J. of KONES* **21** pp 281-8
[5] Steijl R and Barakos G 2008 Sliding mesh algorithm for CFD analysis of helicopter rotor–fuselage aerodynamics *Int. J. for Num. Meth. in Fluids* **58** pp 527-49
[6] Nagata J I, Robbins R D, Skinner G L, Williams R A and Buckanin R M 1981 *Airworthiness and Flight Characteristics Evaluation, UH-60A (Black Hawk) Helicopter* (Edwards AFB, CA: Army Aviation Engineering Flight Activity) USAAEFA-77-17
[7] Nagata J I, Piotrowski J L, Young C J, Lewis W D, Losier P W and Lyle J A 1989 *Baseline Performance Verification of the 12th Year Production UH-60A Black Hawk Helicopter* (Edwards AFB, CA: Army Aviation Engineering Flight Activity) USAAEFA-87-32
[8] Bousman W G and Kufeld R M *UH-60A Airloads Catalog* (Moffet Field, CA: National Aeronautics and Space Administration) NASA/TM-2005-212827
[9] Bousman W G 2001 Out-of-ground-effect hover performance of the UH-60A *UH-60 Airloads Program Occasional Note* **1** pp 1-13
[10] Balch T B 1985 Experimental study of main rotor/tail rotor/airframe interaction in hover *J. of the Amer. Heli. Soc.* **30** pp 49-56
[11] Lorber P F, Stauter R C, Pollack M J and Landgrebe A J 1991 *A Comprehensive Hover Test of the Airloads and Airflow of an Extensively Instrumented Model Helicopter Rotor; Volume 1 - Rotor Airloads and Performance* (USAAVSCOM) TR 91-D-16A
[12] Lorber P F 1991 Aerodynamic results of a pressure-instrumented model rotor test at the DNW *J. of the Amer. Heli. Soc.* **36** pp 66-76
[13] Shinoda P M, Yeo H and Norman T R 2004 Rotor performance of a UH-60 rotor system in the NASA Ames 80-by 120-Foot Wind Tunnel *J. of the Amer. Heli. Soc.* **49** pp 401-13
[14] Bousman W G 2003 *Aerodynamic Characteristics of SC1095 and SC1094 R8 Airfoils* (Moffet Field, CA: National Aeronautics and Space Administration) NASA/TP-2003-212265

IOP Conf. Series: Journal of Physics: Conf. Series **1101** (2018) 012034 doi:10.1088/1742-6596/1101/1/012034

Numerical investigations of an influence of the aspect ratio on the Savonius rotor performance

Krzysztof Sobczak

Institute of Turbomachinery, Lodz University of Technology, 219/223 Wolczanska, 90-924 Lodz, Poland

krzysztof.sobczak@p.lodz.pl

Abstract. An influence of the aspect ratio on the Savonius rotor performance was investigated. Three-dimensional 3D numerical models of the classical Savonius rotor equipped in two semi-circular buckets with an overlap gap and endplates were studied for different aspect ratios. The simulation results were verified against the experimental data and a blockage effect was taken into account. An increase in the power coefficient and the tip speed ratio for the maximal power coefficient with a growth in the rotor aspect ratio was observed. An effect of the wind displacement beyond the rotor was more intensive for lower aspect ratios.

1. Introduction

The most important advantages of a Savonius rotor are its simple design and low manufacturing costs, high starting torque, wind direction independence, low flow and rotation speeds which contribute to low noise, etc. All these aspects make it suitable for applications in urban areas. However, the rotor is characterized by low efficiency with typical values of the power coefficient in the range Cp = 0.15 – 0.20 [1], [2]. Its improvement is necessary to treat the Savonius wind turbine as a reasonable alternative to other wind turbines. Therefore, the turbine is subject to wide experimental and numerical investigations devoted to different aspects of the turbine design.

One of the parameters which influence the Savonius rotor efficiency is its aspect ratio, i.e., a ratio of the rotor height to its diameter (AR = H/D). As indicated in [2], there was a limited number of systematic experimental studies concerning an aspect ratio impact on the Savonius rotor performance. In general, losses at tips of the buckets as well as the wind displacement beyond the rotor lower its efficiency. Therefore, rotors with higher aspect ratios have higher efficiency due to relatively lower tip losses and lower wind displacement. The performance of low aspect ratio rotors can be improved by an application of endplates which limit tip losses and have a similar effect as that produced by a higher aspect ratio. Thus, in the majority of the experiments where the aspect ratio impact was investigated, endplates were used.

Apart from research on low aspect ratio rotors, e.g., at AR = 0.6 – 1.0 in [3], where the highest power coefficient was obtained for AR = 0.7, other studies indicate that the highest efficiency is reached for much higher aspects. In [4], [5] the highest efficiency was attained for the highest aspect under study (AR = 2.5 – 5.0), whereas in [6] the maximum was obtained for AR = 6.0, with a much lower value for AR = 7.0. The investigations presented in [7] show high changes in the maximal value of the power coefficient in the range of AR $<$ 2 and no significant changes above this value. Also in [5] much higher changes were observed for AR $<$ 1 than for higher AR. According to other studies, Savonius rotors with aspect ratios of about 1 – 2 yielded sufficiently good performance. However, the majority of existing Savonius turbines were installed with low aspect ratios (AR $\leq$ 1) due to structural reasons [1], [2]. The aspect ratio of a Savonius rotor can be adjusted to specific requirements. Rotors with a higher diameter generates more torque at lower rotational speed and vice versa [4]. Also, its angular acceleration increases, whereas the rotor moment of inertia decreases with a growth in the aspect ratio [8].

In order to reveal an influence of the aspect ratio on the Savonius rotor performance, a set of numerical simulations was conducted. A classical Savonius rotor equipped in two semi-circular buckets and endplates was studied for its aspect ratio in the range AR = 0.5 – 5. Power coefficient changes and the accompanying flow structure were analyzed as well as.

Apart from the aspect ratio defined above, two dimensionless numbers are used here to describe the Savonius rotor performance for different wind conditions, i.e., a power coefficient:

$$Cp = \frac{T\omega}{0.5\rho v^3 A}$$

and a tip speed ratio:

$$TSR = \frac{\omega R}{v}$$

where: T – torque, ω – angular velocity, R – radius of the rotor, ρ – air density, v – far field wind velocity, and A – rotor projected area (H·D).

2. Definition of simulations

Three-dimensional 3D numerical models of a classical Savonius rotor equipped in two semi-circular buckets with an overlap gap and endplates were studied. Its main dimensions are shown in figure 1. Different values of the turbine aspect ratio were achieved by modification of the turbine height. Taking advantage of the constant section of the rotor and the flow symmetry, its lower half was considered only. A computational domain layout and main dimensions are presented in figure 2. The computational space was divided into two domains. The internal, cylindrical one contained the space around the rotor and it was surrounded by an external, rectangular domain. The external space was divided into two parts to simplify mesh generation, especially when a blockage effect was investigated.

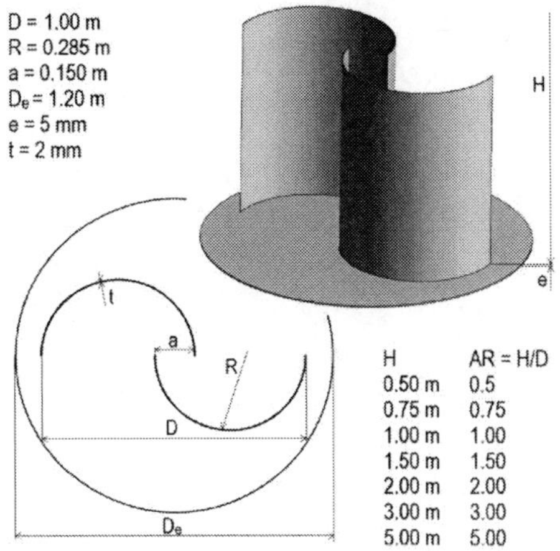

Figure 1. Main dimensions of the Savonius rotors.

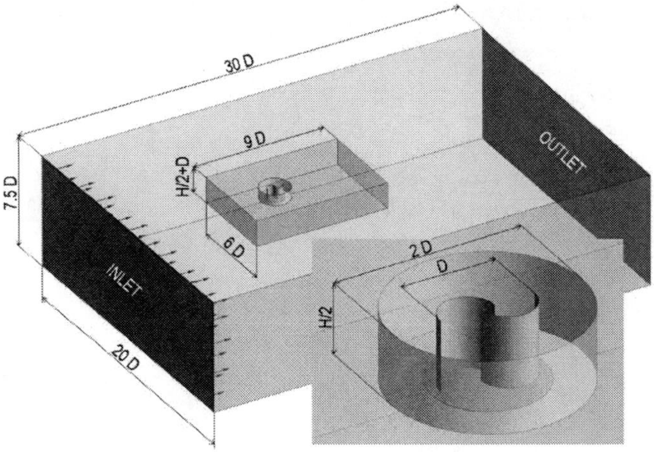

Figure 2. Computational domain arrangement for the blockage of 0.5%.

High quality hexahedral meshes of flow fields around the turbines were generated in ANSYS Meshing with a sweep method in all regions. The mesh refinement in the vicinity of rotor blades and endplates (with the first nodes at the wall y+ < 1) allowed one to fully resolve the flow in boundary layers. The endplate refinement was extended to the rotor surrounding (i.e., in the part of the external domain bounded by the orange box in figure 2) in order to properly resolve the flow in shear zones. Taking into account low changes in flow parameters in the zone far away from the rotor, the mesh outside the orange box was coarsened towards the domain boundaries. The same mesh layout was applied in the planes perpendicular to the rotor axis, however, a number of nodes along the turbine axis was adjusted with respect to the rotor height. Thus, depending on the aspect ratio, the mesh was composed of 8.8 – 13.0 millions of nodes.

Transient simulations of flows in the Savonius rotors were conducted with ANSYS CFX 18.2. 360 time steps per one rotor revolution were selected for all the cases under consideration, thus, the Root Mean Square of the Courant number was RMS Cu = 12 – 15. A further decrease in the time step did not affect the simulation. Due to fact that the Mach number Ma < 0.1, the air was assumed to be incompressible. As indicated above, the computational space was divided into two domains. The internal, cylindrical one (surrounded by the light blue interface in figure 2) was rotating with a constant angular velocity according to the specified tip speed ratio. The external, rectangular domain was stationary. Dimensions of this domain were selected on the basis of numerical tests in such a way that the blockage effect was negligible. Thus, a series of tests for different blockage was conducted for the rotor with AR = 1.5 and for TSR = 0.8. The outcomes of this study are presented in figure 3. A domain for the rotor projected area equal to 0.5% of the domain cross-section was selected for further investigations. In this case, the Cp value differed less than 0.5% from the value estimated for the blockage → 0.

The Shear Stress Transport turbulence model, which was successfully applied in the previous simulations of the Savonius rotor [9], [10], [11], was used. No significant difference was obtained for simulations with other turbulence models/methods (SAS SST and DES) conducted for TSR = 0.8. Preliminary tests revealed that modeling of a transition in the boundary layer has a low impact on the solution results, therefore, the fully turbulent flow was solved. At the domain inlet, the wind velocity of 4 m/s (Reynolds number based on the rotor diameter $Re_D = 2.6 \cdot 10^5$) as well as the low turbulence intensity (1%) and the turbulent to molecular viscosity ratio equal to 1 were imposed. At the outlet, the static pressure equal to the atmospheric one was selected. A half of the rotor was solved, thus, the symmetry condition was applied in the middle of the rotor height. The blade and endplate walls were assumed to be smooth. A transient rotor-stator interface was applied at a transition between the internal and external domain. Two blocks of the external domain mesh were connected by means of the General Grid Interface.

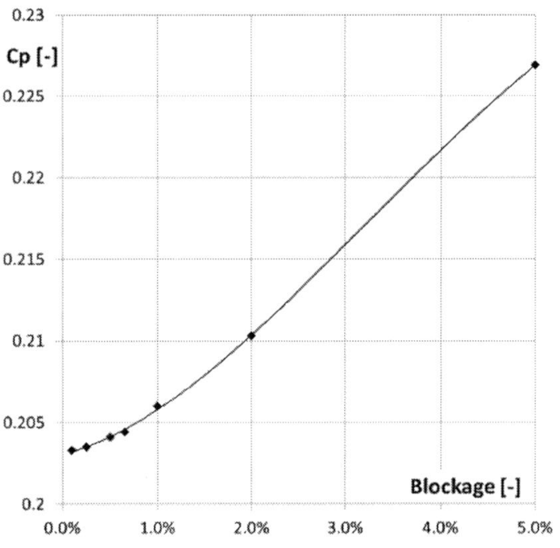

Figure 3. Power coefficient as a function of the domain blockage.

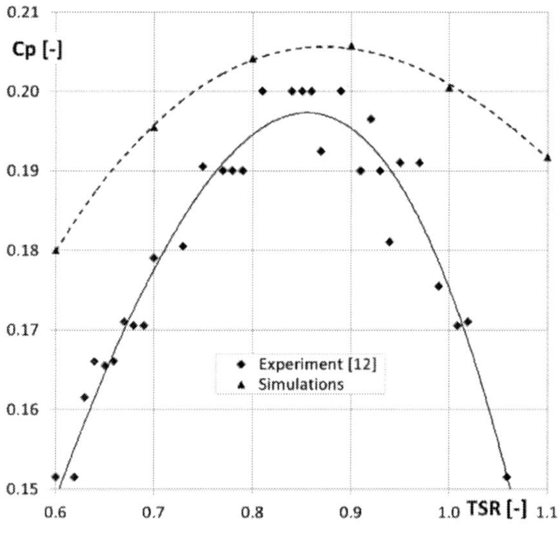

Figure 4. Numerical simulation results versus the experimental data.

Second order time and space discretization schemes were applied. Typically 5 – 8 internal iteration loops were required to reach the prescribed Root Mean Square residual target RMS RES = $1\cdot10^{-5}$ (good solution quality) for each time step. In the majority of cases, the simulations were conducted for 6 – 8 rotor revolutions. It was sufficient to keep the averaged power coefficient Cp changes in the subsequent rotations below 0.1%. However, for TSR ≥ 1.0, the period-averaged value of Cp varied, thus, a higher number of revolutions (10-12) was necessary to average the data.

The simulation results were verified against the experimental data [12]. In the experiment, the natural size Savonius rotor with the dimensions specified in figure 1 and AR = 1.5 was tested under open field conditions. In order to compensate the effects of disturbances due to changes in the wind speed and direction, only the data for wind conditions constant in periods longer than 30 s were taken into account. The power coefficient characteristics of experimental and numerical simulation results are shown in figure 4. A good agreement was reached for the TSRs close to optimal, i.e., for TSR = 0.8 – 0.9, a relative difference with respect to the trend line of the experimental data did not exceed 5%, and it was within 15% for TRS = 0.7 – 1.0. It is not clear if high differences for low (TSR < 0.7) and high (TSR > 1.0) are due to numerical or experimental reasons. Further experimental and numerical studies are planned. However, an agreement for the TSRs close to optimal is sufficient for the needs of this study.

3. Discussion of the results

Three-dimensional 3D numerical simulations of flows in Savonius rotors with aspect ratios in the range AR = 0.5 – 5.0 were conducted for the tip speed ratio TSR = 0.8. A characteristics of the power coefficient Cp for the investigated aspect ratios is presented in figure 5. An increase in Cp with a growth in AR can be observed, but its intensity changes around AR = 1.5. For AR < 1.0, the power coefficient increases rapidly, similarly as presented in [7] and [5]. On the other hand, for AR > 2.0, changes in the coefficient are much less considerable. The 2D simulation result marked in figure 5 with a dashed line indicates the limit of Cp (for AR = ∞). A decrease in the power coefficient with a decrease in the aspect ratio is due to boundary layer effects near the endplates as well as the deflection of a part of the air stream beyond the endplates [7]. Both these effects are very strong for low aspect ratio rotors. With an increase in AR, the relative effect is wakening, but reminds important.

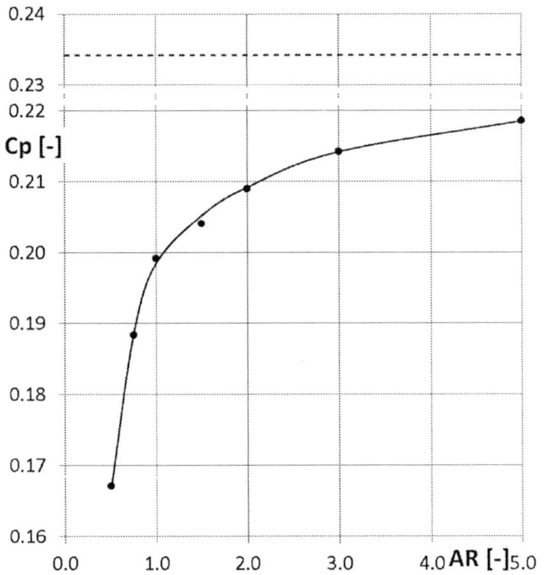

Figure 5. Power coefficient as a function of the aspect ratio for TSR = 0.8.

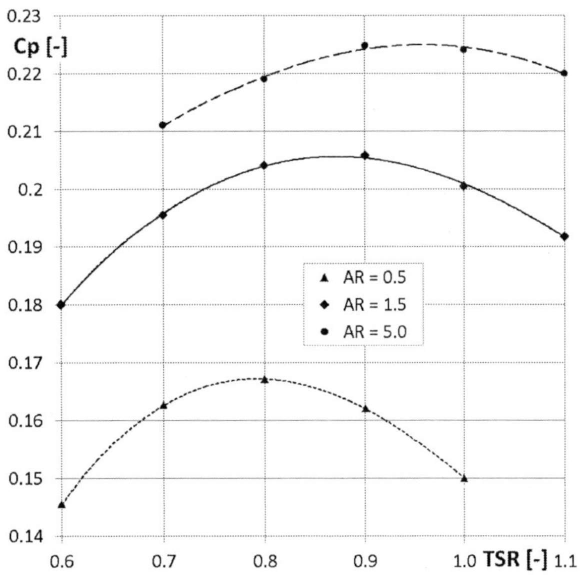

Figure 6. Power coefficient as a function of the TSR for different aspect ratios.

IOP Conf. Series: Journal of Physics: Conf. Series **1101** (2018) 012034 doi:10.1088/1742-6596/1101/1/012034

Additional simulations were conducted for different values of TSR in the case of rotors with AR = 0.5, 1.5 and 5.0. Characteristics of Cp for the investigated AR are presented in figure 6. The results clearly indicate that the tip speed ratio for the maximal power coefficient (Cp_{max}) increases with a growth in the aspect ratio. For AR = 0.5, the power maximum was reached for TSR = 0.8, whereas for AR = 5.0, it increased to TSR = 0.90 – 0.95. Thus, for AR > 1.5, the maximal values of Cp are slightly higher than the values presented in figure 5 (for TSR = 0.8). Such changes were not present in [5] and [6], whereas some shift of Cp_{max} towards higher TSR can be found in [7] and [13]. Also in [3], some impact of AR on TSR for Cp_{max} can be seen. However, limited data available for different Savonius rotor configurations do not allow one to draw clear conclusions.

The friction of endplates consumes a part of the power generated by rotor blades. For all cases under study, almost the same value of power consumption was obtained. Thus, if this value was related to the power output of the turbine, a contribution of friction losses for the low aspect ratio rotors was high. As one can see in figure 7, it exceeded 10% for rotors with AR < 0.75, whereas for AR > 1.5, it was lower than 4%. In all cases, the external side of the endplate was responsible for approximately 70% of the friction torque. However, this assessment of the endplate friction did not take into account effects of the boundary layer development and flow displacement at the blade tips.

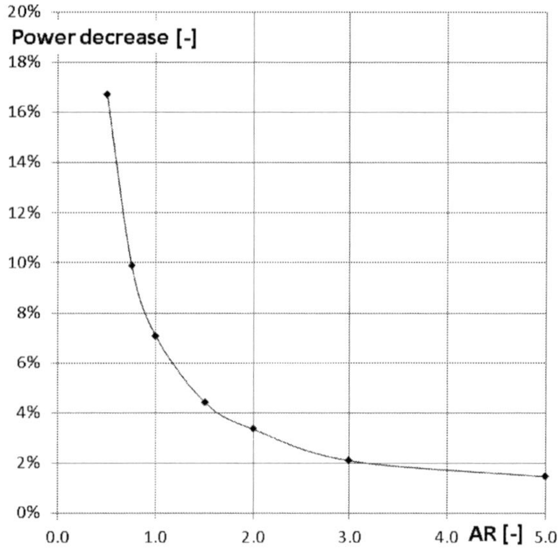

Figure 7. Relative power consumption due to the endplate friction.

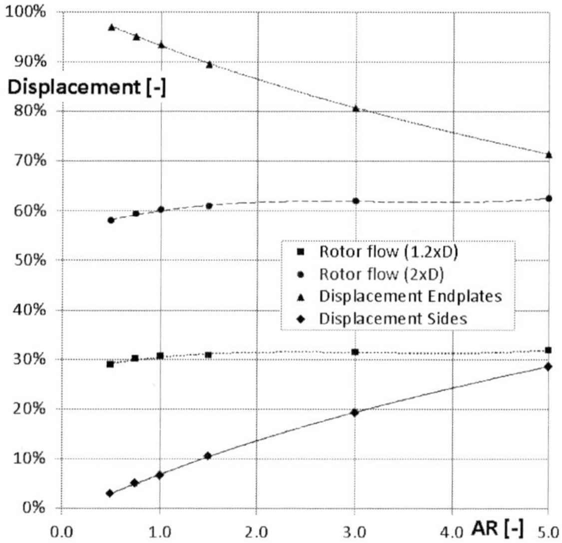

Figure 8. Flow displacement beyond the rotor.

A turbine rotor is a kind of obstacle (bluff/blunt body) which deflects the wind. Due to its specific shape, the Savonius rotor displaces a significant fraction of the wind volume flow rate. The aspect ratio of the rotor influences this displacement as shown in [7]. In figure 8, period-averaged values of the relative air flow indicate the displacement intensity. Data was gathered in the regions located on the plane perpendicular to the wind flow direction and placed on the rotor axis. Rotor flow (1.2xD) and Rotor flow (2xD) data sets indicate the ratios of wind flow rates through the rectangular regions of the rotor height and the width of 1.2 and 2 rotor diameters to the inlet flow rates through the same size regions. Only 30% of the wind stream flew through the region of 1.2xD, whereas this value increased up to 60% if the region was stretched sideways one diameter each side (2xD). A slight decrease in the relative flow rate through these regions were observed with a decrease in AR. The majority of the displaced wind flew behind the endplates (Displacement Endplates data set). Only a small fraction of the fluid was displaced sideways (Displacement Sides data set) for the lowest aspect ratio rotors (less than 5% for AR ≤ 0.75), however, this share increased to 30% for the highest aspect ratio under study AR = 5.0.

4. Conclusions

The aim of this study was to investigate an influence of the aspect ratio of the Savonius rotor on its performance. Transient, 3D numerical simulations of classical Savonius rotors equipped in two semi-circular buckets with an overlap gap and endplates were studied for the aspect ratio AR = 0.5 – 5.0. The numerical model was verified against the experimental data and a good agreement was reached for the tip speed ratios close to optimal. A blockage effect was studied and an appropriate domain size was chosen.

A continuous increase in the power coefficient with a growth of the aspect ratio of the Savonius rotor was observed. Significant changes were present for low aspect ratios (AR < 1.0), whereas for higher aspects, an increase was lower. 2D simulation results indicated that the power coefficient would still increase for AR > 5.0. However, due to some structural reasons, an application of high aspect ratio Savonius rotors is difficult. Therefore, rotors with AR = 1.0 – 2.0 seem to be a good compromise between the rotor performance and compactness.

The effect of wind friction at the endplates was almost the same for all aspect ratios. Thus, a decrease in the power coefficient for low aspect ratio rotors was due to a deflection of higher fraction of the wind stream beyond the rotor (mostly beyond the endplates) and relatively higher thickness of boundary layer zones at the blade tips. The tip speed ratio for the maximal power coefficient was TRS = 0.8 at the lowest aspect ratio under study and it increased up toTSR = 0.9 – 0.95 for the highest aspect ratio.

References

[1] Akwa J V, Vielmo H A and Petry A P 2012 A review on the performance of Savonius wind turbines *Renew. Sust. Energ. Rev.* **16**(5) 3054-64
[2] Kumar A and Saini R P 2016 Performance parameters of Savonius type hydrokinetic turbine–A Review *Renew. Sust. Energ. Rev.* **64** 289-310
[3] Kamoji M A, Kedare S B and Prabhu S V 2009 Experimental investigations on single stage modified Savonius rotor *Appl. Energy* **86** 1064–73
[4] Alexander A J and Holownia B P 1978 Wind tunnel test on a Savonius rotor *J. Wind Eng. Ind. Aerodyn.* **3**(4) 343–51
[5] Mahmoud N H, El-Haroun A A, Wahba E and Nasef M H 2012 An experimental study on improvement of Savonius rotor performance *Alex. Eng. J.* **51**(1) 19–25
[6] Zhao Z, Zheng Y, Xu X, Liu W and Hu G 2009 Research on the improvement of the performance of Savonius rotor based on numerical study *Proceedings of international conference on sustainable power generation and supply (SUPERGEN)* Nanjing China 1–6
[7] Patel V, Bhat G, Eldho T I, and Prabhu S V 2017 Influence of overlap ratio and aspect ratio on the performance of Savonius hydrokinetic turbine *Int. J. Energ. Res.* **41**(6) 829-44
[8] Vance W 1973 Vertical axis wind rotors–status and potential *Proceedings of the conference on wind energy conversion systems* Washington USA 96–102
[9] Kacprzak K, Liskiewicz G and Sobczak K 2013 Numerical investigation of conventional and modified Savonius wind turbines *Renew. Energ.* **60** 578-85
[10] Kacprzak K and Sobczak K 2015 Computational assessment of the influence of the overlap ratio on the power characteristics of a Classical Savonius wind turbine *Open Eng.* **5** 314-22
[11] Ferrari G, Federici D, Schito P, Inzoli F and Mereu R 2017 CFD study of Savonius wind turbine: 3D model validation and parametric analysis *Renew. Energ.* **105** 722-34
[12] Kacprzak K 2016 Numerical and experimental analysis of the flow in Savonius wind turbines (PhD thesis) *Lodz University of Technology* (in Polish)
[13] Jaohindy P, McTavish S, Garde F and Bastide A 2013 An analysis of the transient forces acting on Savonius rotors with different aspect ratios *Renew. Energ.* **55** 286-95

IOP Conf. Series: Journal of Physics: Conf. Series **1101** (2018) 012035 doi:10.1088/1742-6596/1101/1/012035

Experimental study of the aerodynamic perturbations induced by the single normal hot-wire probe positioned normally in the vicinity of a wall

Jacek Sobczyk

Strata Mechanics Research Institute of the Polish Academy of Sciences, Reymonta 27, 30-059 Krakow, Poland

Abstract. The conducted research, though still at the preliminary stage, is a step toward a better understanding of the flow phenomena accompanying hot-wire probes. For the flow in the vicinity of a wall, a potentially non-optimal case was analyzed – a single normal hot-wire probe oriented normally to the surface. According to the results, undisturbed flow, as measured in the absence of a hot-wire probe, was significantly altered in its presence. Differences in the vertical profiles of the velocity magnitude in the upstream direction of up to 40% were observed 0.43 mm from the probe's longitudinal axis of symmetry, and up to 8% at a distance of 3.73 mm. The lower Reynolds numbers were considered the more complex velocity profile perturbations were present. The highest differences in undisturbed vs. disturbed flow were observed at the level of the wire and below, regardless of the distance from the hot-wire probe.

1. Introduction

Hot wire anemometry (HWA) is a well-known technique used on a daily basis in many laboratories around the world. Even though the laser Doppler anemometry (LDA) optical technique seems to be more appropriate nowadays – it is more precise and has a higher temporal resolution – HWA has obvious advantages – it is much cheaper and does not require seeding particles nor sophisticated optical access.

The foundations of HWA were probably laid at the very beginning of the 19th century or possibly even earlier with the papers of Ser [1] and King [2], among others. Aerodynamic perturbations introduced by HWA probes in high-speed flows were recognized as important quite early. However, at low speeds, this problem was, for the first time, raised in 1967 by Hoole & Calvert [3]. They observed the systematic error occurring in mean-velocity measurements, depending on whether the stem of the probe was aligned with or normal to the flow (up to 20%, according to the probe used).

Because of its small size and high temporal resolution, the HWA was acknowledged as a useful tool to investigate not only turbulence in general, but particularly turbulence in the vicinity of walls. However, even bending prongs into a Z-shape (to create the so-called boundary layer probe) did not solve the problem of prong interaction with the wall, while approaching them to the wall at distances comparable to their diameters. An increase of the distance between prongs reduces the influence of the velocity perturbation on the wire and therefore on the measurement results. On the other hand, the decrease of wire length is highly advisable in order to increase the spatial resolution – a possibility to analyze smaller flow structures [4]. What is more, it turned out that the effect of additional cooling from the wall affects the mean-velocity profile [5-6]. According to Bogar & Willmart [7], a strong reduction of the wire diameter down to c.a. 0.5 μm serves as a palliative in this case.

There are numerous other phenomena needed to be taken into account while conducting measurements by means of only a single normal hot-wire probe (SN-probe). They originate from the probe's construction and the materials used [8-9], properties of the media such as fluid composition, pressure and temperature and their stability during the measurement [10-11], extreme velocities [12-13], or effects such as the influence of the wall-normal velocity component on the probe readings [14]. All these issues were solved fully or partially over the years on the basis of theoretical considerations as well as mathematical modelling and experimental research.

An experimental verification of the solutions of the above-mentioned issues, especially regarding complex measurement conditions, until now, typically presupposed the use of another hot-wire probe or LDA at most. In the broadly available literature, the author did not find any attempts of using other optical measurement methods, such as particle image velocimetry (PIV), which allows for a direct and comprehensive study of many flow phenomena. The present paper may serve as an introduction to this task.

2. Measurements

2.1. Measurement stand and conditions

The measurements were carried out in a closed-circuit, low speed wind tunnel at the Strata Mechanics Research Institute of the Polish Academy of Sciences (SMRI PAS) equipped with a 2D PIV measurement set-up based on sCMOS, 5.5 Mpx camera and Nd:YaG, double impulse, 200 mJ per impulse green laser. As seeding, Di-Ethyl-Hexyl-Sebacate (DEHS) was used. In order to obtain the highest possible magnification ratio, a 180 mm micro 1:1 lens was attached to the camera through 2 sets of distance rings. As a result, an area of 6.2 mm x 7.3 mm was projected onto a 2160 px x 2560 px sensor [15]. In the field of view, the edge and the first ~7 mm of the (inclined 1.4°) razor blade surface, with or without the SN-probe oriented normally to the surface, was visible (Figure 1). The wire to wall distance was constant and equal to $h = 0.55$ mm. Details on the used SN-probe: tungsten wire, 5 microns in diameter, 1.3 mm long, welded to two straight, conical, 20 mm long prongs.

The measurements were carried out for five free stream velocities $U_\infty = 0.15$, 0.53, 1.57, 3.9 and 7.59 m/s corresponding to the following Reynolds numbers $Re_d = 1.98$, 7.0, 20.7, 51.4 and 100.1, respectively, to observe possible differences ($d = 0.2$ mm regards to diameter of the prongs' tips).

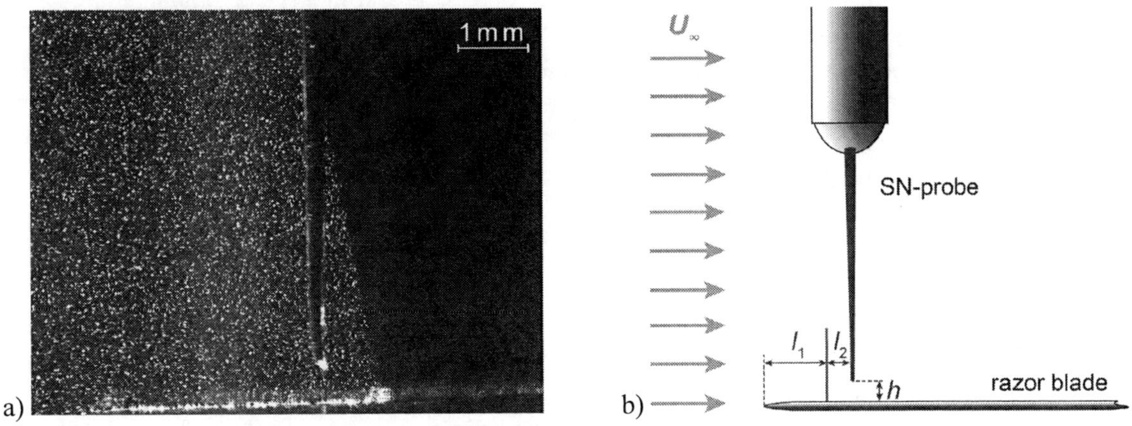

Figure 1. Snapshot from the PIV measurement with the SN- probe present – (a), and sketch presenting the measurement configuration (not in scale) – (b).

Due to demanding conditions related to the small experimental area, simplified lightning conditions were used, i.e. the laser light sheet was positioned above and in front of the SN- probe, outside the wind tunnel measurement chamber. This resulted in a shadow and thus in no measurement results in the downstream part of the measurement area. The flow behavior in this location results mostly from the SN-probe geometry and the first step into its analysis with the use of PIV technique was done in paper [15].

2.2. Observations

Images with such a high magnification ratio tend to expose features and phenomena that are invisible with the naked eyes, which are unexpected or even unwanted. In this case, it was likewise. Image sequences recorded during the measurements revealed, aside from slightly bended tips of prongs, that in higher velocities, the SN-probe and the razor blade were vibrating. These oscillations started being clearly visible for the SN-probe at a velocity of 1.57 m/s, and for the razor blade – at velocity 3.9 m/s, with amplitudes of c.a. 0.01 mm in the case of both. At a velocity of 7.59 m/s, these amplitudes reached a level of c.a. 0.03 and 0.02 mm for the SN-probe and the razor blade, respectively. They result from the natural instability of the measurement set up.

2.3. Measurement results

The tabulation of the resulting ensemble averaged velocity fields is shown in Figure 2. It expresses both qualitatively and quantitatively the main features of the flow over the razor blade without and with the SN-probe present.

A visual comparison of the results leads to an obvious conclusion that, regardless of the Re_d value, the introduction of an SN-probe into the flow, especially in an obstructing way presented here, does disturb the flow significantly. Serious modifications of velocity fields are visible for both components, longitudinal and transverse. The boundary layer formation over the razor blade within the measurement plane is partly disrupted.

Although there were no HWA measurements made for the direct comparison with the PIV results, the presented graphs (Figure 2) clearly indicate that the influence of the SN-probe is visible nearby as well as far in the upstream direction, further than the inflow border of the measurement area. This may explain the high deviation in the velocity estimation observed by Hoole & Calvert [3]. This also suggests that the diameter of the prong was chosen incorrectly as a characteristic dimension for this case. More adequate seems to be the distance between the prongs' outer edges, which is c.a. 1.5 mm at the level of wire and then slightly increases.

From each measurement at the distance $l_1 = 3.00$ mm past the razor blade edge and $l_2 = 0.43$ mm (see Figure 1) in front of the wire (if present), a vertical, mean velocity magnitude profile was extracted. For each Re_d, these profiles were compared in pairs and presented in Figure 3b in terms of relative differences. Additionally, this very procedure was applied to profiles extracted from the position $l_1 = -0.30$ mm and $l_2 = 3.73$ mm (Figure 3a). These graphs enable a more precise comparison of the velocity distribution in the transverse direction.

Two main observations based on the analysis of Figure 3 hold the center stage. Firstly, on average, the lower the Reynolds number, the higher the differences. Secondly, even though conical prongs are an increasing obstacle while moving away from the wire, the highest velocity alternations are observed at the level of the wire and below. This results mainly from the fact that a tiny wire surrounded with a cloud of heated gas is a serious obstacle, which forces the inflowing air stream to move around. What is more, a part of the air stream trying to avoid a couple of prongs and, simultaneously the razor blade surface, increases the mass stream below wire.

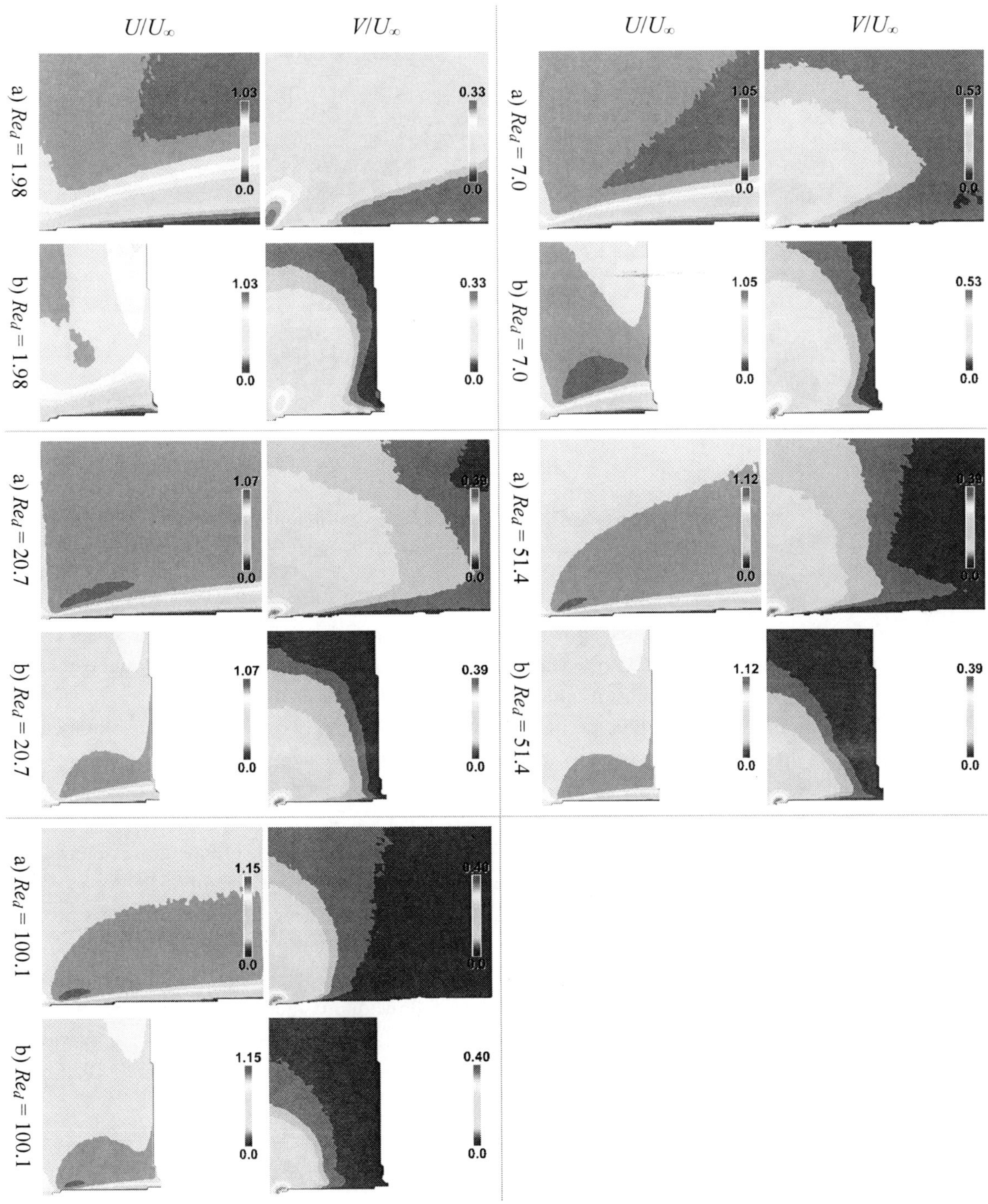

Figure 2. Ensemble averaged velocity fields in the absence of the SN-probe (a) and in the area in front of the SN-probe (b). Presented as normalized scalar fields of their stream-wise and span-wise components.

It is interesting that this trend of highest velocity alternation at the level of the wire and below holds true even far in the upstream direction in front of the razor blade edge. Even more puzzling is the almost mirror reflection of the curve related to $Re_d = 7.0$ versus the one related to $Re_d = 1.98$ in Figure 3a and all others in general. This particular behavior may express certain flow phenomena or a mistake during the measurements. On the other hand, a perfectly regular shape of the curve related to $Re_d = 7.0$ in Figure 3b may suggest the former explanation.

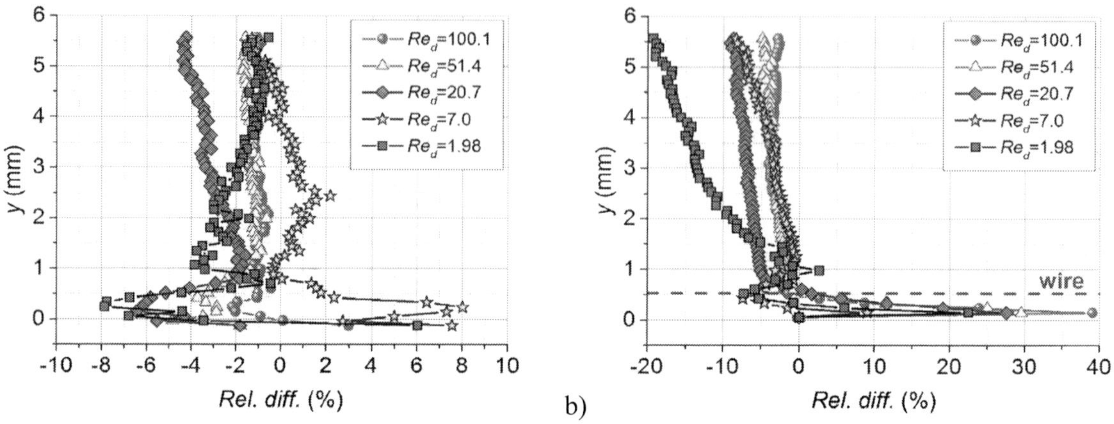

a) b)

Figure 3. Relative differences of the vectors' magnitudes.

3. Conclusions

To conclude, the investigation presented above illustrates the complexity of the flow disturbances caused by placing the SN-probe in the vicinity of a wall. However, the phenomena, described by many authors, was never measured directly by means of optical methods, such as PIV.

Major developments in electronics, the miniaturization of probes and the development of mathematical models made HWA a friendly and relatively cheap measurement technique. A variety of probe shapes allows for more accurate measurements in demanding conditions. However, the SN-probe is the most common and versatile. As the above results indicate, one has to remember that, in measurements where such a probe has to be placed in the close vicinity of one [16] or more objects [17], the flow conditions may vary significantly from those in which the probe was calibrated. As a result, a systematic deviation of the measurement results is expected. What is more, as the probe disturbs the flow, if the latter one is prone to such disturbance, it may exhibit different properties.

The latter statement is not as obvious as it seems. Many users regard the HWA as almost a pointwise technique. They not always have the consciousness of how many factors have an impact on the measurement results and how users modify flow structures by adding the hot-wire probe, in particular. The present paper, being also a continuation of previous ones [15, 18], may serve as a handy illustration for this issue.

Acknowledgments

This publication presents results of the work supported by the Statutory Research of IMG-PAN.

References

[1] Ser L 1888 *Traite de Physique Industrielle* (Paris: Masson) 1, p 142
[2] King L V 1915 *Phil. Mar.* 29 **6**, p 556
[3] Hoole B J, Calvert J R 1967 *J. R. Aeronaut. Soc.* 71 p 511
[4] Ligrani P M, Bradshaw P 1987 *Exp. Fluids* 5 p 407
[5] Richardson E G 1934 *Les Appareils a Fil Chaud.* (Paris: Gauthier Villars) p 67
[6] Alcaraz E, Mathieu J 1975 *C.R. Acad. Sci.* (Paris) 280A p 737

[7] Bogar T J, Willmarth W W 1974 *Bull. Am. Phys. Soc.* 19 **2** p 1144 (Abstr.)
[8] Cierniak W 2009 *Thermoanemometers* (Krakow: Strata Mechanics Research Institute) p 328
[9] Jorgensen F E 2002 *How to measure turbulence with hot-wire anemometers – a practical guide* (Skovlunde: Dantec Dynamics)
[10] Ligęza P 2009 *Simultaneous velocity and temperature measurements in transient gas flows* (Krakow: Strata Mechanics Research Institute) p 132
[11] Noltingk B E 2013 *Measurement of Temperature and Chemical Composition: Jones' Instrument Technology* (Butterworth-Heinemann) p 192
[12] Stainback P C, Nagabushana K A 2018 *Electronic Journal of Fluids Engineering* p 1–54
[13] Andrews G E at al. 1972 *Int J Heat Mass Transfer* 15 p 1765–86.
[14] Dróżdż A, Elsner W 2015 *Open Engineering* **5** 1 p 307-313
[15] Sobczyk J 2018 *EPJ Web of Conferences* **180**, 02094
[16] Gnatowska R 2015 *Pol. J. Environ. Stud.* **24** 4 p 1571–77
[17] Skotniczny P, Sobczyk J 2015 *Transactions of the Strata Mechanics Research Institute* **17** 3–4 p 15–24 (*in polish*)
[18] Bujalski M et al. 2014 *J. Phys.: Conf. Ser.* **530** 012045

IOP Conf. Series: Journal of Physics: Conf. Series **1101** (2018) 012036 doi:10.1088/1742-6596/1101/1/012036

Numerical modelling of flow field within a packed bed of granular material

Marcin Sosnowski[1], Renata Gnatowska[2], Jacek Sobczyk[3], Waldemar Wodziak[3]

[1] Jan Dlugosz University in Czestochowa, Czestochowa, Poland
[2] Czestochowa University of Technology, Czestochowa, Poland
[3] Strata Mechanics Research Institute of the Polish Academy of Sciences, Krakow, Poland

E-mail: m.sosnowski@ajd.czest.pl

Abstract. Computational domain discretization is the pre-processing stage in CFD analysis and it highly contributes to reducing the level of solution error as well as increasing the numerical stability of a model. The discretization of domain representing a packed bed of granular material is a demanding task due to the topology of the model consisting of spherical granules contacting tangentially and narrow regions in the direct vicinity of the contact point. Therefore the aim of the carried out research is to define guidelines to effectively discretize the computational domain representing the packed bed of granular material and validate it using Particle Image Velocimetry (PIV). Two methods of contact point representation between granules (contact point extension, granule radius reduction) and two mesh types (tetrahedral, polyhedral) are examined within the paper. Differences between the analyzed cases are discussed and compared to experimental research results. The obtained results prove that the applied method significantly affects the calculated porosity but does not affect the flow field considerably. Polyhedral mesh turned out to be preferable as the numerical model converges within substantially lower number of iterations and the obtained results are not as affected by the numerical diffusion as in the case of tetrahedral mesh.

1. Introduction

Granular materials are commonly used in many industries and applications; therefore they are the focus of attention of numerous research efforts carried out utilizing both experimental and numerical techniques [1–7]. One of the latter is Computational Fluid Dynamics (CFD), which has proven its applicability as an efficient research tool for numerous industrial sectors and can also be successfully applied in flow analysis within packed beds of granular material [8,9].

The initial and crucial step in CFD analysis is the discretization of the computational domain (meshing) as it influences not only the result accuracy but also the calculation time and numerical stability of the model [10]. Discretization consists of dividing the analyzed geometry into numerous control volumes commonly called elements or cells.

Three mesh types depicted in Figure 1 are used in CFD applications: hexahedral (HEX), tetrahedral (TETRA) and polyhedral (POLY). The first (HEX)is characterized with low numerical diffusivity, but it is difficult or even impossible to be generated in the case of complex geometries. The second (TETRA) is relatively easy to be generated, but it introduces significant errors resulting from its high diffusivity. Therefore, the POLY mesh is a beneficial compromise as it combines acceptable numerical diffusivity with ease of generation. The benefit of POLY results from the fact that each individual cell neighbors with many others. In contrast, TETRA neighbors with only four elements. POLY contributes to better gradient approximations and in consequence to more reliable CFD analysis results . Moreover POLY cells are less sensitive to stretching in comparison to TETRA cells which improves the numerical stability of the model.

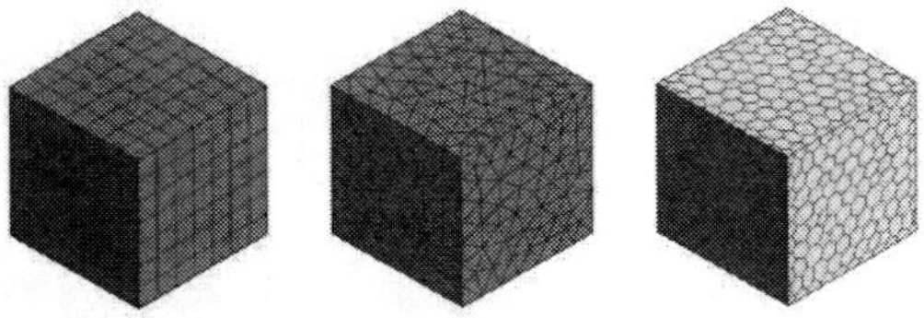

Figure 1. Mesh types: hexahedral – left, tetrahedral – center, polyhedral – right

The mesh quality is significantly important in the case of modelling a flow field in packed beds of granular materials because of the specific topology of the domain, where individual granules of spherical shape contact each other tangentially at a single point. It leads to mesh elements of extremely low quality because the cells in direct vicinity of the contact point are highly skewed. Therefore the aim of the carried out research is to define guidelines to effectively discretize the computational domain representing the packed bed of granular material.

2. Methods

2.1. Experimental research

The experimental research object consisted of three identical spheres of 30mm diameters. The spheres represented granules positioned inline and tangentially contacted. Each of the granules was mounted on a separate post of adjustable height attached to a wooden stand as depicted in Figure 2.

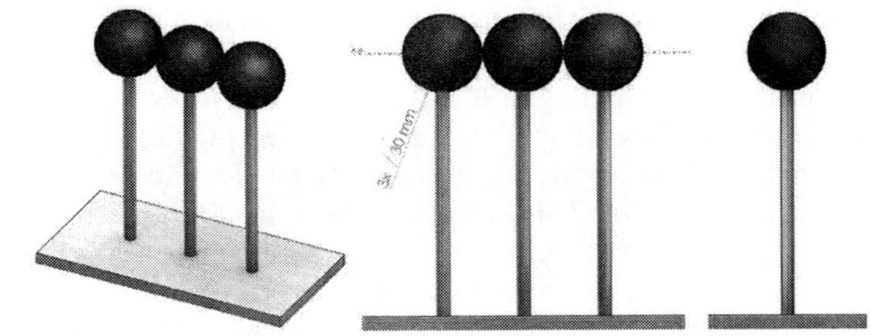

Figure 2. The model of the research object

The experimental research with the application of Particle Image Velocimetry (PIV) was conducted in a low-speed, closed-circuit wind tunnel. The measuring section of dimensions 0.5m x 0.5m x 1.5m was adapted for optical, non-intrusive measurement methods such as PIV (Figure 3). The free stream velocity of the tunnel ranged from 0.1 to 62 m/s and longitudinal free stream turbulence intensity was less than 0.5% as described in [11]. The wind tunnel was equipped with systems designed to control temperature and relative humidity. The utilized PIV system consisted of 5.5Mpx sCMOS camera and Nd:YAG laser emitting a pair of light pulses of energy equal to 0.2J each. Flow was seeded with ~1μm droplets of Di-Ethyl-Hexyl-Sebacate (DEHS). The seeding particles were introduced into the flow in stable concentration adjusted during the preliminary calibration and setting.

The laser light sheet illuminated plane crossing through the centers of the investigated granules. Double frame images were recorded at 15Hz frequency rate and analyzed with the adaptive correlation algorithm. Total number of double frame images taken during the measurement was 250. The research object was mounted centrally along the longitudinal axis of the wind tunnel's measuring section as depicted in Figure 3 and Figure 4. Instantaneous velocity fields were measured within the area marked by the red dotted line in Figure 3 and located in the plane coincident with the centers of the granules. The inflow velocity was equal to 1.81m/s, which corresponds to Re number related to the granule's diameter equal to 3700. The PIV research provided the velocity vector fields with two velocity components in stream-wise and span-wise directions.

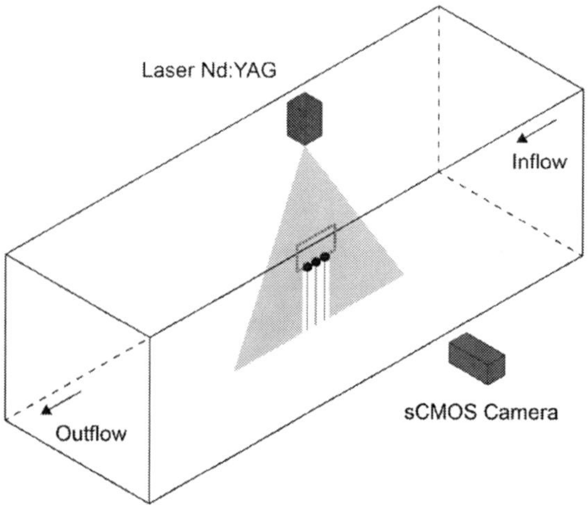

Figure 3.The scheme of the PIV experimental setup

Figure 4.The research object illuminated with the laser light sheet and located in the measuring section of the wind tunnel

2.2. *Computational domain discretization and numerical model parameters*

The pre-processing stage in CFD analysis of granular material domain is a challenging task due to the single-point contact between the two individual granules and very narrow space to be meshed in the direct vicinity of the contact point (Figure 5a). It leads to extremely low-quality mesh elements and in consequence, to numerical instability of the calculations. The common approach to overcome this issue is to reduce the radius of the modelled granules (Granule Radius Reduction - GRR) while maintaining their positions (Figure 5b). This unfortunately leads to the disturbance of flow and heat transfer in the domain. Another approach investigated in [8,12–14] is to extend the contact point between the two individual granules to a cylindrical volume as shown in Figure 5c (Contact Point Extension – CPE). This approach allows generating a high-quality mesh while maintaining contact between granules.

Another important issue influencing the CFD results and associated with mesh is the type of mesh elements used to discretize the domain [15]. As was mentioned earlier, HEX elements are characterized by low numerical diffusion, but the geometric complexity of the packed bed of granular material makes it impossible to apply structured HEX mesh in such a case. Therefore TETRA and POLY elements are analyzed in the paper as they can easily by applied to describe even very complicated geometries [16]. In consequence, four cases were analyzed within the numerical research: two approaches (GRR, CPE) to increase the mesh quality in the contact region and two different mesh types (TETRA, POLY).

The computational domain depicted in Figure 6 represented one quarter of the experimental research object (Figure 2) placed in a virtual wind tunnel domain of dimensions based on best practice guidelines defined in [17,18]. The geometry was created using SolidWorks with ANSYS plugin, which allows the bi-directional associativity between CAD and mesher. It was prepared in two configurations. The first represents a section of the packed bed with granules of reduced diameter (GRR method) – the reduction of granule radius was defined as $\frac{1}{100}r$ (where: r – granule radius) on the basis of findings described in [9,12]. The second configuration represented the full-size granules with contact point extended to the cylindrical volume (CPE method) of radius equal to $\frac{1}{5}r$. Such radius was defined as optimal in [12].

The mesh was generated in ANSYS 18.1 with Workbench Poly Meshing for Fluent 1.8 ACT extension. The mesh sizes on the granule walls were set as $\frac{1}{30}r$ with growth ratio equal to 1.05. The volume mesh size was set to $\frac{1}{10}r$. Such settings were consistent with guidelines provided in [19].

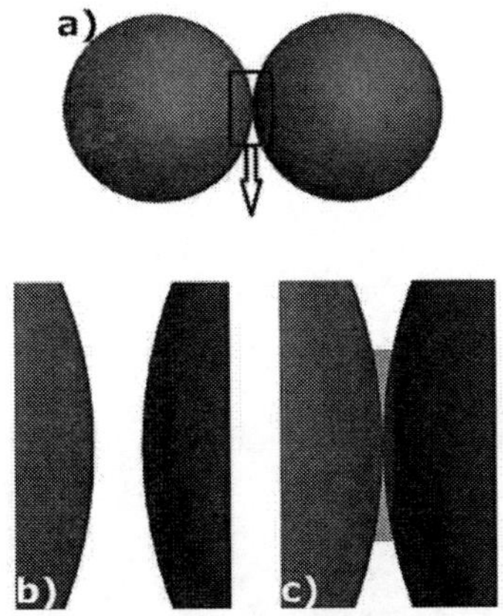

Figure 5. Analyzed methods of granules contact representation: a – contact point, b – GRR method, c – CPE method

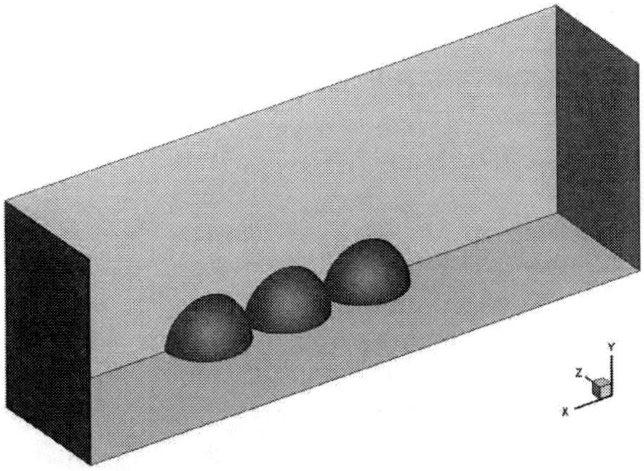

Figure 6. Boundary conditions: wall – red, velocity inlet – green, pressure outlet – blue, symmetry – yellow

The identical computational model (except for the mesh) was used for all cases analyzed within the research. The solver was configured as pressure-based and analysis was performed for a steady state. Standard k-ε viscous model was applied.

Velocity-inlet boundary condition type was assigned to the inlet of the computational domain (marked green in Figure 6) with the velocity magnitude corresponding with the experimental research and equal to 1.81m/s. The outlet of the computational domain (marked blue in Figure 6) was defined as pressure outlet with gauge pressure of 0Pa (operating pressure – 101325Pa), backflow turbulent intensity was set to 5% and backflow turbulent viscosity ratio was set to 10. Symmetry was assigned to the section planes indicated yellow in Figure 6.

Pressure-velocity coupling by SIMPLE algorithm was used as a solution method. This algorithm uses a relationship between velocity and pressure corrections to enforce mass conservation and to obtain the pressure field.

The numerical calculations were performed with ANSYS Fluent 18.1 on Windows 10 workstation equipped with Intel i7 4510U CPU 2.00GHz and 16GB of RAM. The solver was operating in a single mode.

3. Results and Discussion

The experimental research results obtained with PIV concerning time-averaged flow field in the vicinity of the research object are depicted in Figure 7. The boundary layer close to the granules surface is clearly visible although its thickness may be misrepresented due to the resolution of the measuring system equal to 0.712mm for the analyzed region. Moreover slight differences concerning flow patterns in the two regions between the adjacent granules are noticeable.

Another PIV measurement series was performed in order to focus on the flow very close to the contact point between the adjacent granules. Limiting the examined region, resulted in improved measurement resolution equal to 0.085mm. The results concerning the vector plot, as well as time-averaged flow fields in the region between the first and the second granule are shown in Figure 8.

The velocity field is presented as velocity magnitude (Velocity), stream-wise velocity (U) and span-wise velocity (V). As it can be observed, the flow velocity magnitude at the distance of 3mm (20% of granule radius) upwards from the granules' contact point does not exceed 0.02m/s which is approx. 1% of the inflow velocity.

Therefore, it can be concluded that CFD numerical analysis can be effectively performed with the CPE method of discretizing the granular material domain because even in the case of relatively large cylindrical volume radius equal to 3mm, the flow is not disturbed significantly. In contrast, the GRR method affects the main flow stream because of the reduction of the individual granule size.

In addition, the adulteration of the investigated granular material porosity is at the level of 0.03% in the case of CPE and 3.26% in the case of GRR, which directly points to the CPE method as more beneficial.

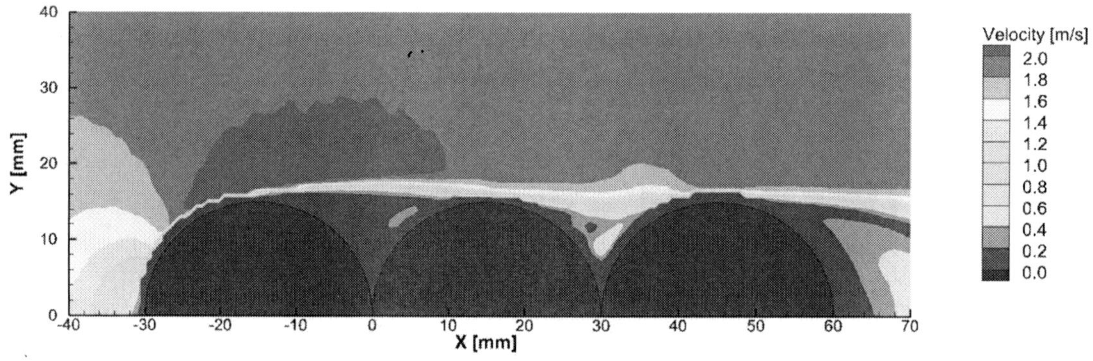

Figure 7. Time-averaged flow field (PIV)

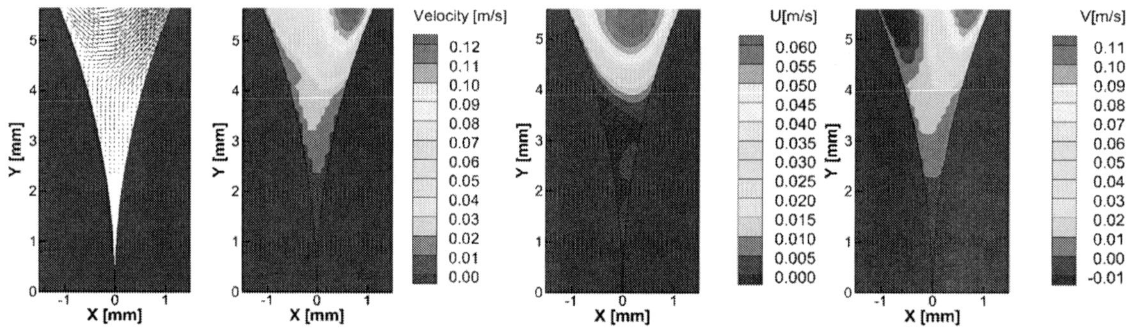

Figure 8. Vector plot and time-averaged flow field (PIV) in the region between the first and the second granule (U – stream-wise velocity, V – span-wise velocity)

As was mentioned earlier, the CFD analyses were performed for four configurations i.e. two discretization methods and two mesh types. Mesh statistics including minimum orthogonal quality (ranges from 0 to 1, where values close to 0 correspond to low quality) and maximum aspect ratio (1 is ideal aspect ratio) for all analyzed cases are listed in Table 1. The above mentioned quality indicators point to a very good quality mesh in the case of POLY and good quality mesh in the case of TETRA. The mesh quality is not affected significantly by the applied method of granule contact representation but only the POLY mesh type indicated preferable results.

The model was assumed as converged where all residuals fell below the value of 1e-3. As can be seen in the last row in Table 1, the number of iterations necessary to assure convergence was almost 2.5 times higher in the case of TETRA mesh in comparison to POLY mesh. It leads to a conclusion

IOP Conf. Series: Journal of Physics: Conf. Series **1101** (2018) 012036 doi:10.1088/1742-6596/1101/1/012036

that the POLY mesh is preferable also in terms of calculation time. The granules' contact point representation method did not affect the number of iterations to convergence.

Table 1. Mesh statistics and iteration to convergence

	POLY CPE	TETRA CPE	POLY GRR	TETRA GRR
Minimum Orthogonal Quality	0.52	0.17	0.68	0.16
Maximum Aspect Ratio	5.53	19.70	5.16	21.86
Iterations to Convergence	153	364	154	364

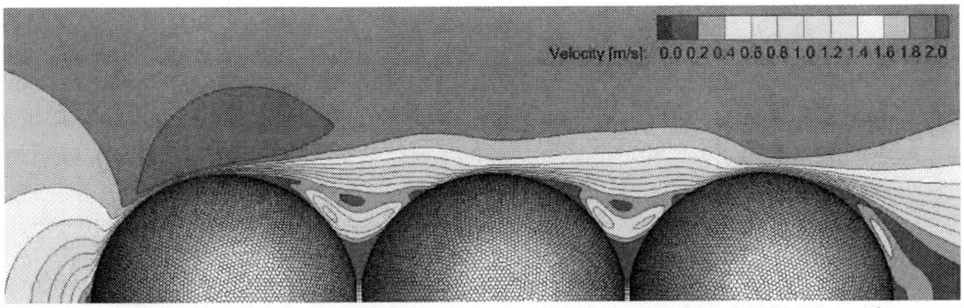

Figure 9. Time-averaged flow field for CPE method and POLY mesh

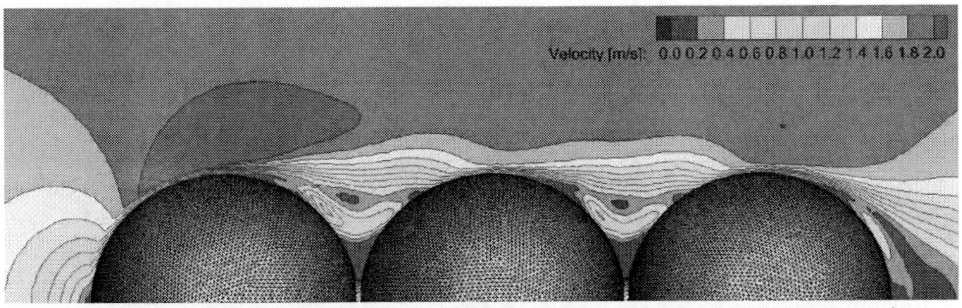

Figure 10. Time-averaged flow field for CPE method and TETRA mesh

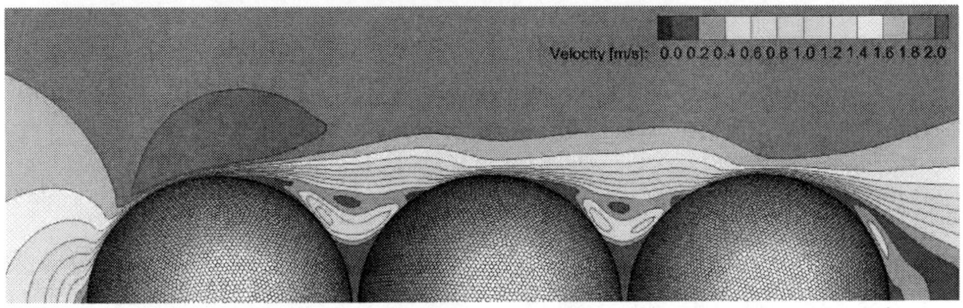

Figure 11. Time-averaged flow field for GRR method and POLY mesh

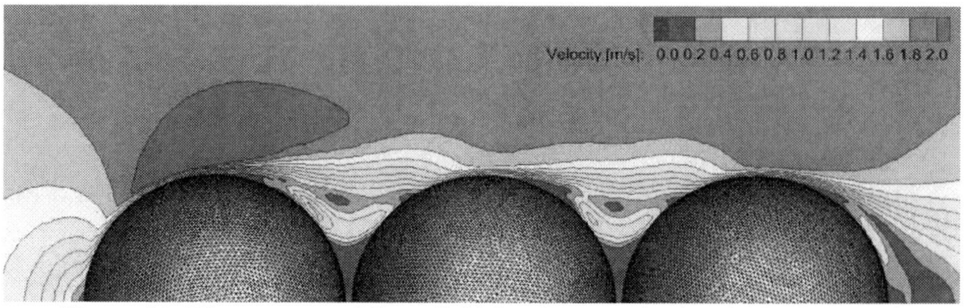

Figure 12. Time-averaged flow field for GRR method and TETRA mesh

Time-averaged flow fields obtained with the use of CFD for the analyzed methods (CPE, GRR) and mesh types (POLY, TETRA) are depicted in Figure 9 - Figure 12. There are no noticeable differences in the flow field near the granules regarding the applied method of contact point representation but a decrease of granule size in the case of GRR method slightly influences the main flow above the granules. The applied mesh type seems to impact the results more significantly as lower numerical diffusion of POLY mesh is reflected in less velocity fluctuations particularly along the vertical axis of the computational domain. This issue is the most significant after the third granule. Therefore the results obtained with POLY mesh are more consistent with the experimental results obtained by means of PIV.

Figure 13 – Figure 16 present the time-averaged flow fields obtained with the use of CFD for the analyzed methods and mesh types zoomed in the region of the contact point between the first and second granule. Velocity magnitude as well as stream-wise and span-wise velocities are depicted in order to compare different cases of CFD and PIV results in detail. It can be noticed that the influence of granules' contact point representation method can be seen only in the region occupied by the additional cylindrical volume generated with the CPE method where the velocity does not exceed 0.02m/s (less than 1% of inflow velocity). However, variations can be observed regarding the mesh type: higher velocity values (more similar to PIV) are observed in the case of TETRA, but POLY represents the boundary layer with better agreement of the experimental research results. In general, CFD underrates velocity in the region between the granules in comparison to PIV. The most significant differences among different cases of CFD simulations are observed between GRR POLY and GRR TETRA, which proves the importance of the applied mesh type.

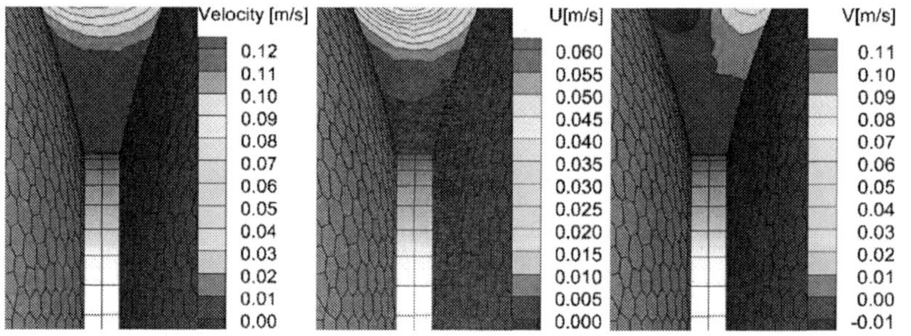

Figure 13. Time-averaged flow fields for CPE method and POLY mesh in the region between the first and the second granule (U – stream-wise velocity, V – span-wise velocity)

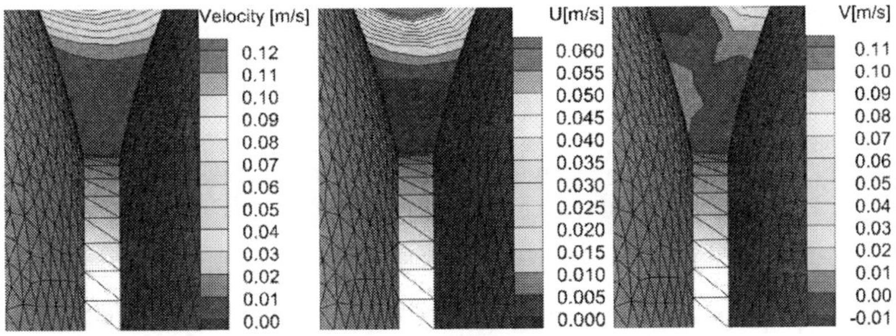

Figure 14. Time-averaged flow fields for CPE method and TETRA mesh in the region between the first and the second granule (U – stream-wise velocity, V – span-wise velocity)

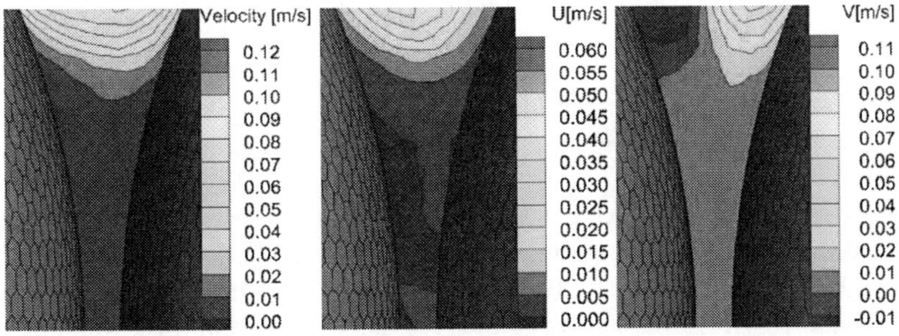

Figure 15. Time-averaged flow fields for GRR method and POLY mesh in the region between the first and the second granule (U – stream-wise velocity, V – span-wise velocity)

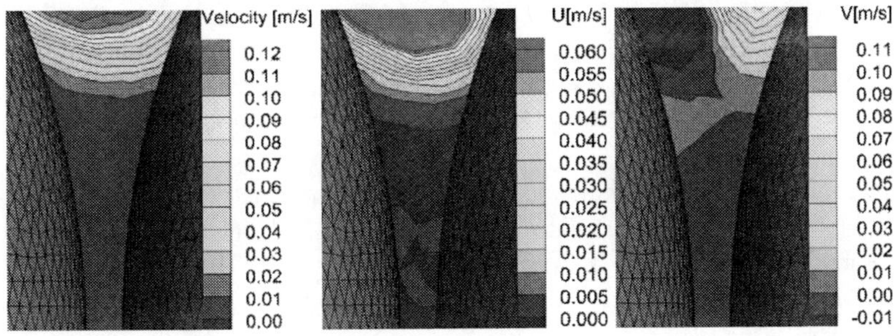

Figure 16. Time-averaged flow fields for GRR method and TETRA mesh in the region between the first and the second granule (U – stream-wise velocity, V – span-wise velocity)

4. Conclusions

The presented results prove that the implementation of the CPE method does not affect the flow above the granules. A reduction of granule diameter in the GRR method slightly affects the flow above the granules but has significant effect on the calculated porosity of the packed bed of granular material which increases 3.26%. In contrast, the CPE method affects the calculated porosity only by 0.03%. The local velocity in the region close to the granules' contact point is underrated by CFD in comparison to PIV.

The applied mesh type influences the flow field more significantly than the applied method of contact point representation. The better characteristics of POLY mesh in terms of numerical diffusivity is revealed in more accurate boundary layer representation and better agreement of CFD and PIV results regarding velocity fluctuations particularly along the vertical axis of the domain. Moreover the results are generated within approx. 2.5 times lower number of iterations in the case of POLY mesh, which may be a major advantage for industrial applications.

Therefore, the recommended method for discretization the computational domain representing a packed bed of granular material is the CPE method with POLY mesh as the preferable mesh type .

Acknowledgements
The work is partially supported by National Science Centre of Poland (Narodowe Centrum Nauki) grant number 2017/01/X/ST8/00019 and 2017/01/X/ST8/00076.

References

[1] Gan J, Zhou Z and Yu A 2017 Effect of particle shape and size on effective thermal conductivity of packed beds *Powder Technol.* **311** 157–66
[2] Szymanek E, Blaszczyk T, Hall M R, Dehdezi P K and Leszczynski J S 2014 Modelling and analysis of heat transfer through 1D complex granular system *Granul. Matter* **16** 687–94
[3] Suekane T, Yokouchi Y and Hirai S 2003 Inertial flow structures in a simple packed bed of spheres *AIChE J.* **49** 10–7
[4] Niegodajew P, Wilczyński M, Marek M, Drobniak S, Asendrych D, Elsner W, Gnatowska R and Stempka J 2018 A study of liquid spreading in laboratory scale random packing column with an optical method supplemented with liquid holdup characteristics *Exp. Therm. Fluid Sci.* **96** 162–8
[5] Niegodajew P and Marek M 2016 Analysis of orientation distribution in numerically generated random packings of Raschig rings in a cylindrical container *Powder Technol.* **297** 193–201
[6] Asendrych D and Niegodajew P 2017 Numerical study of the CO2 absorber performance subjected to the varying amine solvent and flue gas loads *Chem. Eng. Commun.* **204** 580–90
[7] Niegodajew P and Asendrych D 2017 An interfacial heat transfer in a countercurrent gas–liquid flow in a trickle bed reactor *Int. J. Heat Mass Transf.* **108** 703–11
[8] Sosnowski M 2018 Computational domain discretization in numerical analysis of flow within granular materials *EPJ Web of Conferences* vol 180 (EDP Sciences) p 2095
[9] Sosnowski M 2018 Computational domain discretization in numerical analysis of forced convective heat transfer within packed beds of granular materials *Eng. Mech.* 801–4
[10] Gnatowska R and Sosnowski M 2018 The influence of distance between vehicles in platoon on aerodynamic parameters *EPJ Web of Conferences* vol 180
[11] Bujalski M, Gawor M and Sobczyk J 2014 Flow disturbance due to presence of the vane anemometer *Journal of Physics: Conference Series* vol 530 (IOP Publishing) p 12045
[12] Bu S S, Yang J, Zhou M, Li S Y, Wang Q W and Guo Z X 2014 On contact point modifications for forced convective heat transfer analysis in a structured packed bed of spheres *Nucl. Eng. Des.* **270** 21–33
[13] Dixon A G, Nijemeisland M and Stitt E H 2013 Systematic mesh development for 3D CFD simulation of fixed beds: Contact points study *Comput. Chem. Eng.* **48** 135–53
[14] Finn J and Apte S V. 2013 Relative performance of body fitted and fictitious domain simulations of flow through fixed packed beds of spheres *Int. J. Multiph. Flow* **56** 54–71
[15] Sosnowski M, Krzywanski J, Grabowska K and Gnatowska R 2018 Polyhedral meshing in numerical analysis of conjugate heat transfer *EPJ Web of Conferences* vol 180
[16] Sosnowski M, Krzywanski J and Gnatowska R 2017 Polyhedral meshing as an innovative approach to computational domain discretization of a cyclone in a fluidized bed CLC unit *E3S Web of Conferences* vol 14
[17] Gnatowska R, Sosnowski M and Uruba V 2017 CFD modelling and PIV experimental validation of flow fields in urban environments *E3S Web of Conferences* vol 14
[18] Sosnowski M 2018 The influence of computational domain discretization on CFD results concerning aerodynamics of a vehicle *J. Appl. Math. Comput. Mech.* **17** 79–88
[19] Celik I B, Ghia U, Roache P J and Christopher 2008 Procedure for Estimation and Reporting of Uncertainty Due to Discretization in CFD Applications *J. Fluids Eng.* **130** 78001–4

IOP Conf. Series: Journal of Physics: Conf. Series **1101** (2018) 012037 doi:10.1088/1742-6596/1101/1/012037

The effect of heat exchanger geometry on adsorption chiller performance

Marcin Sosnowski[1], Karolina Grabowska[1], Jarosław Krzywański[1], Wojciech Nowak[2], Karol Sztekler[2] and Wojciech Kalawa[2]

[1] Jan Długosz University in Częstochowa, ul. Waszyngtona 4/8, 42-200 Czestochowa, Poland

[2] AGH University of Science and Technology, al. Mickiewicza 30, 30-059 Krakow, Poland

E-mail: m.sosnowski@ajd.czest.pl

Abstract. The adsorption technology applied to cooling production is one of the possible means of waste heat utilization and it therefore contributes to a reduction of power consumption in air-conditioning and desalination facilities. Increasing the performance of the adsorption chiller can be achieved by the intensification of heat transfer in the sorption bed which in turn is directly influenced by heat exchanger geometry. The carried out research utilized numerical methods to define the correlation between the fin-tube heat exchanger design and the main factors influencing adsorption chiller performance such as gradient of heating water temperature, logarithmic mean temperature difference, effective mass factor of silica gel, silica gel average temperature and spatial temperature distribution in the sorption bed. The obtained correlations can be used by sorption bed designers to balance the thermal effectiveness of the device and its overall dimensions and mass.

1. Introduction

The reduction of conventional energy consumption is the main assumption of the sustainable development concept for the global economy, therefore the utilization of low-temperature energy sources i.e. industrial waste heat, solar energy or cogeneration heat, is the intensively investigated area in power engineering. Air-conditioning facilities in highly developed countries, as well as desalination facilities in Middle-East countries are significant energy consumers therefore the reduction of their power consumption is crucial to the reduction of fossil fuel consumption.

The adsorption technology is one of the answers to the above mentioned challenge, as it allows cooling production, as well as desalinated water production, by utilizing low grade thermal energy sources. Moreover, it is an eco-friendly system as it does not use any environmentally hazardous refrigerants.

The adsorption chiller is composed of a sorption bed(s), evaporator and condenser while the first is the most important in terms of overall performance of the chiller. The sorption bed is equipped with a water-sorbent heat exchanger (HE). At least two beds should be used in order to assure continuous and effective operation of the chiller [1,2]. Cooling production with the use of adsorption process is based on periodic changes in temperature and pressure of the sorbent enclosed in said bed. The main factor influencing the performance of the adsorption chiller (described by coefficient of performance - COP) is the heat transfer between the heating/cooling water and the sorbent with the use of a HE [3]. Therefore,

IOP Conf. Series: Journal of Physics: Conf. Series **1101** (2018) 012037 doi:10.1088/1742-6596/1101/1/012037

the optimization of the chiller rated parameters depends on the intensification of heat and mass transfer in the sorbent bed. It is attainable by either modification of the sorbent composition in the bed [4,5], sorbent composition in the vicinity of the HE [6], or adjustment of the HE system design. The latter can be the application of fins characterized with various thickness and spacing [6].

2. Methods

The scope of the research is to analyze the influence of HE fins geometry on the intensification of heat transfer within the sorbent bed, keeping in mind the necessity of limiting the mass and dimensions of the device. Different approaches allowing describing the heat transfer process can be found in [7–10]. Since such analysis require many different configurations to be investigated and evaluated, Computational Fluid Dynamics (CFD) have been used as it allows achieving flawless product design at relatively low cost by combining the prediction of fluid flow, heat transfer and related phenomena [11–18].

The utilized CFD solver (ANSYS Fluent 19.0) employs an algorithm which belongs to a general class of methods called the projection method, where the constraint of mass conservation of the velocity field is achieved by solving a pressure equation. The pressure equation is derived from the continuity and momentum equations in such a way that the velocity field, corrected by pressure, satisfies the continuity. The solution process involves iterations in order to solve the nonlinear and coupled governing equations.

The solver was configured as pressure-based and the analyses were performed for a steady state. The 3D steady RANS equations were solved using standard the k–ε model with standard wall function and turbulence intensity of 5%. The choice of the turbulence model was made based on previous studies [19,20]. The model convergence was monitored and the iterations were terminated when all residuals decreased below $1 \cdot 10^{-3}$.

2.1. The research object

The fin-tube HE of dimensions depicted in Figure 1 was the research object. It consisted of the cuboid filled with sorbent and the heat tube with fins of variable thickness and spacing. The parametrized CAD model with bi-directional associativity between the CAD geometry and numerical mesh generator was prepared as an assembly of three domains: sorbent, heat exchanger and water as depicted in Figure 2. Nine design configurations were built: different fin thicknesses (0.50 mm, 0.75 mm and 1.00 mm) and different fin spacing (2 mm, 3 mm and 4 mm). Moreover the model was prepared as periodic to allow the application of variable number of tubes and minimization of computational domain leading to CPU time reduction.

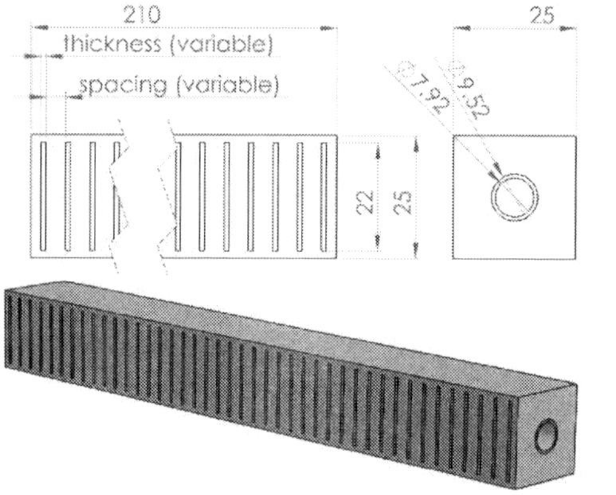

Figure 1. Dimensions of the HE (in mm) with variable thickness and spacing of fins.

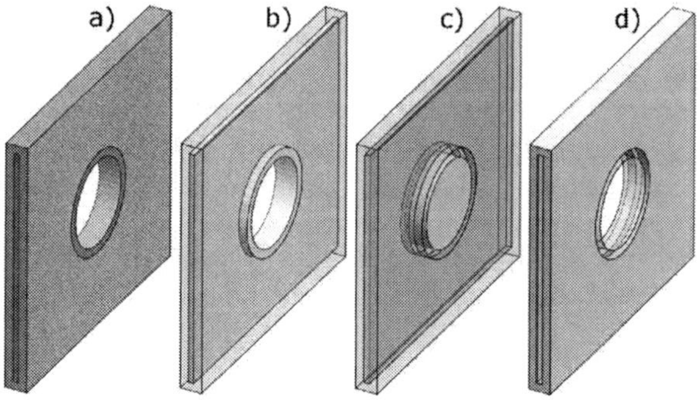

Figure 2. Computational domains: a) all domains, b) heat exchanger, c) water, d) sorbent (only one section of the HE is presented in order to make the figure clear).

2.2. Pre-processing and boundary conditions

The pre-processing was performed using ANSYS Meshing 19.0. Tetrahedral elements were used to discretize the computational domain. Adaptive size function was employed and the element size constrain was set to the value of 0.5 mm. Additional match control settings were applied in order to generate a periodic mesh. The above mentioned meshing settings resulted in the final mesh consisting of approximately four million elements. The mesh resolution was based on mesh sensitivity analysis and generation guidelines described in [21]. The mesh quality was also taken into consideration. Such analysis provides a matrix calculated on the basis of (1) that ranges between 0 and 1 (1 - perfect element, 0 - element of zero or negative volume). The element of the worst quality generated throughout the entire research was characterized by the quality index of 0.17 and therefore it was assumed that the mesh did not affect the numerical stability of the computational model.

$$Q = C \cdot \frac{V}{\sqrt{\left(\sum l^2\right)^3}} \tag{1}$$

where: V – computational domain volume
C – constant characteristic for the applied mesh type
l – the edge length of the analyzed mesh element

The applied boundary conditions correspond with the desorption phase of the adsorption chiller work cycle therefore heating water was the fluid medium. Mass-flow-inlet boundary condition was assigned to the inlet of the heating water with mass flow rate equal 0.02 kg/s and temperature equal 343 K. The outflow boundary condition was defined on the surface of water outflow. The heat transfer between the corresponding subdomains was calculated as conjugate heat transfer. The desorption was modelled as an endothermic process by applying source term of constant and negative value in the whole volume of silica gel.

3. Results and Discussion

According to [22], the heating power of the adsorption chiller is directly proportional to the gradient of heating water temperature (ΔT) calculated as the difference of the temperature of the heating water at the inlet and at the outlet of the HE. Therefore the obtained correlation between ΔT and fin geometry (spacing and thickness) is presented in Figure 3. It can be clearly seen that the temperature difference and resulting heat transfer is proportional to the fin spacing and inversely proportional to fin thickness.

Another important factor defining the influence of HE geometry on adsorption chiller performance is the logarithmic mean temperature difference (LMTD) calculated according to formula (2). The correlation depicted in Figure 4 reveals the strong influence of fin spacing and thickness on LMTD, which is crucial to adsorption chiller performance.

$$LMTD = \frac{\Delta T_{inlet} - \Delta T_{outlet}}{\ln\left(\frac{\Delta T_{inlet}}{\Delta T_{outlet}}\right)} \tag{2}$$

where: ΔT_{inlet} – temperature difference between the heating water and the sorbent in the inlet section of the bed [K]
ΔT_{outlet} – temperature difference between the heating water and the sorbent in the outlet section of the bed [K]

Designers of effective and simultaneously compact adsorption chillers strive to maximize the fraction of silica gel mass in the overall mass of the device. Such dependency can be defined as the effective mass factor (EMF) and it is calculated as the quotient of the silica gel mass to the mass of the device. The carried out research showed that the maximization of the EMF requires minimizing the fin thickness and simultaneously maximizing the spacing between them (Figure 5). Conversely, the average temperature of silica gel also directly influences the sorption kinetics and in consequence the performance of the adsorption chiller and it strongly depends on the geometry of the HE as depicted in Figure 6.

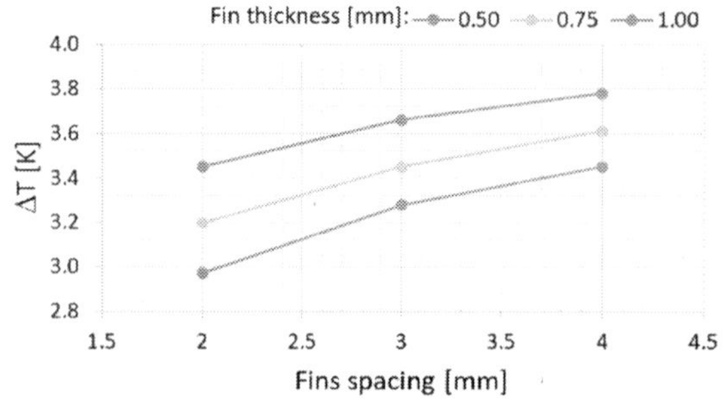

Figure 3. The correlation between heating water temperature gradient and HE geometry.

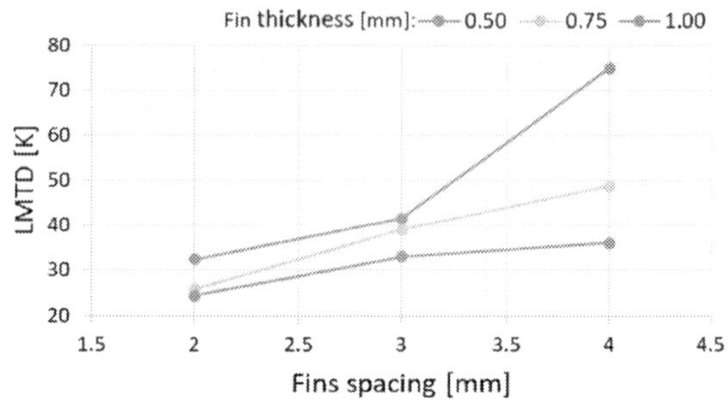

Figure 4. The correlation between logarithmic mean temperature difference and HE geometry.

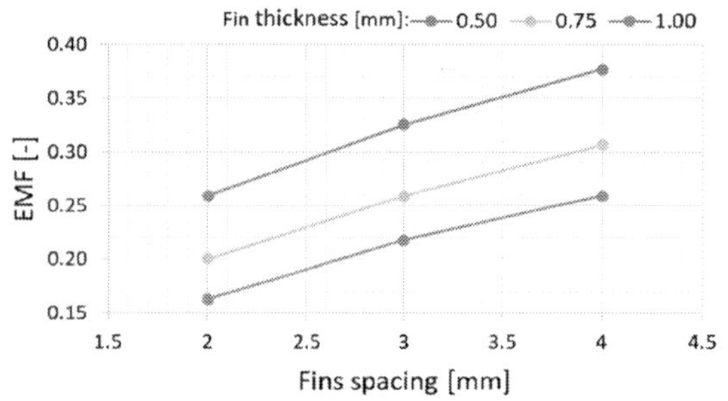

Figure 5. The correlation between effective mass factor and HE geometry.

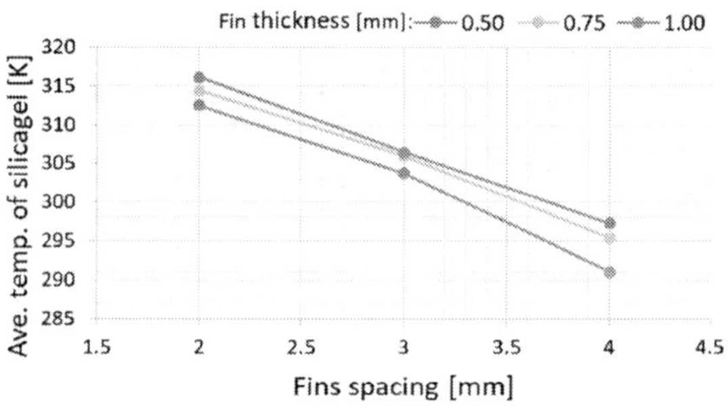

Figure 6. The correlation between silica gel average temperature and HE geometry.

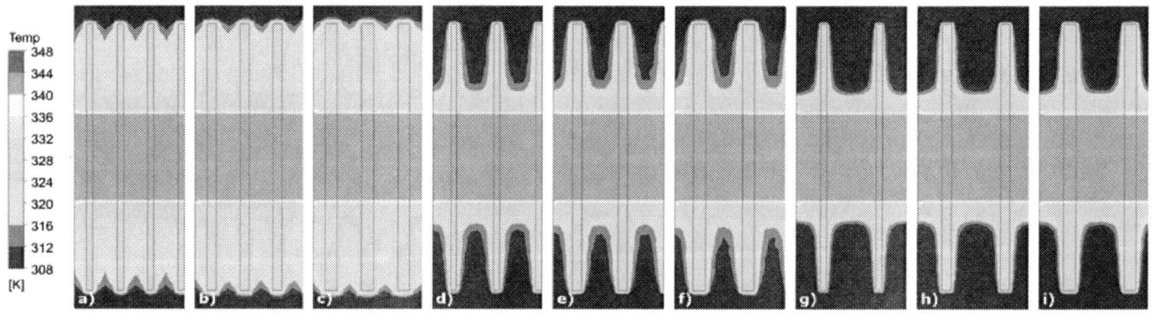

Figure 7. Samples of temperature distribution in the vertical cross-section of the individual HE tube near the inlet obtained for fins spacing equal to 2 mm (a-c), 3 mm (d-f), 4 mm (g-i)) and fins thickness equal to 0.50 mm (a, d, g), 0.75 mm (b, e, h) and 1.00 mm (c, f, i).

Figure 7 presents the samples of temperature distribution in the vertical cross-section of the HE near the heating water inlet, and Figure 8 depicts temperature distribution in the horizontal cross-section of the complete HE consisting of five tubes for nine analyzed cases characterized by different spacing and thicknesses of fins. Figure 6 to Figure 8 prove that the mean silica gel temperature decreases along with the increased fin spacing and decreased fin thickness.

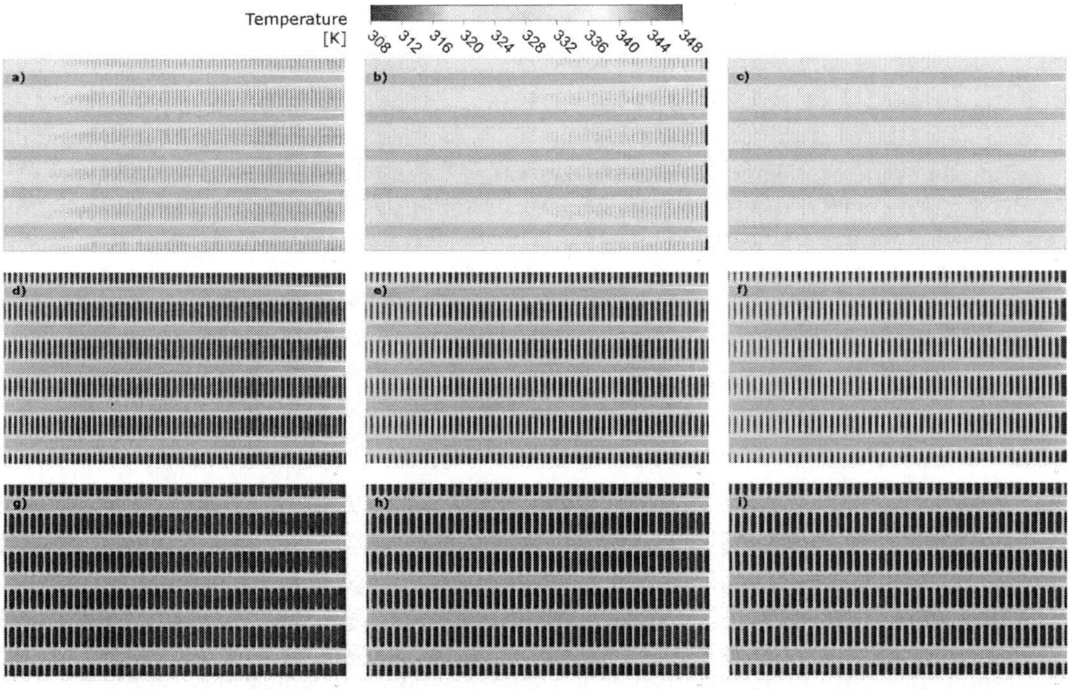

Figure 8. Temperature distribution in the horizontal cross-section of the HE consisting of five heat tubes obtained for fin spacing equal to 2 mm (a-c), 3 mm (d-f), 4 mm (g-i)) and fin thicknesses equal to 0.50 mm (a, d, g), 0.75 mm (b, e, h) and 1.00 mm (c, f, i).

4. Conclusions

The performed analysis allowed to identify the influence of fin spacing and thickness in the HE on the gradient of heating water temperature (ΔT), logarithmic mean temperature difference (LMTD), effective mass factor (EMF) of silica gel, silica gel average temperature and spatial temperature distribution in the sorption bed. All the above mentioned factors directly influence the adsorption chiller performance but the requirement of acceptable mass & dimensions of the sorbent bed also has to be taken into consideration. Therefore the optimal balance between the thermal effectiveness of the device and its overall dimensions has to be individually selected in accordance with specific installation requirements.

The numerical analysis using CFD turned out to be an effective method of comprehensive heat transfer in sorption bed analysis but the applied sorption model has to be validated against experimental data and carefully parametrized in order to improve its reliability.

References

[1] Saha B, Koyama S, Lee J e al, Kuwahara K, Alam K, Hamamoto Y, Akisawa A and Kashiwagi T 2003 Performance evaluation of a low-temperature waste heat driven multi-bed adsorption chiller *International Journal of Multiphase Flow* **29** 1249–63

[2] Krzywanski J, Grabowska K, Herman F, Pyrka P, Sosnowski M, Prauzner T and Nowak W 2017 Optimization of a three-bed adsorption chiller by genetic algorithms and neural networks *Energy Conversion and Management* **153** 313–22

[3] Khan M, Alam K, Saha B, Hamamoto Y, Akisawa A and Kashiwagi T 2006 Parametric study of a two-stage adsorption chiller using re-heat—The effect of overall thermal conductance and adsorbent mass on system performance *International journal of thermal sciences* **45** 511–9

[4] Askalany A A, Henninger S K, Ghazy M and Saha B B 2017 Effect of improving thermal conductivity of the adsorbent on performance of adsorption cooling system *Applied Thermal Engineering* **110** 695–702

[5] Grabowska K, Krzywanski J, Nowak W and Wesolowska M 2018 Construction of an innovative adsorbent bed configuration in the adsorption chiller-Selection criteria for effective sorbent-glue pair *Energy* 317–23
[6] Chang K-S, Chen M-T and Chung T-W 2005 Effects of the thickness and particle size of silica gel on the heat and mass transfer performance of a silica gel-coated bed for air-conditioning adsorption systems *Applied Thermal Engineering* **25** 2330–40
[7] Blaszczuk A and Krzywanski J 2017 A comparison of fuzzy logic and cluster renewal approaches for heat transfer modeling in a 1296 t/h CFB boiler with low level of flue gas recirculation *Archives of Thermodynamics* **38** 91–122
[8] Krzywanski J and Nowak W 2016 Modeling of bed-to-wall heat transfer coefficient in a large-scale CFBC by fuzzy logic approach *International Journal of Heat and Mass Transfer* **94** 327–34
[9] Krzywanski J, Wesolowska M, Blaszczuk A, Majchrzak A, Komorowski M and Nowak W 2018 Fuzzy logic and bed-to-wall heat transfer in a large-scale CFBC *International Journal of Numerical Methods for Heat & Fluid Flow* **28** 254–66
[10] Krzywanski J, Wesolowska M, Blaszczuk A, Majchrzak A, Komorowski M and Nowak W 2016 The non-iterative estimation of bed-to-wall heat transfer coefficient in a CFBC by fuzzy logic methods *Procedia Engineering* **157** 66–71
[11] Jamrozik A, Tutak W, Gnatowski A, Gnatowska R, Winczek J and Sosnowski M 2017 Modeling of Thermal Cycle CI Engine with Multi-Stage Fuel Injection *Advances in Science and Technology. Research Journal* **11** 179–86
[12] Sosnowski M 2017 Computer aided optimization of a nozzle in around-the-pump fire suppression foam proportioning system *Engineering Mechanics 2017* 914–7
[13] Sosnowski M, Krzywanski J and Gnatowska R 2017 Polyhedral meshing as an innovative approach to computational domain discretization of a cyclone in a fluidized bed CLC unit *Energy and Fuels* E3S Web of Conferences vol 14, ed Suwala, W. and Dudek, M. and Leszczynski, J. and Lopata, S.
[14] Gnatowska R, Sosnowski M and Uruba V 2017 CFD modelling and PIV experimental validation of flow fields in urban environments *Energy and Fuels* E3S Web of Conferences vol 14, ed Suwala, W. and Dudek, M. and Leszczynski, J. and Lopata, S.
[15] Gnatowska R and Sikora S 2015 Numerical analysis of wind flow and erosion in flow around the bump terrain *AIP Conference Proceedings* vol 1648 p 850124
[16] Gnatowska R and Rybak T 2015 Numerical analysis of heat transfer around 2D circular cylinder in pulsation inflow *AIP Conference Proceedings* vol 1648 p 850125
[17] Jaminska P, Blazik-Borowa E and Lipecki T 2014 CFD study on wind action on building-scaffolding system *Proc. 6th International Symposium on Computational Wind Engineering*
[18] Jaminska P, Lipecki T and Blazik-Borowa E 2014 CFD study and wind tunnel measurements of flow around prism *Proc. 6th International Symposium on Computational Wind Engineering*
[19] Sosnowski M, Krzywanski J, Grabowska K and Gnatowska R 2018 Polyhedral meshing in numerical analysis of conjugate heat transfer *International Conference on Experimental Fluid Mechanics (EFM)* 589–94
[20] Sosnowski M 2018 Computational domain discretization in numerical analysis of forced convective heat transfer within packed beds of granular materials *Engineering Mechanics 2018* 801–4
[21] Sosnowski M 2018 Computational domain discretization in numerical analysis of flow within granular materials *International Conference on Experimental Fluid Mechanics (EFM)* 582–8
[22] Krzywański J, Nowak W, Grabowska K, Widuch A and Wesołowska M 2017 Minimalizacja rozmiarów chłodziarek adsorpcyjnych wykorzystywanych do produkcji wody lodowej Część I. Chłodziarki adsorpcyjne - charakterystyka *Energetyka Cieplna i Zawodowa* 56–64

IOP Conf. Series: Journal of Physics: Conf. Series **1101** (2018) 012038 doi:10.1088/1742-6596/1101/1/012038

Catalytic utilization of unconventional fuels in a gas turbine

M Stajnke[12] and J Badur[13]
[1]Energy Conversion Department, The Szewalski Institute of Fluid-Flow Machinery PAS-ci, Fiszera 14, 80-231 Gdańsk, Poland

E-mail: [2]mstajnke@imp.gda.pl, [3]jb@imp.gda.pl

Abstract. In this paper, the numerical modelling of acetone utilization in the thermocatalytic reactor based on the intermetallic phase of Ni_3Al has been presented. The experimental results of mixtures containing air contaminated with acetone in a thermocatalytic reactor were used for the preliminary calibration of the Computational Fluid Dynamics "CFD" calculations. The thermocatalytic combustion reaction of acetone has been modelled by employing experimental data related to the active surface area. For more accurate application of current model into the gas turbine regime simulation was performed on three different geometrical cases. Effects regarding flows in microchannel are visible especially when "small" geometry is concerned. One should notice that a decrease of acetone concentration at inlet into package, mainly, two millimetres before the thermocatalytic area, is connected with the diffusion fluxes of the other components, mainly water and carbon dioxide. The commercial CFD code has been expanded by User Defined Functions "UDF's" to include surface chemical reaction rates on the interphase between fluid and solid. Results obtained through numerical calculations were calibrated and compared with experimental data to receive satisfactory agreement.

1. Introduction

There are catalysts like Ni-based that exhibit very versatile spectrum of applications. In case presented in this paper one of this high catalytic potential has been utilised that is extremely high catalytic activity in harmful substances utilization. Along with the reactions with other gasses synthesis and promote the production of carbon nanostructures (mainly carbon nanotubes) [1,2]. The Ni-based, solid-state catalysts that was used in this work as is the Ni_3Al alloy [3], which combines properties of both the structural and functional materials. According to the literature, Ni_3Al intermetallic thin foils exhibit extremely high catalytic properties in hydrocarbon decomposition reactions [4]. Said that the main disadvantage of this material is the relatively high temperature of maximal hydrocarbon conversion as stated in the literature [1,5]. That is where high temperature occurring during normal work of a gas turbine can be utilized. From the thermodynamic point of view such solution would be negative on overall gas turbine performance. But in some instances this utility might be more than useful. As gas turbines are relatively easy to transport they can be used in whole spectrum of situations where this specific function might be used. The thin strips or thin foils based on the Ni_3Al intermetallic phase produced according to technology presented by Jóźwik et al. in works [1,3] could be applied inside gas turbine to fully utilize catalytic potential of mentioned material. The alloy foil package based on the intermetallic phase Ni_3Al created as an prototype used in experiment to define catalytic activity was constructed as a rolled up honeycomb structure (shown in Fig.1a).

Figure 1. a) The alloy foil package based on the Ni_3Al intermetallic phase rolled into a honeycomb; b) the eight single microchannels beginning at inlet into the package and terminating at the outlet from the package taken to numerical simulation; c) the characteristic dimensions of reactor.

1.1. Motivation

While foil packages and experiment data were obtain having very small dimensions in mind (as shown in Fig.1c) there is still question how this catalyst will perform when put in gas turbine which connects with greater overall dimensions of mentioned package. Appropriate method of insertion catalytically active surface inside the hot gas stream will be crucial for performance of gas turbine and catalyst itself. However in this work as a first step to creating full model of gas turbine with catalytically active surfaces the focus was put into examining performance of the alloy foil package for different dimensions. Cases of modelling thermocatalytic processes are becoming increasingly interesting. Besides these works, that describe in general the processes of chemical reactions [6-8] there are articles concerning models of reactors and catalytic micro-reactors [9,10].

This paper is a continuation of the analysis of the presented geometry of a single microchannel [9]. However, different issues have been considered, namely the catalytic surface reactions of the decomposition of acetone and influence of microchannel dimensions on reaction rates. Situation where catalytic surface reactions occurs in the direct vicinity of walls coated in active layers, are often present in technical situations [11,12]. In instances of thermal decomposition reactions, active layers are often built on the base of Ni alloys [2,13,14]. As previously mentioned in this paper, alloy foils based on the intermetallic phase of Ni_3Al can be widely utilized. One of the most promising potential applications is within the field of air purification and removal of harmful substances and the controlled decomposition of hydrocarbons, and extended consideration is given here. The main focus of the this paper is to describe the mathematical modelling the chemical surface reactions of the flow of mixture of air contaminated by acetone. The model has been verified and the performance of the thermocatalytic reactor with microchannels within the air purification regime has been investigated.

2. Model description

A mathematical model used in solver contain basic governing equations that include continuity, momentum, energy, evolution of kinetic energy of turbulence, evolution of dissipation of kinetic energy of turbulence and species transport. Equations were assumed in manner showed below.

2.1. Mass balance (continuity) equation

$$\frac{\partial}{\partial t}\rho + \frac{\partial}{\partial x_i}(\rho v_i) = 0 \tag{1}$$

Where ρ is fluid density, generally dependent on time and location, v_i is fluid velocity (where $i = x, y, z$) and x_i as coordinates used in finite volume method, $i = 1, 2, 3$ (structural mesh) or $i = x, y, z$ (non-structural mesh).

2.2. Momentum balance equation

$$\frac{\partial}{\partial t}(\rho v_i) + \frac{\partial}{\partial x_j}(\rho v_i v_j + p\delta_{ij}) = \frac{\partial}{\partial x_j}(\tau_{ij} + R_{ij}) + \rho b_i \tag{2}$$

Where p is thermodynamic pressure, δ_{ij} stands for Kronecker delta, τ_{ij} for viscous stress flux components, R_{ij} is Reynolds turbulent stress and b_i is mass force of Earth gravity ($b_i = -9.81\delta_{ij}$).

2.3. Energy balance equation

$$\frac{\partial}{\partial t}(\rho e) + \frac{\partial}{\partial x_i}(\rho e v_i + p v_i) = \frac{\partial}{\partial x_i}(q_i + q_i^t + \tau_{ij}v_j + R_{ij}v_j) + \rho b_i v_i \tag{3}$$

Where e represents internal and kinetic energy ($e = u + \frac{\vec{v}^2}{2}$), u – internal energy ($u = c_v T$), q_i is molecular heat flux and q_i^t turbulent heat flux.

2.4. Equation for turbulent kinetic energy evolution k

$$\frac{\partial}{\partial t}(\rho k) + \frac{\partial}{\partial x_j}(\rho k v_j) = \frac{\partial}{\partial x_j}(J_j^k) + \rho S_k \tag{4}$$

2.5. Equation for turbulent dissipation energy evolution ε

$$\frac{\partial}{\partial t}(\rho \varepsilon) + \frac{\partial}{\partial x_i}(\rho \varepsilon k v_i) = \frac{\partial}{\partial x_i}(J_i^\varepsilon) + \rho S_\varepsilon \tag{5}$$

Where J_i^k and J_i^ε are respectively diffusive flux of k and diffusive flux of ε with sources S_k and S_ε.

2.6. Equation for species transport

$$\frac{\partial}{\partial t}(\rho Y_m) + \frac{\partial}{\partial x_i}(\rho Y_m v_i) = \frac{\partial}{\partial x_i}(J_{m,i}) + \rho S_m \tag{6}$$

Where, Y_m is mass fraction of gas component $m = N_2; C_3H_6O; CO; H_2O; CO_2$, S_m is creation/destruction sources of species m and $J_{m,i}$ define flux of m components of mixture. More detailed description of Eq. (1-6) is presented in [6].

2.7. Chemical reaction

The main products of the acetone combustion reaction over the Ni_3Al foils are water and carbon dioxide. The by-products were omitted in this work but are further investigated in works [5,13,14]. For this simplest case focusing on model verification based on experiment results only one equation was implemented:

$$C_3H_6O + 4O_2 \rightarrow 3H_2O + 3CO_2 \tag{7}$$

Some of the by-products, especially water (also present in air), can oxidize the Ni_3Al catalyst surface. As described in the paper [5], where H_2O is the by-product of methanol decomposition and can be used to create the metallic Ni and also can be converted into aluminum hydroxide.

2.8. Acetone combustion source

As it was already mentioned, changes in the components concentration in the mixture due to chemical reactions can be implemented via a volumetric or surface source term, S_m, in the species transport Eq. (1). There are numerous works describing models including flow with chemical reactions [6-8] but it mostly refers to mechanical, thermal and chemical issues of the phenomena occurring in volume. Phenomena occurring on the surface of the combustion chamber are characterized by complete anisotropy. Separately from the volumetric reaction resulting in production of component, S_m, the reaction of the surface absorption also describes chemical kinetics, and therefore they should be treated as some surface source (or discounts) for the gas component [15,16].

In this first approach in modelling complex flow and chemical reactions cases it is assumed in the numerical case that the products of thermocatalytic oxidation are not deposited on the catalyst surface. As it was previously stated all chemical reactions are considered irreversible. In the future, the wall boundary conditions such as temperature and emissivity should be enhanced to account for the effects of this deposit accumulation under the conditions of this reaction. The source term for acetone reaction, $S_{C_3H_6O}$, can be formulated in cells strictly adjacent to the catalyst surface. By the means of UDF (User Defined Functions) we formulated algorithm which can identify finite volumes in direct proximity to the reactor walls. Inside identified in this way cell the source of the acetone oxidation ($S_{C_3H_6O}$) could be implemented which can be expressed via the following equation:

$$S_{C_3H_6O} = W_{C_3H_6O} \nu_{C_3H_6O} (k_{Ni_3Al} [X_{C_3H_6O}] \frac{A_{Ni_3Al}}{V_{cell}}) \tag{8}$$

where:, $\nu_{C_3H_6O\,(\alpha)}$ - the molar stoichiometric coefficient for C_3H_6O and α-th non-reversible reaction and $W_{C_3H_6O}$ molecular mass of acetone, kg/kmol. To obtain the volumetric source term, $S_{C_3H_6O}$, independent from the different mesh sizes occurring in discretization process and only include catalytic surface area, reaction rate should be divided by the height of the computational cell adjacent to the microreactor wall (A_{cat}/V_{cell}). It should be added that the source of acetone combustion $S_{C_3H_6O}$ is then usually given as a function of acetone molar concentration $[X_{C_3H_6O}]$ and the forward catalytic rate constant k_{cat}. There exist no proper data for a full model validation. Partially, it can be done through the acetone conversion to reaction time characteristics diagram that is often available for a particular thermocatalytic micro-reactor.

3. Analysis

3.1. Experiment and constant to calibration

In the examinations the thin Ni_3Al foils with thickness about 50μm and average grain size about 15μm were taken. The stability test for Ni_3Al catalyst in acetone utilization effectiveness been carried out under atmospheric pressure and under the lowest possible temperature, 600 °C, which was sufficient to obtain a full conversion from the initial mole fraction i.e. 2000 ppm. Basic information regarding experiment are shown in table 1. More detailed description of the experiment as well as data obtained can be found in mentioned earlier work [10]. Others parameters have been treated as boundary conditions for CFD calculations.

Table 1. An experimental thermocatalytic decomposition rate of acetone $q_{C_3H_6O}$ over Ni_3Al foil estimated at GHSV = 40000h^{-1} under the lowest possible temperature of the catalysis process which is sufficient to obtain a full conversion from the initial mole fraction i.e. $X_{C_3H_6O}$=2000ppm.

Parameter	Unit	Value
Total mass of catalytic specimens	g	0.5
Conversion of acetone	%	100
Experimental reaction rate of acetone	g / (g_{cat} s)	0.124

3.2. Numerical boundary condition

The recalculation of the experimental reaction rate and diffusion flux were performed under the following assumptions: 1) constant temperature of the process; 2) monolithic structure of the reactor; 3) conversion parameters as in table 1; 4) geometric dimensions as were presented in Fig. 1 c). However to verify the model for gas turbine usefulness the dimensions of analyzed microchannel were changed. To compare changes in reaction effectiveness and flow behavior the geometry was scaled twice, once dimensions in every direction were multiply by two and after that once again multiply by two. This way we obtained 3 geometries of the same microchannel but in different sizes. The results of scaling and its effect on reactor volume and mass flow can be seen in Table 2.

Table 2. Change in parameters (reactor volume and mass flow) for different scale factor for assumed geometry (basic dimensions shown in Fig. 1c).

Parameter	Reactor volume [m^3]	Volume multiple [-]	Mass flow rate [kg/s]	Mass multiple [-]
Scale_1	7,04E-08	1	1,19E-07	1
Scale_2	5,63E-07	8	4,77E-07	4
Scale_3	4,50E-06	64	1,91E-06	16

Microchannels coupled at the inlet and outlet are shown in Fig. 1 b). The velocity was assumed at the level **v** =0,07 cm/s. As known from the experimental data that the lowest possible temperature of the catalysis process which is sufficient to obtain a complete conversion occurred at T =600 °C. Mixture used in simulation consist of air contaminated with acetone in concentration of 250 ppm. For the calculations steady state flow state was assumed and analysis was performed by a CFD commercial solver. This finite volume based code permits one to solve the three-dimensional fluid and heat flow problems concerned with turbulent structures and chemical reactions [17-20]. However, it also allows for the addition of user defined subroutines programmed in C++ for problems that fall outside the capability of the standard version of code. As has been demonstrated in recent works [21-23], in modelling flows with a strong interaction of surface and liquid material, boundary conditions and appropriate closure in mathematical models are essential.

The grid used in the numerical calculations presented here consist of 1 600 000 finite volumes for basic geometry case "Scale_1". For geometry "Scale_2" and "Scale_3" number of finite elements were accordingly increased, respectively to 2 000 000 and 3 400 000 finite elements of structural mesh. This enables the model to maintain a high accuracy of the results, without consuming unnecessary computing power. In the microreactor, a k - ε turbulent approach was applied due to necessity of utilizing the Eddy-Dissipation model which was considered as the simplest one to work through using UDF. It has been assumed that the micro-reactor walls are constantly under influence of a heater, thus assuring a constant temperature conditions. Although the temperature change with this slight acetone concentration will not change much even without additional heater assistance. It was also assumed that the surface structure of the micro-reactor can be treated as homogeneous.

The standard SIMPLE (semi-implicit method for pressure-linked equations) method has been used for pressure-velocity coupling. The second order upwind schemes have been employed for the solution of the convection term in the governing equations. The diffusion terms have been central-differenced with second order accuracy as well.

4. Results and discussion

Thermocatalytic utilization of acetone in a single package micro-reactor is considered as a complicated phenomenon where the main products of the reaction are water and carbon dioxide (see Eq. (7)).

The distribution of the components due to the impact of different processes can be managed by: 1) diffusion effects; 2) mixing effects due to eddy turbulence and 3) thermocatalytic surface utilization. However with relatively small level of air contamination (250 ppm) the diffusion effects are less visible with cases of larger geometry (Fig. 3). This effect can be seen before air stream reaches the reactor and despite that the acetone mole fraction is reduced. For better illustration additional measuring surface was added and placed 2 mm before package inlet.

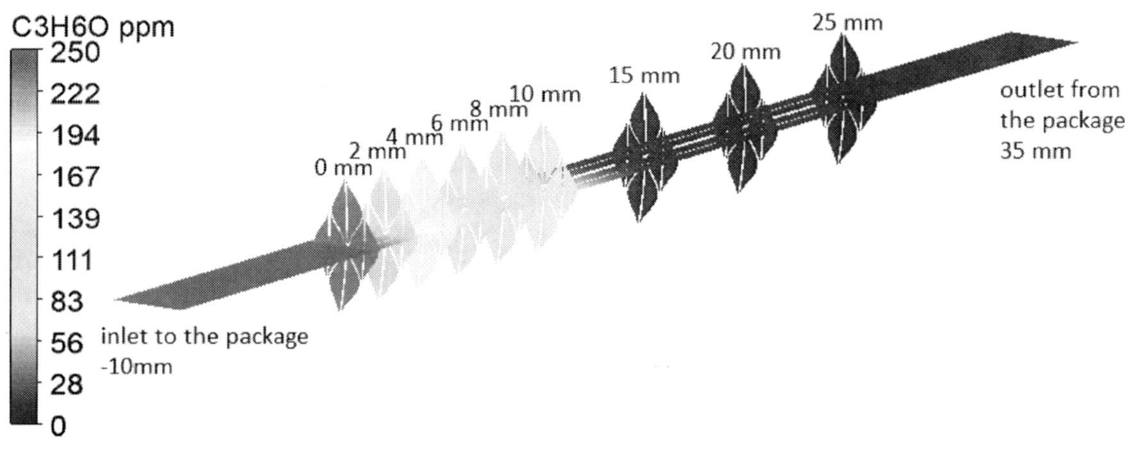

Figure 2. Acetone mole fraction expressed in ppm $X_{C_3H_6O}$ distribution at the eight single microchannels beginning at inlet into the package and terminating at the outlet from the package.

As can be seen in Fig. 2 the three main areas can be distinguished for the whole package: Firstly, the diffusion flux area before microchannels (between $L = -10$ mm and $L = 0$ mm); secondly, the thermocatalytic reaction area (between $L = 0$ mm and $L = 25$ mm) and thirdly, the flow mixing area (between $L = 25$ mm and $L = 35$ mm). As can be seen in Fig. 3 despite significant scale increase the utilization of acetone still is very effective. It should be mentioned that case where harmful substances (in this case acetone) concentration is equal to 250 ppm is quite extreme, but at the same time useful regarding simulation and model preparation.

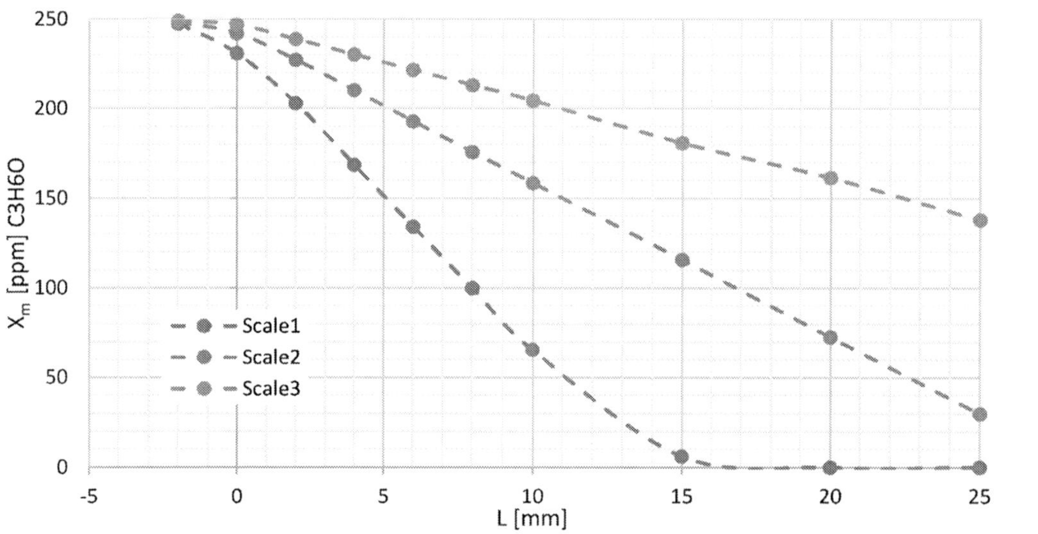

Figure 3. Change of the averaged cross-sectional mole fraction expressed in ppm of acetone as a function of reactor length L.

Although it might seem like different scaling only affect this case in almost non-dimensional way the influence of scale on flow is much more significant. This effect is seen mostly in multicomponent diffusion affecting results of molar concentration of acetone before entering the reactor. For larger scale models and much greater flow speed like in gas turbine the effects of this diffusion will not have such strong influence on gas flow as in micro scale. The important issue in this case would be ability of surfaces to effectively utilize toxic compound without generating much pressure loss.

5. Conclusions

This paper is concerned with of the modelling of acetone utilization which represent harmful substances present in the air in the thermocatalytic reactor based on the intermetallic phase of Ni_3Al. The numerical simulation has been performed via CFD procedure with extensions including the increased chemical reactions rate at the interphase between fluid and solid. Analysis presented should be considered as having demonstrated a promising concept in chemical processing possibilities. Results obtained through numerical calculations were calibrated and compared with experimental data to receive satisfactory agreement, however a more sophisticated approach will be considered which includes the effects of deposit growth on the reaction conditions, increased number of chemical reactions and temperature field at the wall.

Acknowledgments

The work results were obtained in studies co-financed by the National Research and Development Centre in the project PBS 3 ID 246201 titled .: "The development of innovative technology, thin foils of alloys based on intermetallic phase Ni3Al with high activity thermocatalytic in the field of purification of air from harmful substances or controlled decomposition of hydrocarbons."

6. References

[1] Jóźwik P, Grabowski R, Bojar Z, 2010. Catalytic activity of Ni3Al foils in methanol reforming *Mater. Sci. Forum* **636-637** p 895–900.

[2] Olafsen A, Daniel C, Schuurman Y, Raberg L B, Olsbye U, Mirodatos C 2006 Light alkanes CO_2 reforming to synthesis gas over Ni based catalysts *Catal. Today* **115** p 179-185

[3] Jóźwik P, Polkowski W, Bojar Z, 2015 Applications of Ni3Al based intermetallic alloys—current stage and potential perceptivities *Materials* **8** p 2537-2568

[4] Moussa S O, El-Shall M S, 2007 High-temperature characterization of reactively processed nanostructure nickel aluminide intermetallics. *J. Alloys Compd.* **440** p 178-188

[5] Xu Y, Ma Y, Sakurai J, Teraoka Y, Yoshigoe A, Demura M, Hirano T 2014 Effect of water vapor and hydrogen treatments on the surface structure of Ni_3Al foil *Appl. Surf. Sci.* **315** p 475-480

[6] Badur J 2003 *Numerical modelling of sustainable combustion in gas turbines* IFFM Publlishers Gdańsk

[7] Kuo K K and Acharya R 2012 *Applications of turbulent and multiphase combustion* John Wiley & Sons, Inc. New Jersey

[8] Williams F A 1965 *Combustion theory* Addision Wesley Massachussets.

[9] Chen G-B, Chen C-P, Wu C-Y, Chao Y-C 2007 Effects of catalytic walls on hydrogen/air combustion inside a micro-tube *Appl. Catal. A* **332** p 89–97

[10] Jóźwik P, Badur J and Karcz M 2011 *Chemical and Process Engineering* vol 32 (3) p 215-227

[11] Aoki N, Yube K, Mae K, 2007 Fluid segment configuration for improving product yield and selectivity of catalytic surface reactions in microreactors *Chem. Eng. J.* **133** p 105-111

[12] Duran J E, Mohseni M, Taghipour F, 2010 Modeling of annular reactors with surface reaction using computational fluid dynamics (CFD) *Chem. Eng. Sci.* **65** p 1201-1211

[13] Mitani H, Xu Y, Hirano T, Demura M, Tamura R, 2017 Catalytic properties of Ni-Fe-Mg alloy nanoparticle catalysts for methanol decomposition *Catalysis Today* **281** p 669–676

[14] Michalska-Domańska M, Bystrzycki J, Jankiewicz B, Bojar Z 2017 Effect of the grain diameter of Ni-based catalysts on their catalytic properties in the thermocatalytic decomposition of methanol *Comptes Rendus Chimie* **20** p 156-163

[15] Badur J, Ziółkowski PJ and Ziółkowski P 2015 On the angular velocity slip in nano-flows *Microfluid Nanofluid* **19** p 191-8

[16] Badur J, Ziółkowski P, Sławiński D, Kornet S 2015 An approach for estimation of water wall degradation within pulverized-coal boilers *Energy* **92** p 142-152

[17] Ziółkowski P, Stajnke M and Jóźwik P 2018 Modeling of a mixture flow of helium and methanol in thermocatalytic reactor and chemical reactions on the intermetallic phase of Ni_3Al *Transactions IFFM*, accepted to publication

[18] Flaszyński P, Doerffer P, Szwaba R, Kaczyński P, Piotrowicz M 2015 Shock wave boundary layer interaction on suction side of compressor profile in single passage test section. *Journal of Thermal Science* vol 24(6) p 510-515

[19] Badur J, Ziółkowski P, Zakrzewski W, Sławiński D, Kornet S, Kowalczyk T, Hernet J, Piotrowski R, Felicjancik J, Ziółkowski P J 2014 An advanced Thermal–FSI approach to flow heating/cooling *J. Phys.: Conf. Ser.* **530** 10.1088/1742–6596/530/1/012039.

[20] Pianko-Oprych P, Kasilova E, Jaworski Z, 2014 CFD analysis of heat transfer in a microtubular solid oxide fuel cell stack *Chemical and Process Engineering* **35** p 293-304

[21] Badur J, Karcz M, Lemański M, Nastałek L 2011 Enhancement transport phenomena in the Navier-Stokes shell-like slip layer *CMES: Computer Modeling in Engineering & Sciences* **73** p 299–310

[22] Ziółkowski P, Badur J 2018 A theoretical, numerical and experimental verification of the Reynolds thermal transpiration law. International Journal of Numerical Methods for Heat and Fluid Flow vol 28(1) p 64-80.

[23] Lewandowski T, Ochrymiuk T, Czerwińska J 2011 Modeling of heat transfer in microchannel gas flow *ASME Journal of Heat Transfer* **133** 022401-1

IOP Conf. Series: Journal of Physics: Conf. Series **1101** (2018) 012039 doi:10.1088/1742-6596/1101/1/012039

Numerical analysis of an impact of spray characteristics and co-flow temperature on a flame lift-off height

Jakub Stempka[1] and Artur Tyliszczak[1]

[1] Institute of Thermal Machinery, Czestochowa University of Technology, Faculty of Mechanical, Engineering and Computer Science, Armii Krajowej 21, 42-201 Czestochowa, Poland

E-mail: stempka@imc.pcz.pl

Abstract. In the current work the jet spray combustion reflecting an experimental piloted spray burner is simulated numerically. In the simulations the diluted ethanol spray evaporates in the hot surrounding, the gaseous fuel diffuses from spray fuel-rich core into the oxidizer fuel-lean region and autoignites. An in-house high-order compact difference LES solver is used to carry out the computations. The reaction rates are modelled using Implicit LES approach based on the single-step global reaction. The continuous phase is modelled in the Eulerian reference frame while the droplets in the Lagrangian form. The studies are focused on the influence of the co-flow temperature and spray initial Sauter mean diameter (SMD) on the flame lift-off height, autoignition delay and the reaction zone features. The results are compared with the experimental data and show good agreement in terms of lift-off height. It is almost linearly related to the co-flow temperatures. It is found that for the same fuel mass loading, droplets with smaller SMDs significantly shorten the autoignition delay. Moreover the autoignition delay and the lift-off heights are influenced mostly by the droplets SMD rather than by the co-flow temperature.

1. Introduction

The combustion processes are one of the most fundamental ways of global energy conversion. A variety of industrial and small scale devices operate being powered as a consequence of highly turbulent combustion processes that follow the liquid fuel injection, e.g., diesel engines, burners, aircraft propulsion systems, liquid fueled rockets engines and many others. The difficulties of turbulent combustion analysis in two-phase flows arise from a range of interlinked phenomena that cannot be fully captured by the experiment. These are manifested by: the turbulence, droplet dispersion, evaporation, mixing of fuel vapour and chemical reactions. Albeit the experimental and numerical techniques advancement, the two-phase combustion comprises a research field that is far from being fully understood. Numerical studies using both the Direct Navier-Stokes (DNS) [1] and Large Eddy Simulation (LES) [2] approaches enable analysis of two-phase combustion processes with precision unreachable by experiment. The DNS provides a detailed insight into flows nature but is limited mainly to flows in simplified domains and assuming low Reynolds numbers. The LES is the tool which boosts further advancements in spray combustion.

In the framework of LES, the biggest challenge in simulations of turbulent flames is proper modelling of turbulence-flame interactions at sub-grid level that influence the reaction rates. Various strategies for modelling the reaction rates can be employed, such as: Lagrangian Particles, transported PDFs, Eulerian stochastic fields along with Monte Carlo solution techniques. There are also simpler modelling methods, promising from practical point of view, i.e., implicit LES (ILES) [3]. The ILES method is particularly well suited for comparative and parametric studies of complicated reacting flows as it is robust and provides fairly reliable solutions. In the current work it is applied to study the jet spray flame surrounded by the vitiated co-flow, reflecting an experimental configuration investigated originally by the O'Loughlin and Masri [4]. The fuel (ethanol) spray is produced using the ultrasonic nebulizer and is fed to the combustion chamber. The droplets evaporate in the hot surrounding then the gaseous fuel diffuses from spray fuel-rich core into the oxidizer fuel-lean region and autoignites. We analyse the influence of an initial size of fuel droplets and the co-flow temperatures on the flame lift-off heights, autoignition delays and the reaction zone features. The results are compared with the experimental data and show good agreement in terms of lift-off height.

2. Mathematical model

Computations are carried out using an in-house academic LES solver SAILOR [5]. The code solves the LES equations [6] for low Mach number flows disregarding the acoustics modes. The derivatives in Navier-Stokes and energy equations are discretised in Cartesian coordinates using a high-order compact difference scheme [7, 8] on collocated grid arrangement and half-staggered mesh for the pressures nodes. The staggered grid arrangement ensures oscillation-free pressure field since the pressure nodes are decoupled from the values stored in the collocated grid. The pressure field is obtained from the solution of the Poisson equation. The governing equations are integrated in time with the use of predictor-corrector approach. At the predictor step, the second order Adams-Bashforth method is used and at the corrector step the second order Adams-Moulton method is applied. The dispersed phase is modelled in the Lagrangian framework where isolated droplets act on the gas phase through the momentum, mass and energy source terms. The source terms formulations in Navier-Stokes equations were drawn from [9]. Their evaluation is performed by summation over each of the droplet contribution. The equations for droplets positions, velocity and temperature are integrated in time using the 1st order Euler method. The two-way velocity and energy coupling is based on 2nd order approximation of source terms with 4th degree Lagrangian polynomial approximation of the flow field variables at the droplets positions. The numerical code was thoroughly tested and validated for various configurations including both non-reacting and reacting flows [10, 11]. For modelling the species reaction source terms we use the Implicit LES (ILES) approach [3]. The main difference between LES and ILES relies on treatment of the reaction source terms. In ILES the sub-filter scales resulting from the filtration of the convective terms are modelled by the use of eddy viscosity model, but the filtered chemical source terms are modeled directly using the large scale quantities solved on the computational mesh. In this case the reaction rates are obtained from the Arrhenius formulas and in the present study we use simple one-step chemistry for the ethanol oxidation. The reaction rates are tuned with respect to the local equivalence ratio such to reflect correctly the extinction limits.

3. Configuration

The computational domain reflects a jet spray burner configuration from work O'Loughlin and Masri [4]. The burner assembly, showed in figure 1, consists of the ultrasonic nebulizer device (5), which generates spray with specific SMD at the inlet of the central jet nozzle with a diameter of d_j=4.6mm. The spray at temperature 300K is carried downstream the central tube (1) by the co-flowing stream of air at the inlet (6). The annular co-flow assembly surrounds the central

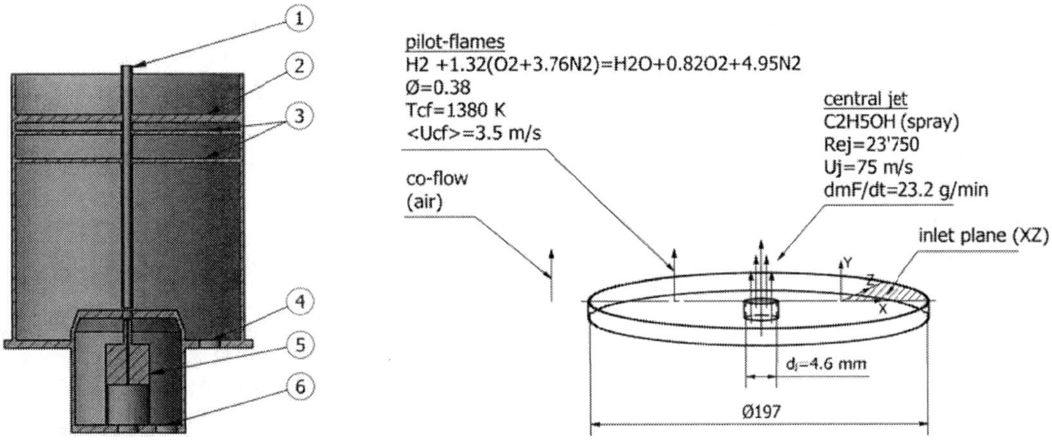

Figure 1. Half-section of the burner (left) and the inlet plane inset (right).

tube. Its diameter is equal to 197mm. The co-flowing mixture is fed in inlet (4), passes through the double glass beads (3) and terminates at the perforated plate (2) with 2200 holes, each with a diameter of 1.58mm. Lean hydrogen/air flames with specified temperature are issued from each of the hole with the velocity equal to 3.5m/s. Experimental data from several spray flames are available for different fuels, mass loadings and co-flow temperatures. As a reference for the current studies an ethanol spray flame is chosen for which the jet spray has bulk velocity of U_j=75m/s, the mass flow of liquid droplets is equal to 23.2g/min, the droplets SMD=40μm and the co-flow temperature T_{cf}=1380K.

The simulation domain comprises of a rectangular volume segment of the burner and is aligned with the jet axis. It starts at the jet exit plane and extends $50d_j$ downstream. The radial dimensions of the domain are equal to $15d_j$. The domain is discretised by $312 \times 200 \times 200$ nodes compressed in the vicinity of the inlet plane and along the jet, i.e., in the region where high scalar gradients are expected. The resulting mesh spacings are as follow: $\Delta x = \Delta z = 1.9 \cdot 10^{-4} \div 1.0 \cdot 10^{-3}$ in the radial directions and $\Delta y = 2.1 \cdot 10^{-4} \div 1.7 \cdot 10^{-3}$ in the axial direction. Except the inlet and outlet, the boundary conditions applied in the simulations are characterised by the zero-gradient Neumann boundary condition. At the outlet the convective boundary condition is prescribed while at the inlet the experimental velocity profile is imposed. Outside the jet inlet region the hot vitiated co-flow of combustion products with temperature and velocity corresponding to the experiment is assumed. Instead of reproducing the exact droplet distribution issuing at multiple points from the jet nozzle the spray is injected from a single randomly picked point, which at each time step has different location. The droplets sizes follow the Rosin-Rammler distribution for which we define 10 classes of diameters with SMD=20, 40 and 60μm. We analyse the flame behaviour depending on SMD and focus on its impact on both global and local flame features during the autoignition and propagation phases. Additionally we compare the lift-off height predictions (L_f) for three different co-flow temperatures T_{cf}=1270, 1330 and 1380K. We verify which parameter (SMD or T_{cf}) has bigger impact on it.

4. Results

4.1. Analysis of spatial and ILES resolutions

To asses the LES quality, the axial variations of resolution parameters based on calculated integral length scale l_0 and Kolmogorov length scale η_K are presented in figure 2. In a round jet, the integral length scale can be approximated by $l_0 \approx d_j (1 + B \cdot z)$ where $B \approx 0.09$ is the expansion rate of the jet [12]. Having the l_0 the Kolmogorov length scale can be computed as $\eta_K = l_0 Re_L^{-3/4}$, for which $Re_L = u' l_0/\nu$ where the RMS of velocity u' is calculated as the spatial average value. The analysis of the ratio of the characteristic mesh spacings in the axial direction Δy across the whole domain shows that spacings between grid points were less than 45% of l_0. At the same time they were not greater than $20\eta_K$ in most part of the domain. It seems that the used grid can be regarded as appropriate for ILES simulations.

The ILES assumptions hold for the perfectly stirred reactor (PSR) conditions, which correspond to the homogeneous sub-grid species and temperatures distributions inside the LES filter size Δ. To ensure PSR conditions the turbulent mixing should occur faster than the chemical reactions. The Damköhler (Da) can be viewed as an indicator of the PSR since Da decreases with decreasing Δ and it is defined as the ratio of subgrid time scale to the chemical time scale $Da_\Delta = \tau_\Delta/\tau_c = S_L/\delta_L \cdot \Delta/u_\Delta$ [3]. The sub-grid velocity fluctuations for Da calculations were estimated as $u_\Delta = \sqrt{2}/c_K \cdot \mu_{sgs}/\rho\Delta$ using computed sub-grid scale viscosity μ_{sgs} and taking the constant $c_K = 0.07$. The sub-grid viscosity was estimated applying the model of Vreman [13]. It is important to note that for PSR the $Da << 1$. However, one can note from figure 2 that there are spatial variations of Da substantially exceeding the value $Da = 1$. They are predominantly linked with the zones that are not resolved well, especially in the outer regions of the jet axis near to the inlet plane and also further downstream, where the grid spacings are bigger. On the other hand, upstream the jet in high heat release regions $Da \approx 5$ or even less. Even though this value is somewhat higher than the one assuring the ILES validity, the simulations still yield correct predictions of the experimental values of the lift-off height, as will be shown in the following subsections.

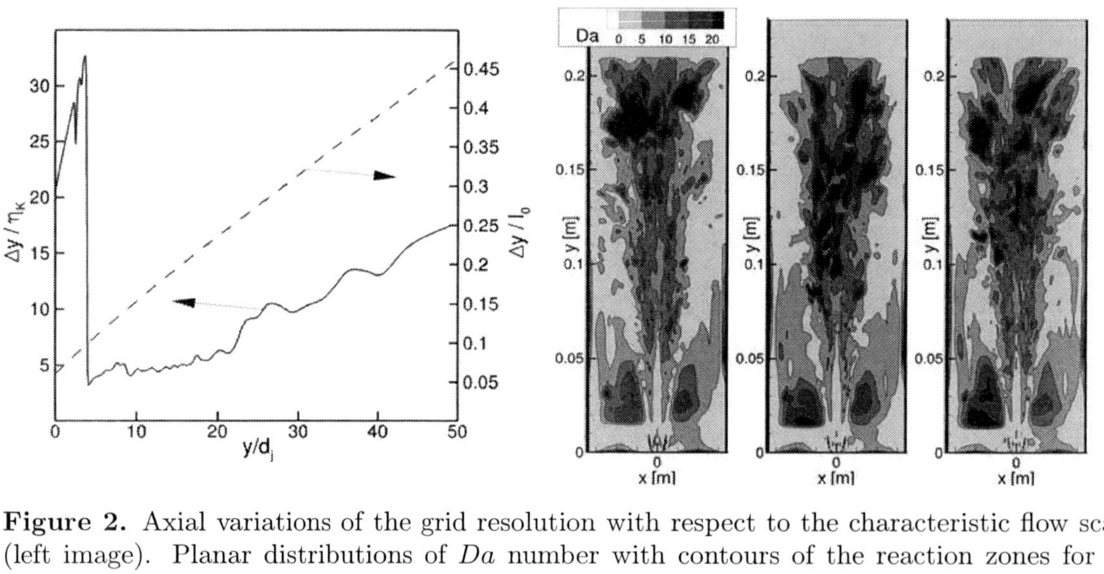

Figure 2. Axial variations of the grid resolution with respect to the characteristic flow scales (left image). Planar distributions of Da number with contours of the reaction zones for T_{cf} equal to: 1270, 1330, and 1380K (from left to right respectively).

4.2. Autoignition delays

Figure 3 shows the autoignition delay times reflected by means of the instantaneous maximum temperature growth. The presented results were obtained for each of the co-flow temperature and initial spray SMD. It can be seen that the autoignition occurs firstly and most rapidly for the smallest droplets (SMD=20μm), regardless the co-flow temperature. The feature of spray with small initial SMD compared to higher SMD for constant fuel loading, is that the smaller one results in larger number of smaller droplets, increasing substantially the droplets surface. This determines the heat diffusion and evaporation rates. The differences in autoignition delays between different co-flow temperatures are the smallest for SMD=20μm. The autoignition for intermediate droplets' SMD=40μm appears almost at the same time instances, being only slightly delayed for low T_{cf}. Substantial differences in the temperature growth rate are evident only for the biggest droplets (SMD=60μm). For the lowest T_{cf}=1270K the autoignition is delayed more than three times, compared with the results for T_{cf}=1380K. The instantaneous maximum fuel mass fraction growth rates for the smallest droplets' SMD for any of the T_{cf}, exceed the fuel growth rates of the other two cases by several times. The evaporation of the bigger droplets (SMD=40 and 60μm) is rather slow and does not present such a substantial jump when the autoignition occurs. It can be seen that the autoignition delay is not directly related to the maximum fuel mass fraction, because the autoignition occurs in each case for completely different $Y_{F,max}$ and always below the stoichiometric mass fraction ($Y_{F,st} = 0.1$). The SMD is in this case the most influential parameter that affects the autoignition delay.

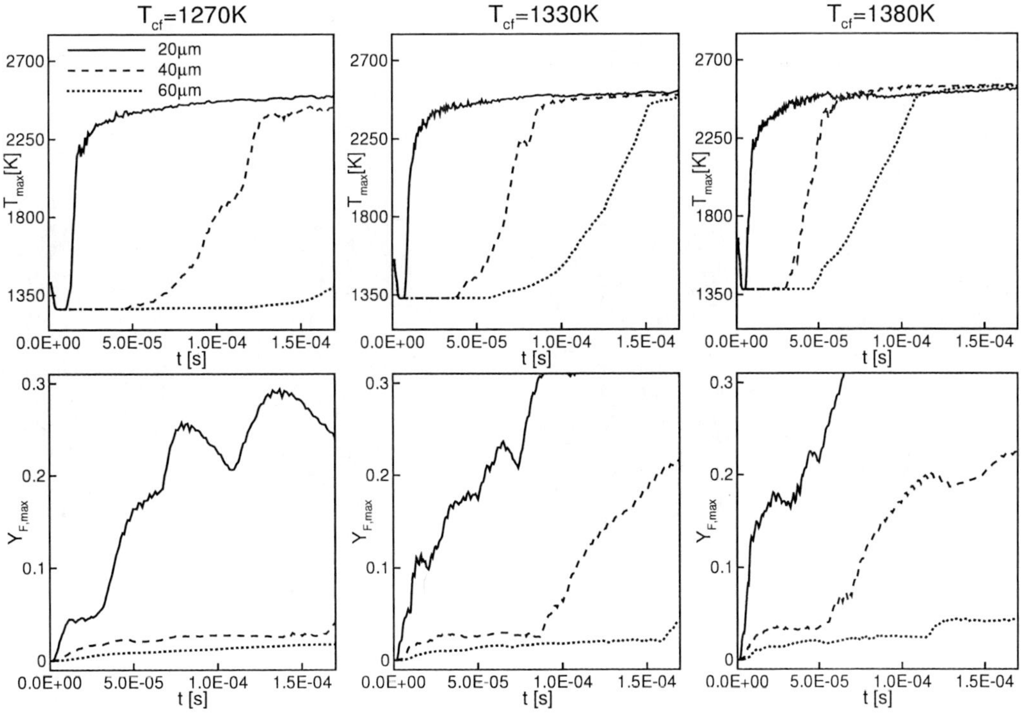

Figure 3. Evolution of maximum temperatures (upper row) and the gaseous fuel mass fractions (lower row) inside the domain.

4.3. Mean and instantaneous temperatures

Comparison of the lift-off heights prediction and the experimental flame depicted in figure 4 shows satisfactory agreement. The measured L_f in the experiment for T_{cf}=1380K was reported as $L_f \approx 15$, similarly as the one obtained presently. Figure 5a) presents the instantaneous distributions of temperatures and the fuel mass fractions and figure 5b) shows the contours of heat release multiplied by the Takeno flame index for each T_{cf} with initial SMD=$40\mu m$. It can be seen that the lift-off height decreases due to increasing co-flow temperature. Inside the highest isotherm envelope there is a broad non-reacting fuel-rich region for the lowest T_{cf}. It is elongated mostly in the downstream direction. From the figure 5a) it is evident that the high temperature reaction zone slides upstream. This is due to the fact that some amount of fuel that evaporated from the spray core, mixes intensively with the oxidizer within the mixing layer that developed at the jet edge. The pockets of high heat release which forerun the autoignition are moved upstream and correspond to the lean fuel regions. However, due to small temperature, lowered by the heat diffusion from the hot co-flow into the spray core, the mixture ignites only further downstream. The Takeno flame index is defined as $\Upsilon = \nabla Y_{O_2} \cdot \nabla Y_{Fuel}$. It distinguishes the premixed combustion regime for which $\Upsilon > 0$ from the diffusion combustion where $\Upsilon < 0$. These correspond to the green and red regions, respectively, showed in figure 5b). The presented values of high heat release regions, multiplied by the Takeno index indicate the combustion mode within the reactive layer. The premixed regime region for the lowest T_{cf} is the smallest one and it extends downstream for increasing co-flow temperatures. The tip of each reaction zone is burning predominantly in the premixed mode while the unburnt fuel is further convected downstream where the combustion occurs mainly in the diffusion mode.

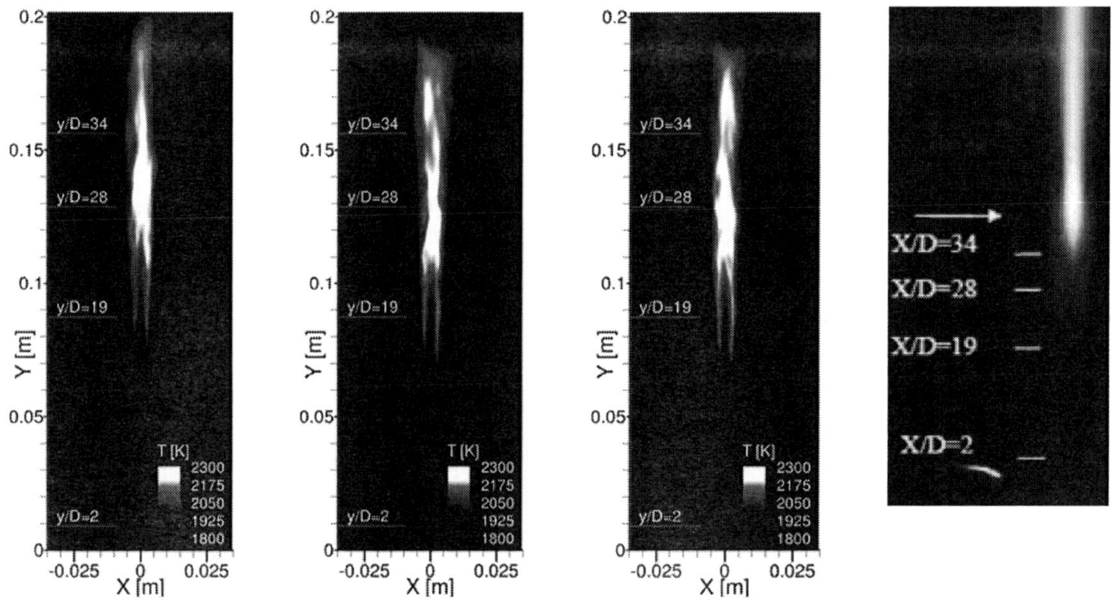

Figure 4. Mean temperature fields obtained for spray characterised by SMD=$40\mu m$ and different T_{cf} (from left to right): 1270, 1330 and 1380K. The last picture shows the experimental results with T_{cf}=1380K.

4.4. Droplet size distributions

Figure 6 shows the droplet size distributions at two distinct positions along the jet axis, i.e., $y/D = 28$ and $y/D = 19$. The bars with different colors denote different classes of the initial

IOP Conf. Series: Journal of Physics: Conf. Series **1101** (2018) 012039 doi:10.1088/1742-6596/1101/1/012039

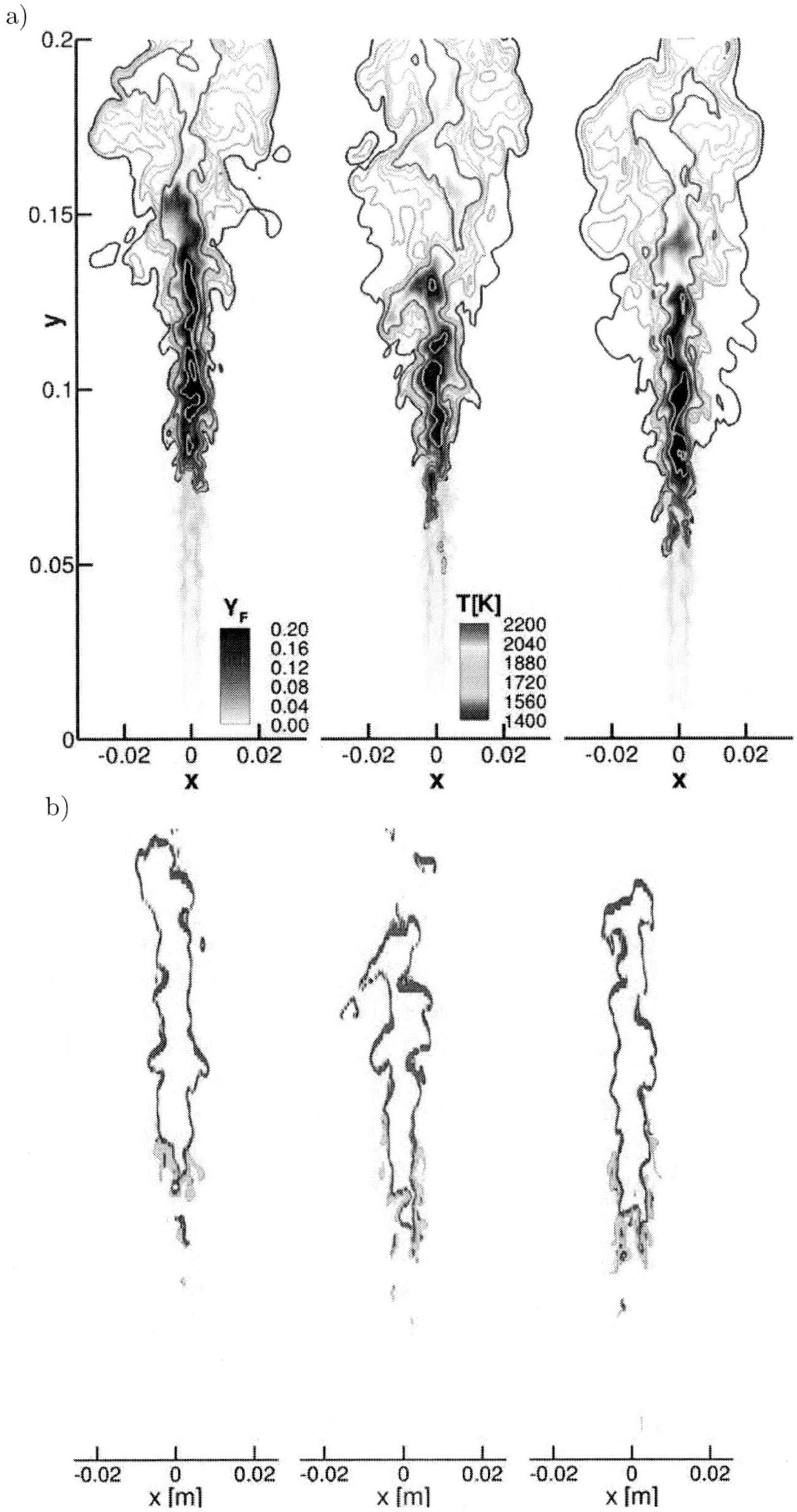

Figure 5. Instantaneous distributions of: a) temperatures (iso-lines) and fuel mass fraction (gray-scale field) and b) iso-contours of high heat release value multiplied by the Takeno index.

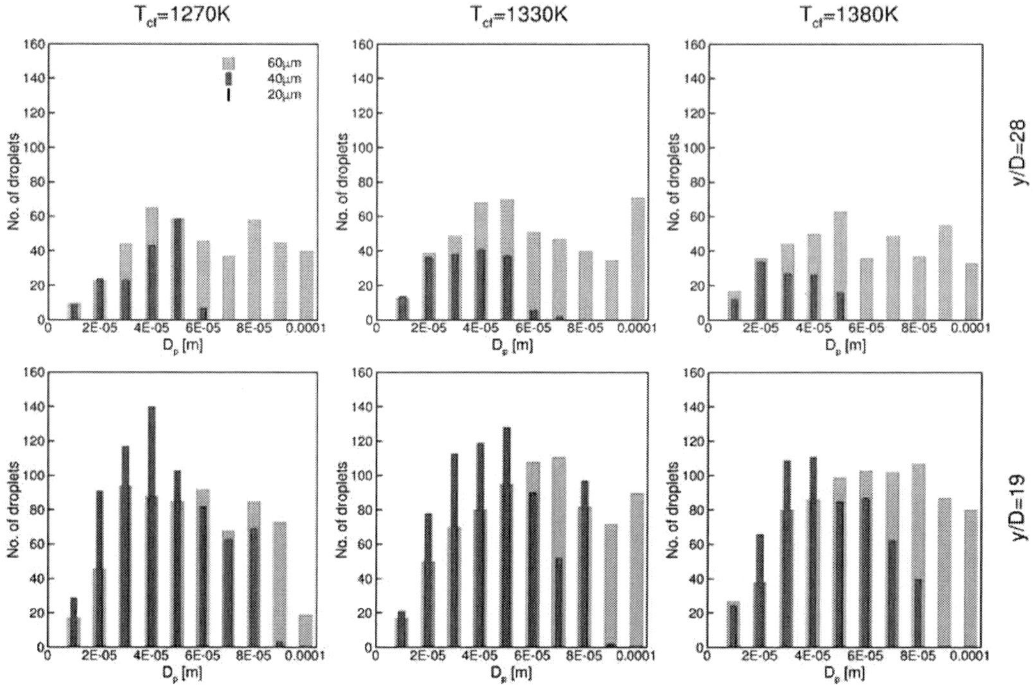

Figure 6. Spatial distributions of droplets for two axial positions of the jet: $y/D = 28$ (upper row) and $y/D19$ (lower row). Columns correspond to T_{cf}: 1270K (left), 1330K (middle) and 1380K (right).

SMDs. One may observe that droplets with the SMD=$20\mu m$ did not prevail to the downstream region and evaporated quickly. Concerning the impact of T_{cf} on droplets diameters one may observe that for its larger values the droplets evaporate quicker, as one could expect. This effect is well visible for droplets with SMD=$40\mu m$. In the case of spray with SMD=$60\mu m$ the influence of T_{cf} is less pronounced but still exists. It is interesting that in this case the total number of droplets in each of the classes is similar at different locations.

4.5. SMD impact

Instantaneous planar snapshots of temperatures distributions with the droplets represented by the spheres coloured by its temperature are presented in figure 7. Because small droplets evaporate quickly for the lowest SMD=$20\mu m$ the flame is attached to the injection nozzle. Significantly different behaviors are observed for the flames with higher SMDs. They are lifted and the lift-off height depends on SMD. Differences in the amount, size and temperature of the drops trapped within the reaction zone are evident. The bigger droplets (SMD=40 and $60\mu m$) prevail almost across the whole flame zone and are not fully consumed. It can be seen that for SMD=$60\mu m$ the flame is thinner and it is anchored further from the inlet ($L_f \approx$ 17D for SMD=$60\mu m$ and $L_f \approx$ 15D for SMD=$40\mu m$). One can also observe that in this case the number of droplets leaving the computational domain is much larger than in the simulations with SMD=$40\mu m$.

5. Conclusions

The current studies concerned the ILES analysis of influence of the co-flow temperatures and the spray size on the combustion process in spray jet flame. It was shown that the lift-off

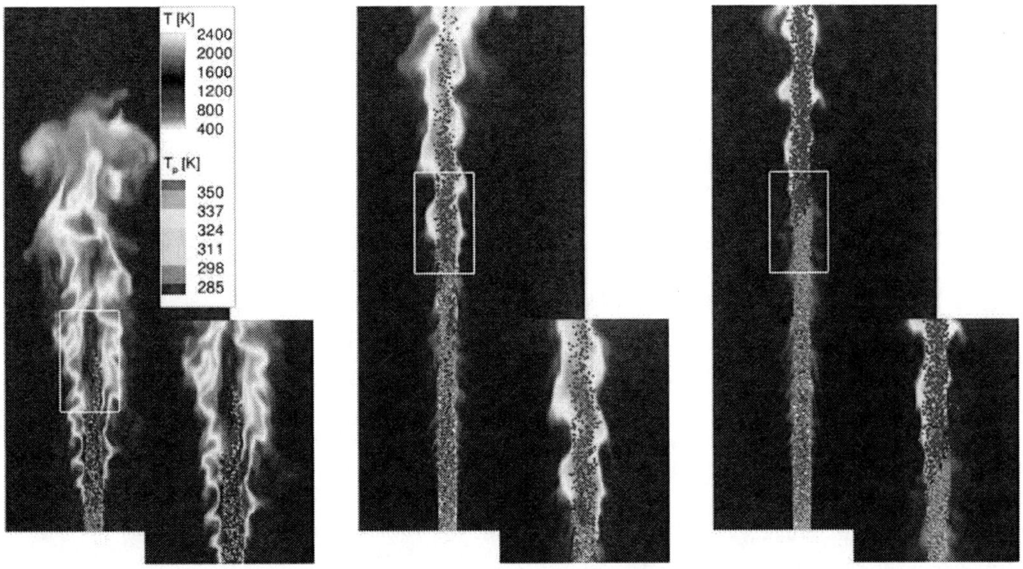

Figure 7. Instantaneous planar temperatures distributions for T_{cf}=1380K and three different SMD=20, 40 and 60μm. The droplets are coloured by its temperatures and scaled by the diameters. Insets show the tip of reaction zone.

height varies slightly with the co-flow temperatures. The co-flow temperature influences the flame by modifying the region of distinct combustion regimes, i.e., the volume of the premixed combustion mode increases with the increasing T_{cf}. The influence of droplet size turned out to be more pronounced than the co-flow temperature variations. Concerning the autoignition time the flame appeared earlier for smaller SMD values. This was connected to the fact that smaller droplets evaporated quicker and produced gaseous fuel more intensively. For the smallest droplets (SMD=20μm) the flame was attached, while for the cases with SMD=40 and 60μm it was lifted. It was observed that the lift-off heights were dependent on SMD and that this dependence was more pronounced than dependence on the co-flow temperature.

Acknowledgments
This work was supported by grant 2015/17/B/ST8/03217 (National Science Centre, Poland) and statutory funds BS/PB-1-103-3010/11/P. PL-Grid infrastructure was used to carry out the computations.

References
[1] Schroll P, Wandel A P, Cant R S and Mastorakos E 2009 *Proc. Combust. Inst.* **32** 2275–82
[2] Prasad V N, Masri A R, Navarro-Martinez S and Luo K H 2013 *Combust. Flame* **160** 2941–94
[3] Duwig C, Nogenmyr K J, Chan C and Dunn M J 2011 *Combust. Theor. Model.* **15** 537–68
[4] O'Loughlin W and Masri A R 2011 *Combust. Flame* **158** 1577–90
[5] Tyliszczak A 2016 *Comput. Fluids* **127** 131–45
[6] Poinsot T and Veynante D 2005 *Theoretical and Numerical Combustion* (R.T. Edwards, Inc.)
[7] Lele S 1992 *J. Comput. Phys.* **103** 16–42
[8] Tyliszczak A 2014 *J. Comput. Phys.* **276** 438–67
[9] Jones W, Lettieri C, Marquis A and Navarro-Martinez S 2012 *Int. J. Heat Fluid Flow* **38** 145–58
[10] Wawrzak A and Tyliszczak A 2018 *Int. J. Hydrogen Energ.* **43** 1–14
[11] Kuban L, Stempka J, Wawrzak A and Tyliszczak A *Proc. 10th Mediterranean Combustion Symp.* 2017 Naples
[12] Chassaing P 2005 *Turbulence en Mecanique des Fluides* (Cepadues-Editions)
[13] Vreman W 2004 *Phys. Fluids* **16** 3670–81

IOP Conf. Series: Journal of Physics: Conf. Series **1101** (2018) 012040 doi:10.1088/1742-6596/1101/1/012040

Cascaded Lattice Boltzmann Method application in forced and natural convection

R Straka[1], K V Sharma[2], D Svyetlichnyy[1] and F W Tavares[2]

[1]Faculty of Metals Engineering and Industrial Computer Science, AGH University of Science and Technology, Krakow, Poland

[2] Department of Chemical Engineering, Federal University of Rio de Janeiro, Rio de Janeiro, Brazil

E-mail: `straka@metal.agh.edu.pl`

Abstract. Lattice Boltzmann Method is applied to forced and natural convection heat transfer from the tube banks. Hot tubes are cooled by flowing or passive air. Two Reynolds numbers (Re=80 and Re=1600) and two Rayleigh numbers (Ra=10^3 and Ra=10^5) of the corresponding heat transfer regime are studied. The method itself is based on the recently derived cascaded collision operator not only for the fluid flow but also for the temperature field. Using this method and moderate space resolution of the lattice we were able to obtain stable and bounded simulations for the non-trivial geometry.

1. Introduction

Lattice Boltzmann Methods (LBM) are numerical methods based on the Boltzmann Transport Equation (BTE) [1]. They solve discrete version of BTE with finite velocity set in order to recover macroscopic variables (density, momentum, temperature) from the underlying distribution functions. The dynamic of the system is described by the collision operator which can be numerically realized in different ways. The most popular are SRT (Single Relaxation Time) and MRT (Multiple Relaxation Time) operators because of their simple form and easy implementation. Cascaded operator [2] can cure some of the numerical artefacts and offers enhanced Galilean invariance of the method [3]. This was the main reason why we derived such an operator for the one conservation law and examined its behavior in case of heat transfer in the recently published article [4]. In this paper we apply cascaded version of a collision operator for D_2Q_5 lattice and one conservation law together with D_2Q_9 cascaded collision operator for fluid flow [2] to forced and natural convection in complex arrays of cylindrical obstacles.

2. Cascaded Lattice Boltzmann method

In LBM, fluid is considered as an ensemble of fictitious particles described by the velocity distribution functions (DFs) denoted as $f(\mathbf{x}, \xi, t)$. The evolution of such DFs obeys Boltzmann's transport equation which reads

$$\frac{\partial f(\mathbf{x}, \xi, t)}{\partial t} + \xi \cdot \nabla f(\mathbf{x}, \xi, t) = \Omega(f, f), \tag{1}$$

where ξ is the microscopic velocity and Ω is the collision operator. Discretization of the velocity space in the above equation by finite set of velocities $\mathbf{c}_i$ leads to the Lattice Boltzmann Equation (written in lattice units)

$$f_i(\mathbf{x}+\mathbf{c}_i, t+1) = f(\mathbf{x}, t) + \mathbb{K}\cdot\mathbf{k} + F_i, \tag{2}$$

where f_i is the velocity distribution function related to the ith characteristic velocity, $\mathbb{K}$ is a collision matrix, $\mathbf{k}$ is a vector of moments of f_i and F_i is a force term. The scheme implementation in the natural convection is detailed in [5]. The macroscopic density ρ and macroscopic velocity $\mathbf{u}$ can be obtained from f_i as

$$\begin{aligned} \rho &= \sum_i f_i, \\ \rho\mathbf{u} &= \sum_i \mathbf{c}_i f_i. \end{aligned} \tag{3}$$

In this work we use two population LBM i.e. we have D_2Q_9 lattice model for fluid flow simulation and D_2Q_5 lattice model for temperature field simulation presented in Fig. 1, where characteristic velocities $\mathbf{c}_i = (c_{i,x}, c_{i,y})$ are defined for both models. The weight factors, w_i for D_2Q_9 and D_2Q_5 are $w_1 = 4/9, w_{3,5,7,9} = 1/9, w_{2,4,6,8} = 1/36$ and $w_1 = 1/3, w_{2,3,4,5} = 1/6$, respectively and sound speed $c_s^2 = 1/3$ for both models.

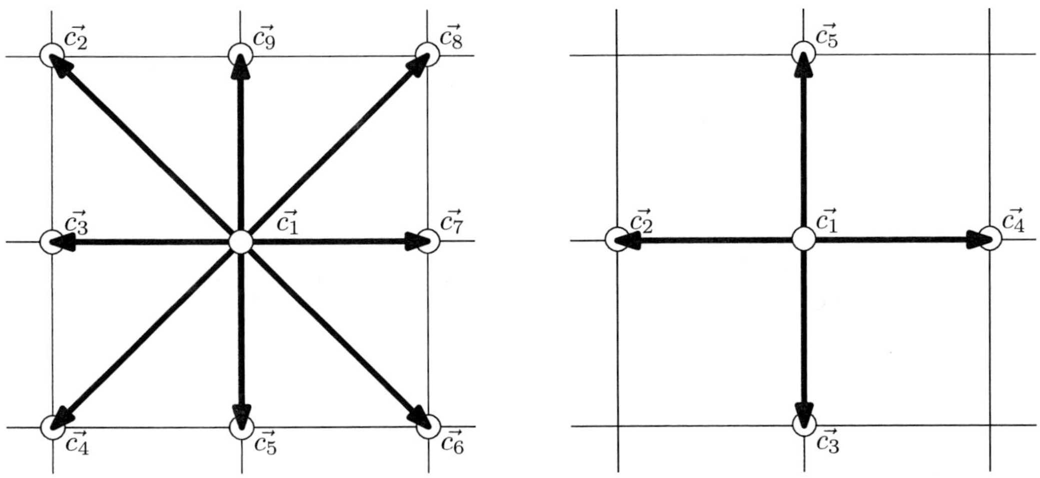

Figure 1. Characteristic velocities (links) for D_2Q_9 (left) and D_2Q_5 (right) lattice models for 2D simulations.

In order to solve equation for the internal energy (temperature) we add another set of DFs and their evolution equation

$$g(\mathbf{x}+\mathbf{c}_i, t+1) = g(\mathbf{x}, t) + \mathbb{Q}\cdot\mathbf{q}, \tag{4}$$

where g_i is the distribution function for temperature field, $\mathbb{Q}$ is a collision matrix and $\mathbf{q}$ is a vector of moments of g_i. The temperature is obtained as a sum of all g_i

$$T = \sum_i g_i, \tag{5}$$

The two matrices used in the above evolution equations are constructed from the monomials of the characteristic velocities [4] and are given as follows

$$\mathbb{K}^T = \begin{bmatrix} 1 & 1 & 1 & 1 & 1 & 1 & 1 & 1 & 1 \\ 0 & -1 & -1 & -1 & 0 & 1 & 1 & 1 & 0 \\ 0 & 1 & 0 & -1 & -1 & -1 & 0 & 1 & 1 \\ -4 & 2 & -1 & 2 & -1 & 2 & -1 & 2 & -1 \\ 0 & 0 & 1 & 0 & -1 & 0 & 1 & 0 & -1 \\ 0 & 1 & 0 & -1 & 0 & 1 & 0 & -1 & 0 \\ 0 & -1 & 0 & 1 & -2 & 1 & 0 & -1 & 2 \\ 0 & 1 & -2 & 1 & 0 & -1 & 2 & -1 & 0 \\ 4 & 1 & -2 & 1 & -2 & 1 & -2 & 1 & -2 \end{bmatrix} \qquad \mathbb{Q}^T = \begin{bmatrix} 1 & 1 & 1 & 1 & 1 \\ 0 & -1 & 0 & 1 & 0 \\ 0 & 0 & -1 & 0 & 1 \\ 4 & -1 & -1 & -1 & -1 \\ 0 & -1 & 1 & -1 & 1 \end{bmatrix}, \qquad (6)$$

where superscript T denotes the transpose and the vectors $\mathbf{k}$ and $\mathbf{q}$ are given by the following relations

$$\begin{aligned}
k_4 &= \frac{1}{12\tau_4}\left[\rho(u_x^2+u_y^2) - f_6 - f_8 - f_4 - f_2 - 2(f_5+f_3+f_7+f_1-\rho c_s^2) + F_4\right], \\
k_5 &= \frac{1}{4\tau_5}\left[f_8 + f_4 - f_6 - f_2 + \rho(u_x^2-u_y^2) + F_5\right], \\
k_6 &= \frac{1}{4\tau_6}(f_7 + f_3 - f_1 - f_5 - u_x u_y \rho + F_6), \\
k_7 &= -\frac{1}{4\tau_7}\Big\{\left[f_5 + f_3 - f_7 - f_1 - 2u_x^2 u_y \rho + u_y(\rho - f_8 - f_4 - f_0)\right] + \\
&\quad + 2u_x(f_7 - f_1 - f_5 + f_3) + F_7\Big\} + \frac{u_y}{2}(-3k_4 - k_5) + 2u_x k_6, \\
k_8 &= -\frac{1}{4\tau_8}\Big\{\left[f_3 + f_1 - f_5 - f_7 - 2u_y^2 u_x \rho + u_x(\rho - f_2 - f_6 - f_0)\right] + \\
&\quad + 2u_y(f_7 - f_1 - f_5 + f_3) + F_8\Big\} + \frac{u_x}{2}(-3k_4 + k_5) + 2u_y k_6, \\
k_9 &= \frac{1}{4}\Big[\frac{1}{\tau_9}(\rho c_s^4 - \kappa_{xxyy}) - 8k_4 - 6k_5(u_x^2+u_y^2) - 2k_5(u_y^2 - u_x^2) + \\
&\quad + 16k_6 u_x u_y - 8k_7 u_y - 8k_8 u_x + F_9\Big], \\
q_2 &= \frac{1}{2\tau_{g,2}}(Tu_x - g_3 + g_1), \\
q_3 &= \frac{1}{2\tau_{g,3}}(Tu_y - g_4 + g_2), \\
q_4 &= -\frac{1}{4\tau_{g,4}}(2c_s^2 T - g_1 - g_2 - g_3 - g_4) + \left(\frac{1}{2\tau_{g,4}} - \frac{1}{2\tau_{g,2}}\right)u_x(Tu_x - g_3 + g_1) + \\
&\quad + \left(\frac{1}{2\tau_{g,4}} - \frac{1}{2\tau_{g,3}}\right)u_y(Tu_y - g_4 + g_2), \\
q_5 &= \frac{1}{4\tau_{g,5}}(g_1 + g_3 - g_2 - g_4) + \left(\frac{1}{2\tau_{g,5}} - \frac{1}{2\tau_{g,2}}\right)u_x(Tu_x - g_3 + g_1) + \\
&\quad + \left(\frac{1}{2\tau_{g,3}} - \frac{1}{2\tau_{g,5}}\right)u_y(Tu_y - g_4 + g_2),
\end{aligned} \qquad (7)$$

where components k_1, k_2, k_3 and q_1 are equal to zero due to the conservation of the density, momentum and internal energy. The implementation of the force terms F_i are described in [5].

The term κ_{xxyy} is a central moment defined by

$$\kappa_{xxyy} = \sum_{i=1}^{9} f_i (c_{i,x} - u_x)^2 (c_{i,y} - u_y)^2. \tag{8}$$

Finally, the parameters τ (called relaxation times) are related to hydrodynamic and non-hydrodynamic modes, those related to hydrodynamic modes are used to setup the transport parameters from the macroscopic partial differential equations and the rest can be used to tune stability and accuracy of the method, in present work we keep all non-hydrodynamic relaxation times equal to unity. In lattice units (superscript "+") we have the following relations for the relaxation times and transport parameters

$$\begin{aligned} \tau_5 = \tau_6 &= \frac{\nu^+}{c_s^2} + \frac{1}{2}, \\ \tau_{g,2} = \tau_{g,3} &= \frac{\alpha^+}{c_s^2} + \frac{1}{2}, \end{aligned} \tag{9}$$

where ν and α denote kinematic viscosity and heat diffusivity (in lattice units). To impose initial and some of the boundary conditions we need equilibrium distribution functions for each characteristic velocity, these are given as follows

$$\begin{aligned} f_i^{eq} &= \rho w_i \left(1 + \frac{\mathbf{u}\cdot\mathbf{c}_i}{c_s^2} + \frac{(\mathbf{u}\cdot\mathbf{c}_i)^2}{2c_s^4} - \frac{\mathbf{u}\cdot\mathbf{u}}{2c_s^2}\right), \\ g_1^{eq} &= (1 - 2c_s^2)T, \\ g_2^{eq} &= (2c_s^2 - 2u_x)\frac{T}{4}, \\ g_3^{eq} &= (2c_s^2 - 2u_y)\frac{T}{4}, \\ g_4^{eq} &= (2c_s^2 + 2u_x)\frac{T}{4}, \\ g_5^{eq} &= (2c_s^2 + 2u_y)\frac{T}{4}. \end{aligned} \tag{10}$$

The initial conditions are given by zero velocity and defined temperature (in lattice units)

$$\begin{aligned} &f_i(\mathbf{x},0) = f_i^{\mathrm{eq}}(\rho_{\mathrm{ini}}, \mathbf{u}_{\mathrm{ini}}),\ g_i(\mathbf{x},0) = g_i^{\mathrm{eq}}(T_{\mathrm{ini}}, \mathbf{u}_{\mathrm{ini}}) \forall \mathbf{x} \in [0, N_{\mathrm{x}}] \times [0, N_{\mathrm{y}}], \\ &\rho_{\mathrm{ini}} = 1,\ \mathbf{u}_{\mathrm{ini}} = (0,0), \end{aligned} \tag{11}$$

where N_{x} and N_{y} are lattice dimensions of the computational domain.

Boundary conditions needed during the simulations are Dirichlet for inlet, extrapolation for outlet and no-slip conditions for the velocity at walls, Dirichlet, extrapolation and adiabatic conditions for the temperature at inlet, outlet and at walls. The no-slip and adiabatic conditions can be simulated by the bounce-back approach [1]

$$f_{\bar{i}}(\mathbf{x}_{\mathrm{bb}}, t+1) = f_i^*(\mathbf{x}_{\mathrm{bb}}, t), \quad g_{\bar{i}}(\mathbf{x}_{\mathrm{bb}}, t+1) = g_i^*(\mathbf{x}_{\mathrm{bb}}, t), \tag{12}$$

where the star denotes postcollision state i.e. the right hand sides in equations 2 and 4, $\mathbf{x}_{\mathrm{bb}}$ is the position of a wall adjacent fluid site, $f_{\bar{i}}$ and $g_{\bar{i}}$ correspond to DFs which characteristic velocity points in the reflected direction i.e. $\mathbf{c}_{\bar{i}} = -\mathbf{c}_i$. This results in no-slip and/or adiabatic wall. In the case of Dirichlet conditions (i.e. for wall with given temperature) we use "anti-bounce-back" conditions to setup the temperature of the walls

$$g_{\bar{i}}(\mathbf{x}_{\mathrm{w}}, t+1) = -g_i(\mathbf{x}_{\mathrm{w}}, t) + c_s^2 T. \tag{13}$$

For the inlet boundary condition we use the equilibrium DF

$$f_i(\vec{x}_{\mathrm{in/d}}, t) = f_i^{\mathrm{eq}}(\rho_{\mathrm{in}}, \vec{u}_{\mathrm{in}}),$$

where $\vec{x}_{\mathrm{in/d}}$ is the location of the inlet. The outlet is modeled by a first-order extrapolation

$$f_i(\vec{x}_{\mathrm{out}}, t+1) = f_i^*(\vec{x}_{\mathrm{out}} - \vec{n}, t), \quad g_i(\vec{x}_{\mathrm{out}}, t+1) = g_i^*(\vec{x}_{\mathrm{out}} - \vec{n}, t)$$

where $\vec{x}_{\mathrm{out}}$ is location of the outlet, $\vec{n}$ is an outer normal and i is equal to directions that are pointing to the inside of the domain.

Macroscopic equations solved by the aforementioned schemes are the following [3, 4]

$$\begin{aligned}
&\frac{\partial u_x}{\partial x} + \frac{\partial u_y}{\partial y} = 0 \\
&\frac{\partial u_x}{\partial t} + \frac{\partial {u_x}^2}{\partial x} + \frac{\partial u_x u_y}{\partial y} = -\frac{\partial p}{\partial x} + \nu \left[\frac{\partial}{\partial x}\left(\frac{\partial u_x}{\partial x}\right) + \frac{\partial}{\partial y}\left(\frac{\partial u_x}{\partial y}\right)\right] + \Gamma_x \\
&\frac{\partial u_y}{\partial t} + \frac{\partial {u_y}^2}{\partial y} + \frac{\partial u_y u_x}{\partial x} = -\frac{\partial p}{\partial y} + \nu \left[\frac{\partial}{\partial x}\left(\frac{\partial u_y}{\partial x}\right) + \frac{\partial}{\partial y}\left(\frac{\partial u_y}{\partial y}\right)\right] + \Gamma_y, \\
&\frac{\partial T}{\partial t} + \frac{\partial T u_x}{\partial x} + \frac{\partial T u_y}{\partial y} = \alpha \left(\frac{\partial^2 T}{\partial x^2} + \frac{\partial^2 T}{\partial y^2}\right).
\end{aligned} \tag{14}$$

with ν and α being the kinematic viscosity and heat diffusivity of the fluid. These material parameters can be set through the relaxation times of the LBM, which are located inside the expressions for the $\mathbf{k}$ and $\mathbf{q}$ vectors. The force term is present only when the natural convection flows are solved and is given by the Boussinesq approximation

$$\Gamma = \mathbf{g}\beta(T - T_{\mathrm{ref}}), \tag{15}$$

with $\mathbf{g}$ being the vector of gravitational acceleration and β is the linear isobaric thermal expansion coefficient defined by

$$\beta = -\frac{1}{\rho}\left(\frac{\partial \rho}{\partial T}\right)_p. \tag{16}$$

The accuracy and stability of the presented method was studied by the authors and results can be found in the recent articles [4] (forced convection) and [5] (natural convection).

3. Description of the problem

We solve forced and natural convection in the vicinity of the 5x5 hot tube banks. For the case of forced convection, the tubes are inserted into the adiabatic channel and in the case of natural convection the tubes are inside the adiabatic box. The geometry of the forced convection problem is presented in Fig. 2. Walls are no-slip and adiabatic, dimensionless numbers are Pr=0.71, Re=80 and Re=1600 based on the tube's diameter. Dimensionless temperature of the fluid at inlet (left open boundary) and the surface of the tubes are 0 and 1 respectively.

The geometry of the natural convection problem is presented in Fig. 3. Walls are no-slip and adiabatic, dimensionless numbers are Pr=0.71, Ra=10^3 and Ra=10^5 based on the tube's diameter. Dimensionless temperatures of the fluid and tubes are 0 and 1 respectively.

IOP Conf. Series: Journal of Physics: Conf. Series **1101** (2018) 012040 doi:10.1088/1742-6596/1101/1/012040

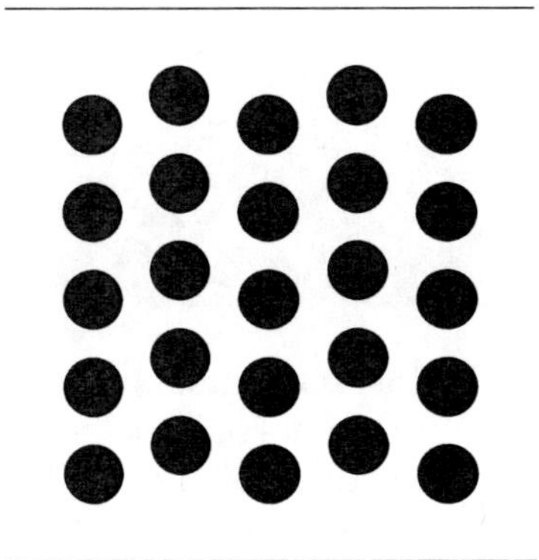

Figure 2. Geometry of the channel with tube banks.

Figure 3. Geometry of the adiabatic box with tube banks.

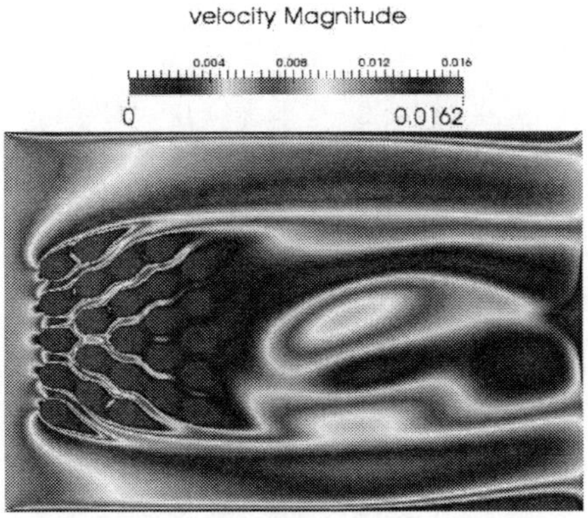

Figure 4. Instant velocity magnitude (in lattice units) for forced cooling of tube banks at Re=80, Pr=0.71 at lattice time $t^+=25\cdot10^4$.

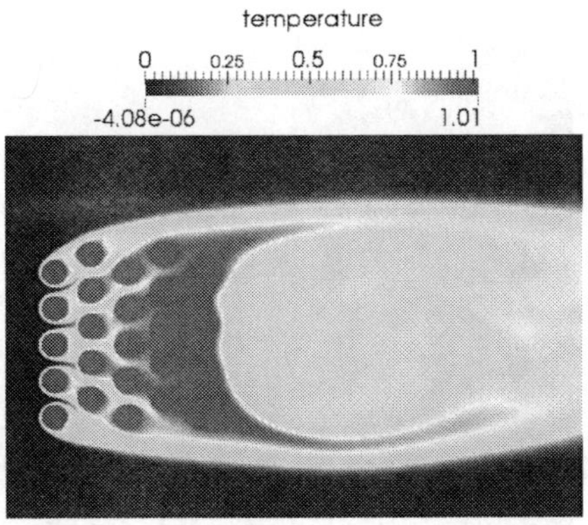

Figure 5. Instant temperature (in lattice units) for forced cooling of tube banks at Re=80, Pr=0.71 at lattice time $t^+=25\cdot10^4$.

4. Numerical simulations

Numerical setup for forced convection is following; the domain is divided into L_x=768 and L_y=512 lattice sites, the tube diameter in lattice units is D^+=32, the Mach number is set to Ma=0.01 and the rest of the parameters (ν^+ and α^+) are computed from the dimensionless numbers. We simulate $t^+=25\cdot10^4$ dimensionless timesteps. Temperature and velocity magnitude fields in lattice units are presented in Figs. 4-7.

Numerical setup for natural convection is following; the domain is divided into L_x=L_y=512 lattice sites, the tube diameter in lattice units is D^+=32, the Mach number is set to Ma=0.01 and the rest of the parameters (ν^+ and α^+) are computed from the dimensionless numbers. We simulate $t^+=25\cdot10^4$ dimensionless timesteps. Temperature and velocity magnitude fields in

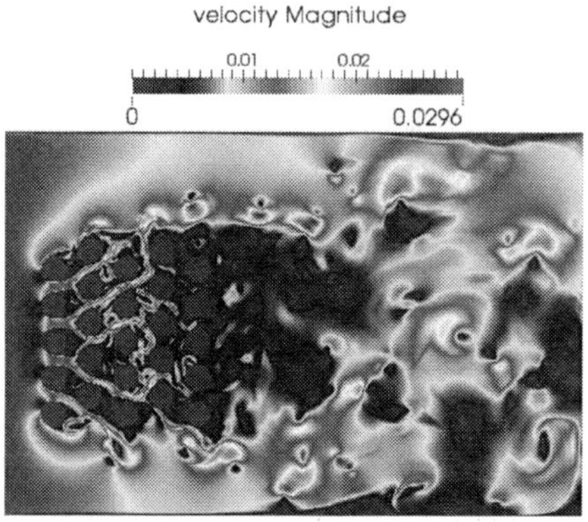

Figure 6. Instant velocity magnitude (in lattice units) for forced cooling of tube banks at Re=1600, Pr=0.71 at lattice time $t^+=25\cdot10^4$.

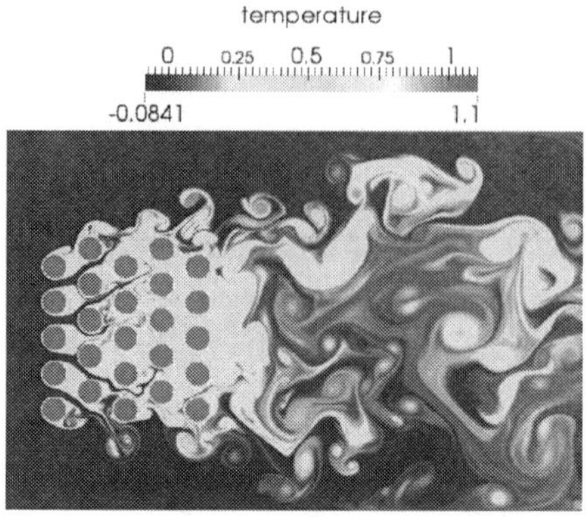

Figure 7. Instant temperature (in lattice units) for forced cooling of tube banks at Re=1600, Pr=0.71 at lattice time $t^+=25\cdot10^4$.

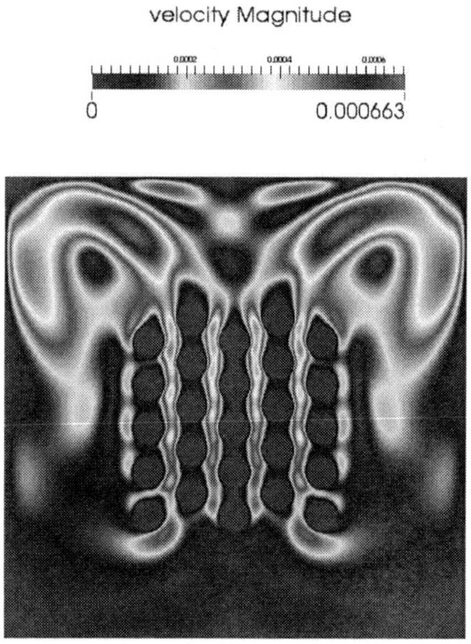

Figure 8. Instant velocity magnitude (in lattice units) for natural convection from tube banks at Ra=10^3, Pr=0.71 at lattice time $t^+=25\cdot10^4$.

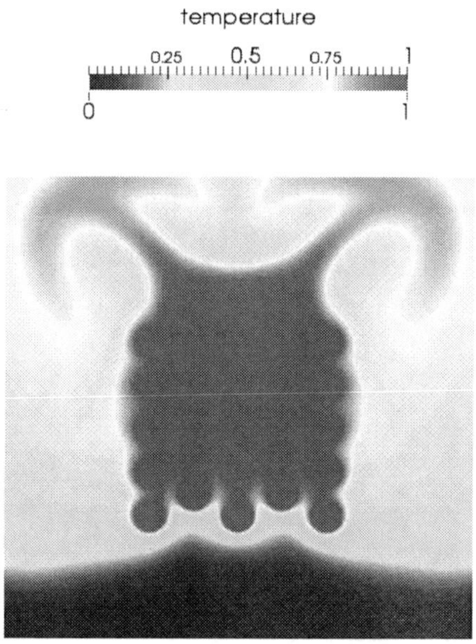

Figure 9. Instant temperature (in lattice units) for natural convection from tube banks at Ra=10^3, Pr=0.71 at lattice time $t^+=25\cdot10^4$.

lattice units are presented in Figs. 8-11.

5. Discussion

Comparing the forced convection for two different Re numbers, one can observe that for lower Re number big recirculation zone exist, right behind the bank (the flow is in the laminar regime but with unstable wake), limiting the heat transfer in tthe wo last columns of the bank. The

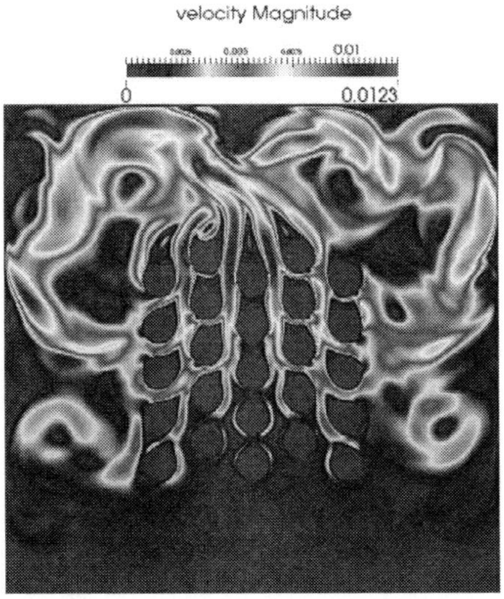

Figure 10. Instant velocity magnitude (in lattice units) for natural convection from tube banks at Ra=10^5, Pr=0.71 at lattice time $t^+=25 \cdot 10^4$.

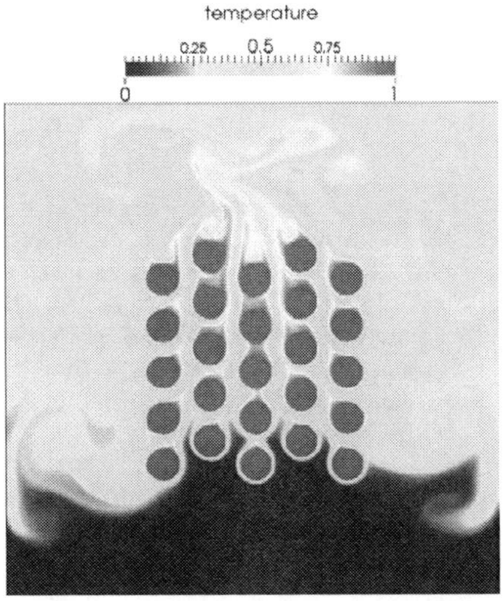

Figure 11. Instant temperature (in lattice units) for natural convection from tube banks at Ra=10^5, Pr=0.71 at lattice time $t^+=25 \cdot 10^4$.

bypass flow above and under the bank confines the temperature wake into the band area which is nearly of the height of the bank. The flow in the bank is close to laminar and for the first two columns causes better heat transfer as can be inspected from the temperature field, where last two columns are very poorly cooled. For higher Re flow (which lies in the subcritical regime), the flow and temperature patterns are more complex. One can clearly see the vortex street instabilities, and also flow pattern inside the bank is unstable. In this case the heat transfer is more intense as the flow between tubes are transitional. One recirculation zone does not exist in this case and therefore heat is removed more efficiently from the last two columns.

During the natural convection, the situation is similar, for the low Ra number, flow is laminar and slow, which affects the heat transfer from the tubes. Two recirculation zones exist at the top of the cavity on both its sides and the temperature inside the bank is equal to the tubes temperature. The heat is removed from the sides and mainly from the top of the bank as can be seen from temperature fields. For higher Ra number, the flow become unstable and the heat is removed more efficiently from the bank. The unstable wakes behind the cylinders in the upper half of the bank is clearly visible now. As in the previous case, the first few rows transport heat more efficiently than the rest of the bank (the gradients of temperatures near cylinders are bigger).

6. Conclusion

Lattice Boltzmann Method with cascaded collision operators for both fluid flow and heat transfer was applied to thermal flows. We simulated forced and natural convection from the hot tube banks to the cold flowing and passive air. The method was able to deliver stable and bounded results. The cascaded operator for the case of one conservation law is still not well studied and we are currently working on its deep analysis in terms of accuracy and stability. The next step is the extension of the scheme to the three dimensions and exploration its ability to simulate industrial flows.

Acknowledgments
Acknowledgments The work was partially supported by the Czech Science Foundation project No. 18-09539S. K.V. Sharma would like to thank CAPES, CNPq, and Petrobras-ANP for providing research scholarship and for supporting this research.

References
[1] Kruger T, Kusumaatmaja H, Kuzmin A, Shardt O, Silva G and Viggen E M 2017 *The Lattice Boltzmann Method: Principles and Practice* 1st ed Graduate Texts in Physics (Springer International Publishing)
[2] Geier M, Greiner A and Korvink J 2006 *Phys. Rev.* A **73** p 066705
[3] Geier M, Schonherr M, Pasquali A and Krafczyk M 2015 *Comput. Math. Appl.* **70** pp 507–547
[4] Sharma K, Straka R and Tavares F 2017 *Int. J. Therm. Sci.* **118** pp 259–277
[5] Sharma K, Straka R and Tavares F 2018 *Int. J. Therm. Sci., submitted*

IOP Conf. Series: Journal of Physics: Conf. Series **1101** (2018) 012041 doi:10.1088/1742-6596/1101/1/012041

Prediction of helicopter rotor noise in hover using FW-H analogy

T Suresh, O Szulc, P Flaszynski and P Doerffer

Institute of Fluid-Flow Machinery, Polish Academy of Sciences, Fiszera 14, 80-231 Gdansk

E-mail: thanu@imp.gda.pl

Abstract. The integral formulation of Ffowcs Williams and Hawkings (FW-H) analogy developed by Farassat is implemented to understand the sound generation and propagation of a rotating slender body like the helicopter rotor blade in hover. Using the linear acoustic theory, the thickness and loading noise terms are implemented in a flexible, new, in-house aero-acoustic tool. The pressure distribution on the blade surface is obtained from CFD simulation which sufficiently captures the correct flow characteristics of the rotor blade. Although, there is an improved estimation of the acoustic pressure signal in comparison to existing numerical noise signal results, there is still an under prediction of the peak amplitude compared to experimental values. The code is validated against the experiment for the UH-1H model helicopter rotor equipped with two rectangular blades having NACA0012 section.

1. Introduction

Noise emission is a major concern in most industries such as automotive, aerospace, wind energy and turbomachines. The global initiative to reduce noise to meet environmental targets brings focus to the study and development of aero-acoustics. This has led to noise control now becoming a design parameter rather than a post production fix. Taking advantage of the advancements made in computers, development of a new, in-house acoustic tool to study and investigate sound generated by rotating bodies such as helicopter rotors and wind turbine rotors is the main objective here. The sound producing mechanisms are far too many and complicated for rotating bodies. For example, in the case of helicopter rotor in subsonic motion with no shocks in the flow field, there exists sound generated due to thickness, loading, blade-vortex interaction, turbulence, blade-wake interaction and high-speed impulsive noise. For a preliminary study, linear noise theory is implemented to study thickness and loading noise. Since there exist experimental data-sets for the sound produced by the UH-1H model rotor in [1], it was used as one of the validation cases for the tool developed. Acoustic fields can be analyzed by either the direct method where the flow equations are classically solved or through hybrid methods where the sound source region is solved through classical CFD simulations and the far-field sound propagation is solved through acoustic analogies. Due to the criticality imposed by the acoustic scales which are at least a magnitude lower than the flow scales, the direct methods are cumbersome and expensive. Hybrid methods offer elegant solutions by separating sound generation and sound propagation mechanisms. This has led to development of several formulations over the years with a varying degree of success in predicting the sound signal matching the experimental data. This depends

IOP Conf. Series: Journal of Physics: Conf. Series **1101** (2018) 012041 doi:10.1088/1742-6596/1101/1/012041

on the non-linearity effects of the sources being included in the model or not. However, they give useful insights into the specific areas of major noise generation on a rotor and help in devising noise control methods. Since there is no clear consensus on which formulation to be considered while studying different noise mechanisms, it is necessary to understand the physics and governing principles of sound generation and propagation.

2. Theoretical modelling

Sound generation is captured by the non-zero source terms in the right hand side of the in-homogeneous wave equation. This linear wave equation is obtained by re-arranging the conservation of mass and momentum equations. Lighthill first developed a theory to predict noise produced by a turbulent flow field in an unbounded medium during his study of jet noise [3]. Ffowcs Williams and Hawkings (FW-H) developed an equation to predict noise while including the presence of solid bodies having arbitrary motion in the flow region [4].

This equation is an exact rearrangement of flow governing equations. FW-H derived the equation using the idea of embedding the flow problem in a larger unbounded domain using generalized Green's functions to solve the original acoustic flow case. They used the ordinary derivatives as generalized derivatives while keeping the conservation laws in differential form valid. The moving surface is represented as $f(\vec{x},t) = 0$ such that $\nabla f = \hat{n}$, where $\hat{n}$ is the unit outward normal as shown in figure 1.

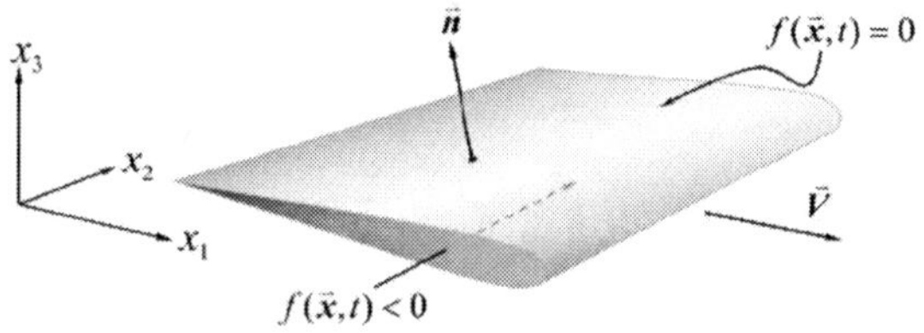

Figure 1. Definition of moving surface implicitly as $f(\vec{x},t) = 0$ with $\nabla f = n$ [2].

This implies that $f < 0$ is the volume enclosed by the surface. This volume is replaced by the undisturbed medium and surface distribution of sources. The FW-H equation from [4] is expressed as :

$$\Box^2 p' = \frac{\partial}{\partial t}\left[\rho_0 v_n \delta\left(f\right)\right] - \frac{\partial}{\partial x_i}\left[p n_i \delta\left(f\right)\right] + \frac{\partial^2}{\partial x_i \partial x_j}\left[H\left(f\right) T_{ij}\right] \tag{1}$$

with $T_{ij} = \rho u_i u_j - \sigma_{ij} + (p' - c^2\rho')\delta_{ij}$. p', c, ρ_0, v_n, T_{ij}, σ_{ij}, δ_{ij}, H(f) and $\delta(f)$ are the acoustic pressure, speed of sound, density, normal velocity, Lighthill stress tensor, viscous stress tensor, Kronecker delta, Heaviside and Dirac delta functions in the undisturbed medium respectively. The solution to this equation is obtained by first confining all the flow non-linearities into a control volume V which is bounded by a control surface S over which the pressure field is integrated to get the acoustic signal. These non-linearities are preserved in the Lighthill stress tensor. When a body is in motion, it produces pressure disturbances in the fluid region which are felt as sound by an observer in the medium. One important approximation employed is that the body does not disturb the medium appreciably and hence the non-linearities are neglected. The turbulence generated is of low intensity and also is a very inefficient noise generating mechanism. However, the fluctuating pressure produced by the turbulence is indirectly accounted for by the loading term in the equation. Here, the viscous shear forces acting over the blade are neglected. Thus converting a non-linear problem to a linear one. But this holds true only for low Mach numbers. Also, the discontinuities of the flow like the shock waves are not taken into account in the FW-H formulation. Both compact source and non-compact source theories have been developed for this equation. The three terms on the right side of the equation are known as monopole, dipole and quadrupole respectively. Monopole (thickness term) which is defined as the time derivative of elementary mass source, accounts for the displacement of the fluid produced by the body. Dipole (loading term) is defined as the doublet of equal elementary force sources of opposite phase that account for the unsteady loading exerted by the body on the fluid, quadrupole (volume term) is defined as the double doublet in opposite phase accounting for all non-linearities and noise due to turbulence.

Farassat [5] derived the integral solution known as Formulation 1 for the FW-H equation with surface sources in sub-sonic motion. This solution was proposed in two frames of reference: the observer space-time variable (**x**,**t**) which is always kept fixed and the source space-time system (**y**,τ) as seen in figure 2.

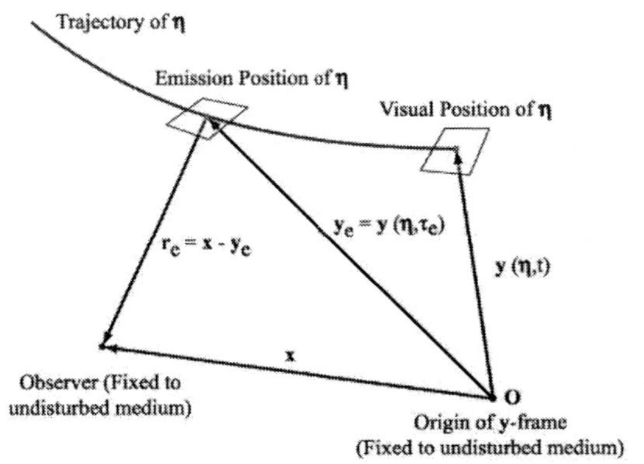

Figure 2. Trajectory of a source point η as seen by an observer fixed to the medium(x or y frame) and the definitions of visual and emission source positions [5].

For a study of the helicopter rotor noise, the blade is represented as a surface in a frame, the η-frame fixed relative to the surface. The motion of the blade is given by the trajectory of a source on the surface described by the fixed η in a y-frame. Applying convolution to Green's functions, G $= \frac{\delta(g)}{r}$ where $g = t - \tau - \frac{r}{c}$ and r = |x - y|, the volume integrals are converted to surface integrals by using the properties of δ function. After rigorous mathematical operations, the equation for retarded time τ is obtained:

$$g = \tau - t + \frac{\mid x - y(\eta, \tau) \mid}{c} = 0 \tag{2}$$

For a subsonic source motion, there is only one emission time τ_e for a given source point η. This source point has an emission position $y_e = y(\eta, \tau_e)$ and emission distance r_e, $r_e \equiv c(t - \tau_e)$ to the observer position x. This emission time is found usually numerically by a shooting technique described in [5]. The source position and trajectory is known at the observer time t. This is known as the visual position of η. To find the emission time, probing is done in small time steps in the backward direction of the trajectory for $\tau < t$ so that it satisfies the retarded time equation. The Formulation 1 as proposed by Farassat in [5] is of the form:

$$4\pi p^{'}(x,t) = 4\pi(p_T^{'}(x,t) + p_L^{'}(x,t)) =$$
$$= \frac{\partial}{\partial t}\int_{f=0}\left[\frac{\rho_0 v_n}{r\left(1-M_r\right)} + \frac{pcos\theta}{cr\left(1-M_r\right)}\right]_{ret} dS + \int_{f=0}\left[\frac{pcos\theta}{r^2\left(1-M_r\right)}\right]_{ret} dS \tag{3}$$

The critical factor defined by $r \mid 1 - M_r \mid$ seen in the above equation contains the source motion information. It indicates the difference in the time between the sound received by the observer and the time at which the same sound signal was emitted by the source. M_r represents the source Mach number in the direction of the observer. The singularity of this formulation arises when the M_r reaches 1. This leads to non-linearity which affects the quadrupole term and indirectly the dipole term. This is an important limitation of this formulation. Compact source body condition is a critical criterion of this formulation as demonstrated in [5]. Numerically, when the blade surface is divided into many small source cells, it is necessary that the compact source condition is satisfied by every single source cell. It implies that for each source cell, the acoustic field should be related to the global parameters of the rotating body, for example the thickness noise to the net rate of mass injection, loading noise to the net force on the body. If the sources are non-compact instead of the net force contributing to the noise signal, only the local pressure would be accounted for. These compact conditions also depend on the observer time and position making it quite challenging to numerically quantify them for each and every source cell and at each time interval.

3. Methodology

The integral formulation of Farassat has been implemented in the Tecplot 360 macros and scripting language. Tecplot 360 is a visualization and analysis software with built-in macros for scripting and also provides python scripting options. The helicopter rotor blade surface is discretized into small cells which act as acoustic sources. Various parameters of equation 3 are calculated for each source cell at the emission time. Then the corresponding observer time is obtained using the retarded time equation 3. The steady pressure distribution for the loading term is supplied from CFD simulation. The surface integration is obtained through a mid-point formula which is a second-order approximation. The differentiation scheme used for observer time derivatives is also a second-order scheme. The blade surface given as input is taken at time instant t=0 and it is rotated through small time steps to complete one period of revolution. All the parameters are computed for each and every single source cell at each time step. This is known as the acoustic potential for each source at emission time. Then their corresponding observer time is computed satisfying the retarded time equation. At given observer time instant, all the source panels with the correct emission time, contributing to the noise signal are added to get the acoustic pressure pulse. This tool provides flexibility in terms of studying the different types of noise terms individually, in terms of analyzing the individual parameters of the equation thus providing opportunities to implement noise control methods. Also enables to study noise contribution from different blade portions such as the leading edge, trailing edge and blade tips and at various observer locations aiding in better understanding of sound generation and propagation.

4. Results and discussion

The analysis was conducted for 1/7-scale model of the UH-1H helicopter rotor in hover, having two rectangular blades with NACA0012 profile with tip Mach number of 0.8 and radius of 1.045m with the observer located at a distance of 3R from the blade tip in the rotor plane, same as in the experimental investigations of Boxwell [1]. The blades are untwisted and un-tapered with aspect ratio of 13.71 as shown in figure 3. The pressure data was obtained from the CFD simulations. The case was set up similar to the CFD computational model as presented in [9]. These simulations were conducted using a cell-centered, block-structured, parallel code SPARC with a Spalart–Allmaras turbulence model with a mesh of 3 million cells. The acoustic signal obtained had to be flipped due to the reverse direction of probing of the acoustic parameters in the implemented formulation.

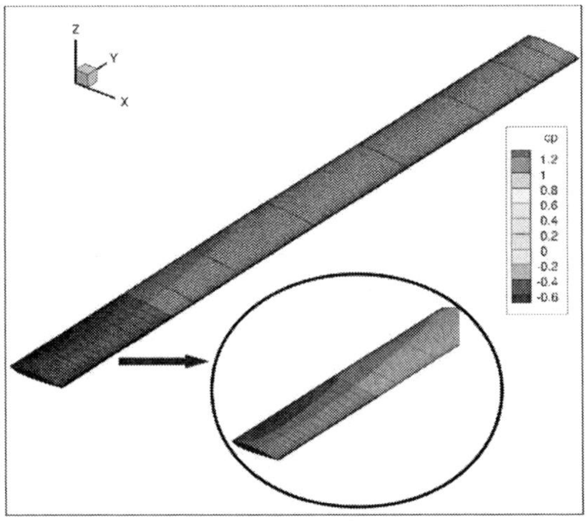

(a)

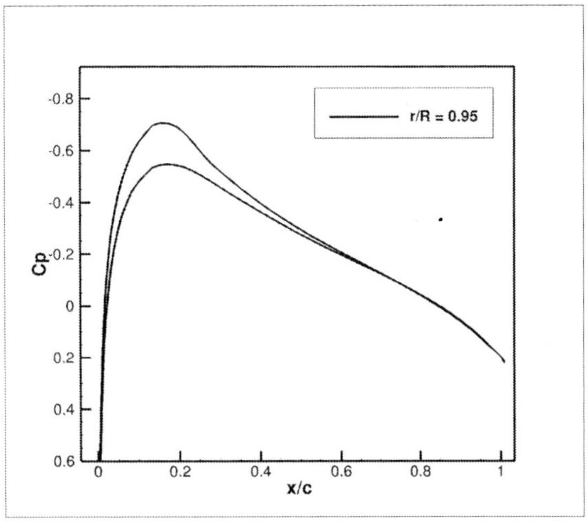

(b)

Figure 3. UH-1H model helicopter rotor a) grid with local refinement of $3.0.10^6$ volumes (7208 surface cells) (b) pressure coefficient (Cp) distribution at r/R = 0.95.

4.1. Grid dependency

A grid dependency test was conducted for varying number of grids from coarse 584 to 26708 source panels as seen in figure 4. The pressure distribution used for the computation of the loading term for all grids is constant steady data from the fine grid case of 26708 cells. With increasing number of cells, the peak negative amplitude decreases and also the small positive peaks on eitherside of the negative peak change nature. This is because of the varying distribution of source cells on the blade surface. Loading noise computed represents both the near-field and the far-field loading terms added together and it decreases the amplitude of the thickness noise peak. Thickness noise signal and the loading noise signal are added to get the final acoustic pressure pulse. It is seen that the peak amplitude decreases with increasing cell number rather in a counter intuitive way. This could be because of the sharp slopes that are seen in the fine grids in the loading term gets smoothed out as well as the peak thickness amplitude decreases, thereby reducing the overall peak amplitudes.

4.2. Time step dependency

Time step plays a crucial role in predicting noise as the acoustic potential of the sources were differentiated with respect to the time taken. Tests were conducted from a large time step such as 1° to 0.25° of the azimuth. The thickness peak amplitude increases with increasing time step thereby increasing the overall acoustic peak amplitude as seen in figure 5.

4.3. Higher order of differentiation

The default differentiation scheme in Tecplot 360 is a second order approximation. To improve the peak amplitude, higher order observer time derivatives - 4th and 6th order of central differencing scheme are used. The 4th order scheme improves the signal by 1 Pa but there is no difference between the 4th and 6th order schemes as seen in figure 6. Hence, 4th order scheme is preferred for further analysis.

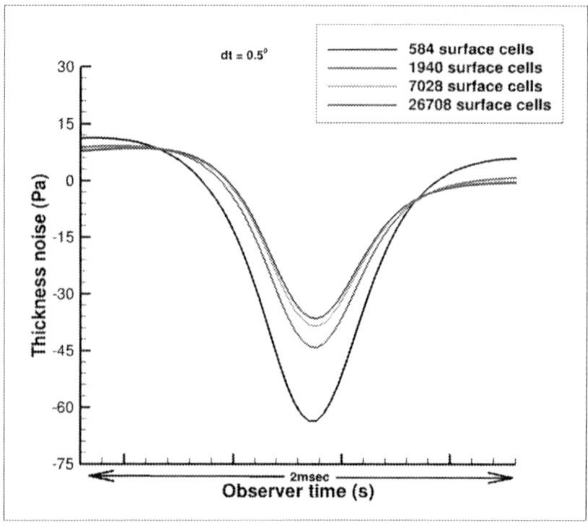

(a)

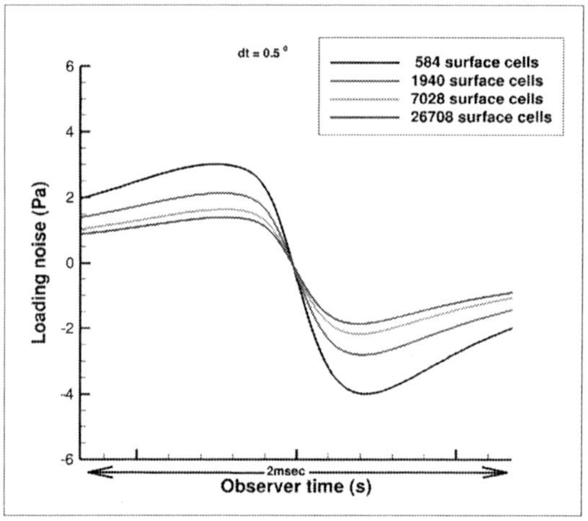

(b)

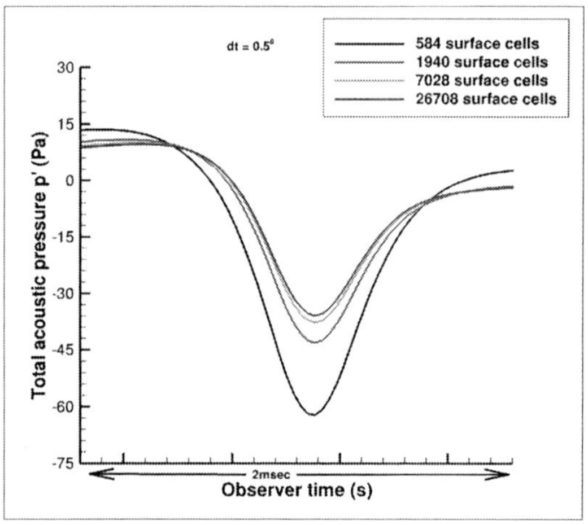

(c)

Figure 4. Grid dependency study a)thickness noise b)loading noise c)total acoustic pressure.

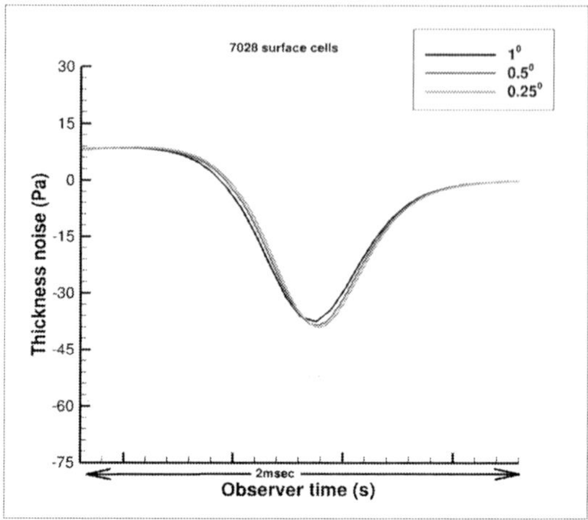

(a)

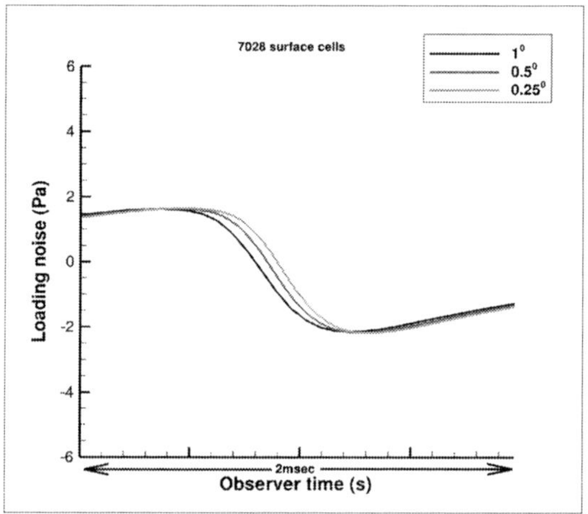

(b)

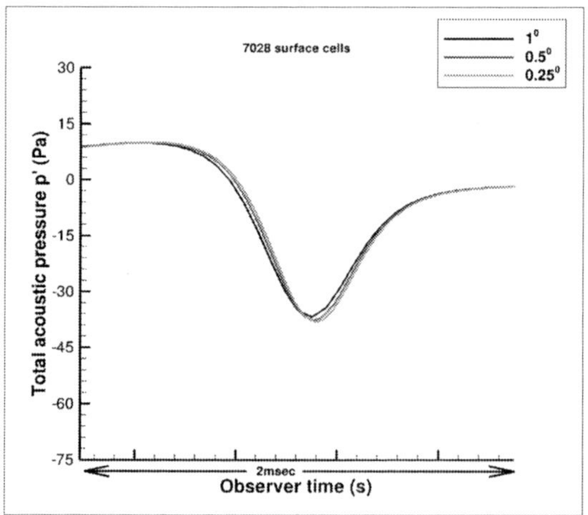

(c)

Figure 5. Time step dependency study a)thickness noise b)loading noise c)total acoustic pressure.

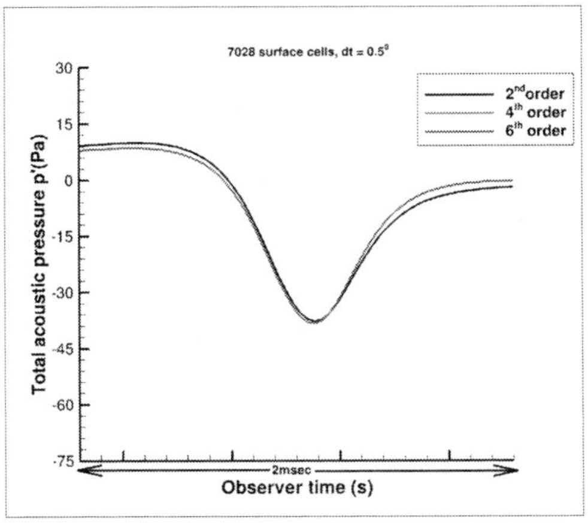

Figure 6. Higher order schemes of differentiation (7208 surface cells, $\Delta t = 0.5°$).

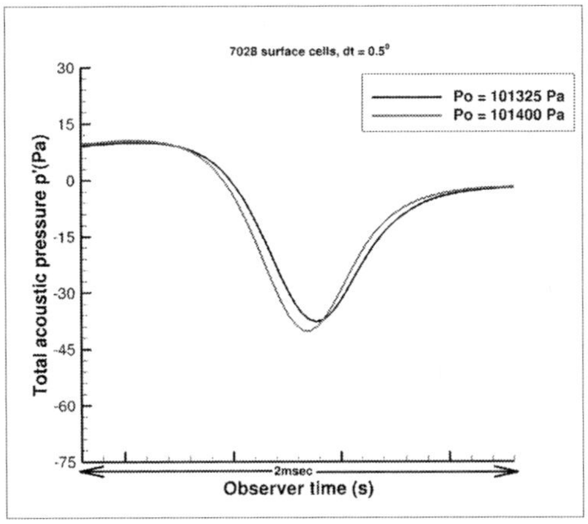

Figure 7. Precise experimental input data (7208 surface cells, $\Delta t = 0.5°$).

4.4. Precise experimental input data

The ambient conditions like the reference pressure, density and temperature recorded during the experiment is unknown. Taking the average conditions during the time of recording in the wind tunnel, the atmospheric pressure is taken as 101400 Pa. The angular frequency (ω) is extracted from the experimental plot by measuring the period of one revolution and using the expression $\omega = \frac{2\Pi}{T}$, the new ω is calculated. Using this value in Mach number formula, the ambient temperature is found to be $T_0 = 285.3K$. Using the ideal gas law, density is also computed $\rho = 1.238$ kg/m^3. These new values are then implemented and this improves the peak amplitude of the signal by about 2.3 Pa approximately as seen in figure 7.

5. Validation

The final acoustic signal is computed with higher order of differentiation scheme, with the correct experimental input data for the reference grid of 7028 surface cells and $\Delta t = 0.5°$ as seen in figure 9. This signal is compared with the experimental data from [1]. It is observed that the final peak amplitude obtained using the new in-house tool has a value of 42 Pa approximately when compared to the experimental value value of 60 Pa as seen in figure 8. It is improved compared to the 30 Pa approximately achieved in the theoretical modeling in [1] and also the numerical analysis conducted in [10]. It still underpredicts the experimental value. Other factors like the twisted blade, pre-cone angle are not implemented here and the experimental investigations shows an uncertainity in the value of about +/- 10 Pa.

6. Conclusions

The correct shape of the acoustic signal is captured very well by the tool. The numerical results underpredict the experimental values which points towards the limitations of the formulation itself and the assumptions of the linear theory. The sound due to turbulence encaptured by quadrupoles is not modelled here as at subsonic motion and low Mach numbers their contribution is insignificant. The underprediction in the validation raises the question if the non-linearity effects must also be considered in the formulation. Considering the effects of quadrupoles, non-linearity in higher Mach flows or in forward motion of helicopter rotors is understandable but their significance in low speeds is questionable as pointed out in [8]. Another important criterion is the acoustic compact source conditions. It is necessary that each and every cell satisfies the

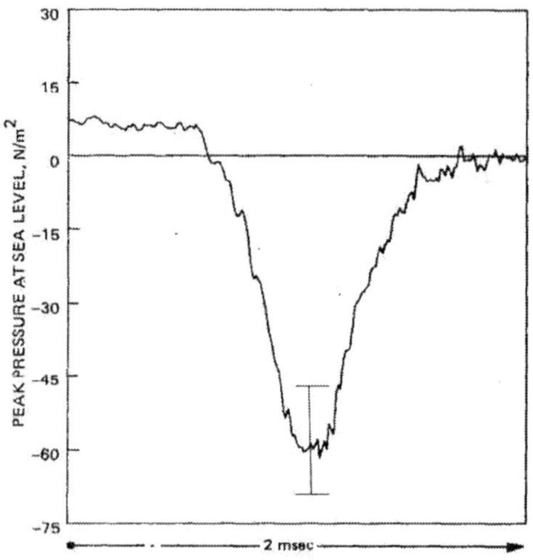

Figure 8. Acoustic pressure pulse recorded at inplane location of r/D = 1.5 ($M_T = 0.8$) [1].

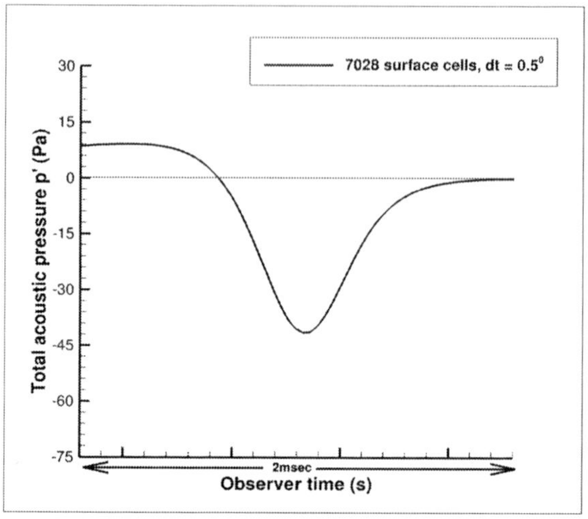

Figure 9. Acoustic pressure pulse computed at inplane location of r/D = 1.5 ($M_T = 0.8$).

acoustic compactness condition and at every observer time step which should be quantified.

In the implementaion of the equation 3, it is observed that the acoustic signal is very sensitive to the number of source cells on the blade surface and their distribution. The case with fewer cells and higher time steps seem to produce results more closer to the experiment than the fine grid with smaller time step cases. The location of the positive peaks in the signal which corresponds to sink and source strength and the time shift in the arrival of these signals towards the observer makes the discretization of the blade surface very crucial. This raises the question of the upper and the lower limits of discretization that is allowed while implementing the integral solution of the FW-H equation. The surface integration is a critical step in the computation and since the errors present in this step only gets amplified with numerical differentiation, it is imperative to reduce these numerical errors. This forms the basis of future work to understand the reasons for underprediction at low Mach number helicopter hover cases.

Acknowledgments

This research was supported by the European Union's Horizon 2020 research and innovation programme under the Marie Sklodowska-Curie grant agreement No 722401- SmartANSWER and in part by the CI TASK and PL-Grid Infrastructure.

References

[1] Boxwell D A, Yu Y H and Schmitz F H 1978 *NASA CP-2052*
[2] Farassat F 2007 *NASA TM-214853*
[3] Lighthill M J 1962 *Proc. Roy. Soc. London* A **267** 147–82
[4] Ffowcs Williams J E and Hawkings D L 1969 *Phil. Tran. Roy. Soc.* A **264** 321–42
[5] Farassat F 1975 *NASA Tech. Rep.* R **451**
[6] Schmitz F H 2016 *J. of the Amer. Hel. Soc.* **61** 041001
[7] Schmitz F H Sep 7-9, 1977 *Third European Rotorcraft and Powered Lift Aircraft Forum* Aix-en-Provence
[8] Farassat F and Brentner K 1987 *J. of the Amer. Soc.* **33 (1)**
[9] Szulc O, Doerffer P and Tejero F 2016 *J. Phys.: Conf. Series* **760** 012031
[10] Kusyumov A N, Mikhailov S A, Garipova L I, Batrakov A S and Barakos G 2015 *EPJ Web of Conf.* **92** 02042

IOP Conf. Series: Journal of Physics: Conf. Series **1101** (2018) 012042 doi:10.1088/1742-6596/1101/1/012042

Numerical study of potential application of active suction and blowing through blade tip perforation to reduction of helicopter rotor thickness noise

O Szulc and P Doerffer

Institute of Fluid-Flow Machinery, Polish Academy of Sciences, Fiszera 14, 80–231, Gdansk

E-mail: oskar.szulc@imp.gda.pl

Abstract. The article explains a new concept (called here ATNC – active thickness noise control) of the application of surface transpiration to the reduction of helicopter main rotor harmonic noise. The ATNC method is based on an introduction of four cavities covered by perforated plates (connected to pressure reservoirs) at the leading and trailing edges of the outer 20% of blade span. For an exemplary, two-bladed (NACA 0012 airfoil) model helicopter rotor of Purcell in low-lifting, hover conditions (tip Mach number $Ma_{\mathrm{T}} = 0.66$ and collective $\theta = 1.5°$) the results of numerical simulations, based on the SPARC code (Spalart–Allmaras turbulence and Bohning–Doerffer transpiration models), prove that the acoustic pressure fluctuations are significantly reduced by ATNC in the near-field of blade tip. The peak amplitude is attenuated by more than 45%, with a reduction of the Overall Sound Pressure Level *OASPL* by 3.4 dB. It is shown also that the venting has not only a large impact on the acoustic radiation, but also on the aerodynamic performance of the rotor, expressed in terms of the torque penalty of 38%. The ATNC method proves to be a promising candidate as a mean of helicopter rotor thickness noise control, but not in the current arrangement.

1. Introduction

In the helicopter rotor tip-path plane the acoustic signal has a distinct, symmetrical shape, with a large negative pressure pulse that radiates mostly in the direction of forward flight, often dominating all other sources of rotorcraft noise [1]. From the point of view of civil or military operations it is highly undesirable since the community annoyance is increased and the detection distance of an approaching aircraft is compromised. For helicopters operating with the advancing tip Mach numbers Ma_{AT} in the 0.75 – 0.85 range (many contemporary designs) the shape of the impulse can be explained by considering only the monopole (thickness) term of the Ffowcs Williams–Hawkings solution of the inhomogeneous wave equation proposed by Lighthill (FWH analogy) for cases with arbitrary motion of the source surfaces (e.g. helicopter rotor blades) [2]:

$$p'(x,\,y,\,z,\,t) \;=\; \frac{1}{4\pi}\,\frac{\partial}{\partial t}\iint_S \left[\frac{\rho_0\,U_{\mathrm{n}}}{r\,|1-Ma_r|}\right]_{\tau} dS\;. \tag{1}$$

At a fixed observer position $(x,\,y,\,z)$ the acoustic pressure value $p' = p - p_0$ [Pa] (p – static pressure, p_0 – ambient pressure) at any given instant of time t is computed as a time derivative of the sum of all acoustic sources distributed over the entire surface of the blade S.

The acoustic potential (integrand) of each source panel dS is evaluated at the retarded (emission) time: $\tau = t - r/a$ (a – speed of sound), and is dependent on four parameters: the density of surrounding fluid ρ_0, the local normal velocity of the panel due to blade motion U_n, the distance between the panel and the observer r, and finally the Mach number of the panel in the observer direction Ma_r. As can be seen in equation 1 the in-plane acoustic pressure signal is determined by the blade shape and kinematics only, being independent of the aerodynamics. In effect an almost symmetrical acoustic signal is predicted, as shown in figure 2.

It is obvious from equation 1 that to reduce the thickness noise it is necessary to consider all mentioned components of the monopole term. For modern helicopter rotors it is primarily achieved by changing the blade tip shape (S), introducing thinning and sweep, accompanied by a lower hover tip Mach number Ma_T (limiting both, U_n and Ma_r). If the rotational speed of the rotor and the blade shape are considered as given, the only choice is to influence the normal velocity component U_n of the source panel dS. It may be achieved by active suction and blowing through perforated surfaces placed directly at the blade tip. The proposed method of ATNC is effective in damping the main source of thickness noise, but introduces an important coupling since the transpiration influences not only the acoustic radiation, but also the surrounding flow-field, modifying the magnitude of the aerodynamic forces acting on the blade.

2. Mechanism of active thickness noise control (ATNC) by surface ventilation

The presented arrangement of the ATNC method is based on the application of four cavities (connected to low or high pressure supplies) covered by perforated plates and distributed at the rotor blade tip (see figure 1). Two identical (symmetrically placed with respect to the chord line), short leading edge cavities covered by perforated plates of equal, high porosity are responsible for sucking the air in. The remaining two (also identical and symmetrically positioned), extended trailing edge cavities covered by perforated plates of equal, low porosity are responsible for blowing out. A large pressure difference between both sides of each plate induces choking of the flow in the holes. Therefore, the resulting transpiration velocity U_t distribution is practically independent of the external flow conditions above the surface of the ventilated wall, and may be considered as constant. The surface coverage of the perforation is chosen based on a location of the maximum values of normal velocity U_n during rotation, to counteract the main source of the thickness noise. So far, only constant suction and blowing rates have been considered, leading to a uniform noise reduction efficiency in the entire rotor plane (no directivity pattern present), meaning lack of dependence on the azimuth angle of emission. The suction and blowing transpiration streams are imbalanced with approximately 2:1 ratio.

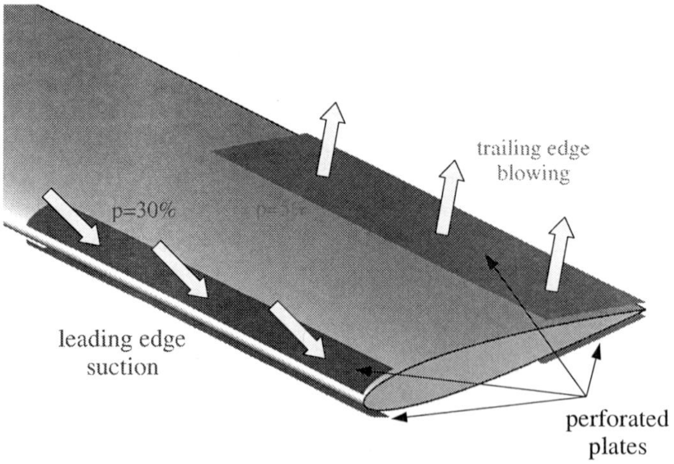

Figure 1. A helicopter rotor blade tip equipped with ATNC.

From the flow point of view the leading edge suction alone improves the boundary layer state. On the contrary, the trailing edge blowing disturbs highly the velocity profile leading to a negative impact on the aerodynamic performance. In forward flight the most intensive acoustic signal is emitted from the advancing blade, in a narrow azimuth range, and with strong directivity forward and near the rotor plane. In such conditions the ATNC method would need to be activated only partially, within a narrow range of azimuth (centred at approx. 90°), having a much lower impact on the performance than in the studied hover conditions. What is important, in contrast to other means of thickness noise reduction (tip thinning and lowering of tip speed), it does not require any modification of the operating conditions nor the blade outer shape. Instead, a high drag penalty has to be paid, but the control may be easily deactivated in conditions when noise emission is not a critical issue. The economical aspects of the implementation of ATNC and technological limitations of its usage are not considered here.

3. Physical and numerical modelling

The present investigation has been carried out with a cell-centred, block-structured, parallel CFD code SPARC [3]. It solves numerically the compressible, Favre-averaged Navier–Stokes equations with several 1- and 2-equation eddy viscosity models. The turbulence field is fairly well predicted by a low-Reynolds closure of Spalart–Allmaras (SA) in reference and flow control conditions [4]. The numerical algorithm is based on a semi-discrete scheme, utilizing the finite-volume formulation for the spatial discretization (central, 2nd order) and the implicit dual time-stepping approach (supplemented by the explicit Runge-Kutta method) for the integration in time (2nd order). To damp numerical oscillations and to assure a high resolution of discontinuities the artificial dissipation model SLIP is implemented. In order to increase the convergence rate the local time-stepping, implicit residual averaging, and full-multigrid techniques are incorporated in the solver. The equations are solved in a global coordinate system where the rotation is imposed by an entire grid movement using an unsteady mode (URANS). A perforated wall boundary condition, based on the empirical, Bohning–Doerffer (BD) transpiration law [5] and designed towards the modelling of surface ventilation, has been implemented in the solver [6], validated for various internal and external flows [7] (including the shock wave–boundary layer interaction control), and applied to the helicopter rotor high-speed impulsive noise reduction [8] in the past. Moreover, the SPARC code has proven to deliver reliable aerodynamic results in hover and forward flight conditions for many rotor flows. In terms of the acoustic analysis only near-field observer locations are considered. Inherent dissipation and dispersion errors, introduced by the CFD algorithm, are therefore minimised, allowing for comparison of pressure signals extracted for the reference and flow control configurations.

4. Numerical simulation of ATNC method

The initial, testing phase of the active thickness noise control (ATNC) method was performed with a two-bladed model of the UH-1H helicopter main rotor, investigated by Purcell in low-thrust hover conditions [9]. The blades were untwisted, rectangular, and had the NACA 0012 section (aspect ratio of 13.71). The tip Mach number of $Ma_{\mathrm{T}} = 0.66$ and the tip Reynolds number of $Re_{\mathrm{T}} = 1.6 \cdot 10^6$ (collective angle $\theta = 1.5°$) were used, in-line with the experimental study of a similar rotor system conducted by Sargent and Schmitz at the University of Maryland (2012) [10]. This investigation was focused on the development of a new, on-blade acoustic cancellation method, based on the active tip jet blowing control of in-plane rotor harmonic noise. The details of the computational model (i.e. mesh topology, boundary conditions, convergence criteria, grid and time step dependency studies) are similar to the description specified in [8] for the hovering Caradonna–Tung rotor, and for this reason are omitted here. The presented URANS simulations were performed using a coarser mesh, but still consisting of an adequate $3.0 \cdot 10^6$ of control volumes per blade (in comparison with $7.0 \cdot 10^6$)

thanks to a lack of shock waves and flow separation in the investigated subsonic conditions.

The reference blade was equipped with two suction systems at the leading ($0 < x/c < 0.1$, c – chord length) and two blowing systems at the trailing edge ($0.65 < x/c < 1.0$), covering the outer 20% of blade span. The forward cavities were closed by perforated plates of high porosity of 30%. On the contrary, the rearward cavities were covered by perforated plates of low porosity of 5%. These values lay within the range of applicability of the BD model which, apart from the transpiration flow characteristics, delivers the maximum mass flow rate achievable in chocked conditions. The pressure in the connected supply system was kept at a sufficiently low (or high) level to reach the sonic velocity in the holes. This way the resulting transpiration (normal) velocity U_t was independent of the local conditions near the blade surface, exhibiting constant rates of suction ($60\,\mathrm{m\,s^{-1}}$) and blowing ($10\,\mathrm{m\,s^{-1}}$). Taking into account the surface area subjected to bleeding the final mass flow rate of suction was 2.4 times larger than for blowing.

Near-field (at $1.5\,c$ from the tip) acoustic pressure fluctuations p'/p_0 (p_0 – ambient pressure) in time t/T (T – period of rotation), recorded in the rotor plane, at 1.11 R (R – rotor radius), are depicted in figure 2 for three investigated configurations. The reference signal (ATNC off) exhibits a typical, almost symmetrical shape of the thickness noise pulse measured in wind tunnels at similar operating conditions. When the air is sucked-in through two leading edge perforated plates (ATNC on, suction only) the peak negative pressure is reduced by 25%, while the Overall Sound Pressure Level *OASPL* is limited by 1.4 dB. In these almost non-lifting conditions the penalty is mainly considered in terms of an increase of the rotor shaft torque which is rather small and equal to 4%. Addition of blowing-out through two trailing edge perforated plates (ATNC on, suction/blowing) significantly improves the peak amplitude attenuation to 46%, leading to a drop of *OASPL* by 3.4 dB. It is well known that the blowing has a detrimental effect on the aerodynamic performance, that's why an increase of torque as high as 38% is observed which is considered as a price of the activation of the noise cancellation device.

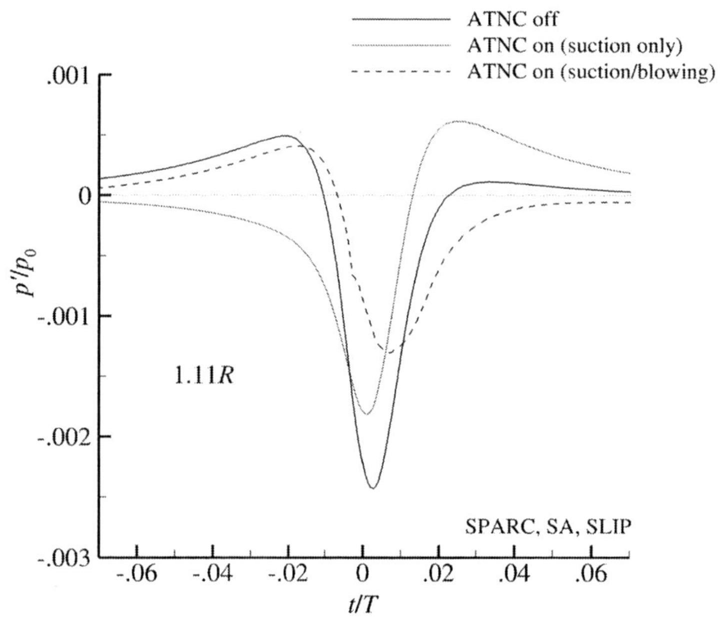

Figure 2. Impact of the ATNC method on the near-field acoustic pressure variation p'/p_0 in time t/T at an in-plane location of 1.11 R.

The Overall Sound Pressure Level *OASPL* acoustic parameter is defined by a relation:

$$OASPL\,\mathrm{dB} = 20\,\log\frac{p'_{\mathrm{rms}}}{p'_{\mathrm{ref}}}\,\mathrm{dB}\ ,\qquad p'_{\mathrm{rms}} = \sqrt{\frac{1}{N}\sum_{i=1}^{N} p_i'^{\,2}} \tag{2}$$

where p'_{rms} is the root mean square of pressure fluctuations p', p'_{ref} is the threshold of human hearing (20 µPa), and N is the number of samples (here 1440 per period of rotation T). It must be stressed out that 3 dB in *OASPL* corresponds to a reduction of p'_{rms} by 50% that is directly perceived by a human ear as a two times lower noise level. The blowing induces a strong disturbance to the boundary layer, but it does not lead to flow separation for the investigated case. In forward flight the advancing blade emits the most intensive acoustic pulse when it is approaching 90° of azimuth and the maximum inflow Mach number is present (the advancing tip Mach number Ma_{AT}). Close to this position the angle of attack at the tip is close to 0° (for twisted blades) – a necessary requirement to balance the overproduction of lift. On the contrary, in hover (with the tip Mach number $Ma_{\mathrm{T}} = Ma_{\mathrm{AT}}$) the inflow conditions are steady in time, hence the acoustic signal is of much higher amplitude. Therefore, the application of the ATNC method (in the current form) in forward flight allows for a limitation of the azimuthal coverage of the surface transpiration, which would reduce the torque penalty considerably.

5. Conclusions

The article explains the details of a new concept of the application of surface transpiration to the reduction of helicopter rotor in-plane, harmonic noise. The proposed ATNC method is based on an introduction of four cavities covered by perforated plates, through which air is sucked-in or blown-out, located at the leading and trailing edges of the outer 20% of the blade span. For the exemplary model helicopter rotor in low-lifting, hover conditions the results of numerical simulations, based on the SPARC code (SA closure and BD transpiration model), prove that the acoustic pressure fluctuations are significantly reduced by ATNC in the near-field of the blade tip. The peak amplitude of p' is attenuated by 46%, with a reduction of *OASPL* by 3.4 dB. It is shown also that the tip venting has not only a large impact on the acoustic radiation, but also on the aerodynamic performance of the rotor expressed in terms of the shaft torque penalty of 38%. The ATNC method proves to be a promising candidate as a mean of helicopter rotor thickness noise control, but not in the current arrangement.

It has been demonstrated that the proposed flow control strategy (active surface bleed) may have a radical impact on the radiated acoustic pulse amplitude and shape. The imposed transpiration velocity U_{t} directly counteracts the normal velocity U_{n} of the blade surface element due to rotation that is a main source of the thickness noise. Still, many questions remain unanswered, and worth further investigation. Is it possible to separate the outer flow-field from the disturbance and drag introduced by the surface ventilation? What is the impact of the ATNC method on other sources of rotary noise (e.g. loading) in out-of-plane locations? At least two possibilities arise. Instead of the applied uniform transpiration velocity for each perforated plate, U_{t} could be non-uniformly distributed along the chordwise and radial directions, adopting its intensity to the values of U_{n} present at the surface (known a priori). It may be easily achieved by a proper choice of the distribution and diameter of the perforation holes, reflected in a spatially varying aerodynamic porosity of the plate. Moreover, the suction and blowing in an unsteady manner would give an another option for reaching an improvement of the efficiency of ATNC. It is evident from equation 1 that it is not the absolute value of the monopole integral that is responsible for the noise generation, but its time derivative. It also seems (based on the results published in [8]) that it takes some substantial time for the external flow-field to adapt to a suddenly appearing and disappearing transpiration, with the delay reaching approximately 5° – 7° in terms of the azimuthal angle. It may suggest an application of a series of short bursts that cannot be perceived by the outer flow. It could still provide a sufficient anti-noise pulse strength that would significantly attenuate the main source of thickness noise, without a severe impact on the rotor performance. Additionally, a supplemental valve system would allow for a convenient activation and deactivation of ATNC, leaving it operational only in conditions when it is really needed from the point of view of mission requirements. Finally, the ATNC method is general (not assigned to rotorcrafts) and may be used for any rotating machinery that is know to emit strong thickness noise (e.g. propellers).

Acknowledgments

This research was supported by CI TASK and in part by PL-Grid Infrastructure.

References

[1] Schmitz F H 2016 *J. American Helicopter Society* **61** 4 041001
[2] Ffowcs Williams J E and Hawkings D L 1969 *Phil. Trans. Roy. Soc. Lond.* **264** 1151
[3] Magagnato F 1998 *TASK Quarterly* **2** 2 pp 217–70
[4] Spalart P R and Allmaras S R 1992 *Proc. AIAA 30th Aerospace Sciences Meeting and Exhibit* Reno 92-0439
[5] Doerffer P P and Bohning R 2000 *Aerosp. Sci. Technol.* **4** pp 525–34
[6] Doerffer P and Szulc O 2011 *Int. J. Engineering Systems Modelling and Simulation* **3** 1-2 pp 64–73
[7] Szulc O, Doerffer P, Flaszynski P and Suresh T 2018 *Proc. 10th Int. Conf. on Computational Fluid Dynamics* Barcelona
[8] Szulc O, Doerffer P and Tejero F 2016 *J. Phys.: Conf. Series* **760** 1 012031
[9] Purcell T W 1988 *Proc. 14th European Rotorcraft Forum* Milano 2
[10] Sargent D C and Schmitz F H 2012 *Proc. AIAA 33rd Aeroacoustics Conf.* Colorado Springs 2012-2140

IOP Conf. Series: Journal of Physics: Conf. Series **1101** (2018) 012043 doi:10.1088/1742-6596/1101/1/012043

Experimental and numerical research on heat and air flow through a granular material

E Szymanek[1], **A Tyliszczak**[1]

[1] Faculty of Mechanical Engineering and Computer Science, Czestochowa University of Technology. Al. Armii Krajowej 21, 42-201 Czestochowa, Poland

E-mail: ewaszym@imc.pcz.pl, atyl@imc.pcz.pl

Abstract. The paper is devoted to experimental and CFD (Computational Fluid Dynamics) studies of heat and mass transfer processes in granular material layers. The 3D numerical model is based on a high-order computational code combined with the immersed boundary method (IB) for modelling the flow and temperature distributions in geometrically complex systems. The temperature measurements are performed using thermocouples placed inside the layers closed in a climatic chamber. Besides providing data for validation of the numerical results one of the key tasks of the experiment was to examine an impact of an external temperature changes on the temperature distribution in the interior of the layers depending on their structures and physical properties. The obtained results reveal very complex flow and heat transfer inside the layers. It is shown that the temperature inside the layers can be effectively changed and controlled by the parameters of air flowing through them.

1. Introduction

The research related to the flow and heat transfer in granular materials is motivated by a wide range of its engineering and industrial applications, e.g. in the fermentation process, for storage of grains, in systems for production and recovery of crude oil, in fluidized beds, in conversion and thermal storage systems (catalytic reactors, energy storage systems). These various applications enforce the need for better understanding of the basics of transport phenomena in granular layers. Their structures are complicated and specification of the configuration of the material's arrangement inside them is impossible in real conditions. The configurations of the layers for the needs of mathematical and numerical models are usually idealised. Although such descriptions are fraught with errors and will not accurately reflect real structures of the layer they enable analysis of the phenomena inside the layers using discrete/finite elements type methods [1, 2, 3, 4] or continuous and fractional models [6]. These approaches and models are widely used by many researchers but they are not sufficient to accurately describe all processes in granular layers.

In this paper the flow and heat transfer was modelled using the immersed boundary technique based on a volume penalization method (IB-VP)[7, 8], which allows the use of Cartesian meshes for objects with very complex shapes. Although, in case of the granular layers this would translate to arbitrary form of granulates in the present studies the layers were modelled in a simplified manner assuming them as regions of packed spheres. The numerical code used in this research was an in-house code called SAILOR. It is based on the unsteady equation of heat conduction (3D) and the Navier-Stokes equations. The analysed problem was treated as unsteady and the flow was assumed to be laminar. The experimental research consisted on temperature measurements and was performed using LTCL600 Series 3 climatic chamber (TASLtd., WestSussex, Great Britain). The purpose of the experiment was to determine internal temperature distributions and their variability in respect to a changing temperature outside the layers. These results were then used also as the exemplary data to validate the numerical solutions. We focused on the flow behaviour inside the granular layers and analysed their evolution in between the granulates. We investigated how the flow movement changes the temperature distribution inside the layers.

2. Mathematical modelling

Variable density, variable temperature low Mach number flows are described by the continuity equation, the Navier-Stokes equations and the energy equation defined as:

$$\partial_t \rho + \nabla \cdot (\rho \mathbf{u}) = 0 \tag{1}$$

$$\rho \left(\partial_t \mathbf{u} + (\mathbf{u} \cdot \nabla)\mathbf{u}\right) + \nabla p = \nabla \cdot \boldsymbol{\tau} + \mathbf{f}^{\mathrm{IB}} \tag{2}$$

$$\rho C_p \left(\partial_t T + (\mathbf{u} \cdot \nabla) T\right) = \nabla \cdot (\kappa \nabla T) + \mathrm{f}_T^{\mathrm{IB}} \tag{3}$$

where ρ stands for the density, T - temperature, p - hydrodynamic pressure, $\mathbf{u}$ - velocity vector, C_p - heat capacity and κ - heat conductivity. The set of Eqs. (1)-(3) is complemented with the equation of state $p_0 = \rho R T$, where p_0 denotes the thermodynamic pressure and R - is the specific gas constant. In open flows with inlet/outlet boundaries p_0 is constant in space and time [14] and in this work it is assumed to be 101325Pa. The molecular viscosity (μ) within the viscous stress tensor $\boldsymbol{\tau}$ is computed from the Sutherland law.

IB-VP source term. The source terms $\mathbf{f}^{\mathrm{IB}}$ and $\mathrm{f}_T^{\mathrm{IB}}$ originate from the immersed boundary volume penalization approach (IB-VP). They role is to penalize a difference between the actual and assumed velocity and temperature of the solid body. They act on a fluid in such a way as if there were solid objects immersed in the flow domain and they are defined as:

$$\mathbf{f}^{\mathrm{IB}} = -\frac{\rho}{\eta}\Gamma(\mathbf{x})(\mathbf{u} - \mathbf{u}_s), \quad \mathrm{f}_T^{\mathrm{IB}} = -\frac{\rho C_p}{\eta}\Gamma(\mathbf{x})(T - T_s) \tag{4}$$

where $\eta \ll 1$ is the so-called penalization parameter with dimension of time unit and Γ - the phase indicator defined as:

$$\Gamma(\mathbf{x}) = \begin{cases} 0, & \text{for } \mathbf{x} \in \Omega_f \\ 1, & \text{for } \mathbf{x} \in \Omega_s \end{cases} \tag{5}$$

where Ω_f and Ω_s are the regions of fluid and solid part of the computational domain. For $\Gamma(\mathbf{x}) = 1$ discrete forms of Eq. (2) and (3) with the time step Δt reduce to $\mathbf{u}^* \approx \Delta t \mathbf{u}_s/(\eta + \Delta t)$ and $T^* \approx \Delta t T_s/(\eta + \Delta t^n)$ which for $\eta \ll \Delta t$ leads to $\mathbf{u}^* \approx \mathbf{u}_s$ and $T^* \approx T_s$. Thus, the forcing terms enforce the no-slip boundary conditions and set the required temperature of the solid objects. Simplicity of the IB-VP method has, however, direct consequences in lower solution accuracy. Similarly, as in the classical IB method with a stepwise approach (i.e. without the interpolation [9]) the formal order of the IB-VP method is at most equal to one [8, 10, 11].

Solution algorithm. The solution algorithm for Eqs. (1-3) is formulated in the framework of a projection method [12] for pressure-velocity coupling. The time integration is based on a predictor-corrector approach (Adams-Bashforth/Adams Moulton) and the spatial discretisation is performed using 6th/5th order compact difference and WENO (Weighted Essentially Non-Oscilatory) schemes on half-staggered meshes [13, 14].

3. Experimental facility

The experimental research were performed using a climatic chamber (LTCL600 Series 3, TASLtd., WestSussex, Great Britain), which allowed setting the outside temperature and its exact control during the whole experiment. A sample layer consisted of granular material (expanded clay) filled with air is shown in Fig. 1(a). Its effective heat diffusivity was equal to $\alpha_g = 8.0 \times 10^{-7}\mathrm{m}^2/\mathrm{s}$. The length ($L_x$), height ($L_y$) and width ($L_z$) of the layer were equal to 0.12m, 0.3m, 0.3m, respectively, and it was insulated from the top and bottom ($y = \pm Ly/2$) and at $z = \pm L_z/2$. The temperature measurements were carried out using thermocouples placed between the granules. They were mounted on specially constructed racks allowing their positioning inside the layer at the selected locations, as shown schematically in Fig. 1(b). The goal of the experimental research was to provide the temperature distribution inside the layers, which could be than used to verify the proposed numerical model.

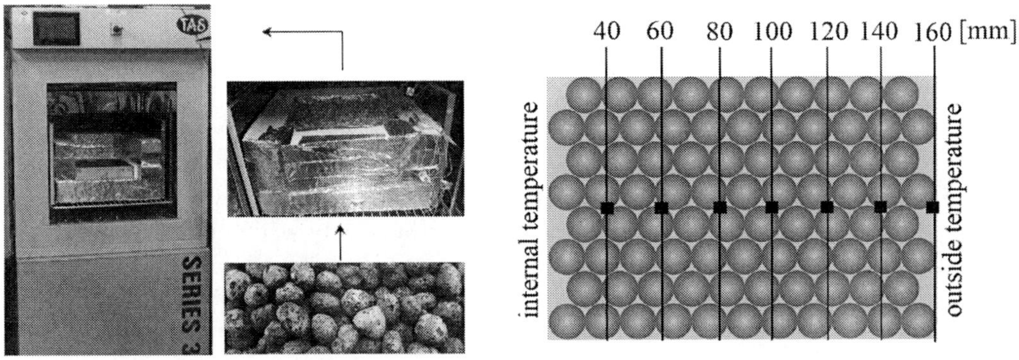

(a) (b)

Figure 1. Research test stand: a) climatic chamber with a layer of granular material in a close-up, b) the locations of thermocouples in the layer

4. Results

4.1. Assessment of the IB-VP accuracy

This section shows how the theoretical limitation of the solution accuracy of IB-VP method translates on the results obtained for the cases including the heat and mass transfer in the flows in channels and in between spheres. The test cases included: (1) an isothermal flow in a channel filled with six layers of spheres and (2) a flow through a layer $10 \times 10 \times 10$ spheres.

In first test case (1) we considered the flow through the rectangular channel in which six layers of spheres ($D = 28mm$) were placed in the corners, as shown in Fig. 2. The size of the domain was $1D \times 1D \times 15D$. The simulations were performed for Reynolds numbers $Re = UD/\nu = 28.88$ (U - inlet velocity, ν - viscosity), $Re = 59.78$, $Re = 105.57$ and $Re = 204.74$. The numerical results were compared with experimental data [15] and with the results obtained using the ANSYS Fluent code with 2nd order discretisation. An important factor affecting the solution accuracy is the density of the mesh. To analyse this issue for the present test case the computations were performed for the meshes with significantly different number of nodes. For the SAILOR code the meshes were uniform and consisted of $40 \times 160 \times 40$ (mesh M1), $80 \times 160 \times 80$ (M2) and $80 \times 320 \times 80$ (M2) nodes. In the simulations performed using the ANSYS Fluent code the body-fitted meshes precisely matched the shapes of the spheres and they consisted of approximately $2.5 \cdot 10^6$ (mesh M1) and $7 \cdot 10^6$ (M2) tetrahedral cells. Preparation of the meshes in the ANSYS software required cutting of small cylinders in locations in which the spheres were in contact (see Fig. 2), otherwise the mesh generation resulted in error. Comparisons of the streamwise velocity

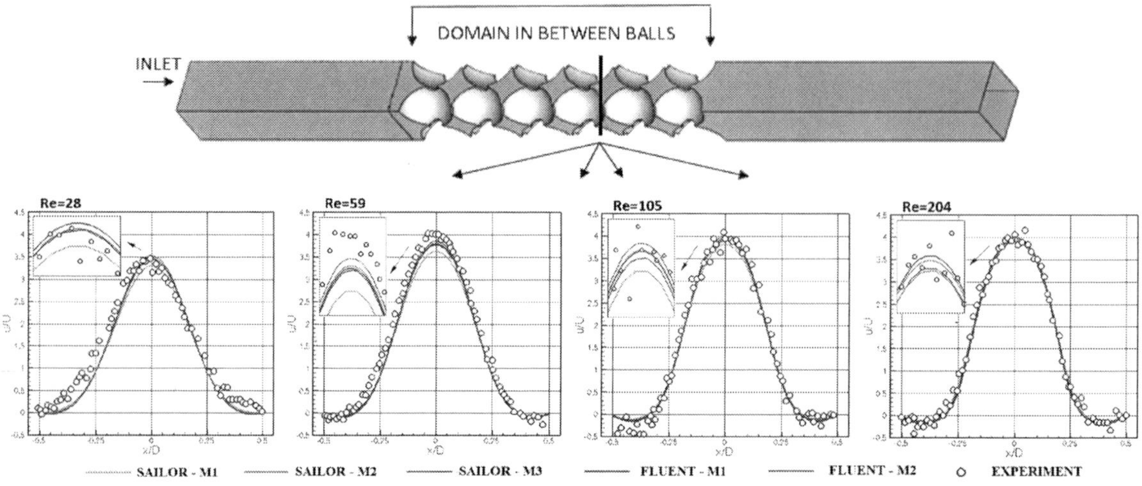

Figure 2. Isothermal flow in a channel filled with six layers of spheres

profiles along the x-direction from the location in between 5th and 6th sphere are shown in Fig. 2. The obtained results agree with the experimental data relatively well and the accuracy increases along with the Reynolds number. It seems that the numerical solutions slightly underpredict the measured values in the locations $x/D > 0.25$ for the case with Re $= 28$ and at $x/D = 0.0$ for Re $= 59$. For the higher Reynolds numbers the agreement is very good. It should be noted that the results obtained from the SAILOR and ANSYS Fluent codes practically overlap and are virtually independent of the grid density, which means that the accuracy of the IB-VP method is very good. It seams that lowering the formal order accurace of IB-VP method to 1st order has no such negative consequences as one could initially presume.

The configuration of the second test case was a simplified model of the granular layer. The computational domain was a rectangular box $0.1 \times 0.3 \times 0.1\text{m}^3$ in which ten layers of 10×10 hot spheres ($T_s = 400$K, $D = 0.008$m) were placed in the central part of the domain starting from $y = 0.05$m, as shown in Fig. 3. The gaps in between particular spheres were equal to 0.0002m. At the lower boundary we assume uniform velocity 0.1959m and temperature 300K ($Re = UD/\nu = 100$). The computations were performed with the use of a uniform mesh consisting of $181 \times 240 \times 181$ for the SAILOR code and body-fitted mesh with approximately 11×10^6 cells for the ANSYS Fluent code. The comparison of the results is presented in Fig. 3(right figures) showing the velocity and temperature profiles at the locations at $y = 0.06$m (the second row of the spheres) and $y = 0.15$m (above the layer) along the x-coordinate on the lines placed in between the spheres. It can be seen that discrepancies between the results obtained with the help of the SAILOR and ANSYS Fluent codes are small and only quantitative. The locations of maxima/minima of the velocity and temperature are predicted very well and only its absolute values are slightly under-predicted or over-predicted, depending on the localisation and the compared quantity.

4.2. Heat and mass transfer in the granular layer

It has been shown that IB-VP approach can be applied for modelling of the heat and fluid flow in complex domains with good accuracy. In this subsection we analyse the heat transfer inside the granular material layer studied experimentally. In the simulations the model of the layer was taken smaller in the vertical (y) and horizontal (z) direction, i.e., $L_y = 0.04$m, $L_z = 0.04$m, and the domain boundaries at $y = \pm L_y/2$ and at $z = \pm L_z/2$ were assumed to be adiabatic.

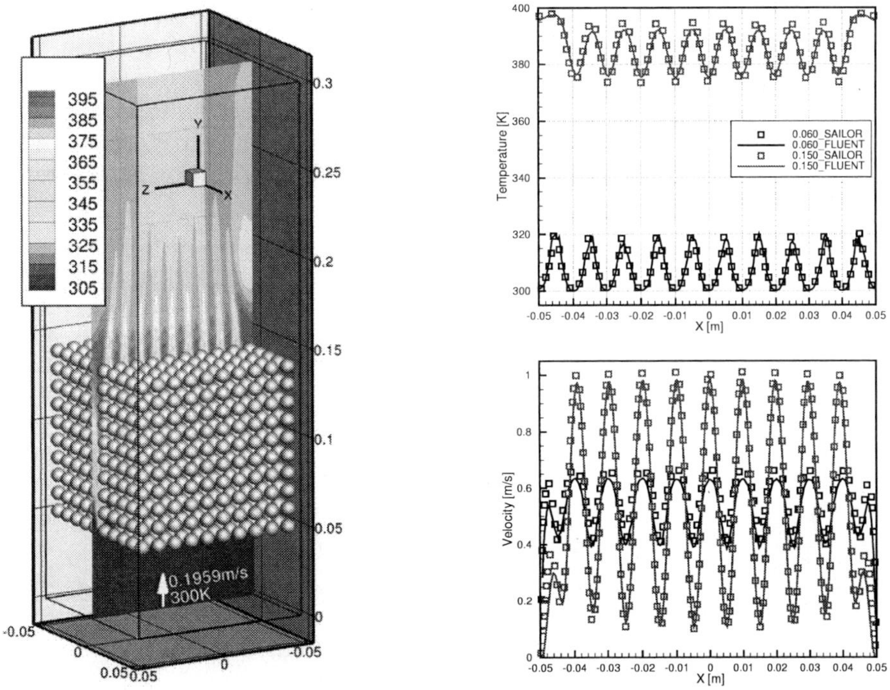

Figure 3. Left figure: the computational domain with the contours of temperature in the central cross-section. Right figures: temperature and velocity profiles at $y = 0.06$m and $y = 0.15$m

Taking the smaller layer dimensions for the simulations was necessary because of very high computational costs that would be required for the full scale model. The mesh was uniform and consisted of $256 \times 208 \times 84$ nodes in the x, y and z direction, respectively. The preliminary tests have shown that such a mesh was dense enough to obtain qualitatively consistent solutions. We considered three cases: (1) substitute layer treated as uniform solid material (no flow) with the effective heat diffusivity $\alpha_g = 8.0 \times 10^{-7}\mathrm{m}^2/\mathrm{s}$); (2) non-uniform layer (no flow) represented by spherical $13 \times 5 \times 5$ orderly packed spheres (D = 10mm, $\alpha_g = 5.3 \times 10^{-7}\mathrm{m}^2/\mathrm{s}$) with the empty spaces filled with air ($\alpha_{\mathrm{air}} = 1.9 \times 10^{-5}\mathrm{m}^2/\mathrm{s}$); (3) as the configuration 2) but with the air injected from the lower side of the layer.

Configuration 1. This was the configuration studied experimentally. It provided the temperature distribution inside the layer, which allowed for an additional verification of the applied numerical method. The test was performed as follow. Initially, the temperature of the layer was uniform and equal to 298K. Then, the outside temperature on the right side of the layer (see Fig. 1(b)) started to decrease linearly ($T(t) = 298 - 2.083 \times 10^{-3}t$ [K]) such that at the time 16100s it fallen down to 265K. Figure 4(a) presents a comparison of numerical results obtained from the SAILOR code and with experimental data as well as the simulation results obtained using the ANSYS Fluent code. It can be seen that the convergence of the numerical solutions is very good. Only slight differences between these results and experimental data prove high precision of the applied codes and solutions algorithms.

Configuration 2. This test case is very similar to the previous one and the only difference is that the granules and air are treated separately. As can be seen in Fig. 4(a) in this case, the differences between the results obtained for configuration 1 are large and increase across the

IOP Conf. Series: Journal of Physics: Conf. Series **1101** (2018) 012043 doi:10.1088/1742-6596/1101/1/012043

layer in the x-direction. Due to larger heat diffusivity of the air the changes in the temperature distribution occur faster. At the end of the layer at position $x = 0.04$m at time 12000 the temperature is about 10K lower compared to the case with the effective diffusivity. It seems that in this particular case modelling of the heat flow trough the layer assuming it as homogenous is more accurate than treating it as compound of solid spheres and empty spaces. It can be caused by incorrect specification of the solid material properties provided by the producer.

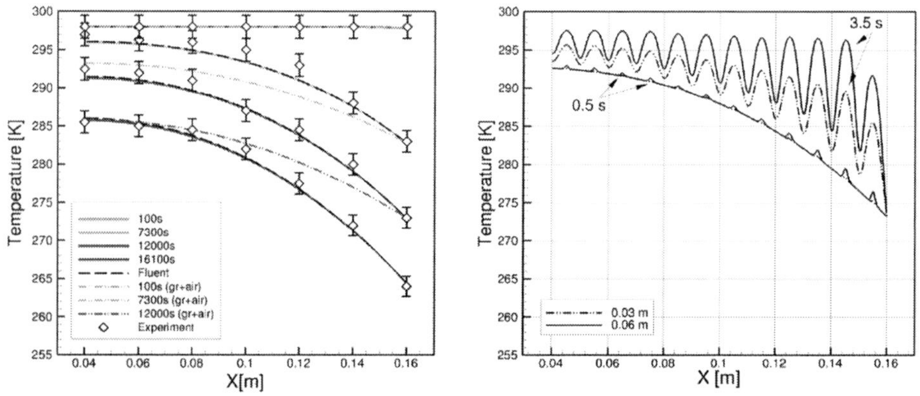

(a) stagnant air (b) flowing air

Figure 4. Comparison of the numerical and experimental results for the granular layer

The complex structure of the layer and thus the problems with the exact placement of measuring devices may be the reason for the slight deviations in the experimental results. Additionally, the case in which the layer was considered as a system composed of solid granulate with the air filling the intergranular spaces (dashed line designated as "gr+air") was analyzed.

Configuration 3. For this configuration two buffer layers were added to the domain, i.e., 0.02m from the bottom and 0.04m from the top, as shown in Fig. 5. They were required to specify the inlet and outlet boundary conditions and to minimise their impact on the flow in between the spheres. As the initial solution we took the results at the time instant 12000s presented in Fig. 4(a). The air with the temperature 300K was injected from the bottom side of the domain with a uniform velocity equal to 0.1m/s. It was higher than the temperature of the layer and this implied that the spheres were heated. Figure 5 on the left side shows the temperature iso-surface inside the layers and vertical velocity contours and velocity vectors in the main cross-section plane for the time moment 4 seconds from the beginning of the air flow. Figure 5(a,b) show the time evolutions of the temperature and velocity fields in the same plane. The words "SPHERE" in the zoomed figure showing the velocity vectors point the locations of the spheres which were removed for a better visibility of the results. It can be seen that the flow structures and temperature distributions are very complex and non-uniform in space. There are recirculation regions (negative velocity), which occur both in between the spheres as well as above the layer. Such an inhomogeneity of the flow around individual spheres enhance the heat exchange process. As one could expect, because of the smaller heat diffusivity of the spheres, they are heated slower than the spaces in between them. Figure 4(b) shows the temperature variations across the layer in two different vertical locations and at two time instances. A wavy character of the profiles is readily apparent and the maxima correspond to the places in between the spheres.

Figure 6 presents the velocity contours in $x-z$ cross-sections planes and $y-z$ cross-sections at $x = 0.065$ (across the spheres) and $x = 0.06$ (in between). It can be seen that the maxima

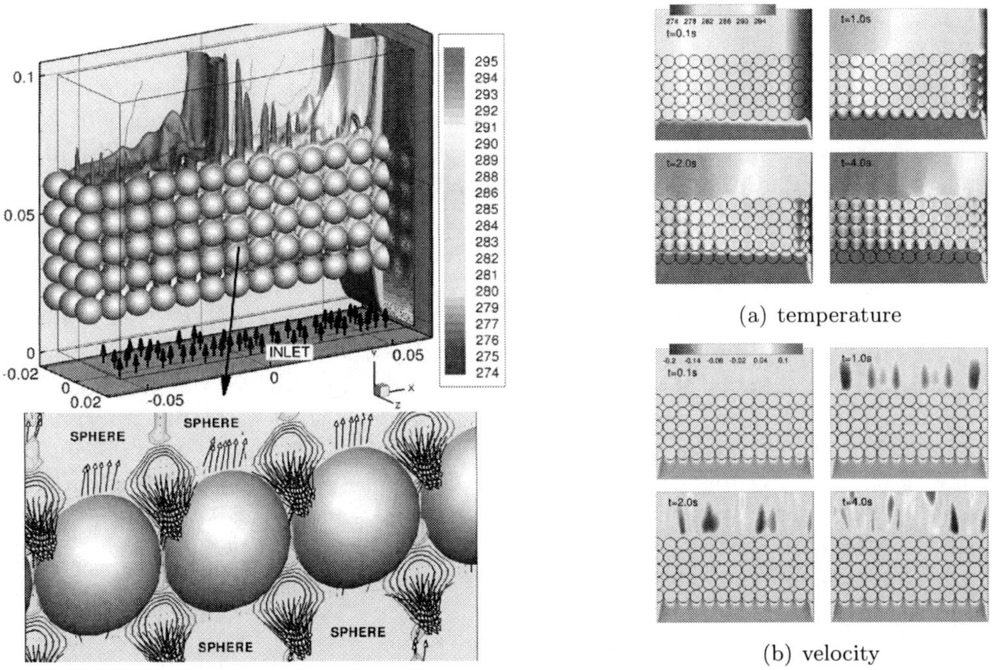

(a) temperature

(b) velocity

Figure 5. Left figures show the solution at 4.0s from the beginning of the air flow: temperature iso-surfaces (upper figure) and velocity contours with vectors (lower figure, zoom). Figures on the right side show the temperature and velocity contours at the selected time instances.

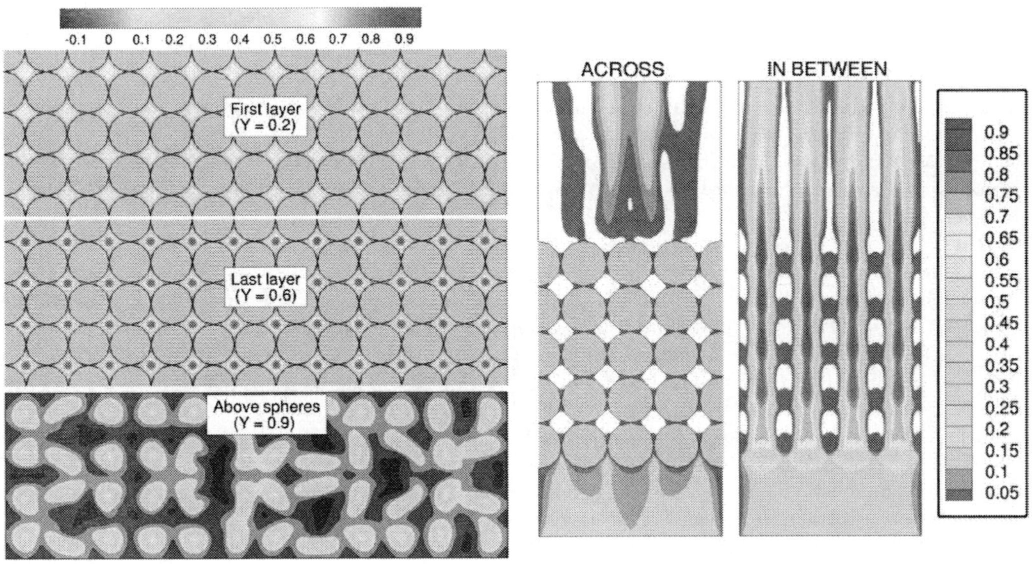

Figure 6. Velocity contours in $x - z$ cross-sections (left figures) and $y - z$ cross-sections at $x = 0.065$ (across the spheres) and $x = 0.06$ (in between).

of the velocity occur in the centres of the regions in between the spheres. In these locations the flow strongly accelerates and compared to the inlet conditions the velocity increases almost

10 times. The streams of the air with high velocity extend even above the layers, as can be seen in Fig. 6(right figures). The complex structure of the flow behind the spheres (plane at $x = 0.065$ - across the spheres) manifests by recirculation regions denoted by white colour. It can be seen that these regions appear already behind the first row and their shapes do not change significantly across the layer. This means that the flow structure inside the layer at different vertical locations is similar, and hence, its impact on the heat transfer process on particular row of the sphere is also similar.

5. Summary

This paper presented numerical and experimental studies on the heat and flow in granular layers. The IB-VP method applied in the research was proven to be sufficiently accurate to predict the inhomogeneities of the granulates and complex mass and heat transfer phenomena in between the inter-granular spaces. This was demonstrated by comparisons with the experimental data and the numerical data obtained using the ANSYS Fluent code. The results obtained for the granular layers revealed existence of various size of recirculation regions developing behind the granulates (spheres). Their presence intensifies the heat exchange process between the solid and gas phases. It was shown that the temperature inside the layers can be effectively changed and controlled by the parameters of air flowing through them. Note that such air can be taken from the air-conditioning system, for instance.

Acknowledgements

This work was supported by National Science Centre, Poland (Grant no. 2017/27/N/ST8/02318) and statutory funds of Czestochowa University of Technology under BS/MN 1-103-301/2018/P. PL-Grid infrastructure was used to carry out the computations.

References

[1] Vargas W L and McCarthy J J 2001 Heat Conduction in Granular Materials *AIChE J* **47** 1052-1059
[2] Tian K S and Shu H J 2009 Progress in porous media research *Nova Science Publishers Inc. New York*
[3] Amhalhel G A and Furmanski P 1997 Problems of modeling flow and heat transfer in porous media *Biuletyn Instytutu Techniki Cieplnej Politechniki Warszawskiej* **85**
[4] Sosnowski M, Krzywanski J, Grabowska K and Gnatowska R 2017 Polyhedral meshing in numerical analysis of conjugate heat transfer *Proceedings of International Conference on Experimental Fluid Mechanics (EFM)* **589-94**
[5] Breugem W P, van Dijk V and Delfos R 2014 Flows Through Real Porous Media: X-Ray Computed Tomography, Experiments, and Numerical Simulations *J. Fluids Eng.* **136(4)** 040902
[6] Szymanek E, Blaszczyk T, Hall M R, Dehdezi P K and Leszczynski J 2014 Modelling and analysis of heat transfer trough 1D complex granular system *Granular Matter* **16(5)** 687-694
[7] Mittal R and Iaccarino G 2005 Immersed boundary methods *Ann. Rev. Fluid Mech.* **37** 239-261
[8] Khadra K, Angot P, Parneix S and Caltagirone J P 2000 Fictitious domain approach for numerical modelling of Navier-Stokes equations *Int. J. Numer. Meth. Fluids* **34** 651-684
[9] Fadlun E A, Verzicco R, Orlandi P and Mohd-Yusof J 2000 Combined Immersed-Boundary Finite-Difference Methods for Three-Dimensional Complex Flow Simulations *J. Comp. Phys.* **161** 35-60
[10] Kadoch B, Kolomenskiy D, Angot P and Schneider K 2012 A volume penalization method for incompressible flows and scalar advection diffusion with moving obstacles *J. Comp. Phys.* **231** 4365-4383
[11] Tyliszczak A and Ksiezyk M 2018 Large eddy simulations of wall-bounded flows using a simplified immersed boundary method and high-order compact schemes *Int. J. Num. Meth. Fluids* **87** 358-381
[12] Fletcher C A J 1991 Computational Techniques for Fluid Dynamics *Springer-Verlag*
[13] Tyliszczak A 2014 A high-order compact difference algorithm for half-staggered grids for laminar and turbulent incompressible flows *J. Comp. Phys.* **276** 438-467
[14] Tyliszczak A 2016 High-order compact difference algorithm on half-staggered meshes for low Mach number flows *Comput. Fluids* **127** 131-145
[15] Suekane T, Yokouchi Y and Hirai S 2003 Inertial flow structures in a simple-packed bed of spheres *AIChE J.* **49** 10-17.

IOP Conf. Series: Journal of Physics: Conf. Series **1101** (2018) 012044 doi:10.1088/1742-6596/1101/1/012044

Discrete-continuous optimisation of an axial flow blood pump

K Tesch and K Kaczorowska-Ditrich
Gdansk University of Technology, ul. G. Narutowicza 11/12, 80-233 Gdansk, Poland
E-mail: krzyte@pg.edu.pl

Abstract. This paper presents results of discrete-continuous optimisation of an axial flow blood pump. Evolution Strategies (ES) are used as a global optimisation method in order to localise the optimal solution in relatively short time. The whole optimisation process is fully automated. This also applies to geometry modelling. Numerical simulations of the flow inside the pump is performed by means of the Reynolds-Average Navier-Stokes approach. All equations are discretised by means of the finite volume method and the corresponding algebraic equation systems are solved by the open source software for CFD, namely OpenFOAM. Finally, optimisations results are presented and discussed. The objective function to be maximised is simply pressure increase. The higher pressure increase the lower angular velocities required. This makes it possible to minimise the effect of haemolysis because it is mainly caused by high shear stress which are related, among others, to angular velocities.

1. Introduction

Since advanced medical treatment is usually not enough to prevent further decline of patients with heart failure, two treatments of patients with such a disease can be distinguished, namely heart transplantation and artificial heart blood pumps. The former is somewhat difficult because of relatively high costs and, what is even more important, lack of donor organs [5], not to mention organ rejection and mortality rates. The latter approach to heart failure problem, i.e. artificial heart pumps, gain popularity due to constantly improved design and features.

Several design of artificial hearts are know. The most complicated aim to replace the ailing heart whereas simpler pumps are design in order to support it. The supporting devices fall into three groups: LVAD (left ventricular assist device), RVAD (right ventricular assist device) and BIVAD (bi-ventricular assist device) [3]. Despite the fact that the human heart nature is pulsatile, continuous flow pumps are the most popular solutions. This is because of the simplicity reason and size of the device. Furthermore, we can distinguish centrifugal and axial blood pumps. The latter being smaller in comparison with the former. However, axial blood pumps require significantly higher angular velocities in order to increase outlet pressure. This may lead to blood damage, i.e. thrombosis and haemolysis [10, 9, 1] in particular. Haemolysis is mainly caused by high shear stress [2]. In order to minimise the effect of haemolysis an optimisation process of the axial pump is undertaken. The optimisation results leads to improvement in reduction of the wall shear stresses.

IOP Conf. Series: Journal of Physics: Conf. Series **1101** (2018) 012044 doi:10.1088/1742-6596/1101/1/012044

2. Geometry description

The whole geometry consists of a rotor and stator and is described by fourteen parameters (design variables) $\mathbf{x} = \{x_1, \ldots, x_{14}\}$ listed in table 1. First of all, the rotor blade shape is a non-linear helix given by

$$x(t) = R\cos t, \tag{1}$$
$$y(t) = R\sin t, \tag{2}$$
$$z(t) = f(t) \tag{3}$$

where R stands for the radius of rotor blades. Further, the parameter $t \in [0; 2\pi x_1]$ is related to rotor blade pitch $2\pi x_1$. The non-linearity $f(t)$ of the helix (1) is described by one parameter x_2 shown in figure 1.

Two of fourteen design variables are discrete, i.e. the number of rotor and stator blades. The former is named x_3 and the latter x_8. Next, the shape of the shaft is described by four points (x_4, x_5), (x_6, x_7) of the spline shown in figure 1. Stator blades are given by the so called camber line and the blade thickness. The former is represented by two variables (one points) (x_{11}, x_{12}) of the spline whereas the latter by two variables (two points) $(0, x_{13})$, $(1, x_{14})$, see figure 2. Finally, the two remaining variable x_9 and x_{10} are the stator blade upper and lower twist angles, respectively. The considered rotor radius R is 8 mm and the length of the shaft is 4.5 R.

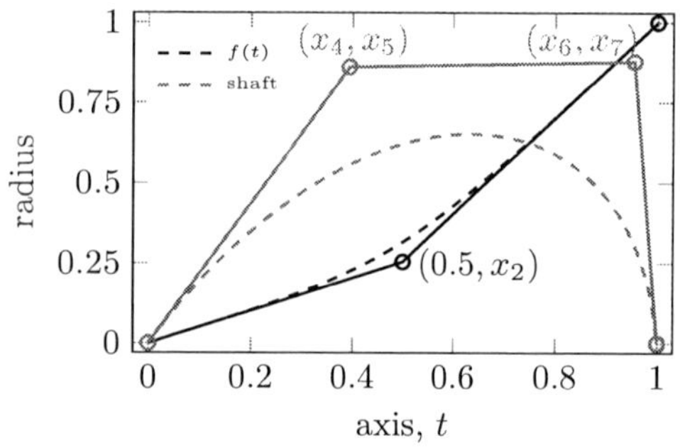

Figure 1. Shaft and helix splines.

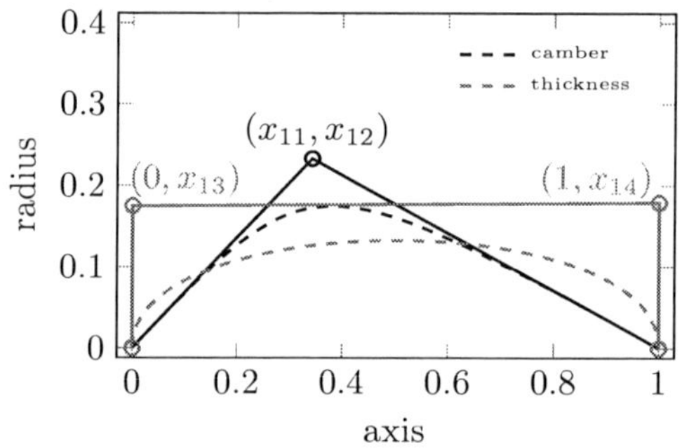

Figure 2. Stator splines.

The whole optimisation process is fully automated. This also applies to geometry modelling.

3. CFD

3.1. Governing equations and discretisation

Numerical simulations of the flow inside the pump is performed by means of the Reynolds-Average Navier-Stokes approach. A closed system of equations [14] for incompressible fluid and two additional transport equation for the two-equation SST [6] model is solved.

All the equations are discretised by means of the finite volume method and the corresponding algebraic equation systems are solved by the open source software for CFD, namely OpenFOAM [8]. Divergence schemes include both convection and other diffusive terms and involve Gauss integration. The discretised convection terms need to be interpolated by means of cell centred values because the values are located at the face centroids. Limited linear interpolation is used, being second order accurate. Further, the discretised diffusive terms involving surface normal gradients are evaluated at a cell face that connects two cells. In order to maintain second order accuracy for non-orthogonal meshes, apart from orthogonal schemes, a non-orthogonal correction is considered.

IOP Conf. Series: Journal of Physics: Conf. Series **1101** (2018) 012044 doi:10.1088/1742-6596/1101/1/012044

The SIMPLE algorithm is used in order to solve pressure-velocity coupling and the pressure equation is solved by means of GAMG solver with DIC smoother. For the velocity fields and turbulent quantities standard solvers using a GS smoother are utilised. Under-relaxation factors are used in order to improve stability of a solution. This is particularly important when solving steady-state flows. The assumed factors are 0.3 for pressure, 0.7 for velocity and 0.5 for turbulent quantities k and ω.

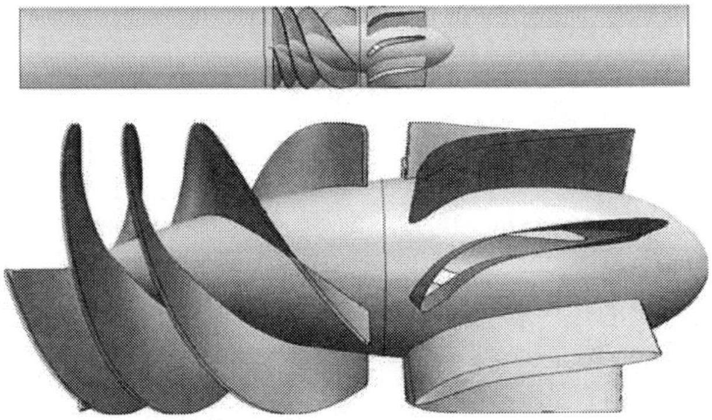

Figure 3. Flow domain (top) and optimal pump geometry (bottom).

3.2. Mesh and boundary conditions

The flow domain is divided into three parts, see figure 3. Apart from the rotating rotor, two additional steady pipes are considered. All three domains are discretised separately and merged by the so called arbitrary mesh interface. The total number of nodes is 1 553 806 and total number of volumes 1 378 033 where 1 269 555 of them are hexahedra. Furthermore, the mesh used may be classified as Cartesian, see figure 4.

Figure 4. Mesh.

In order to make certain that flow near walls is properly resolved thin layers around the physical walls are generated. The quality of the mesh near the walls is inspected in terms of the maximal y^+ values. Maximal values of y^+ are below 3 for all considered walls including blades.

The main boundary conditions are:

- Inlet. The specified constant volumetric flow rate $\dot{V} = 3$ dm/min is directed perpendicularly to inlet surface accompanied by the zero normal gradient pressure. Low turbulence intensity is also considered. This means that turbulence intensity is at the level of 1% and viscosity ration $\nu_t/\nu = 1$;
- Outlet. The constant pressure distribution is assumed here together with zero gradient velocity for flow out of the domain. This is because the outlet surface is located relatively far from the rotor;

- Walls. The so called no-slip conditions is applied meaning that impermeability and adhesion requirements are forced. Rotating wall velocity $n = 6000$ rev/min is considered in the rotating frame of reference. The flow in the near wall region is modelled by means of the scalable wall function;
- Interfaces. In order to allow for coupling between stationary pipes and rotating part of the pump the so called cyclic arbitrary mesh interfaces are considered. A steady state approach is used rather than full transient rotor-stator interaction. This is crucial for time consuming optimisation processes since CFD calculations need to be repeated hundred of times. This approach is commonly known as a multiple reference frame simulation;

4. Optimisation

4.1. General remarks

A discrete-continuous optimisation approach is needed to find the optimal shape. This is because two of fourteen design variables are discrete, namely the number of rotor and stator blades. Remaining variables such as the shaft, rotor and stator shapes as well as their thickness are continuous.

Evolution Strategies are used as a global optimisation method to localise the optimal solution in relatively short time [7, 11]. ES can be further classified as multi-point (population based) optimisation algorithms. Recombination, mutation and survival are adapted in order to evolve better solution. Most importantly, no additional information about the objective function is required. Additionally, randomisation is introduced through the probability of crossover and mutation. This makes it possible to efficiently explore the design space and escape local minima.

4.2. Objective function

The optimisation problem is to find a maximal pressure increase Δp. Since most optimisation algorithms are designed for minimisation of objective function, the pressure increase is considered with a minus sign

$$\Delta p_0 = \min_{\mathbf{x}\in\Omega\subseteq\mathbb{R}^D} \left(-\Delta p(\mathbf{x})\right). \tag{4}$$

Furthermore, the argument of the global minimum value of the objective function is expressed as

$$\mathbf{g} = \arg \min_{\mathbf{x}\in\Omega\subseteq\mathbb{R}^D} \left(-\Delta p(\mathbf{x})\right) \tag{5}$$

where $D = 14$ stands for the dimension of constraint space Ω or simply the so called optimisation domain. Thus, the objective function is subjected to box constraints listed in table 1

$$\Omega = \left\{\mathbf{x} \in \mathbb{R}^D : L_i \leq x_i \leq U_i\right\} \tag{6}$$

where L_i and U_i are lower and upper bounds, respectively. Box constraints are regarded as a special case of inequality constraints. This type of constraints is commonly met in optimisation problems and does not need any special treatment.

4.3. Algorithm

The ES algorithm in the pseudocode from [11] is shown in figure 5. Firstly, the initial population is randomly generated by means of a random variate of a continuous uniform distribution (lines 1-3). Lines 7 and 8 represent parent selection steps. This is achieved by means of tournament selection (lines 7 and 8) where $T = 3$ stands for the tournament size out of a parent population of N members. Once parents are selected, the arithmetical crossover is performed (line 11).

Input: p_c, p_m, T, N, n_{max}, $\mathbf{L}$, $\mathbf{U}$
Output: $\mathbf{g}$
1 **for** $i := 0$ **to** $N-1$ **do**
2     $\mathbf{x}_i := \mathbf{L} + (\mathbf{U} - \mathbf{L}) \circ \mathcal{U}(0,1)$;
3     $\mathbf{y}_i := \mathbf{0}$;
4 $\mathbf{g} := \arg\min_{\mathbf{x}_i} f(\mathbf{x}_i)$;
5 **for** $n := 1$ **to** $n_{max} - 1$ **do**
6   **for** $i := 1$ **to** $N-1$ **do**
7         $a :=$ Tournament $(\mathbf{x}, T)$;
8         $b :=$ Tournament $(\mathbf{x}, T)$;
9         $\mathbf{p}_1 := \mathbf{x}_a$;
10         $\mathbf{p}_2 := \mathbf{x}_b$;
11         $(\mathbf{c}_1, \mathbf{c}_2) :=$ Crossover $(\mathbf{p}_1, \mathbf{p}_2, p_c)$;
12         $\mathbf{y}_i :=$ Mutation $(\mathbf{c}_1, i, p_m)$;
13         $\mathbf{y}_{i+1} :=$ Mutation $(\mathbf{c}_2, i, p_m)$;
14         $i := i + 2$;
15     $\mathbf{l} := \arg\min_{\mathbf{x}_i} f(\mathbf{x}_i)$;
16     $\mathbf{g} := \arg\min \{f(\mathbf{g}), f(\mathbf{l})\}$;
17     $\mathbf{x} = \text{Selection}(\mathbf{x}, \mathbf{y})$;

Figure 5. Evolutionary strategy pseudocode.

Namely, two parents $\mathbf{x}_1$ and $\mathbf{x}_2$ are crossed with probability p_c. Two offspring vectors $\mathbf{y}_1$ and $\mathbf{y}_2$ are produced according to

$$\mathbf{y}_1 := a\,\mathbf{x}_1 + (1-a)\mathbf{x}_2, \tag{7a}$$

$$\mathbf{y}_2 := a\,\mathbf{x}_2 + (1-a)\mathbf{x}_1 \tag{7b}$$

where $a := \mathcal{U}(0,1)$. Next, the integer variables (the number of rotor and stator blades) are uniformly mutated (lines 12-13) if $\mathcal{U}(0,1) < p_m$ as opposed to floating point variables. The latter are non-uniformly mutated if $\mathcal{U}(0,1) < p_m$ according to

$$x_{ik} := \begin{cases} x_{ik} + (U_k - x_{ik})\,\Delta, & \text{if } \mathcal{U}\{0,1\} = 0; \\ x_{ik} - (x_{ik} - L_k)\,\Delta, & \text{otherwise} \end{cases} \tag{8}$$

where

$$\Delta := 1 - \mathcal{U}(0,1)^{\left(1-\frac{n}{n_{max}}\right)^2}. \tag{9}$$

In order to damp mutation as the algorithm approaches to an end the value of Δ should decrease.

The selection step (line 17) passes best solutions onto next generations. The so called $(\mu+\lambda)$ strategy is taken into consideration. Here μ denotes the total number of parent vectors $\mathbf{x}_i$ whereas λ is the number of offspring vectors $\mathbf{y}_i$. In this case $\mu = \lambda = N$. Simply, the considered strategy creates the next parent vector generation with the best μ vectors from the combined $(\mu+\lambda)$ parent and offspring population.

4.4. Results

The EA population size was set as $N = 20$ and the number of generations as $n_{max} = 20$. This results in 400 CFD calculations. Convergence is easily achieved after about 10 generations, see figure 6. The dashed line denotes the average pressure increase for the entire population, while solid line is the global maximum for the entire optimisation process.

Figure 7 shows how the mutation and the crossovers count varies during the optimisation process for the probabilities $p_m = 0.7$ and $p_c = 0.15$. Moreover, figure 8 makes it possible to monitor the population variability as the so called discrepancy is such a measure [13, 12]. Discrepancy D^* takes on values in the neighbourhood of 0 for random populations and close to 1 for uniform populations.

Table 1. Constraints of variables and optimal values.

Name	Constraints	Optimum	Description
x_1	$[0.2; 1]$	0.823	rotor blade pitch
x_2	$[0.1; 0.5]$	0.254	non-linearity of the helix
x_3	$\{2, \ldots, 4\}$	4	number of rotor blades
x_4	$[0; 0.5]$	0.393	two points of shaft's spline
x_5	$[0.7; 1]$	0.860	
x_6	$[0.7; 1]$	0.954	
x_7	$[0.7; 1.2]$	0.877	
x_8	$\{4, \ldots, 8\}$	8	number of stator blades
x_9	$[0; 45°]$	31.99°	stator blade upper twist angle
x_{10}	$[0; 30°]$	15.26°	stator blade lower twist angle
x_{11}	$[0.2; 0.6]$	0.342	point of stator shape blade spline
x_{12}	$[0; 0.3]$	0.232	
x_{13}	$[0.05; 0.3]$	0.174	point of stator thickness blade spline
x_{14}	$[0.05; 0.2]$	0.179	

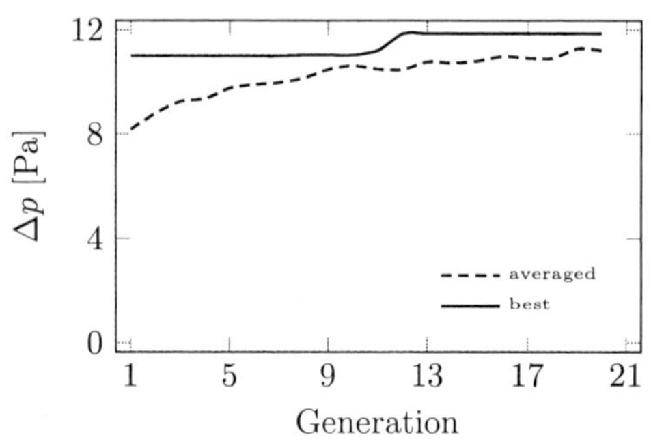

Figure 6. Convergence.

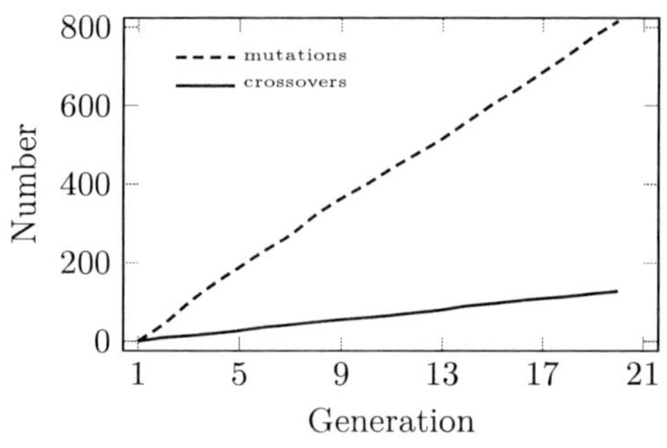

Figure 7. Mutations and crossovers.

The optimal values of **g** are listed in table 1. Lower part of figure 3 presents the optimal pump shape according to table 1. It may be observed that the optimal geometry consists of 4 rotor blades and 8 stator blades being the upper accessible ranges. As for the remaining variables, they are located inside the considered box constraints.

The average shape evolution for selected generations (iterations) of ES, reflecting convergence process, is presented in figure 9. It is interesting that the first shape (iteration 1) is simply an arithmetical average of purely random shapes. This is because the initial population is randomly generated. At the same time, the last shape (iteration 20) is similar to the optimal shape since

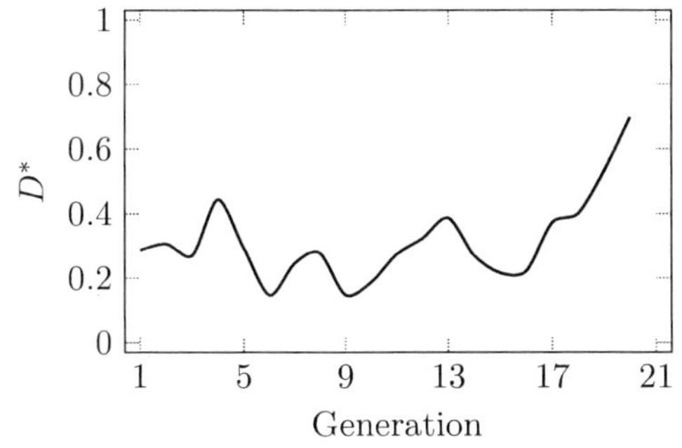

Figure 8. Discrepancy D^*.

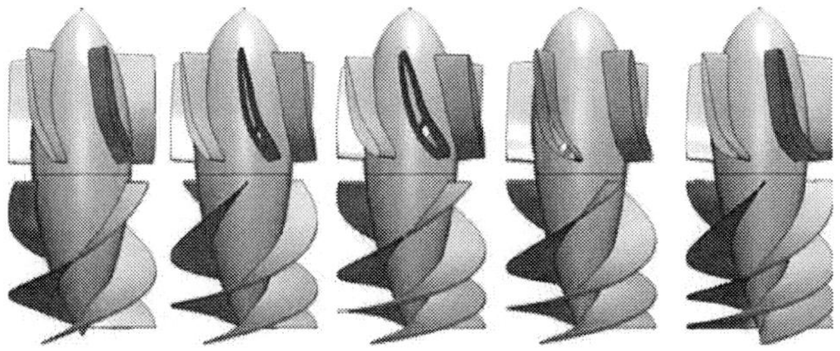

Figure 9. Average shape evolution (iteration number 1, 3, 5, 10 and 20).

the population here is nearly uniform according to the discrepancy value, see figure 8. Finally, since these shapes are arithmetical averages of individual generations, it should be noted that none of these has been subject to any CFD calculations.

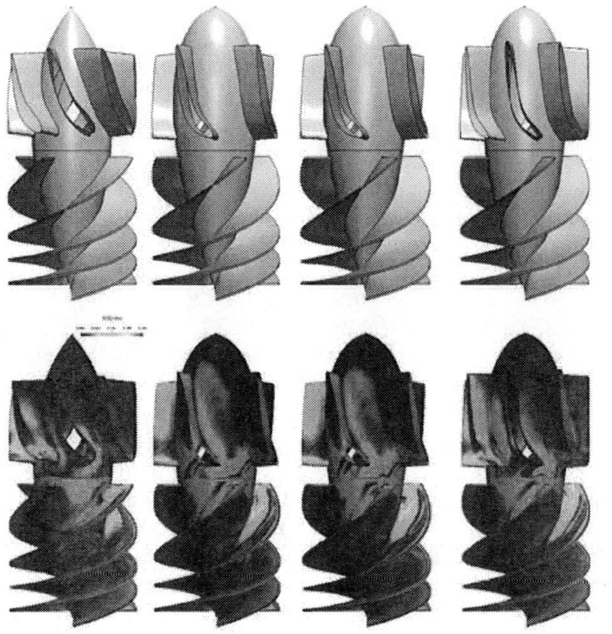

Figure 10. Best shape evolution and corresponding wall shear stresses distributions (iteration number 7, 10, 11 and 20).

Upper part of figure 10 displays the best shape evolution resulting from the shape optimisation process. The last shape (iteration 20) is the optimal solution shown also in figure 3. Interestingly, all shapes apart from first (iteration 7) are similar in terms of rotor configurations. What distinguishes the first geometry from the remaining three is the shape of the shaft and stator blades. Lower part of figure 10 presents corresponding wall shear stresses distributions. The higher the iteration number the smoother the distribution on the shaft and blades. One has to keep in mind, however, that high shear stresses are related primarily to angular velocities.

5. Conclusions

- Results of a discrete-continuous global optimisation process of an axial flow blood pump are presented. The optimal results are achieved by means of ES in relatively short time. Since 2 of 14 design variables are discrete (integer) special care of ES is necessary;
- A simple and effective method of geometry modelling is proposed. This makes it possible to make the whole optimisation process fully automated. There is no need for any intermediate CAD software since the geometry is created directly be means of GNU Octave [4] scripts;
- Figure 10 presents the wall shear stress distribution which is responsible for haemolysis. It may be observed that the highest values are localised, among other, on the rotor blade tips. The higher pressure increase the lower angular velocity required. Thus the optimisation process leads to improvement in reduction of the wall shear stresses and the effect of haemolysis. This is because haemolysis is mainly caused by high shear stress which are related, among others, to angular velocities;

References

[1] Aaronson K D et al. 2012 Use of an intrapericardial, continuous-flow, centrifugal pump in patients awaiting heart transplantation *Circulation* **125**(25) 3191
[2] Behbahani M at al. 2009 A review of computational fluid dynamics analysis of blood pumps *Euro. J. of Applied Mathematics* **20**(4) 363
[3] Carr C M et al. 2010 CT of Left Ventricular Assist Devices *RadioGraphics* **30**(2) 429
[4] Eaton J W et al. 2017 GNU Octave version 4.2.1 manual: a high-level interactive language for numerical computations, https://www.gnu.org/software/octave/doc/v4.2.1/
[5] Kafagy D and Gitano-Briggs H 2013 Design of axial blood pumps for patients with dysfunctional fontan physiology: computational studies and performance testing *Trends Biomater. Artif. Organs* **27**(3) 124
[6] Menter F R 1994 Two-equations eddy-viscosity turbulence models for engineering applications *AIAA-Journal* **32**(8)
[7] Michalewicz Z 1996 Genetic algorithms + data structures = evolution programs (Berlin, Heidelberg, New York: Springer)
[8] OpenFOAM user guide 2015, OpenFOAM Foundation Ltd.
[9] Rogers J et al. 2017 Intrapericardial left ventricular assist device for advanced heart failure *N Engl J Med* **376** 451
[10] Slaughter M et al. 2013 HeartWare ventricular assist system for bridge to transplant: combined results of the bridge to transplant and continued access protocol trial *J Heart Lung Transpl* **32**(7) 675
[11] Tesch K 2016 *Continuous Optimisation Algorithms* (Gdansk: GUT Publishers)
[12] Tesch K and Kaczorowska K 2016 Arterial cannula shape optimization by means of the rotational firefly algorithm *Engineering Optimization* **48**(3) 497
[13] Thiémard E 2001 An algorithm to compute bounds for the star discrepancy *Journal of Complexity* **17** 850
[14] Wilcox D C 1994 Turbulence modeling for CFD (California: DCW Industries)

IOP Conf. Series: Journal of Physics: Conf. Series **1101** (2018) 012045 doi:10.1088/1742-6596/1101/1/012045

Analysis of the fuel spray diversity in the opposed-piston engine

T Tulwin[1] and P Karpiński[1]

[1] Lublin University of Technology, Faculty of Mechanical Engineering, Department of Thermodynamics, Fluid Mechanics and Aviation Propulsion Systems, Nadbystrzycka 36 Str., 20-618 Lublin, Poland

t.tulwin@pollub.pl

Abstract. Experimental analysis of combustion in CI engines is difficult due to the very short duration of this process and the unfavourable thermodynamic conditions prevailing in the combustion chamber (high pressure and temperature). An alternative is to perform numerical calculations using the CFD method based on a defined combustion model. Such an approach was used in this paper to investigate the influence of different fuel injection configurations and different swirl ratio on selected parameters of the combustion process in the designed two-stroke opposed piston engine. The combustion chamber of the tested engine was formed of two movable surfaces of pistons. The AVL Fire software was used to perform the simulation. As a result of the calculations, the dependence of rate of heat release, burnt mass fraction, mean pressure in cylinder and heat transfer coefficient on defined calculation cases was obtained. In addition, a spatial distribution of the combustion stream of injected fuel was obtained. On the basis of the results obtained, conclusions regarding the impact of the fuel injection configuration and the swirl ratio on the obtained engine performance were presented.

1. Introduction

In CI engines a strong gradient of the air-fuel ratio inside the combustion chamber results in the formation of self-ignition sources and further propagation of the flame front. Numerous works have been written describing the formation and evolution of the fuel stream over time. An example is the paper [1], which presents the Eulerian-Lagrangian Spray Atomization (ELSA) model and the paper [2], which shows the fuel stream model based on the Lagrangian-Droplet and Eulerian-Fluid (LDEF) methodology. Furthermore, the nature of the combustion process is significantly influenced by the injection timing and the auto-ignition delay period, which should be as short as possible in order to avoid excessive pressure build-up during the kinetic combustion process. The authors of the paper [3] using the CFD method showed the relative influence of chemical mechanism on spray/flame structure in terms of ignition delay and also ignition and flame stabilization sites. In the paper [4] the phenomenon of the ignition delay period under steady and variable conditions for supercharged heavy-duty diesel engine was analysed on the basis of experimental results. It has been shown that under the established conditions the proposal presented by Watson in his paper [5] provides the same accuracy as the correlation presented in the paper. In the paper [6], CFD simulation of diesel engine operation was performed, in which the auto-ignition delay period was investigated depending on the cetane number of fuel and engine load. The obtained simulation results were compared with experimental results. The combustion process can be divided into three key stages: kinetic combustion, diffusion combustion and afterburning. The duration of the individual stages determines the nature of the combustion process. The parameters that characterize the combustion process are the rate and amount of heat dissipation and heat transfer coefficient. The analysis of these parameters in two-stroke engines is particularly important because of the high heat load of the piston crown and cylinder liner caused by the ignition of the mixture at each turn of the crankshaft. A commonly used method of investigating the combustion process is the CFD (Computational Fluid Dynamics) method. It is based on solving partial differential equations on the basis of an adopted model characterizing the combustion process and defined boundary conditions. This model should take into account the transport equation for the flow and the heat transfer equation. The unquestionable advantage of this method is the speed of obtaining results, low price and low level of complexity of tests compared to bench tests. An example of its application is the paper [7], in which the combustion process in an airplane star engine was studied. This method can also be used to analyze heat exchange, as shown in [8-10], where heat exchange in the aircraft engine cylinders was studied. An example of a work in which the CFD method was used to create a diesel engine model that takes into account diesel sprays, ignition and combustion is work [11]. To model the flow, the RNG k-ε turbulence model was used. The engine model parameters have been validated using a single-cylinder version of a heavy duty truck engine. In work [12] three-dimensional simulation tests of a diesel engine for different shapes of the combustion chamber were made. The resulting flow was for a limited number of measurements. As a result of the simulation, the air flow characteristics were obtained during the intake and the compression stroke. Another example of research in which the CFD method was used is the study presented in work [13], in which the combustion process was analyzed in a two-stroke opposed-piston diesel engine. For the defined geometry of the combustion chamber, the influence of the swirl intensity of the injected fuel and the diameter of the injector's hole on the distribution of the mixture inside the cylinder was investigated. It has been found that smaller diameters result in better distribution of fuel to droplets. In addition, greater turbulence improves the process of mixing fuel with air, but its too high value results in the creation of areas with a richer mixture. The optimal value of the diameter and the swirl coefficient for the tested engine was determined. In the paper [14] the effects of various ambient and injection parameters on the flame "lift-off" length have been investigated under typical diesel conditions in a constant-volume combustion vessel. As a result of the simulation, a fuel stream distribution was obtained for different injection pressures, nozzle hole sizes, ambient temperatures, and ambient densities. Very good convergence was obtained for liquid length and flame lift-off length between simulation results and experimental results. The purpose of the work was to examine how the fuel delivery method (various injection configurations) and the swirl intensity affect the selected parameters of the combustion process and the performance (average cylinder pressure and thermal efficiency) of the tested two-stroke opposed-piston engine. The combustion chamber model was created in the CATIA V5 program. Various types of injectors were modeled, differing in the number of holes, diameter and setting angle that were used in the combustion chamber model. Based on the created cases, simulation calculations were made in the FAME Hex module in the AVL Fire program.

2. Analysis and modelling

The subject of this study is a two-stroke reciprocating piston CI engine with an opposed cylinder system. The process of rinsing the cylinders and filling them with fresh charge was carried out with the use of a mechanical compressor Rotrex C30-64 combined with an intercooler. Diesel was assumed as the type of fuel. The basic technical characteristics of the engine under test are shown in Table 1, and in order to improve the filling of cylinders, the operating phases of crankshafts are different. The crankshaft on the inlet side is assumed to exceed the crankshaft on the outlet side by 14°. The research consisted of two parts. The first part examines how the method of fuel delivery affects the heat release coefficient, the combustion rate of the injected fuel and the average cylinder pressure as a function of the crank angle. The process of heat exchange with the walls was also analysed by comparing heat transfer coefficient (HTC) values and thermal efficiency was calculated for defined cases on the basis of a closed indicative diagram. Two single injectors and two double injectors were compared. In the second part of the study we analysed how the turbulence factor affects the parameters mentioned above. The calculation cases are shown in Table 2. The calculations are based on the crankshaft rotation angle in the range of 610-800° and the rotational speed of 4000 rpm. It is engine speed which corresponds to the nominal power.

These operating conditions can account for up to 50% of the engine's working time during the flight [16]. It is assumed that the fuel injection starts at a shaft rotation angle of 700° and ends at a shaft rotation angle of 730°. The time step at the time of fuel delivery was 0.2°CA. It was assumed that the temperature of the fuel supplied was 317 K. The weight of the fuel supplied was identical in all calculation variants. The basic technical parameters of the injectors used are presented in Table 3.

Table 1. Basic technical parameters of the tested engine.

Piston bore	65.5 mm
Piston stroke	76.4 mm
Compression ratio	24:1
Take-off power	100 kW (4200 rpm)
Maximum continuous power	86 kW (4000 rpm)

The size of the holes was selected so as to maintain the initial velocity of the fuel stream equal to approx. 350 m/s. The tests were performed for a charge swirl factor of 1 (swirl 4000 min^{-1} at 4000 rpm) and for an absolute pressure at the beginning of the simulation of 194 634 Pa. The area of active surface was approximately the same for all injector models and amounted to about 0.06 mm2. Initially, the air density was defined as 1.19 kg/m3 and the temperature was 428 K, turbulence kinetic energy was 30 m2/s2, turbulence length scale was 0.0013 m, and turbulence dissipation rate 20769.2 m2/s3.

Table 2. Calculation cases.

Variant	Compared injectors	Swirl ratio (SR)
1 a)	Single 4h and 8h	1
1 b)	Single 4h and double 2h	1
1 c)	Double 2h and double 3h	1
2	Double 2h	0.5, 1, 1.5, 3

It has been assumed that the mass of fuel injected during one cycle is 13.4 mg. On the basis of the defined injector characteristics, the program automatically scaled the mass flow characteristics as a function of the crank angle, so that the integral of the mass flow by the injection time was equal to the assumed mass of injected fuel during one cycle.

Table 3. Injector parameters.

Injector	Start of injection (°)	End of injection (°)	Injected fuel mass m_I (mg)	Active area (mm^2)	Number of holes	Diameter of holes (µm)	Angle between the holes (°)
Single, 4h	700	730	26.8	0.0615752	4	140	15
Single, 8h			26.8	0.0628318	8	100	30
Double, 2h			13.4	0.0615752	2	140	15
Double, 3h			13.4	0.0678584	3	120	15

To perform the research, a geometrical model of the combustion chamber in the CAD program was developed. The best volume to surface ratio has a spherical shape. Thus, the more spherical shape of the combustion chamber means less heat loss to the piston and cylinder walls. Therefore, the designed combustion chamber is characterized by a spherical shape and has curved asymmetrical edges to induce tumble flow. The asymmetrical shape of the transition from the flat gap to the spherical bowl of the combustion chamber results in transverse vortices. Transverse vortices are designed to limit the flow of hot gases into the gaps between the pistons. In addition, the use of radial turbulence results in a symmetrical and even distribution of the mixture and the propagation of the flame front. The shape and dimensions of the chamber are shown in Figure 1a. The basic technical parameters of the piston-cylinder system for the tested engine are shown in Table 4. As a combustion model, the 3-Zones Extended Coherent Flame Model (ECFM-3Z+) was adopted with the TKI (Tabulated Kinetic Ignition) auto-ignition model. The model allows to define the fuel with one chemical species or several components. Transport equations are solved for the averaged quantities of chemical species O_2, N_2, CO_2, H_2, H_2O, O, H, N, OH and NO. Average values of these quantities are the global values for the three mixing zones [15]. This combustion model includes post-flame chemistry treatment and the mechanisms of oxidation and formation of soot. The selected self-ignition model is based on a data table containing a cool flame heat releaase and auto-ignition progressive variable rates, which are linearly interpolated based on the amount of fresh gas. The k-ζ -f model, taking into account the total enthalpy and the phenomenon of compressibility, was assumed as the turbulence model. It is a model based on kinetic turbulence energy k, normalized velocity scale ζ and elliptic relaxation function f. The model of heat exchange through the walls has been defined as the Standard Wall Function Model. [15]

a) b)

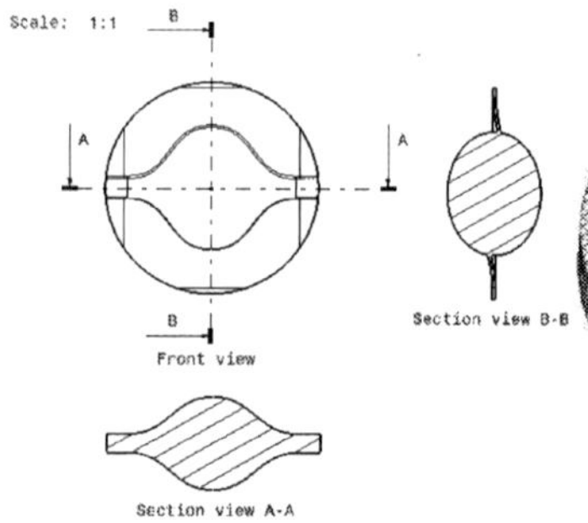

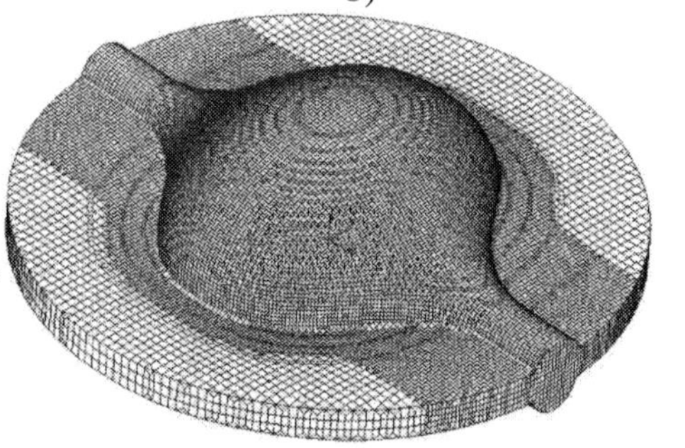

Figure 1. Section of the two planes of combustion chamber geometry at the mean top dead center of pistons (a) and calculation grid of the combustion chamber at 700 °CA (b).

Table 4. Basic technical parameters of the piston-cylinder system of the test engine.

Piston diameter	76 mm
Piston displacement	75 mm
Displacement volume	514 607 mm^3
Combustion chamber volume	22 562 mm^3
Compression ratio	23.81
Crankshaft rod length	125 mm
Crankshaft axes distance	326.6 mm
Crankshaft offset	14°

The calculated grid at the beginning of the fuel injection process (700 °CA) consisted of 403 524 elements is presented in Figure 1b. The model uses an adaptive mesh that automatically adapted the grid resolution to the defined injection configuration.

3. Results and discussion

Influence of the fuel injection method on the combustion process parameters

Figure 2a shows the rate of heat release as a function of the shaft rotation angle for various fuel delivery methods with a constant injection pressure and swirl ratio. The rate of heat release (ROHR) corresponds to the amount of heat released in the fuel combustion process in relation to the crank angle. The use of a single four-hole injector results in higher ROHR values during injection (710-725 °CA) compared to a single eight-hole injector. Starting from a shaft rotation angle of about 735 degrees, ROHR adopts higher values for the four-hole injector due to the release of more energy in the diffusion combustion process. The use of a double injector, both two-hole and three-hole, results in a significant increase (by nearly 50%) in the ROHR value immediately after ignition, which indicates better mixing of fuel with air, resulting in more energy being released during the combustion process. For the double-hole injector ROHR takes higher values in the range of shaft rotation angles 710-722° compared to the tri-hole injector. This is due to the fact that the fuel stream supplied by the two-hole injector creates a larger angle, and therefore the injected fuel is atomized more accurately. This translates into a larger combustion area and more efficient heat generation. Figure 2b shows the portion of fuel burned (BMF - burned mass fraction) as a function of the crankshaft rotation angle for the various fuel delivery modes. It was assumed that the volumetric ratio of all injected fuel into the combustion chamber was 4%. The use of a single eight-hole injector results in slower fuel combustion in the diffusion combustion phase (725-743°) than in the case of other injection methods. This is due to the shorter penetration of the injected fuel into the combustion chamber. A single four-hole injector provides comparable fuel combustion rate as a double-hole or three-hole injector. For all configurations,

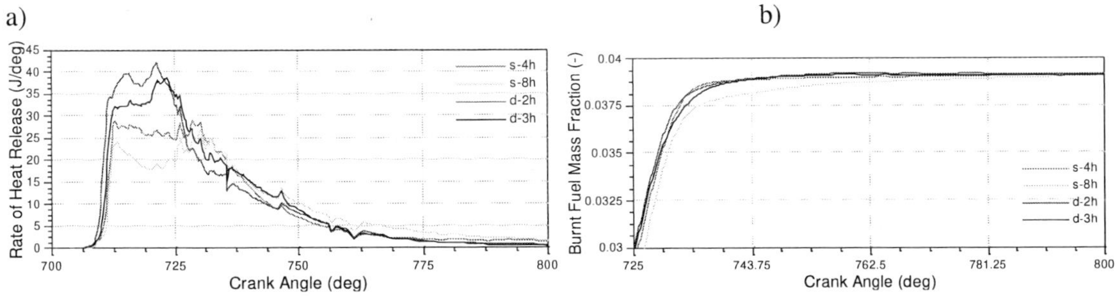

Figure 2. ROHR (a) and BFM (b) as a °CA function for defined injection configurations.

Figure 3a shows the average cylinder pressure as a function of the crankshaft rotation angle for the various fuel delivery modes. The highest average pressure in the cylinder equal to about 15 MPa was obtained using a double-hole injector. The use of single injectors: four- and eight-hole, resulted in a decrease of the maximum average pressure in the cylinder to about 9 and 13 MPa, respectively.

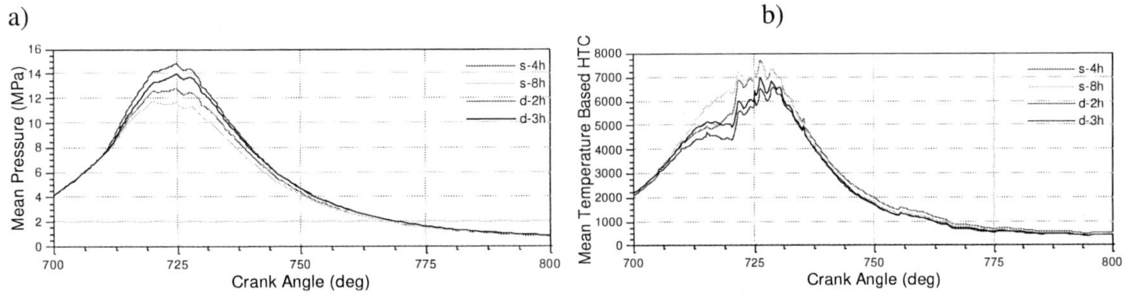

Figure 3. Average cylinder pressure (a) and HTC (b) as a function of °CA for the defined injection configurations.

Figure 3b shows the heat transfer coefficient (HTC) related to the average combustion chamber temperature as a function of the crankshaft rotation angle for the various fuel delivery modes. For the angle range of 710-720 °CA with a single 8-hole injector the HTC is higher than that of other injectors. The highest HTC value (approx. 7800 W/(m2K)) is obtained for single injectors at an angle of 728 °CA. HTC takes the lowest values in the whole CA angle range for the injection configuration d-2h. Starting at an angle of approx. 730°, the HTC progress is similar for all injection configurations. Figure 4 shows the iso-surface temperature defined as 8% by mass of fuel burned for 715 °CA and SR = 1 for defined injection configurations. In the case of the s-4h configuration, the combustion range of the injected fuel stream is greater than in the case of the s-8h configuration, where the flame reaches the symmetrical axis of the combustion chamber at an angle of 715 °CA, however, it is inclined towards the chamber wall. In all configurations the maximum combustion temperature on the defined iso-surface was about 2900 K. For the s-4h configuration, the high temperature area was located near the injection point, while for the s-8h configuration, this area was located closer to the centre of the chamber at the corrugated flame front. In the case of d-2h configuration there are four smaller streams of burnt fuel reaching the centre of the combustion chamber. Compared to the configuration with a single injector, the average temperature on a given surface was higher. For the d-3h injection configuration two opposite streams of combusted fuel were created, reaching the center of the geometric combustion chamber, formed of three smaller streams. A higher temperature occurred near the fuel injection points. The thermal efficiency for the s-4h injection configuration was assumed as the reference efficiency. Thermal efficiency for the s-8h, d-2h and d-3h configurations in comparison with the reference efficiency was 2.3% smaller, 5.3% higher, 4.7% higher, respectively.

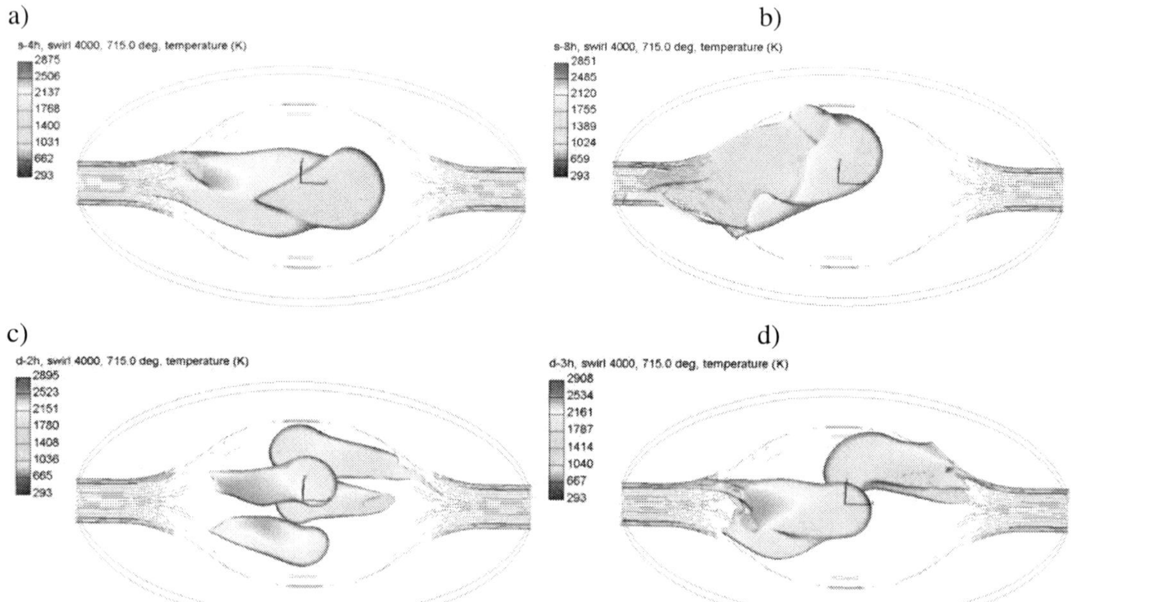

Figure 4. Iso-surface temperature defined as 8% by mass of the fuel burned criterion at 715°CA and SR = 1 for the defined injection configurations, isometric projection of the combustion chamber.

The influence of the swirl ratio on the parameters of the combustion process

The d-2h configuration provided the highest ROHR values, the highest average cylinder pressure, the smallest HTC values as a function of the crank angle and the highest thermal efficiency among the considered configurations. Therefore, on its basis, the second part of the calculations was performed, aimed at analyzing the influence of the swirl ratio on selected parameters of the combustion process. Figure 5a shows the rate of heat release as a function of the crank angle for the injection configuration d-2h for defined SR values. The low value of swirl results in a worse mixing of the fuel with the air, and thus a smaller ROHR in the kinetic combustion phase. The maximum ROHR value for SR = 0.5 is nearly 30% less than ROHR for SR = 3. Figure 5b shows the burnt fuel mass fraction as a function of the crank angle for the injection configuration d-2h for defined SR values. It was assumed that the mass fraction of all injected fuel into the combustion chamber is 4%. With increasing SR, the combustion rate of the fuel injected increases. However, for SR = 3, 98.5% of the total mass of the injected fuel was combusted, while for the remaining SR values, approx. 96.9%. Figure 6a shows the mean cylinder pressure as a function of the crank angle for the injection configuration d-2h for defined SR values. For SR = 0.5, the smallest maximum pressure value was 14 MPa. In addition, for this SR value the pressure curve is below the other curves, and therefore the indicated work for this case is the smallest. The highest values of the maximum pressure were obtained for SR = 1.5 and SR = 3, respectively 15.6 MPa and 15.8 MPa. Figure 6b shows the heat transfer coefficient related to the average temperature in the combustion chamber as a function of the crank angle for the injection configuration d-2h for defined SR values. A small increase in the turbulence, for example from SR = 0.5 to SR = 1.5, results in a small increase in the value of

HTC in the entire range of angles at which combustion occurs. Too much turbulence (SR = 3) results in a significant increase in the value of HTC in the second part of the injection (715-730 °CA). For the CA equal to 715° HTC is higher by almost half, therefore, thermal losses to the walls of the combustion chamber will increase significantly.

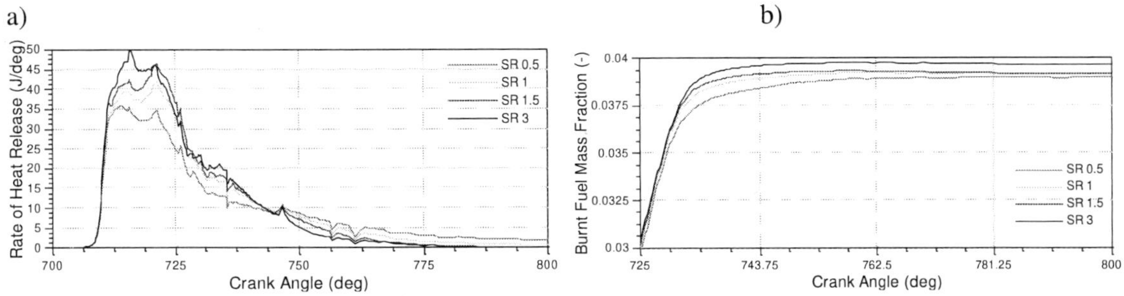

Figure 5. ROHR (a) and BFM (b) as a function of crank angle for the injection configuration d-2h for defined SR values.

Figure 7 shows the iso-surface temperature defined as 8% of mass of the fuel burned for the injection configuration d-2h at 715°CA and defined SR values. At low swirl (SR = 0.5), at the half of the injection time (715°CA) there is a joining of opposite streams of fuel burned. Increasing the SR to 1 results in a reduction of the contact surface of opposite streams. For SR = 1.5, four streams remain separated - there is no contact between them. For a large swirl (SR = 3) there is a significant deformation of the flame front and directing it towards the walls of the combustion chamber. Maximum temperatures in all analyzed cases amounted to approximately 2900 K. The temperatures in the considered iso-surfaces were higher in places further away from the geometrical center of the combustion chamber.

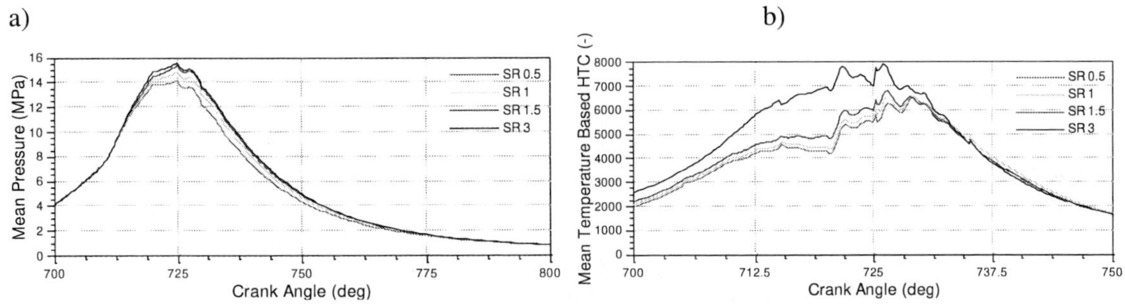

Figure 6. The average pressure in the cylinder (a) and HTC (b) as a function of crank angle for the injection configuration d-2h for defined SR values.

The thermal efficiency for SR = 1 was assumed as the reference efficiency. Thermal efficiency for the SR = 0.5, SR = 1.5 and SR = 3 configurations was 4.2% smaller, 3.9% higher, and 7.6% higher, respectively. The paper presents the results of numerical calculations using CFD method for investigating the combustion process in two-stroke opposed piston engine for defined injection configurations and defined values of swirl ratio. The most advantageous configuration from the point of view of achieved performance and thermal efficiency is d-2h configuration (double-hole injector). For this configuration, the highest thermal efficiency, the highest indexed pressure, the lowest heat transfer coefficient and the fastest rate of heat release were achieved. Single injector configurations do not ensure even combustion of the resulting mixture, which translates into lower average cylinder pressure, higher HTC and lower thermal efficiency. Too little turbulence (SR = 0.5) results in a significant decrease in heat release rate and thus a slower combustion rate of the injected fuel. This results in a lower average cylinder pressure. An increase in the swirl intensity (SR ≥ 1.5) results in faster heat release and thus an increase in the average cylinder pressure and a small increase in the HTC. Too much swirl intensity (SR = 3) causes the front of the flame to be directed towards the walls of the combustion chamber and a significant increase (even 50%) of the HTC value, which results in an increase in heat losses to the cooling system. Despite the high thermal efficiency for the SR = 3, the high value of HTC makes the case where the SR = 1,5 turbulence is more advantageous.

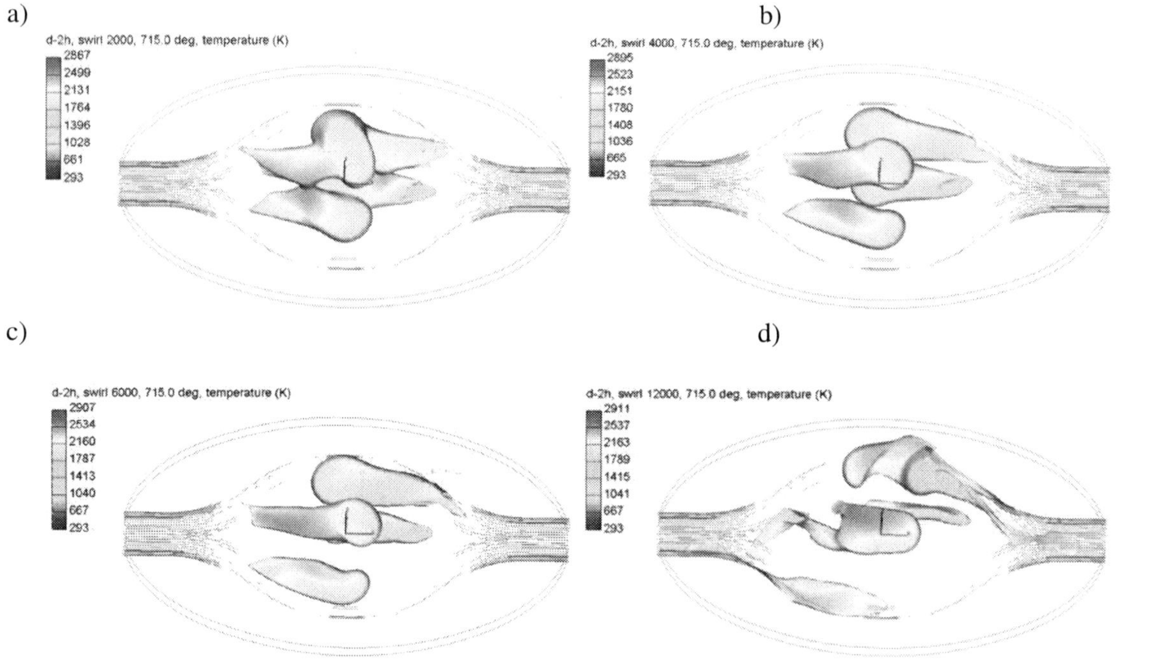

Figure 7. Temperature on the iso-surface defined as 8% of the mass of the fuel burned for the injection configuration d-2h for 715°CA and SR equal to: (a) 0.5, (b) 1, (c) 1.5, (d) 3, isometric view of the combustion chamber.

4. Conclusions

In further studies, it is possible to analyze geometric parameters (the position of injectors in relation to the geometrical center of the combustion chamber, the diameter of the holes, the angle between the holes) of the d-2h configuration for further optimization of thermal efficiency.

Acknowledgment
This work has been realized in the cooperation with The Construction Office of WSK "PZL-KALISZ" S.A." and is part of Grant Agreement No. POIR.01.02.00-00-0002/15 financed by the Polish National Centre for Research and Development.

References

[1] Hoyas S Pastor J Khuong-Anh D Mompó-Laborda J and Ravet F 2011 Application and Evaluation of the Eulerian-Lagrangian Spray Atomization (ELSA) Model on CFD Diesel Spray Simulations *SAE Technical Paper* 2011-37-0029
[2] Abani N Kokjohn S Park S W Bergin A M Ning W Sun Y and Reitz R D 2008 An Improved Spray Model for Reducing Numerical Parameter Dependencies in Diesel Engine CFD Simulations *SAE Technical Paper* 2008-01-0970
[3] Novella R García A Pastor J M and Domenech V 2011 The role of detailed chemical kinetics on CFD diesel spray ignition and combustion modelling *Mathematical and Computer Modelling* vol 54 no 7-

8 pp 1706-1719
[4] Assanis D N Filipi Z S Fiveland S B and Syrimis M 2003 A Predictive Ignition Delay Correlation Under Steady-State and Transient Operation of a Direct Injection Diesel Engine *J. Eng. Gas Turbines Power* vol 125 no 2 pp 450-457
[5] Watson N Pilley A D and Marzouk M 1980 A Combustion Correlation for Diesel Engine Simulation *SAE Technical Paper* 800029
[6] Huang M Gowdagiri S Cesari X M and Oehlschlaeger M A 2016 Diesel engine CFD simulations: Influence of fuel variability on ignition delay *Fuel* vol 181 pp 170–177
[7] Pietrykowski K 2014 The CFD Model of the Combustion Process in a Radial Engine *SAE Technical Paper* 2014-01-2578
[8] Grabowski L Czyz Z and Kruszczynski, K 2014 Numerical Analysis of Cooling Effects of a Cylinders in Aircraft SI Engine *SAE Technical Paper* 2014-01-2883
[9] Pietrykowski K and Tulwin T 2015 Aircraft Radial Engine CFD Cooling Model *SAE Int. J. Engines* vol 8 no 1 pp 82-88
[10] Tulwin T 2016 A coupled numerical heat transfer in the transient multicycle CFD aircraft engine model *Procedia Engineering* vol 157 pp 255-263
[11] Reitz R D and Rutland C J 1995 Development and testing of diesel engine CFD models. *Progress in Energy and Combustion Science* vol 21 no 2 pp 173-196
[12] Payri F Benajes J Margot X and Gil A 2004 CFD modeling of the in-cylinder flow in direct-injection Diesel engines *Computers & fluids* vol 33 no 8 pp 995-1021
[13] Zhang Z Zhao C Xie Z Zhang F and Zhao Z 2014 Study on the effect of the nozzle diameter and swirl ratio on the combustion process for an opposed-piston two-stroke diesel engine *Energy Procedia* vol 61 pp 542–546
[14] Senecal P K Pomraning E Richards K J Briggs T E Choi C Y McDavid R M Patterson M A 2003 Multi-Dimensional Modeling of Direct-Injection Diesel Spray Liquid Length and Flame Lift-off Length using CFD and Parallel Detailed Chemistry *SAE Technical Paper* 2003-01-1043
[15] AVL Fire v2014.2 - Combustion Module Manual
[16] Czyż Z Grabowski Ł Pietrykowski K Czarnigowski J Porzak M 2018 Measurement of flight parameters in terms of toxic emissions of the aircraft radial engine ASz62-IR *Measurement* vol 113 pp 46-52

Extended Bubble Cavitation Model to predict water hammer in viscoelastic pipelines

Kamil Urbanowicz, Mateusz Firkowski
West Pomeranian University of Technology Szczecin, Faculty of Mechanical Engineering and Mechatronics, Piastów Avenue 19, 70-310 Szczecin, Poland

E-mail: kamil.urbanowicz@zut.edu.pl, mateusz.firkowski@zut.edu.pl

Abstract. The equation of continuity describing the two-phase bubble flow was modified to take into account the influence of the viscoelasticity of the pipe walls. The modified equation of continuity together with the equation of motion was solved using the method of characteristics. Simulations carried out using in-house written program in Matlab indicate a high correspondence between simulated and experimental runs. The modified Shu model is characterized by simple construction, thanks to which it is easy to quickly implement this solution in commercial computer programs used to model transient states in pipes. It takes into account three basic phenomena accompanying these flows, namely: frequency-dependent hydraulic resistance, cavitation and viscoelastic effect off the pipe wall. Further work on its extension is underway.

1. Introduction

Today polymers are extensively used in pressurized pipe systems, due to their temperature, chemical and abrasion resistance, high pressure ratings, light weight, easy and fast installation, and low-price. The viscoelastic behaviour of polymers, which is characterized by a time-dependent strain, significantly influences the pressure response during transient events by attenuating the maximum or minimum pressure variations in the pipeline and by increasing the dispersion of the pressure wave [1]. Phenomena such as cavitation [2–4], frequency-dependent hydraulic resistance [5–8] or fluid structure interaction [9–11], commonly studied in metal conduits, are also found in plastic pipes.

In these pipes the equation of continuity is commonly used in the form presented by Rieutord and Blanchard [12, 13]. The equation of motion has the same form as for metal pipes. Güney [14–16] was first who solved this set of hyperbolic partial differential equations. The classical method of characteristics was used to obtain ordinary differential equations, and next they were numerically solved by means of a finite difference technique. The creep compliance $J(t)$ function was obtained from dynamic tests carried out on a Rheovibron apparatus. From the frequency dependence of $E'(\omega)$ and $E''(\omega)$ determined experimentally the values of creep compliance coefficients (J_i) and retardation times (T_i) was found. This experimentally obtained creep function for the polyethylene pipe, many researchers are using today [17–20]. The experimental results of the water hammer obtained by Güney at that time are also used by the researcher to validate they own computer programs [17–21].

Covas presented other interesting papers [1, 22–24]. She noticed that the nonlinear elastic behaviour of pipe walls is significant in plastic pipes and it cannot be ignored. The creep compliance function was approximated as in earlier Güney works by a generalized Kelvin-Voigt model. Covas assumed that, the fluid velocity is negligible compared with the pressure wave speed, this allowed to neglect all convective derivatives. The final simplified solution was based on a rectangular method of characteristics computational grid.

Both Güney and Covas solutions were obtained using the same technique, namely: the method of characteristics. However, one can find papers in which general equations were solved by other methods. Hadj-Taïeb and Hadj-Taïeb [17] neglected convective terms, and following the technique described by Payret and Taylor [25] solved the set of partial differential equations with use of the two-step Lax-Wendroff finite difference scheme. Other authors [26–28] extended 2-D Pezzinga's model [29] to simulate the nonlinear behaviour of pipe walls. Ferrante and Caponi [30, 31] implemented the viscoelastic component in a frequency domain model which reduces the computational time.

Only a few authors studied the transient flow in plastic pipes with additional effects. Authors of [18, 19, 32] investigated the influence of fluid structure interaction in plastic pipes, and in the papers [17, 33–35] the influence of cavitation on the flows in plastic pipes was investigated.

Hadj-Taïeb and Hadj-Taïeb [17] used the Henry's law to calculate the instantaneous liquid volume fraction when pressure fall below the vapour pressure p_v. The vapour phase was assumed to satisfy an ideal gas law. Consequently, the instantaneous mixture density was modelled.

Soares et al. [34,35] used popular discrete vapour cavity model (DVCM) based on the column separation hypothesis. The flow of liquid in the tube is instantaneously and completely separated by its vapour phase when the cavity is formed. Cavities were allowed to form at any of the computational sections if the pressure is computed to be below the vapour pressure. Pure liquid with a constant pressure wave speed was assumed to occupy the reach in between two computational sections. The absolute pressure in a cavity was set equal to the vapour pressure, then the upstream and downstream discharges at a cavity were computed ignoring mass transfer during cavitation. Having an instantaneous grid node discharges a cavitation volume were calculated with use of this model. Soares lumping the mass of free gas at computing sections also used the discrete gas cavity model (DGCM). In this model each isolated small volume of gas expands and contracts isothermally as the pressure varies, in accordance with the perfect gas law.

Keramat et al. [33] also studied the cavitational flow in plastic pipes using a DVCM model of column separation similar to [34], but with different numerical implementation. The preliminary conclusion was that column separation (even for the fast closure of a valve) can hardly result in pressures higher than the Joukowsky pressure. The second conclusion was that the simplest model of column separation, DVCM, although sensitive to the numerical implementation, provided acceptable results for cavitating flow in viscoelastic pipes. In addition to simulation research, it is also worth distinguishing experimental research on the occurrence of cavitation in polyethylene and PVC pipelines made by Mitosek [36–38].

Retarded strain in all discussed papers is calculated in the same way, namely from the convolutional integral, which is the product of the local derivative from pressure and the weighting function, which is a derivative of the creep function. Today, most authors [30, 39–47] use the Pezzinga-Scandura [48] and Covas [22–24] calibration method to determine coefficients describing the creep function to which experimental runs of water hammer in a conduit from a given type of polymer is needed. In this paper we will use creep function experimentally obtained by Güney.

Significant changes in the modelling approach can be noted in two recent works. Keramat et al. [21] developed a mathematical model taking the time dependency of Poisson's ratio into account for linear viscoelastic pipes. Poisson's ratio was written in terms of relaxation function and bulk modulus which is assumed to be constant. The relaxation function is obtained from creep function given as the viscoelastic property data of pipe material. This model it's recommended by its authors as more accurate than the conventional viscoelastic models but is much more complicated for numerical modelling. Ferrante and Caponi [30] presented the generalized Maxwell model which in the opinion of it authors performs slightly better than the well-known standard linear solid model, for both HDPE and PVC-O pipes.

In this work, which is the continuation of the previous two papers [20, 49], the cavitational bubble flow model [50] will be extended so that the influence of retarded strain is properly taken into account in polymeric pipes.

2. Mathematical model

Starting from the equation of motion in the form of:

$$\frac{\partial}{\partial t}\left(\rho_m v_m A\right)+\frac{\partial}{\partial x}\left(\rho_m v_m^2 A\right)+A\frac{\partial p}{\partial x}+\pi D\tau+g\rho_m A\sin\gamma=0, \tag{1}$$

where: t– time $[s]$, x – distance along the pipe $[m]$, A – pipe cross-sectional area $[m^2]$, v_m – mixture velocity $[m/s]$, p – pressure $[Pa]$, D – inner pipe diameter $[m]$, τ – wall shear stress $[Pa]$, g – acceleration due to gravity $[m/s^2]$, γ – pipe slope angle $[°]$, α – volumetric fraction of liquid phase $[-]$, ρ_v – density of vapour phase $[kg/m^3]$, ρ_l – density of liquid phase $[kg/m^3]$ and ρ_m – mixture density

$$\rho_m=\alpha\rho_v+(1-\alpha)\,\rho_l. \tag{2}$$

After differentiation, one comes to the following form:

$$\frac{\rho_m v_m}{A}\frac{dA}{dt}+\rho_m\frac{dv_m}{dt}+\rho_m v_m\frac{\partial v_m}{\partial x}+v_m\frac{d\rho_m}{dt}+\frac{\partial p}{\partial x}+\frac{4}{D}\tau+g\rho_m\sin\gamma=0. \tag{3}$$

In this paper, for simplicity, a horizontal pipeline ($\gamma=0$) was assumed, thus the last term on the left-hand side of the above equation will disappear. Then, using the continuity equations written separately for the gas phase:

$$\frac{\partial}{\partial t}\left(\rho_v(1-\alpha)A\right)+\frac{\partial}{\partial x}\left(\rho_v(1-\alpha)Av_v\right)=0, \tag{4}$$

where v_v – average velocity of vapour flow $[m/s]$. And liquid phase:

$$\frac{\partial}{\partial t}\left(\rho_l\alpha A\right)+\frac{\partial}{\partial x}\left(\rho_l\alpha Av_l\right)=0, \tag{5}$$

where v_l – average velocity of liquid flow $[m/s]$. By adding the equation (4) and (5), one will get:

$$\frac{\partial}{\partial t}\left(\rho_v(1-\alpha)A+\rho_l\alpha A\right)+\frac{\partial}{\partial x}\left(\rho_v(1-\alpha)Av_v+\rho_l\alpha Av_l\right)=0. \tag{6}$$

Now assuming that the gas phase has the same velocity as the liquid phase (non-slip flow) $v_v=v_l=v_m$:

$$\frac{\partial}{\partial t}A\rho_m+\frac{\partial}{\partial x}\left(Av_m\rho_m\right)=0. \tag{7}$$

After differentiation, one will get:

$$A\frac{d\rho_m}{dt}+\rho_m\frac{dA}{dt}+A\rho_m\frac{\partial v_m}{\partial x}=0. \tag{8}$$

Dividing equation (8) by $A\rho_m$:

$$\frac{1}{\rho_m}\frac{d\rho_m}{dt}+\frac{1}{A}\frac{dA}{dt}+\frac{\partial v_m}{\partial x}=0. \tag{9}$$

However, multiplying the equation (8) with v_m and dividing by A, one gets another useful form:

$$v_m \frac{d\rho_m}{dt} + \frac{\rho_m v_m}{A}\frac{dA}{dt} + v_m \rho_m \frac{\partial v_m}{\partial x} = 0, \tag{10}$$

what makes three terms disappear in the equation (3), and the resulting final form of the equation of motion is as follows:

$$\rho_m \frac{dv_m}{dt} + \frac{\partial p}{\partial x} + \frac{2}{R}\tau = 0. \tag{11}$$

From [23, 51] it is known that in plastic pipes:

$$\frac{1}{\rho_l}\frac{d\rho_l}{dt} = \frac{1}{\varepsilon_l}\frac{dp}{dt} \text{ and } \frac{1}{\rho_v}\frac{d\rho_v}{dt} = \frac{1}{\varepsilon_v}\frac{dp}{dt} \text{ and } \frac{1}{A}\frac{dA}{dt} = \frac{D\theta}{Ee}\frac{dp}{dt} + 2\frac{d\varepsilon_r}{dt}. \tag{12}$$

The total derivative of ρ_m is calculated from the following equation:

$$\frac{d\rho_m}{dt} = \frac{\partial \rho_m}{\partial t} + v_m \frac{\partial \rho_m}{\partial x}. \tag{13}$$

Substituting ρ_m from (2) and after differentiation:

$$\frac{d\rho_m}{dt} = \alpha \frac{d\rho_l}{dt} + (\rho_l - \rho_v)\frac{d\alpha}{dt} + \frac{d\rho_v}{dt}. \tag{14}$$

Introducing solution (14) into (9) and using the last relationship presented in (12), one will get:

$$\left[\rho_m \frac{D\theta}{Ee} + \frac{\alpha\rho_l}{K_l} + \frac{(1-\alpha)\rho_v}{K_v}\right]\frac{dp}{dt} + (\rho_l - \rho_v)\frac{d\alpha}{dt} + 2\rho_m \frac{d\varepsilon_r}{dt} + \rho_m \frac{\partial v_m}{\partial x} = 0. \tag{15}$$

As in the method of characteristics the pressure wave speed c must be constant, the assumption in this work (also J.J. Shu assumed it) is that it value will be calculated as pre-transient. At that moment, only liquid flow occurs so for starting time $\alpha = 1$, and the expression in bracket of above equation reduces to:

$$c^{-2} = \left[\rho_l \left(\frac{D\theta}{Ee} + \frac{1}{K_l}\right)\right]. \tag{16}$$

Then, finally the equation of continuity has such form:

$$\frac{1}{c^2}\frac{dp}{dt} + (\rho_l - \rho_v)\frac{d\alpha}{dt} + 2\rho_m \frac{d\varepsilon_r}{dt} + \rho_m \frac{\partial v_m}{\partial x} = 0. \tag{17}$$

In non-slip flows, the share of the dispersed phase of a statistical nature, i.e. the volume α and mass c_m concentration, are equal to the relevant proportions of dynamic nature, i.e. transport concentration and the degree of dryness [52]. The following relationship applies then: $v_m = \frac{v}{\alpha}$, where v is superficial velocity of liquid phase.

From paper [20] it follows that:

$$\frac{\partial \varepsilon_r(t)}{\partial t} = \frac{\theta D}{2e}\int_0^t \frac{\partial}{\partial t}\left(p(t-u) - p(0)\right) \cdot \left(\sum_{i=1}^{k} \frac{J_i}{T_i} e^{-\frac{u}{T_i}}\right) du, \tag{18}$$

where: ε_r – retarded strain $[-]$, e – pipe wall thickness $[m]$, θ – parameter reporting support condition of pipe $[-]$, J_i – creep compliance of the i-th spring of the Kelvin-Voigt element $[Pa^{-1}]$,

T_i – retardation time of the dashpot of i-th element $[s]$. Recognizing the analogy of the above equation with the convolution integral representing the wall shear stress, let us write:

$$\sum_{i=1}^{k} \frac{J_i}{T_i} e^{-\frac{u}{T_i}} = w_J(u). \tag{19}$$

Equations (17) and (11), after taking into account the formulas for the wall shear stress and the derivative of the retarded strain, form the following system:

$$\begin{cases} \frac{1}{c^2}\frac{dp}{dt} + (\rho_l - \rho_v)\frac{d\alpha}{dt} + \rho_m \frac{\partial}{\partial t}\left(\frac{v}{\alpha}\right) + \rho_m \Xi \int_0^t \frac{\partial p}{\partial t}(u) \cdot w_J(t-u)du & = & 0, \\ \rho_m \frac{d}{dt}\left(\frac{v}{\alpha}\right) + \frac{\partial p}{\partial x} + \frac{2\rho_m}{R}\left(\frac{f v|v|}{8\alpha^2} + \frac{2\mu_m}{\rho_m R}\int_0^t \frac{d}{dt}\left(\frac{v}{\alpha}(u)\right) \cdot w_{UF}(t-u)\right) & = & 0, \end{cases} \tag{20}$$

where: $\Xi = \frac{2R}{e}\theta$ – enhanced θ parameter $[-]$, w_J – creep weighting function $\left[Pa^{-1} \cdot s^{-1}\right]$, w_{UF} – unsteady friction weighting function $[-]$, μ_m – Dukler's [53, 54] two-phase mixture dynamic viscosity $[Pa \cdot s]$. Using the numerical solution of convolution integral presented by Schohl [55] to solve equation (18), we obtain:

$$\frac{\partial \varepsilon_r}{\partial t}(t+\Delta t) \approx \frac{D\theta}{2e}\sum_{i=1}^{k} \underbrace{\left(x_i(t) \cdot e^{-\frac{\Delta t}{T_i}} + \frac{J_i}{\Delta t}\left[1 - e^{-\frac{\Delta t}{T_i}}\right]\left(p_{t+\Delta t} - p_{(t)}\right)\right)}_{x_i(t+\Delta t)}, \tag{21}$$

where Δt – a constant time step in method of characteristics $[s]$. The above equation can be written in the following simple form:

$$\frac{\partial \varepsilon_r}{\partial t}(t+\Delta t) = p_{(t+\Delta t)}F - G(t), \tag{22}$$

$$F = \frac{D\theta}{2e}\sum_{i=1}^{k} M_i,\ G(t) = \frac{D\theta}{2e}\sum_{i=1}^{k}\left(M_i \cdot p_{(t)} - x_i(t) \cdot N_i\right),\ M_i = \frac{J_i}{\Delta t}\left[1 - e^{-\frac{\Delta t}{T_i}}\right],\ N_i = e^{-\frac{\Delta t}{T_i}}. \tag{23}$$

The solution of the convolutional integral describing the wall shear stress is as follows [5, 56]:

$$\begin{aligned} \tau(t+\Delta t) \approx\ & \frac{\rho_m f v_{(t)}|v_{(t)}|}{8\alpha_{(t)}^2} \\ + & \frac{2\mu_m}{R}\sum_{i=1}^{3} \underbrace{\left[A_i y_i(t) + \eta B_i\left[\frac{v_{(t)}}{\alpha_{(t)}} - \frac{v_{(t-\Delta t)}}{\alpha_{(t-\Delta t)}}\right] + [1-\eta]\, C_i\left[\frac{v_{(t-\Delta t)}}{\alpha_{(t-\Delta t)}} - \frac{v_{(t-2\Delta t)}}{\alpha_{(t-2\Delta t)}}\right]\right]}_{y_i(t+\Delta t)} \end{aligned} \tag{24}$$

where:

$$\eta = \frac{\int_0^{\Delta \hat{t}} w_{class.}(u)du}{\int_0^{\Delta \hat{t}} w_{eff.}(u)du},\ A_i = e^{-n_i \Delta \hat{t}},\ B_i = \frac{m_i}{\Delta \hat{t} n_i}\left[1 - A_i\right],\ C_i = A_i B_i. \tag{25}$$

The system of partial differential equations (20) was solved in this work using the method of characteristics and finite differences method:

$$\begin{cases} \frac{p_D}{c\rho_l} - \frac{\kappa p_v}{c\rho_l} + \frac{v_D}{\alpha_D} + \frac{c}{2}\ln\frac{\rho_m D}{\rho_l} + p_D c \Xi F \Delta t & = & C_A, \\ -\frac{p_D}{c\rho_l} + \frac{\kappa p_v}{c\rho_l} + \frac{v_D}{\alpha_D} - \frac{c}{2}\ln\frac{\rho_m D}{\rho_l} - p_D c \Xi F \Delta t & = & C_B, \end{cases} \tag{26}$$

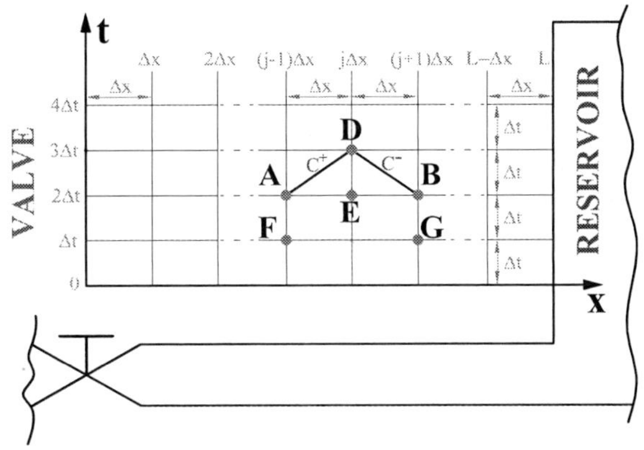

Figure 1. Rectangular grid in method of characteristics

where:

$$\begin{cases} C_A &= \frac{v_A}{\alpha_A} + \frac{p_A - \kappa p_v}{c\rho_l} - \frac{f_A \Delta t v_A |v_A|}{4R\alpha_A^2} + \frac{c}{2}\ln\frac{\rho_{m_E}\rho_{m_F}}{\rho_l \rho_{m_A}} + G_E(t)\Xi c\Delta t, \\ C_B &= \frac{v_B}{\alpha_B} - \frac{p_B - \kappa p_v}{c\rho_l} - \frac{f_B \Delta t v_B |v_B|}{4R\alpha_B^2} - \frac{c}{2}\ln\frac{\rho_{m_E}\rho_{m_G}}{\rho_l \rho_{m_B}} - G_E(t)\Xi c\Delta t \end{cases} \tag{27}$$

and $\kappa = 1 + c^2\rho_l \Xi F \Delta t$. From the above equations, solutions for the internal node of the characteristic grid were obtained (Fig. 1) for averaged cross-section velocity:

$$v_D = \frac{\alpha_D (C_A + C_B)}{2}, \tag{28}$$

and for the instantaneous pressure value:

$$p_D = \frac{(C_A - C_B)\, c\rho_l}{2\kappa} + p_v - \frac{c^2\rho_l}{2\kappa}\ln\frac{\rho_{m_D}}{\rho_l}. \tag{29}$$

From the analysis of the above formulas it follows that in order for $p_D > p_v$ (then $\rho_{m_D} = \rho_l$), the condition that $C_A \geq C_B$ must be satisfied. When $C_A \geq C_B$, then $\alpha_D = 1$ and $p_D = \frac{(C_A - C_B)c\rho_l}{2\kappa} + p_v$. And if $C_A < C_B$ cavitation occurs, then $p_D = p_v$ and mixture density is calculated $\rho_{m_D} = \rho_l e^{\left(\frac{C_A - C_B}{c}\right)}$. Having the instantaneous value of the density of the mixture ρ_{m_D}, the vapor density ρ_v and the density of the liquid ρ_l, one should determine from the equation (2) the instantaneous value of the liquid phase concentration α_D:

$$\alpha_D = \frac{\rho_l e^{\left(\frac{C_A - C_B}{c}\right)} - \rho_v}{\rho_l - \rho_v}. \tag{30}$$

3. Exemplary simulation

In this section, exemplary simulation studies will be carried out. Received results will be compared with the experimental results presented by Güney [14–16]. The experimental setup build in INSA Lyon (France) was composed of a horizontal low-density polyethylene test pipe (LDPE, Poisson's ratio $\nu_P = 0.38[-]$), a constant pressure reservoir and a quick-closing piston valve at the downstream end of the pipe (Fig. 2). The pipe length was $L = 43.1$ $[m]$, inner diameter $D = 0.0416$ $[m]$, wall thickness $e = 0.0042$ $[m]$, and characteristic polymer pipe roughness size $k_s = 1.5 \cdot 10^{-6}$ $[m]$ were assumed. The fluid was water with temperature 13.8 $[°C]$, density 999.3 $[kg/m^3]$ and kinematic viscosity $\nu = 1.17 \cdot 10^{-6}$ $[m^2/s]$. Initial steady-state flow ($v_0 = 1.28$ $[m/s]$) was determined using a simple method involving measuring volumes of water collected in a fixed time. It is not clear, as the author of the experimental setup do not mention about it, but according to the schematic diagram (Fig. 2) it looks like that in all cases the liquid flow into the atmosphere. The absolute reservoir pressure can be then derived from following formula:

$$p_R = p_{atm.} + hg\rho, \tag{31}$$

where g – acceleration due to gravity $[m/s^2]$ and according to table from Güney's PhD [14] h=2.88 [m]. The experimental uncertainty was estimated within ± 3 $[kPa]$ for pressure and $\pm 1\%$ for the initial flow velocity. In analyzed case the closure time of the valve was $T_c = 0.012$ $[s]$. The creep function is an filtered experimental function characterized by following creep and retardation time coefficients: ($J_1 = 0.637 \cdot 10^{-9}$, $J_2 = 0.871 \cdot 10^{-9}$) $[Pa^{-1}]$ and ($T_1 = 0.0166$, $T_2 = 1.747$) $[s]$.

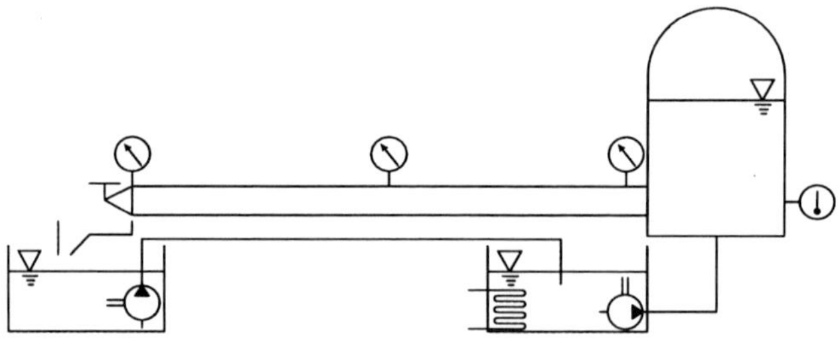

Figure 2. Schematic diagram of Guney's experimental test stand

Due to the fact that the Reynolds number in the stationary flow preceding the unsteady state was $Re_0 = 45511$ $[-]$, in this study examined will be the effect of calculated resistance. Four cases were examined:

a) weighting function calculated for laminar flow (the shape of the filtered simplified function similar to the classic weighting function according to Zielke [57], the coefficients were calculated using the analitycal method described in [7]),
b) weighting function calculated for turbulent flow (for initial condition, coefficients were scaled using the procedure discussed in the paper [58]),
c) variable values of weighting function coefficients (constant values during the single dimensionless time step, followed by a step change to the new values calculated on the basis of the instantaneous Reynolds number),
d) resistances calculated from quasi-steady formula (convolutional integral not taken into account).

In all numerical simulations the pipeline was divided for 64 equally long elements ($N = 64$ $[-]$, $\Delta x = 0.6734$ $[m]$). With that assumption and knowing the value of calculated pressure wave speed $c = 300$ m/s, a constant time step were calculated $\Delta t = 2.245 \cdot 10^{-3}$ $[s]$ ($\Delta\hat{t} = 6.07 \cdot 10^{-6}$ $[-]$). The graphical representations of simulated transient states are presented on Fig. 3 and Fig. 4.

From the results shown (Fig. 3 and Fig. 4) it can be seen that:

- transient cavitation runs are simulated with a modified bubble cavitation model with sufficient accuracy,
- in the runs taking into account the frequency-dependent hydraulic resistance, the selection of the weighting function necessary to calculate the values of frequency-dependent resistances were not crucial (Fig. 4),

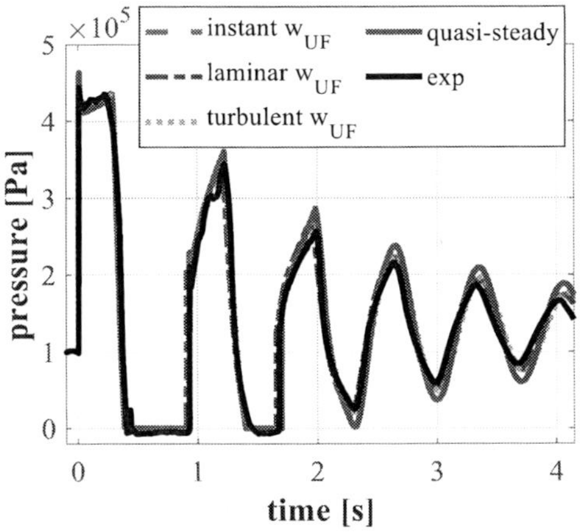

Figure 3. Comparison of simulated and experimental results ($Re_0 = 45511$)

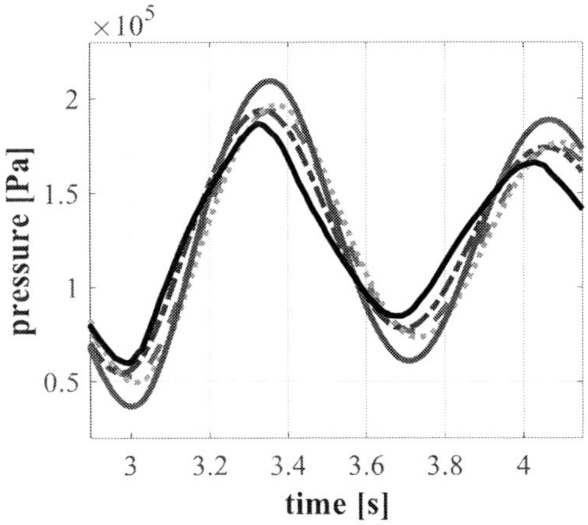

Figure 4. Enlarge of selected time range

- the simplest weighting function, which is nominally dedicated to laminar flow, gived the best results,
- the use of a quasi-steady friction model resulted in results that were clearly different from those simulated (Fig. 4)

Simulation results obtained for other initial conditions will be presented during the conference.

4. Conclusions

The presented modified bubble cavitation flow model, which takes into account the retarded deformation of the plastic pipe walls, very accurately models the analyzed transient states. The numerical simulation carried out showed that the inclusion of frequency-dependent hydraulic resistance approaches the simulation results to the experimental one. However, the significance of frequency-dependent friction is much smaller than the creep function. Its derivative determined in this work as a weighting function w_J has a fundamental influence on the modeled pressure changes runs.

New modified model is characterized by simple construction, thanks to which it is easy to quickly implement this solution in commercial computer programs used to model transient states in pipes. It takes into account three basic phenomena accompanying these flows, namely: frequency-dependent hydraulic resistance, cavitation and viscoelastic effect off the pipe wall material.

Due to the fact that two other models of cavitation flow in plastic pipes are known from the literature, namely the discrete vapor cavity model (DVCM) and the discrete gas cavity model (DGCM), a comparative analysis of all these models is necessary.

References

[1] Covas D, Stoianov I, Ramos H, Graham N, Maksimovic C and Butler D 2004 *Urban Water J.* **1** 177
[2] Zarzycki Z and Urbanowicz K 2006 *Chem. Process Eng.* **27** 915
[3] Adamkowski A and Lewandowski M 2009 *J. Fluids Eng.* **131** 071302
[4] Adamkowski A and Lewandowski M 2012 *J. Hydraul. Eng.* **138** 935
[5] Urbanowicz K 2015 *Proc. 12th Int. Conf. Press. Surges* 113
[6] Urbanowicz K and Zarzycki Z 2015 *J. Theor. App. Mech.-Pol.* **53** 295
[7] Urbanowicz K 2017 *J. Theor. App. Mech.-Pol.* **55** 1029
[8] Urbanowicz K 2017 *Pol. Marit. Res.* **24** 68

[9] Henclik S 2014 *J. Phys.: Conf. Ser.* **530** 012034
[10] Henclik S 2018 *J. Sound Vib.* **417** 245
[11] Henclik S 2018 *J. Fluids Struct.* **76** 469
[12] Rieutord E and Blanchard A 1972 *C. R. Acad. Sci. Paris* **274** 1963
[13] Rieutord E and Blanchard A 1979 *J. Hydraul. Res.* **17** 217
[14] Güney M S 1977 *Contribution à l'étude du phénomène de coup de bélier en conduite viscoélastique* Ph.D. thesis Université de Lyon I
[15] Gally M, Güney M S and Rieutord E 1979 *J. Fluids Eng.* **101** 495
[16] Güney M S 1983 *Proc. 4th Int. Conf. Press. Surges* 189
[17] Hadj-Taïeb L and Hadj-Taïeb E 2009 *Int. J. Model. Simul.* **29** 206
[18] Achouyab E and Bahrar B 2011 *C. R. Mecanique* **339** 262
[19] Bahrar B, Rieutord E and Morel R 1998 *La Houille Blanche* **1** 26
[20] Urbanowicz K and Firkowski M 2018 *Springer Proc. Math. Stat.* **249** DOI: 10.1007/978-3-319-96601-4_35
[21] Keramat A, Kolahi A G and Ahmadi A 2013 *J. Fluids Struct.* **43** 164
[22] Covas D 2003 *Inverse transient analysis for leak detection and calibration of water pipe systems modelling special dynamic effects* Ph.D. thesis Imperial College of Science, Technology and Medicine (University of London)
[23] Covas D, Stoianov I, Ramos H, Graham N and Maksimovic C 2004 *J. Hydraul. Res.* **42** 517
[24] Covas D, Stoianov I, Ramos H, Graham N and Maksimovic C 2005 *J. Hydraul. Res.* **43** 56
[25] Peyret R and Taylor T D 1983 *Computational Methods for Fluid Flow* (Springer-Verlag Berlin Heidelberg)
[26] Duan H F, Ghidaoui M, Lee P J and Tung Y K 2010 *J. Hydraul. Res.* **48** 354
[27] Shamloo H and Mousavifard M 2015 *J. Fluids Eng.* **137** 111203
[28] Wahba E M 2017 *J. Fluids Struct.* **68** 113
[29] Pezzinga G 1999 *J. Hydraul. Eng.* **125** 676
[30] Ferrante M and Capponi C 2017 *J. Hydraul. Res.* **55** 1
[31] Ferrante M and Capponi C 2018 *J. Hydroinform.* **20** 1
[32] Keramat A, Tijsseling A S, Hou Q and Ahmadi A 2012 *J. Fluids Struct.* **28** 434
[33] Keramat A, Tijsseling A S and Ahmadi A 2010 *IOP Conf. Ser.: Earth Environ. Sci.* **12** 012081
[34] Soares A K, Covas D, Ramos H M and Reis L F R 2009 *Int. J. Fluid Mach. Syst.* **2** 269
[35] Soares A K, Covas D and Carriço N J G 2012 *J. Hydraul. Res.* **50** 228
[36] Mitosek M 1997 *Plast. Rub. Compos. Pro.* **26** 324
[37] Mitosek M and Roszkowski A 1998 *Plast. Rub. Compos. Pro.* **27** 436
[38] Mitosek M 2000 *J. Hydraul. Eng.* **126** 904
[39] Soares A K, Covas D and Reis L F R 2008 *J. Hydraul. Eng.* **134** 1389
[40] Bergant A, Hou Q, Keramat A and Tijsseling A S 2011 *Proc. 4th Int. Meet. Cavitation Dyn. Probl. Hydraul. Mach. Syst.* 27
[41] Meniconi S, Brunone B and Ferrante M 2012 *J. Fluids Struct.* **33** 44
[42] Apollonio C, Covas D, de Marinis G, Leopardi A and Ramos H M 2014 *Urban Water J.* **11** 160
[43] Bergant A, Hou Q, Keramat A and Tijsseling A S 2013 *J. Hydraul. Struct.* **1** 23
[44] Pezzinga G, Brunone B, Cannizzaro D, Ferrante M, Meniconi S and Berni A 2014 *J. Hydraul. Eng.* **140** 04014036
[45] Keramat A and Haghighi A 2014 *J. Hydraul. Eng.* **140** 04014058
[46] Weinerowska-Bords K 2015 *J. Fluids Eng.* **137** 054501
[47] Yao E, Kember G and Hansen D 2016 *J. Eng. Mech.* **142** 04016063
[48] Pezzinga G and Scandura P 1995 *J. Hydraul. Eng.* **121** 802
[49] Urbanowicz K, Firkowski M and Zarzycki Z 2016 *J. Phys.: Conf. Ser.* **760** 012037
[50] Shu J J 2003 *Int. J. Pres. Ves. Pip.* **80** 187
[51] Milan Enel-Cris 2000 *Hydraulic Transients with Water Column Separation: IAHR Working Group 1971-1991 Synthesis Report* (IAHR Secretariat)
[52] Hsu Y Y and Graham R W 1976 *Transport Processes in Boiling and Two-Phase Systems: Including Near-Critical Fluids* (Washington, D.C., Hemisphere Publishing Corp.; New York, McGraw-Hill Book Co.)
[53] Dukler A E, Wicks M and Cleveland R G 1964 *AIChE J.* **10** 38
[54] Dukler A E, Wicks M and Cleveland R G 1964 *AIChE J.* **10** 44
[55] Schohl G 1993 *J. Fluids Eng.* **115** 420
[56] Urbanowicz K 2018 *Z. Angew. Math. Mech.* **98** 802
[57] Zielke W 1968 *J. Basic Eng.* **90** 109
[58] Urbanowicz K, Zarzycki Z and Kudźma S 2012 *J. Theor. App. Mech.-Pol.* **50** 889

On the structure of the boundary layer under adverse pressure gradient on an inclined plate

V Uruba[1,2], **P Procházka**[1] **and V Skála**[1]

[1]Institute of Thermomechanics of the CAS, v. v. i., Dolejškova 5, Praha 8, Czech Republic
[2]University of West Bohemia, Faculty of Mechanical Engineering, Department of Power System Engineering, Universitní 8, Plzeň, Czech Republic
E-mail: uruba@it.cas.cz

Abstract. The presented study is focused on experimental investigation of a boundary layer on a flat plate in adverse pressure gradient. The flat plate is placed in regular flow, the pressure gradient is generated by the plate inclination. The study [9] deals with structure of the wake behind the plate, the presented study concentrates on structure of the flow close to the suction surface of the plate. Dynamical behaviour of the flow structures is studied in details with respect to the streamwise topology changes. In spite of the fact that the time-mean flow field is 2D, constant along the span, the instantaneous structures topology is fully 3D.

1. Introduction

The presented study is dedicated to experimental investigation of a boundary layer on a flat plate in adverse pressure gradient. The flat plate is placed in regular flow, the pressure gradient is generated by the plate inclination. The study [9] was focused on structure of the flow in the transversal planes, the presented paper studies the topology in streamwise direction close to the suction surface of the plate. In this region streamwise vortices emerge due to instability mechanism within the boundary layer subjected to adverse pressure gradient.

The presence of streamwise vorticity has been proved experimentally without any doubt – see e.g. [6, 7, 8]. The vortices are present in the suction region just above the plate, they touch the plate surface. However appearance of the vortices in this region is random both in space (position along the span) and time.

The aim of the present study is to get the relevant information about the topology of the vortices especially in the streamwise direction.

2. Experimental setup

Flat plate inclined with angle of attack (AOA) 7 degrees has been placed in a uniform low turbulent stream. The blow-down facility produces a jet with uniform velocity distribution, mean velocity about 5 m/s, intensity of turbulence less than 0.2%. The plate of thickness 2 mm has rounded edges, chord is 100 mm and span 450 mm. The Reynolds number based on plate chord is about 33 thousands. Schematics of the experiment is in figure 1.

The plane of measurement is located parallel to the plate surface in the distance from 2 up to 8 mm, defined as the constant Z value. The Cartesian coordinate system is introduced with X axis in streamwise direction parallel to the plate surface, Y in spanwise direction and Z perpendicular to the plate surface.

Origin of the coordinates is fixed on the upper plate surface at the centre of the leading edge (red dot in figure 1)

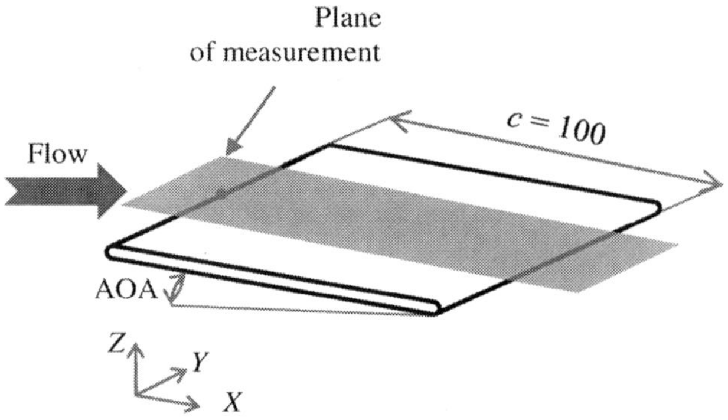

Figure 1. Schematics of the experiment.

3. Methods

The experimental methods and methods of data analysis will be described in details.

3.1. Instrumentation

The time-resolved PIV method was used for the experiments.

The measuring system DANTEC consists of the double-pulse laser with cylindrical optics and the CMOS camera. For the presented experiments the frequency was 2 kHz and 4000 double-snaps in sequence corresponding to 2 s of record for mean evaluation was acquired. The plane of measurement has been located parallel to the plate suction surface – see figure 1 in green. More details on measuring technique could be find in [8].

However the laser sheet thickness was approximately 1 mm, representing uncertainty of the measurement location. If the measuring plane is well defined in a distinct distance from the surface within the boundary layer viscous sublayer, then the value of the measured velocity is proportional to the local skin friction. The results should be considered as indicators only of the flow topology close to the surface, similar as results of surface flow visualisation using surface paintings. The vectorlines could be supposed to follow the skin friction on the surface.

More details on the near-wall PIV results interpretation see e.g. [1].

3.2. Analysis methods

The acquired velocity fields are to be subjected to further analysis. The classical statistical method are used for determining time mean and variances. Vorticity *Z*-component is evaluated as well. To demonstrate the dynamics of flow-field fluctuations topology the Proper Orthogonal Decomposition and Oscillation Pattern Decomposition methods are used.

The decomposition methods are based on idea Hilbert space, which is defined by the snapshots, forming the basis of the space. The goal of the decomposition methods is to find different, optimized basis with clear physical meaning.

The Proper Orthogonal Decomposition method (hereinafter POD) looks for the orthonormal modes basis maximizing the variance. Each mode consists of spatial pattern (Topos), evolution in time (Chronos) and amplitude, which is square root of kinetic energy (or enstrophy for vorticity) of the mode. The modes are ordered according to decreasing amplitude.

The POD method represents the classical approach to analyze the fluctuating flow topology on the basis of energetic content. Historically, it was introduced in the context of turbulence by Lumley [2] as an objective definition of what was previously called big eddies and which is now widely known as coherent structures.

IOP Conf. Series: Journal of Physics: Conf. Series **1101** (2018) 012047 doi:10.1088/1742-6596/1101/1/012047

The Oscillation Pattern Decomposition method (hereinafter OPD) evaluates the basis representing oscillating modes. The OPD method is based on stability assessment of fluctuation patterns in the flow under study. The method results into set of modes. Each mode is characterized by its frequency, damping factor called e-folding time and the topology. The periodical behavior is supposed with the given frequency and decaying amplitude, the mean time period for its decay by factor *e* is given by the e-folding time. Periodicity value is defined as the e-folding time to the mode period ratio, thus it is a dimensionless quantity. The mode topology is defined by a complex spatial mode. Its real and imaginary parts characterize fluctuating flow patterns shifted by a quarter of period, respectively.

More details on the OPD method could be find e.g. in [3,4,5].

4. Results

All presented results are given in dimensionless form, coordinates are divided by the plate chord defining the system (xyz). Velocities are divided by the inlet velocity U_e = 5 m/s defining dimensionless velocity components U and V. The OPD and POD modes are non-dimensional from definition.

The results evaluated in measuring planes in 4 distances from the suction plate surface $z = 0.02$, 0.04, 0.06 and 0.08, however detailed analysis will be shown for the measuring plane closest to the surface $z = 0.02$.

4.1. Statistics

Time mean flow-field will be shown first. In figures 2 and 3 the dimensionless streamwise mean velocity component U and sum of the dimensionless velocity components U and V variances sv distributions are shown respectively in the whole inspected domain. In figures 4 and 5 the same quantities are depicted within the single measuring plane $z = 0.02$. The spanwise mean velocity component V approaches zero value in the whole domain.

In figures 4 and 5 the lines $x = 0$ and $x = 1$ indicate positions of the leading and trailing edges respectively. In figures 2 and 4 the dark blue color indicates negative streamwise velocity component meaning back-flow region caused by the flow separation on the leading edge and reattachment close to the position $x = 0.25$. Above the separation zone there is a region of streamwise velocity overshoot, the velocity exceeds the free stream value (i.e. it is over 1).

The results indicate that the time-mean flow-field is more or less 2D, distributions are independent on the position y across the span.

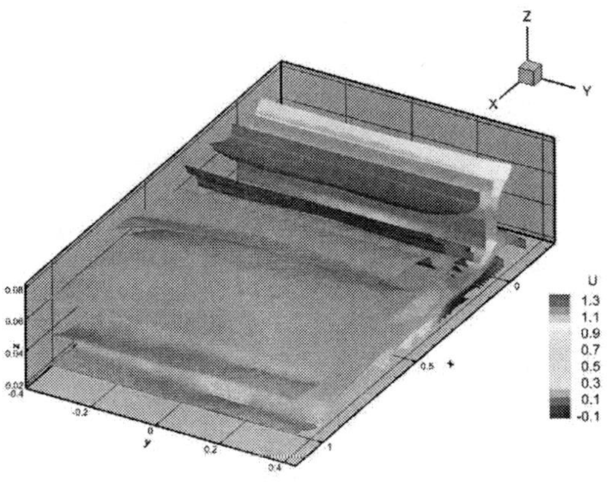

Figure 2. Iso-surfaces of dimensionless mean streamwise velocity component U.

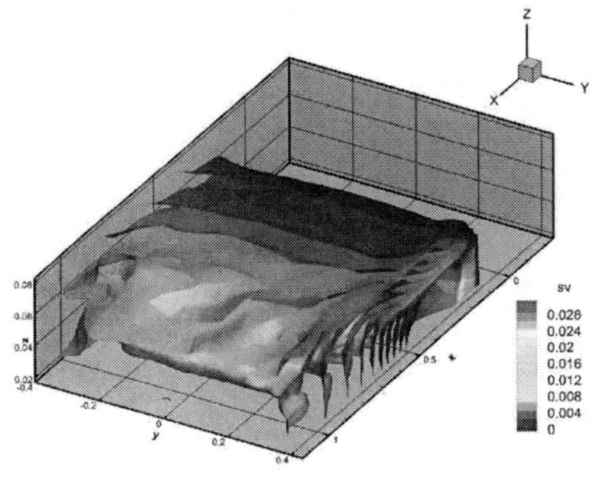

Figure 3. Iso-surfaces of sum of dimensionless velocity component variances sv.

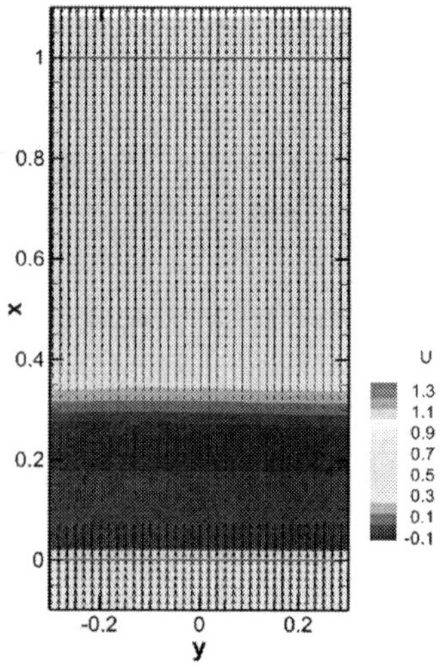

Figure 4. Dimensionless mean velocity U distribution.

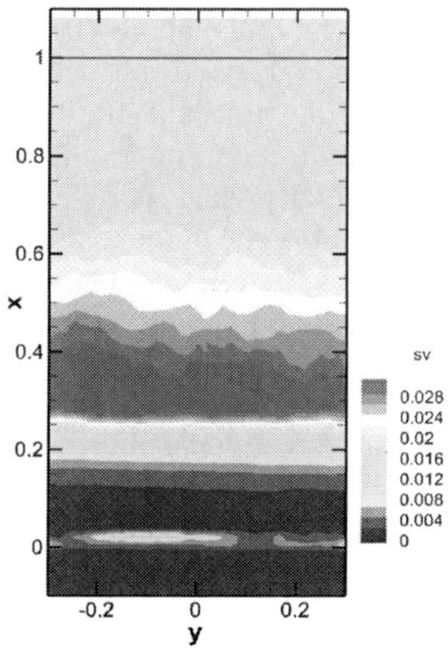

Figure 5. Sum of dimensionless velocity components variances sv distribution.

Maximal dynamical activity region is located just downstream the reattachment point – see the red color in figures 3 and 5.

For the subsequent dynamical analysis the domain was in the streamwise direction limited to the region with attached boundary layer $0.25 < x < 1$, i.e. between the reattachment point and the trailing edge.

4.2. OPD analysis of velocity fields

The OPD method has been applied on velocity vector distribution time-resolved data in the measuring plane $z = 0.02$, the 10 OPD modes has been evaluated.

Parameters of all OPD modes are given in table 1, in figure 6 there are the modes in the frequency – periodicity plain. The 3 modes no. 1, 5 and 8 with the largest value of periodicity were selected for presentation. Those modes are true oscillating ones represented by traveling patterns in the streamwise direction. The modes with low periodicity value or even 0 are of pulsating or decaying nature, not moving in space.

Table 1. OPD modes.

Mode	Frequency [Hz]	e-folding time [ms]	Periodicity [-]
1	104.0	23.3	2.42
2	61.8	22.3	1.38
3	34.8	22.2	0.77
4	29.9	21.6	0.64
5	117.5	21.2	2.49
6	42.6	20.6	0.88
7	76.6	17.7	1.36
8	133.0	15.2	2.02
9	0	14.8	0
10	0	12.0	0

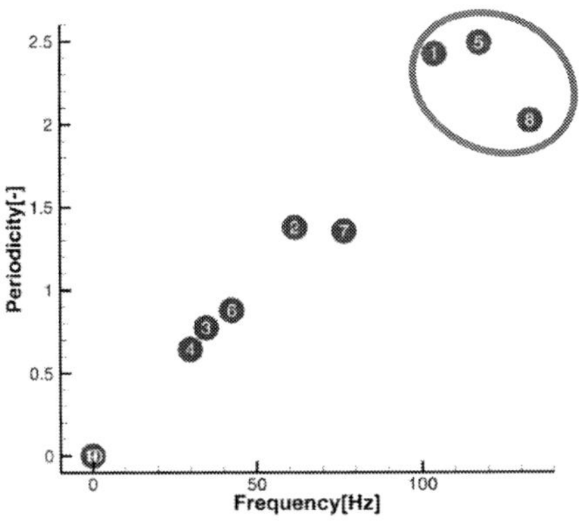

Figure 6. OPD spectrum, periodicity and frequencies of the OPD modes.

The spatial patterns will be shown for the selected modes, in figures 7 and 8 there are real and imaginary parts of the OPD mode 1 respectively. The mode topology is represented by the dimensionless velocity vector pattern and colour field representing dimensionless vorticity component perpendicular to the surface: red positive, blue negative values (the legend applies for figures 7 to 10).

The patterns in real and imaginary part of the OPD mode 1 are very similar, there is only difference in position of the patterns in streamwise direction and intensity. The intensity difference means that the amplitude of patterns changes periodically. The shift in space is connected with movement of patterns in the streamwise direction. Spacing of the structures is about 30 mm, divided by the mode period in time results in pattern velocity about 3 m/s, representing about 62% of incoming velocity.

For the OPD modes 5 and 8 only real parts of the topology will be shown in figures 9 and 10, the corresponding imaginary part is again shifted in x direction resulting in streamwise velocity of the pattern.

The studied OPD modes consist of shear layers, visible as vorticity concentration, in zigzag configuration.

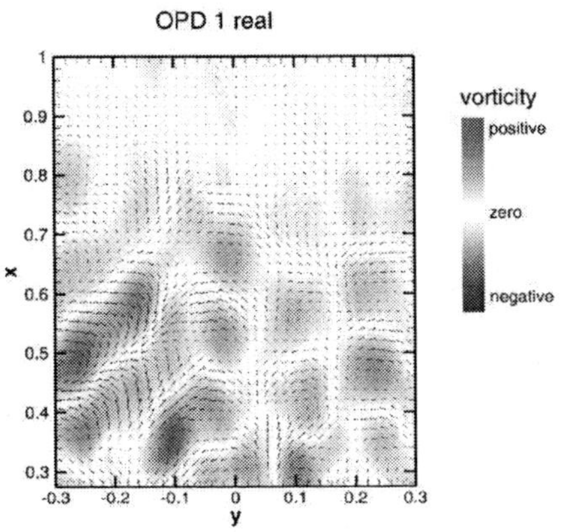

Figure 7. Real part of the OPD 1 mode.

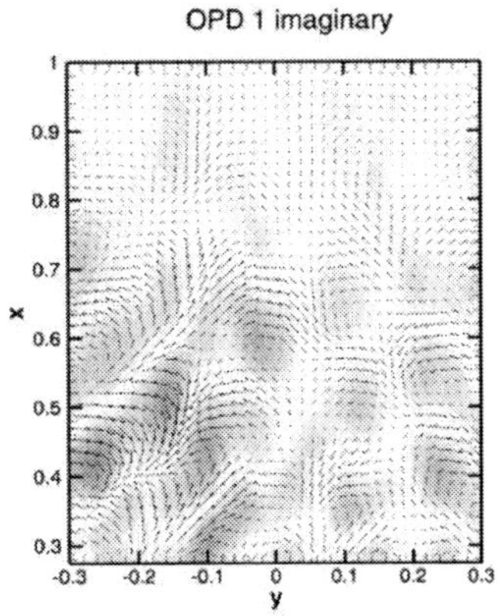

Figure 8. Imaginary part of the OPD 1 mode.

IOP Conf. Series: Journal of Physics: Conf. Series **1101** (2018) 012047 doi:10.1088/1742-6596/1101/1/012047

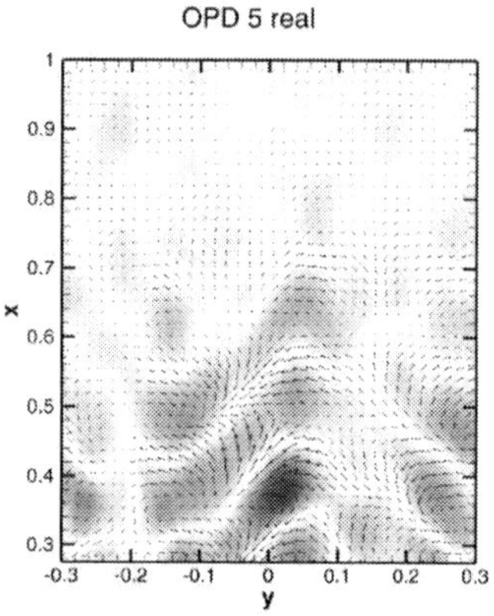

Figure 9. Real part of the OPD 5 mode.

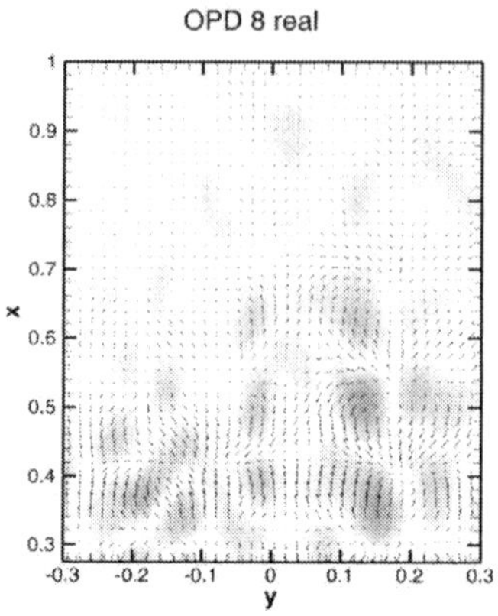

Figure 10. Real part of the OPD 8 mode.

4.3. POD analysis of vorticity fields
The vorticity z-component has been evaluated from all individual vector velocity fields. In figures 11 and 12 there is distribution of mean value and variance of vorticity, respectively. Mean vorticity forms streamwise streaks. Maximum of fluctuating activity is located just behind the reattachment, decaying in the streamwise direction.

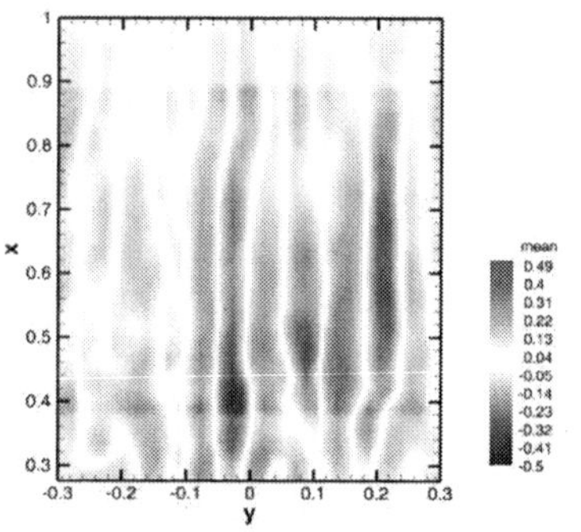

Figure 11. Mean dimensionless vorticity distribution.

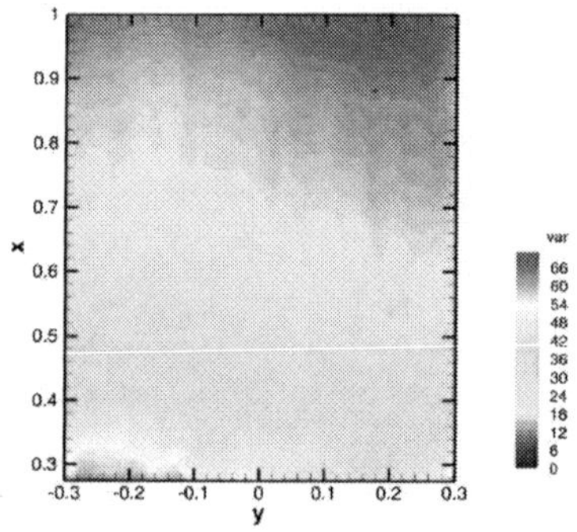

Figure 12. Dimensionless vorticity variance distribution.

The resulting vorticity fields have been subjected to the POD analysis to determine the modes explaining the vorticity variance, which could be considered as an enstrophy part. In figure 13 there is the enstrophy fraction and accumulated enstrophy of the vorticity POD modes depicted for all 2000 modes. The first POD mode contains 1.3% of the total enstrophy, while first 10 modes cover about 11%.

Selected POD modes topologies are to be shown. In figure 14 the Topos of the POD mode 1 is shown as vorticity fluctuations distribution (the legend apply for figures 6-18 as well). In figure 15 there is corresponding Topos representing time evolution of the mode 1, obtained as projection of the mode topology on the velocity vector field time series.

The higher order lower energy Toposes are shown in figure 16 number 4, in figure 17 number 40 and in figure 18 the POD mode number 99. The patters consist of vorticity strips in streamwise or oblique directions, filling the region behind the point of reattachment. Higher order modes fill the studied region regularly with smaller and more randomly distributed structures. The Topos representing time evolution of the mode 99 is in figure 19. Comparing with time evolution of the mode 1 (figure 15), the presence of higher frequencies is visible.

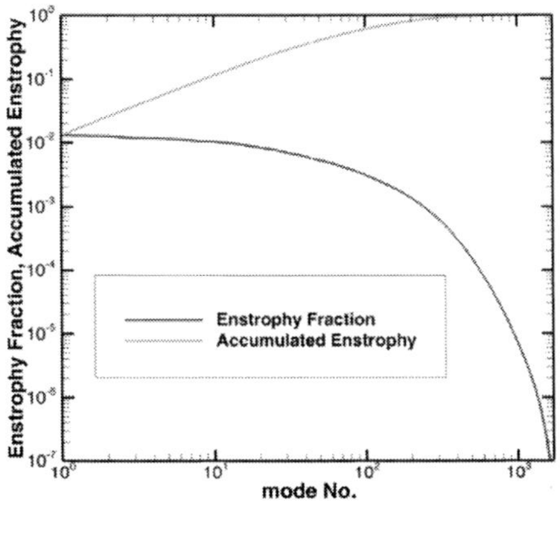

Figure 13. Enstrophy fraction and accumulated enstrophy of the vorticity POD modes.

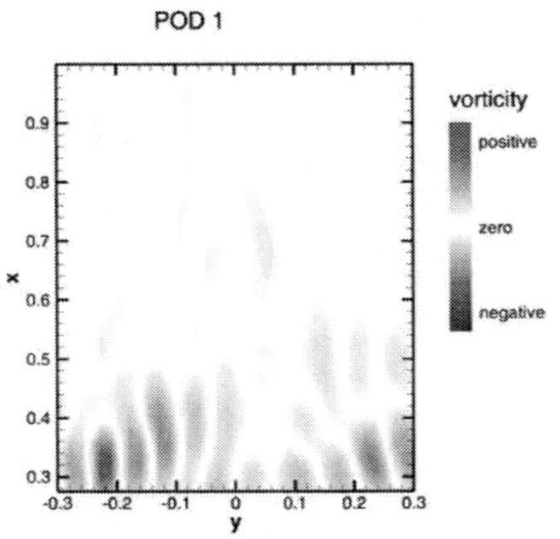

Figure 14. Topos of the POD 1 mode.

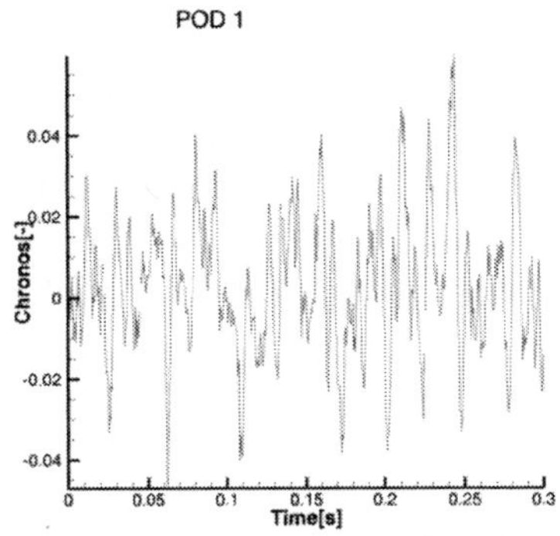

Figure 15. Chronos of the POD 1 mode.

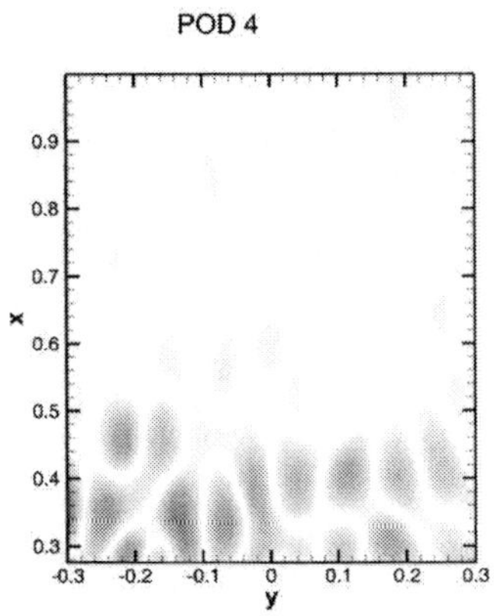

Figure 16. Topos of the POD 4 mode.

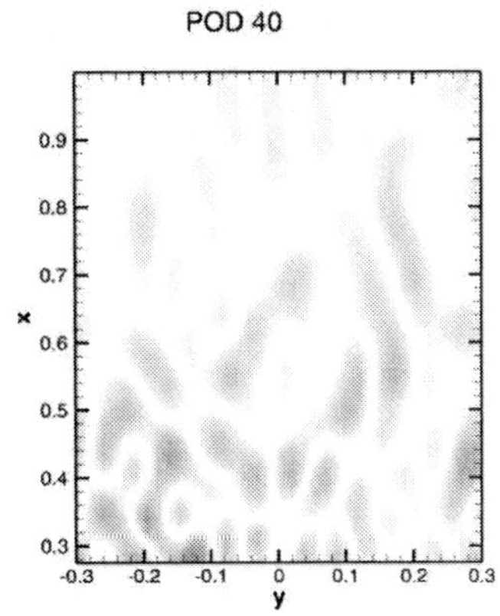

Figure 17. Topos of the POD 40 mode.

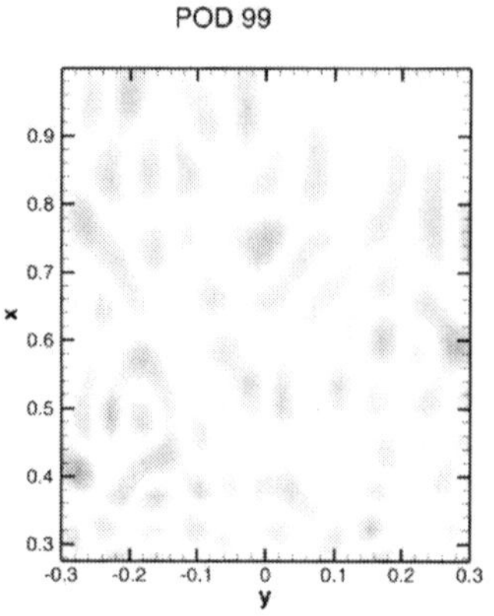

Figure 18. Topos of the POD 99 mode.

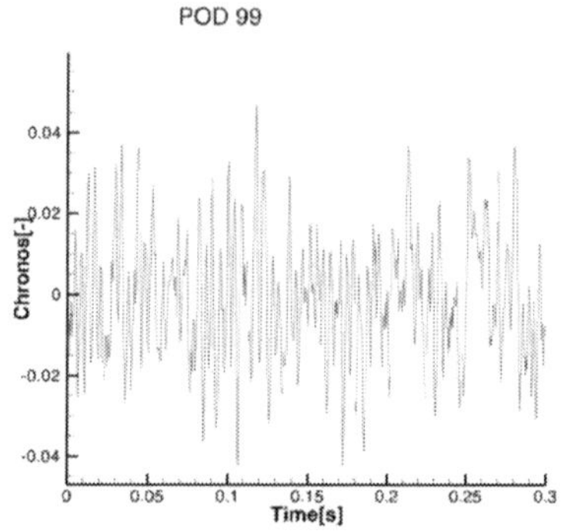

Figure 19. Chronos of the POD 99 mode.

5. Conclusions

Topology of the flow in the vicinity of suction surface of the flat plate in regular stream with angle of attack 7° was studied experimentally. Time resolved PIV method has been applied with measuring plane located close to the plate surface parallel to it. Local separation of the boundary layer has been detected on the first quarter of the surface. Both velocity and vorticity fields dynamics in the attached region have been explored in details. While the time-mean flow-field is close to 2D structure, the dynamical structures are highly 3D. Typical structures topology in the form of streamwise streaks and zigzag patterns have been shown. Maximal dynamical activity is located just behind the reattachment point decaying in the streamwise direction gradually.

Acknowledgements

This work was supported by the Grant Agency of the Czech Republic, project No. 17-01088S.

References

[1] Depardon, S., Lasserre, J.J., Boueilh J.C., Brizzi, L.E., Borée J., Skin friction pattern analysis using near-wall PIV, *Experiments in Fluids* **39**, 805–818, 2005.

[2] Lumley, J.L., The structure of inhomogeneous turbulent flows, Atm.Turb. and Radio Wave Prop., Yaglom and Tatarsky eds., Nauka, Moskva, pp. 166-178, 1967.

[3] Hasselmann, K., PIPs and POPs: The Reduction of Complex Dynamical Systems Using Principal Interaction and Oscillation Patterns, *Journal of Geophysical Research* **93**, D9, 11.015-11.021, 1988.

[4] Uruba, V., Decomposition Methods in Turbulence Research, *EPJ Web of Conferences* **25**, 01095, 2012.

[5] Uruba, V., Near wake dynamics around a vibrating airfoil by means of PIV and Oscillation Pattern Decomposition at Reynolds number of 65000, *Journal of Fluids and Structures* **55**, 372–383, 2015.

[6] Uruba, V., Pavlík, D., Procházka, P., Skála, V. and Kopecký, V., On 3D flow-structures behind an inclined plate, *EPJ Web of Conferences* **143**, 02137, 2017.

[7] Uruba, V., Pátek, Z., Procházka, P., Skála, V., Zacho, D. and Kulhánek, R., Flow Structure behind a Wing at High Reynolds Numbers, *EPJ Web of Conferences* **180**, 02111, 2018.

[8] Uruba, V., Procházka, P. and Skála V., On 3D Flow Structure of the Boundary Layer on the Suction Side of a Plate, *EPJ Web of Conferences* **180**, 02112, 2018.

[9] Uruba, V., Procházka, P. and Skála, V., On the 3D structure of the flow-field in the vicinity of inclined plate, to be published.

IOP Conf. Series: Journal of Physics: Conf. Series **1101** (2018) 012048 doi:10.1088/1742-6596/1101/1/012048

Turbulence Behind 3D Multi-Scale Sparse Grids

Syed M. Usama[1], Jackson Tellez-Alvarez[2], Jacek Kopec[3], Kamil Kwiatkowski[3], Jose-Manuel Redondo[2], Nadeem A. Malik[1*]

[1]Department of Mathematics & Statistics, King Fahd University of Petroleum and Minerals, P.O. Box 5046, Dhahran 31261, Saudi Arabia

[2]Department of Physics, Technical University of Catalonia, BarcelonaTech., Barcelona 08034, Spain

[3]Interdisciplinary Centre for Mathematical and Computational Modelling, Faculty of Physics, University of Warsaw ul. Prosta 6900-838, Poland

E-mail: nadeem_malik@cantab.net

Abstract. A fundamentally new idea in grid generated turbulence is the 3D Sparse Grid (3DS) concept [N. A. Malik. Sparse 3D Multi-Scale Grid Turbulence Generator. US Patent No. US 9,599,269 B2 (2017)] which reduces the effective blockage ratio compared to the 2D flat fractal grids, $\sigma_{3DS} \ll \sigma_{2DF}$, and possess a much greater parameter space which could allow further optimization of the turbulence as compared to the 2D fractal (2DF) grids. Here, we report on some theoretical results regarding blockage ratio reduction in a 3-frame 3DS system, and some results from Direct Numerical Simulations comparing the turbulence characteristics generated by 3DS with Regular (RG) and 2DF grids cases.

1. Introduction to 3D Sparse Grids

One of the most important properties of turbulence is its ability to enhance mixing. This has enumerable applications in chemical and industrial contexts and in many natural systems where turbulence is present, such as drag reduction. The control of turbulence is therefore desired. For example, 2D flat (2DF) square fractal grids have been shown to alter the turbulence

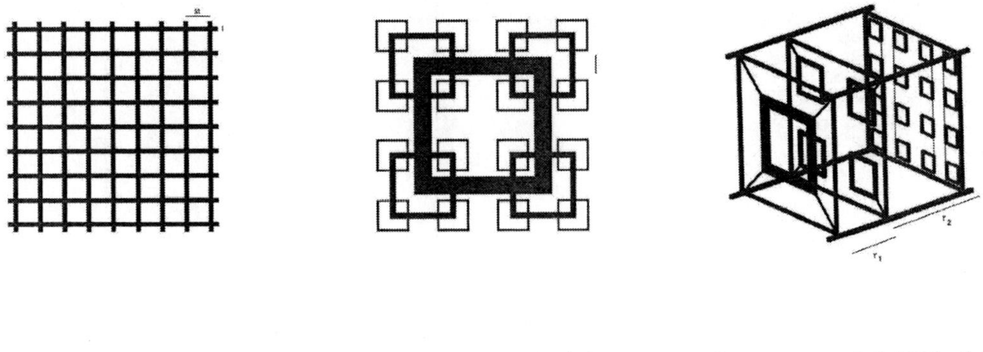

Figure 1: Left to rght: (a) 2D regular grid (RG). (b) 2D fractal square grid (2DF). (c) 3D sparse grid (3DS).

characteristics compared to the regular grids, it is believed that the turbulence intensity is enhanced for the same blockage ratio.

Grid generated turbulence is the most common way of studying turbulence in laboratory experiments. Until recently regular grids (RG), Figure 1(a), with bars of constant thickness producing open cells of constant width for flow passage, have been used. As fractals became popular in the 1990s a new type of grid with bars of different thicknesses and different lengths in a flat 2D fractal arrangement (2DF) was developed, Figure 1(b). The essential idea here is that different scales of turbulence are generated spontaneously in the same plane. The turbulence generated has different characteristics compared to the RGs, with the turbulence intensity peaking at a higher level [1].

A recent innovation in grid generated turbulence, the 3D Sparse Grid (3DS) [2, 3], called the Sparse 3D Multi-Scale Grid Turbulence Generator, or 3D sparse grid (3DS) for short, has excited interest in the turbulence community because of its potential to alter turbulence characteristics downstream of the grid. The 3DS goes further than the 2DF construction by separating each generation of length scale of turbulence grid elements in to its own frame in

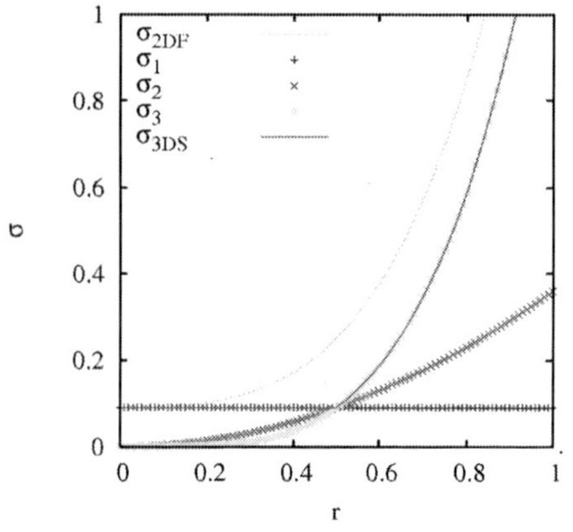

Figure 2: $a = 0.5$, $w_1 = 0.05$

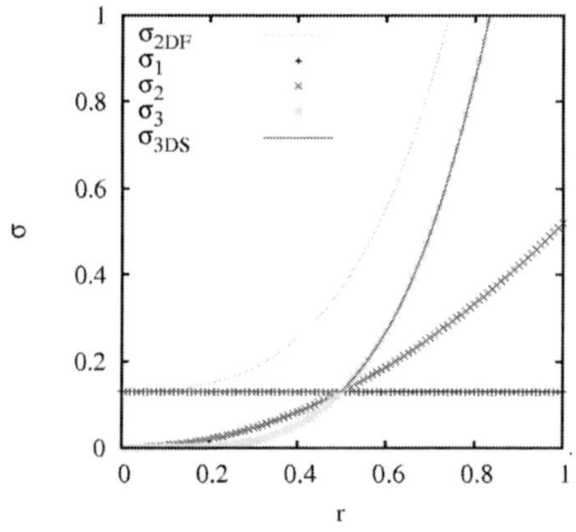

Figure 3: $a = 0.7$, $w_1 = 0.05$

Solidity (blockage ratios) against the scaling factor r. σ_{2DF}, and σ_{3DS}, and the scale-by-scale, $\sigma_1, \sigma_2, \sigma_3$ in the 3DS are shown.

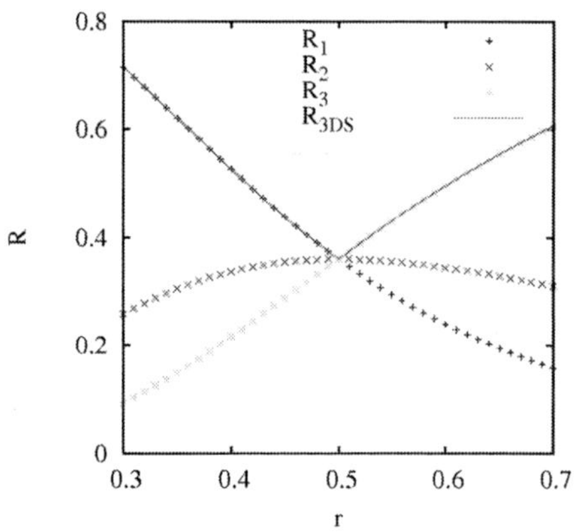

Figure 4: $a = 0.5$, $w_1 = 0.05$

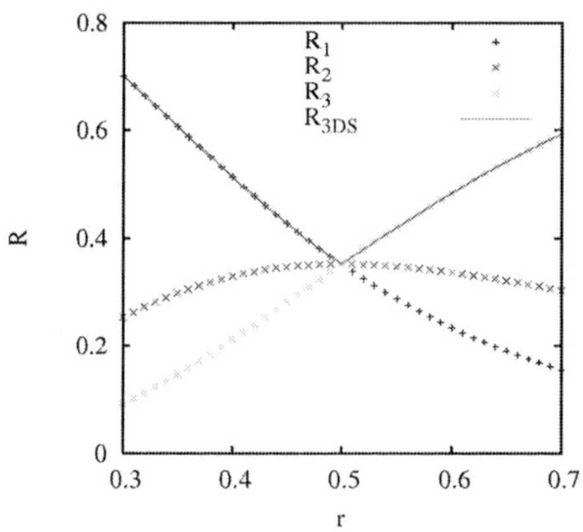

Figure 5: $a = 0.7$, $w_1 = 0.05$

Reduced solidity against the scaling factor r. The scale-by-scale, R_1, R_2, R_3, and R_{3DS} in the 3DSGT are shown.

IOP Conf. Series: Journal of Physics: Conf. Series **1101** (2018) 012048 doi:10.1088/1742-6596/1101/1/012048

overall co-planar arrangement, Figure 1(c), which produces a 3D sparse grid system. Each generation of grid elements produces a turbulent wake pattern that interacts with the other wake patterns downstream. The length scale of the grid elements from frame to frame can be in any multiscale ratios, although a fractal pattern is a popular choice.

Here, we report on some theoretical results regarding mass flow rates in 3DS in Section 2, and some results from Direct Numerical Simulations (DNS) that demonstrate the performance of the 3DS in a conduit compared to the classical flat 2D Fractal Grid (2DF) arrangement.

2. Some Properties of 3D Sparse Grids

A direct consequence of separating scales in the 3DS is that the effective blockage ratio (solidity), is greatly reduced compared to the 2DF. Furthermore, the parameter space is increased because the distances between successive frames are additional parameters.

In the 3DS arrangement, the optimal reduction in the blockage ratio as compared to the 2DF flat fractal grid, $R_{3DS}(r) = \sigma_{3DS}(r)/\sigma_{2DF}$, occurs when the geometric ratio between successive length scales is, $r = 0.5$, see Figures 4 and 5. The mass flow rate is increased for the same pressure gradient due to the lower blockage.

Consider a 3-generation 2DF, and a corresponding 3DS sparse arrangement. In both cases, the grids are contained inside a square conduit of height and width equal to H, to give a cross-sectional area of,

$$T = H \times H \tag{1}$$

Suppose that in the first generation (largest scale) the fractal square grid has length $L_1 = aH$, with $a < 1$. The thickness of the grid bars is W_1. The second generation of grid scales is produced from the first generation by scaling with scaling factor r with $0 < r < 1$; thus $L_2 = rL_1$, and $W_1 = rW_0$. In the third generation, $L_3 = r^2 L_1$ and $W_3 = r^2 W_1$.

Then, the 2DF flat grid in Fig. 1, has the total cross-sectional area,

$$A_{2DF} = (L_1 W_1 - W_1^2)(4 + 16r^2 + 64r^4) - W_1^2(8r + 32r^3) \tag{2}$$

This is an exact expression, taking in to account the small overlap areas between the different generations. Laziet & Vassilicos [1] derive an approximate expression in terms of the perimeter

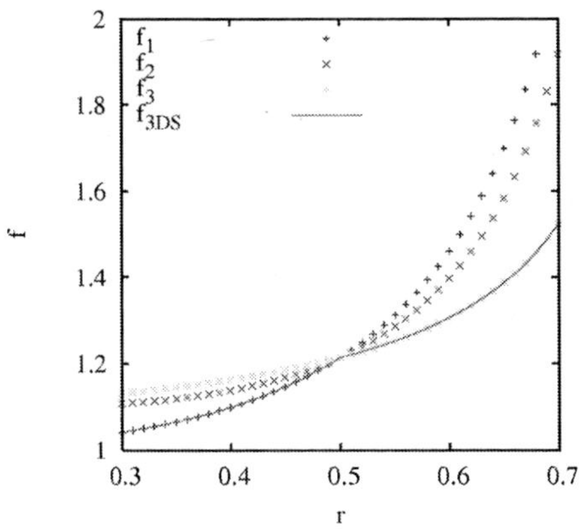

Figure 6: $a = 0.5$, $w_1 = 0.05$

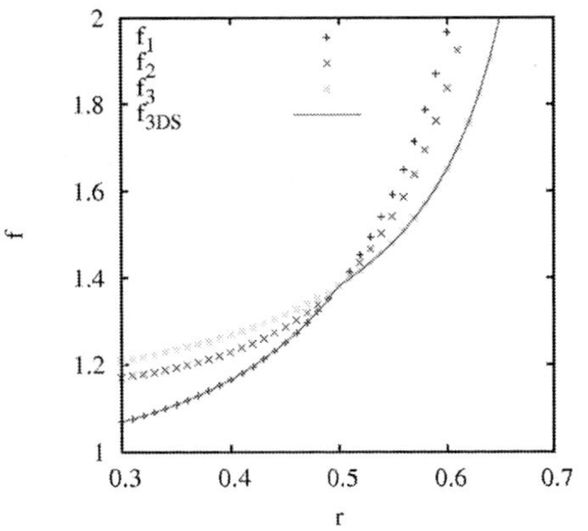

Figure 7: $a = 0.7$, $w_1 = 0.05$

Enhanced mass flux against the scaling factor r. The scale-by-scale, f_1, f_2, f_3, and f_{3DS} in the 3DSGT are shown.

of the square fractal arrangement. The cross-sectional areas for each generation of grid bars in the 3DS separately, are

$$A_1 = 4(L_1W_1 - W_1^2) \tag{3}$$
$$A_2 = 16r^2(L_1W_1 - W_1^2) = 4r^2A_1 \tag{4}$$
$$A_3 = 64r^2(L_1W_1 - W_1^2) = 4r^2A_2 = 16r^4A_1 \tag{5}$$

We non-dimensionalise all lengths by H to obtain the non-dimensional parameters, $l_1 = L_1/H$, and $w_1 = W_1/H$. Then, the blockage ratio (solidity) in the 2D flat grid is given by $\sigma_{2DF} = A_{2DF}/T$,

$$\sigma_{2DF} = (aw_1 - w_1^2)(4 + 16r^2 + 64r^4) - w_1^2(8r + 32r^3) \tag{6}$$

and the solidity of each of grid frame separately in the 3DS are,

$$\sigma_1 = 4(aw_1 - w_1^2); \; \sigma_2 = 4r^2\sigma_1; \; \sigma_3 = 16r^4\sigma_1 \tag{7}$$

The effective solidity of the 3DS arrangement is therefore smaller than the corresponding 2DF,

$$\sigma_{3DS} = Max(\sigma_1, \sigma_2, \sigma_3) \tag{8}$$

The solidity is a function of three non-dimensional variables,

$$\sigma_{3DS} = \sigma_{3DS}(w_1, a, r) < \sigma_{2DF} \tag{9}$$

Figures 2 and 3 show the $\sigma's$ against r for $w_1 = 0.05$, and for $a = 0.5$ and 0.7 respectively.

From equations (7)-(9), for $r = 0.5$ we have $\sigma_1 = \sigma_2 = \sigma_3$. This is because the increase in the number of self-similar squares competes with the decrease in the per unit area from generation to generation. At $r = 0.5$ these two effects exactly balance and lead to the optimal reduction at $r = 0.5$.

We define the reduction in solidity (blockage ratio) from 2DF to 3DS to be,

$$R_\sigma(w_1, a, r) = \frac{\sigma_{3DS}}{\sigma_{2DF}}. \tag{10}$$

We also define scale-by-scale solidity reduction by,

$$R_1 = \frac{\sigma_1}{\sigma_{2DF}}; \; R_2 = \frac{\sigma_2}{\sigma_{2DF}}; \; R_3 = \frac{\sigma_3}{\sigma_{2DF}}. \tag{11}$$

Figures 4 and 5 shows the solidity reductions R as a function of r, for $w_1 = 0.05$, and for $a = 0.5$ and 0.7 respectively. R_σ shows a minimum of $R_\sigma \approx 0.36$ at $r = 0.5$.

If we approximation the mass flux rate through a grid with blockage ratio σ to be,

$$M = (1 - \sigma)F(Re; \rho, U, \mu, L) \tag{12}$$

where F is some function (for example, F could be proportional to the pressure gradient). ρ is the fluid density, U is the mean flow velocity, ν is the kinematic viscosity Re is the Reynolds number. The increase in mass flow rate from the 2DF flat to the 3DS sparse grids is,

$$f_\sigma(w_1, a, r) = \frac{M}{M_{2DF}} = \frac{1 - \sigma}{1 - \sigma_{2DF}} \tag{13}$$

and similar expressions f_1, f_2, f_3 for each generation can be obtained.

Figures 6 and 7 shows the enhanced mass flow rates, $f's$, against r for the same cases as in Figures 4 and 5 respectively. We obtain $f_\sigma = 1.21$ which is an increase of 21%. Although f_σ continues to increase for $r > 0.5$, the solidity approaches 1, and $1 - \sigma \to 0$, and the flow rate approaches zero. Thus $r = 0.5$ is the optimal enhancement in mass flux rates.

IOP Conf. Series: Journal of Physics: Conf. Series **1101** (2018) 012048 doi:10.1088/1742-6596/1101/1/012048

3. Direct Numerical Simulations

The aim here is to compare the turbulence characteristics produced by RG, 2DF, and 3DS grids. We use DNS for this purpose.

The domain is a cuboid of dimensions $460.8 \times 115.2 \times 115.2d^3_{min}$ where d_{min} is the thickness of the smallest square. Thus, the height and width of the channel is $H = 115.2d_{min}$. We consider three different grids, regular grid RG, 2D flat fractal grid 2DF, and 3D sparse grid 3DS. The blockage ratios in the RG and 2DF is equal to 32%, and the constant effective mesh size in the RG is $M_{ef} = 13.33d_{min}$. The bars have length $115.2d_{min}$ in the RG with thickness $2.6d_{min}$, the same as in [1].

The bars in the 3DS have the same lengths and thicknesses as in the 2DF. All lengths are henceforth non-dimensionalized by d_{min}.

The 2DF has non-dimensionlized lengths and widths $\{l_i, d_i\}$, in generation $i = 0, 1, 2,$. The geometric ratio for the large grid scales is, $r = 0.5$, and $a = 0.5$. Thus, $l_0 = 57.6 = 0.5h$, $l_1 = 0.5l_0$, $l_2 = 0.5l_1$. The geometric ratio for the bar thicknesses is $1/\sqrt{8.5} = 0.34$, thus $d_0 = 8.5$, $d_1 = 2.9d_2$, $d_2 = 1$.

The time scale is defined by $t_2 = d_{min}/U_\infty$ where U_∞ is the inlet velocity set equal to 1.

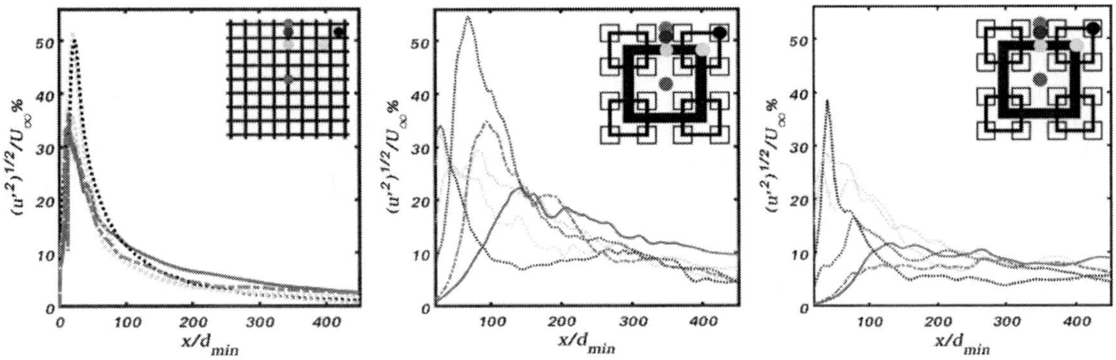

Figure 8: Turbulence intensity along pencils in the channel x-directions. Left: RG; Centre: 2DF; Right: 3DS.

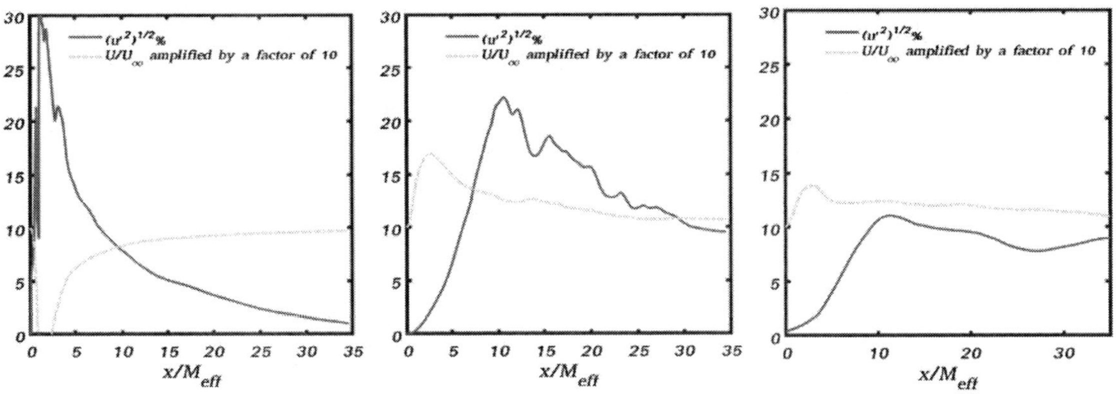

Figure 9: Turbulence intensity along pencils in the channel x-directions. Left: RG; Centre: 2DF; Right: 3DS.

The 3DS has the same lengths and thickness as the 2DF in [1], however each generation is held in a frame separated from the next by non-dimensional distances, $r_1 = x_1 - x_0 = 17$, and $r_2 = x_2 - x_1 = 8.5$, and $x_0 = 10$, where $x_i' s$ are the non-dimensionalised x-coordinates of the ith frame.

The solidity (blockage ratio) in the 2DF is 32%, and the solidity in the 3DS is 15%.

To resolve down to the smallest scales, a numerical grid size of one fifth of the thickness of the smallest bar is created, $\Delta x = 0.2 d_{min}$. This creates a grid of $N_x \times N_y \times N_z = 2304 \times 576 \times 576$. The RG and 2DF turbulence grids lie in the plane $x_0 = 10$ downstream of the channel inlet. Periodic boundary conditions are applied on the walls in the y and z directions; and inlet-outlet boundary conditions were applied in the x-direction. The initial condition is a uniform inflow velocity $U_\infty = 1$. The Reynolds number is, $Re = \frac{U_\infty d_{min}}{\nu} = 300$.

OpenFOAM, (OFoam), an opensource CFD toolbox, is chosen for the simulation. OFoam uses finite volume discretization is performed using Pressure Implicit Splitting of Operator Algorithm (PISO). Time discretization using Backward Euler method, whereas gradient and Laplacian term discretization using Gauss linear method are performed. Divergence term discretization is done using Gauss cubic method which is a third order scheme. Interpolation and other terms are discretized using Gauss Linear schemes. The resulting linear systems are solved by preconditioned conjugate gradient method with diagonal incomplete Cholesky preconditioner for pressure solution whereas iterative solver is used with symmetric Gauss-Siedel as the smoother to calculate velocities. Tolerance is set at 10^{-6}. Simulation time step is $\Delta t = 0.015 d_{min}/U_\infty$ which corresponds to a Courant number of 0.75. Probes and pencils are placed at various locations and 100 complete field snapshots have also been recorded in the time range from $300 t_2$ to $600 t_2$. The flow statistics have also been averaged over this time.

Figures 8-10 shows comparison of turbulence characteristics from RG, 2DF, and 3DS grids from the current DNS simulations. The RG and 2DF plots are close to the results of Laziet & Vassilicos [1], which validates the current DNS for these calculations.

Figure 8 compares the turbulence intensities u'/U_∞ along the channel length in the x-directions – pencils taken at various (y, z) locations as indicated by the different colors. The peak intensities in the 3DS is lower than in 2DF. After about $x/d_{min} = 100$ the intensities in the 3DS is higher and sustained for longer than from the RG, but lower than in 2DF. The latter is probably due to the lower blockage in the 3DS (15%) compared to the 2DF (32%).

Figure 9 shows the plots of the turbulence intensity and the mean flow along the central pencil $(y, z) = (0, 0)$. The mean flows in the 2DF and 3DS are close, and both are significantly

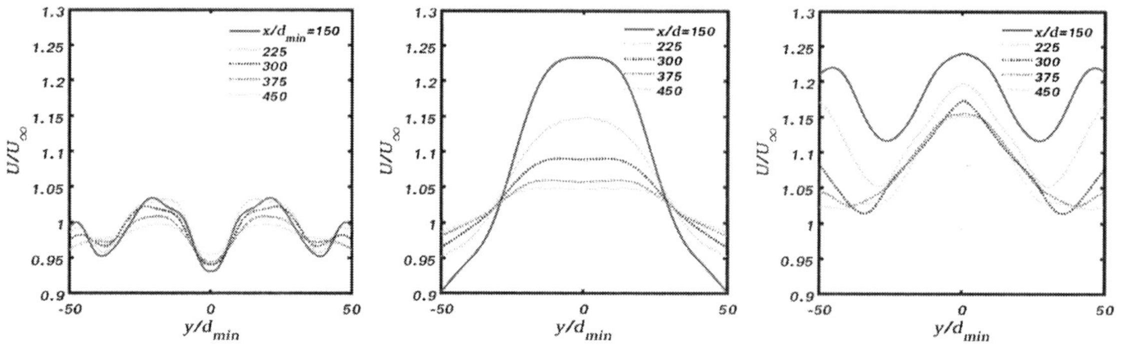

Figure 10: Turbulence intensity along pencils in the channel x-directions. Left: RG; Centre: 2DF; Right: 3DS.

higher than in the RG grid.

Figure 10 shows the mean x-velocity along the y-axis, $U(y)/U_\infty$, taken at different stations along the channel. There is significant difference between the 2DF and 3DS grids. The 2DF grid produces smooth non-oscillating profiles that peak and are uniform in the central region and eventually flatten out further downstream. The results from the 3DS is similar to that from the RG in so far as it shows an oscillating profile across the y-direction, however it shows peak levels similar to the 2DF of about 1.25, compared to the RG where the average is about 1 at all stations along the channel. The 3DS does not flatten out as fast as the 2DF being sustained on average at a higher level further downstream.

The overall impression from the results from the 3DS, with the given configuration, is that the 3DS is roughly in between the RG and 2DF cases for the turbulence characteristics shown.

However, it is important to remember that the blockage ratio in the 3DS is about half of the RG and 2DF so that the mass flow rate in the 3DS turbulence here is about half of that in the other two; to obtain the same flow rate in the RG and 2DF the pressure gradient would have to be doubled, so the mixing efficiency in the 3DS may yet to be greater.

These are the first results from DNS that we have obtained. The aim of this work is a parametric study of 3DS for different r_1 and r_2, and for different blockage ratios, and to address to question of mixing efficiency.

4. Conclusions

The 3DS [2, 3] has been investigated, and some comparisons with the Regular RG and flat 2D fractal grid 2DF has been made. Theoretically, the 3DS admits a higher mass flow rate for the same pressure gradient because of the lower blockage ration (solidity), equation (13). An optimal value for the geometric ratio was found to be 0.5.

DNS was first validated against the previous work of Laziet & Vassilicos [1]. The DNS was then used to simulate RG, 2DF, and 3DS with frame separations of $r_1 = 17d_{min}$ and $r_2 = 8.5d_{min}$ and a blockage ratio of 15%, compared to 32% for the RG and 2DF. It was found that overall the turbulence characteristics generated by 3DS was generally in between the RG and 2DF grids cases; the peak turbulence intensities were lower than in 2DF, and downstream the intensities were also lower than in 2DF. But the intensities were higher than in RG and were sustained for longer downstream.

A critical question is what happens if the blockage ratio of the 3DS is increased towards the 2DF value of 32%. Another is, does 3DS lead to greater mixing efficiency? These issues are currently under investigation numerically. In addition to the velocity field, we are looking at the vorticity field, the pressure field, and diffusing scalar fields.

Acknowledgements

This work has been carried out partly in the framework of European High-Performance Infrastructures in Turbulence (EuHIT) project Trans National Access program at the EuHIT grant, Turbulence Generated by Sparse 3D Multi-Scale Grid (M3SG), 2017.

References

[1] S. Laizet & J. C. Vassilicos. DNS of Fractal-Generated Turbulence. Flow Turbulence Combust 87:673-705 (2011).
[2] N. A. Malik. Sparse 3D Multi-Scale Grid Turbulence Generator. US Patent No. US 9,599,269 B2 (2017).
[3] N. A. Malik. Sparse 3D Multi-Scale Grid Turbulence Generator. EPO Patent no. 2965805 (2017).

IOP Conf. Series: Journal of Physics: Conf. Series **1101** (2018) 012050 doi:10.1088/1742-6596/1101/1/012050

Analysis of species diffusion and methanol decomposition source in thermocatalytic reactor based on the intermetallic phase of Ni_3Al for low Reynolds numbers

P Ziółkowski [1,4], M Stajnke[1], P Jóźwik[2], Z Bojar[2], PJ Ziółkowski[3] and J Badur[15]
[1] Energy Conversion Department, The Szewalski Institute of Fluid-Flow Machinery PAS-ci, Fiszera 14, 80-231 Gdańsk, Poland
[2] Faculty of Advanced Technologies and Chemistry, Military University of Technology, Warszawa, Poland
[3] Faculty of Civil and Environmental Engineering, Gdańsk University of Technology, Gdańsk, Poland

E-mail: [4]pziolkowski@imp.gda.pl, [5]jb@imp.gda.pl

Abstract. In this paper, the numerical modelling of hydrogen production by methanol decomposition in the thermocatalytic reactor based on the intermetallic phase of Ni_3Al has been presented. The experimental results of flowing mixtures containing helium and methanol in a thermocatalytic reactor incorporating microchannels were used for the preliminary calibration of the Computational Fluid Dynamics "CFD" calculations. The thermocatalytic decomposition reaction of methanol has been modelled by employing experimental data related to the active surface area. It should be understood that a decrease of methanol concentration at inlet into package, mainly, ten millimetres before the thermocatalytic area, is connected with the diffusion fluxes of the other components, mainly hydrogen and carbon monoxide. The commercial CFD code has been expanded by User Defined Functions "UDFs" to include surface chemical reaction rates on the interphase between fluid and solid. The data extrapolated via the implemented numerical model have made it possible to assess the minimal length of the micro-reactor channels which predicts an optimal dimension at the system outlet. Results obtained through numerical calculations were calibrated and compared with experimental data to receive satisfactory agreement.

1. Introduction

Ni-based catalysts exhibit extremely high catalytic activity in methanol decomposition and other gasses synthesis and promote the production of carbon nanostructures (mainly carbon nanotubes) [1,2]. One of the Ni-based, solid-state catalysts is the Ni_3Al intermetallic phase and its alloys [3], which belongs to a sort of multifunctional materials, combining properties of both the structural and functional materials. According to the literature, Ni_3Al intermetallic thin foils exhibit extremely high catalytic properties in hydrocarbon decomposition reactions [4]. On the other hand, also based on the literature, the relatively high temperature of maximal hydrocarbon conversion is the main disadvantage of this material [1,5]. Nevertheless, the high temperature of the process can be utilized by placing a regenerative heat exchanger downstream of the reactor [3]. Therefore, the thermocatalytic reactor with thin strips or thin foils based on the Ni_3Al intermetallic phase produced according to technology presented by Jóźwik et al. in works [1,3] appears as an innovative and extremely promising technology in the making. One

IOP Conf. Series: Journal of Physics: Conf. Series **1101** (2018) 012050 doi:10.1088/1742-6596/1101/1/012050

example of an alloy foil package based on the intermetallic phase Ni_3Al, constructed as a rolled up honeycomb structure, is shown in Fig.1a). The micro-reactor generally consists of microchannels which are separated by a thin solid wall, as is highlighted schematically in Fig.1b). The cross section of the micro-channels with characteristics dimensions in the honeycomb set is presented in Fig. 1 c).

Figure 1. a) The alloy foil package based on the Ni_3Al intermetallic phase rolled into a honeycomb; b) the eight single microchannels beginning at inlet into the package and terminating at the outlet from the package taken to numerical simulation; c) the characteristic dimensions of reactor.

1.1. Motivation

The appropriate selection of the dimensions of the device, in such way that the contact time of the flow across the catalytically active surface of the substrate providing the reaction of the target product, is very important. In addition to experimental studies, work focused on the modelling of thermocatalytic processes is becoming increasingly interesting. Besides these works, that describe in general the processes of chemical reactions [6-8] there are articles concerning models of reactors and catalytic micro-reactors [9,10].

This paper is a continuation of the analysis of the presented geometry of a single microchannel [9]. However, different issues have been considered, namely the catalytic surface reactions of the decomposition of methanol and the volume reaction of shifting or methanation inside a microchannel. Generally, it is widely known that the catalytic surface reactions, in the direct vicinity of walls coated with active layers, are very often present in technical situations [11,12]. For thermal decomposition reactions, these active layers are often built on the base of Ni alloys [2,13,14]. As previously mentioned in this paper, one of the most promising potential applications of the thin alloy foils, based on the intermetallic phase Ni_3Al with high thermocatalytic activity, is within the field of air purification and removal of harmful substances and the controlled decomposition of hydrocarbons, and extended consideration is given here. In this study, the thermocatalytic reactor with microchannels has, with respect to the production of hydrogen, been numerically considered. The main focus of the this paper is to describe the mathematical modelling the chemical surface reactions of the flow of a helium and methanol mixture. The model has been verified and the performance of the thermocatalytic reactor with microchannels within the hydrogen production regime has been investigated.

2. Model description

A mathematical model of the thermocatalytic micro-reactor is based on the elementary balance equations solved for fluids to include continuity, momentum, energy, evolution of kinetic energy of turbulence, evolution of dissipation of kinetic energy of turbulence and species transport in the frame of finite volume method (FVM). The governing equations for the whole device model can be given in the following compact with CFD form:

$$\frac{\partial}{\partial t}\begin{Bmatrix} \rho \\ \rho\mathbf{v} \\ \rho e \\ \rho k \\ \rho\varepsilon \\ \rho Y_m \end{Bmatrix} + \operatorname{div}\begin{Bmatrix} \rho\mathbf{v} \\ \rho\mathbf{v}\otimes\mathbf{v} \\ \rho e\mathbf{v} \\ \rho k\mathbf{v} \\ \rho\varepsilon\mathbf{v} \\ \rho Y_m\mathbf{v} \end{Bmatrix} + \operatorname{div}\begin{Bmatrix} 0 \\ p\mathbf{I} \\ p\mathbf{v} \\ 0 \\ 0 \\ 0 \end{Bmatrix} = \operatorname{div}\begin{Bmatrix} 0 \\ \mathbf{t}^t \\ \mathbf{t}^t\mathbf{v} + \mathbf{q}^t \\ \mathbf{J}_k \\ \mathbf{J}_\varepsilon \\ \mathbf{J}_m \end{Bmatrix} + \begin{Bmatrix} 0 \\ \rho\mathbf{S}_\mathbf{v} \\ \rho S_e \\ \rho S_k \\ \rho S_\varepsilon \\ \rho S_m \end{Bmatrix} \tag{1}$$

In the above $\rho = \rho(\mathbf{x}, t)$ is the mixture density that depends, in general, on gas component m, time t and location $\mathbf{x}$, $\mathbf{v} = v_i\mathbf{e}_i$ - velocity vector, $\mathbf{e}_i$- versor in direction i , v_i- value of vector. Next, in the second row $\rho\mathbf{v}$, are the bulk momentum density vectors, p is thermodynamic pressure, $\mathbf{I} = \delta_{ij}\mathbf{e}_i\otimes\mathbf{e}_j$ defines unit tensor, δ_{ij} is Kronecker's delta. $\mathbf{t}^t = \mathbf{t}^{lam} + \mathbf{t}^{tur}$ takes into account total diffusive momentum flux (flux components of viscous stress, namely, laminar and turbulent, respectively), $\rho\mathbf{S}_\mathbf{v}$ includes all momentum sources. Additionally, the total specific energy $e = u + \frac{1}{2}\mathbf{v}^2$ is a sum of specific internal u and specific kinetic energy $\frac{1}{2}\mathbf{v}^2$, $\mathbf{q}^t$ represents the total diffusive heat flux, and S_e are the energy sources. Furthermore, in the evolution part of kinetic energy of turbulence k and dissipation of the kinetic energy of turbulence ε, occur diffusive fluxes of k and diffusive flux of ε, i.e. $\mathbf{J}_k$, $\mathbf{J}_\varepsilon$, respectively, and sources, namely, the source of the kinetic energy of turbulence S_k together with the source of the dissipation of the kinetic energy of turbulence S_ε. Finally, Y_m is mass fraction of gas component $m =$ He; CH_3OH ; CO; H_2; CH_4; H_2O; CO_2, S_m is creation/destruction sources of species m and $\mathbf{J}_m$ define flux of m components of mixture. More detailed description of Eq. (1) is presented in [6].

2.1. Chemical reactions

The main products of the methanol decomposition reaction over the Ni_3Al foils are the hydrogen, carbon monoxide and solid carbon deposits. The by-products are carbon dioxide, methane and water [5,13,14]. In the simplest case, methanol decomposition may be described by the following equation [14]:

$$CH_3OH \rightarrow CO + 2H_2 \tag{2}$$

If we consider the thermocatalytic microreactor fed by methanol via the decomposition reaction, it should be noticed that carbon monoxide and hydrogen, which are the products, can be consumed in the reacting flow via the following methanation reaction:

$$CO + 3H_2 \rightarrow CH_4 + H_2O \tag{3}$$

The carbon monoxide component is converted into carbon dioxide via shift reaction. We decided to employ a simplified single-step non-reversible reaction in the form of:

$$CO + H_2O \rightarrow CO_2 + H_2 \tag{4}$$

Carbon deposition can be considered through the Boudouard reaction:

$$2CO \rightarrow C + CO_2 \tag{5}$$

It should be mentioned that some of the by-products, especially water, can oxidize the Ni_3Al catalyst surface. Based on the paper [5], where H_2O is the by-product of methanol decomposition and can be used to create the metallic Ni and also can be converted into aluminium hydroxide. It is worth noting that the appearing Ni nanoparticles can be oxidized by H_2O according to the reaction, and it can then also take part in the subsequent reaction of spinel formation.

2.2. Methanol decomposition source

As has been already mentioned, changes of gas component, m, concentration in the mixture due to chemical reactions can be implemented via a volumetric or surface source term, S_m, in the species transport Eq. (1). At this point it should be emphasized that numerous problems of the gas-dynamic boundary layer increase as a result of combination with the chemical phenomena that are taking place in the vicinity of solids. The flow model of chemical reactions, discussed in many works [6-8] refers to mechanical, thermal and chemical issues of the phenomena occurring in volume. Phenomena occurring on the surface of the thermocatalytic reactor (or the combustion chamber) are characterized by complete anisotropy. Apart from the volume production of component, S_m, the reaction of the surface absorption also describes chemical kinetics, and therefore they should be treated as some surface source (or discounts) for the gas component [15,16].

It is assumed in the numerical case that there is no coke creation during micro-reactor operation and the products of thermocatalytic oxidation are not deposited on the catalyst surface. Hence, to provide a complete CFD model of the catalytic micro-reactor, it will be necessary to consider the coke deposition phenomena. In the future, the wall boundary conditions (temperature and emissivity) should be corrected to account for the effects of this deposit accumulation under the conditions of this reaction. Examples of publications that consider ash deposition and temperature fields, may be found as [17-19], and could be helpful in the future works including the growth of carbon nanotubes. In this paper, it is also assumed that there are no site species involved in the surface reactions and hence only gas phase species are modelled. For modelling purposes only the methanol decomposition reaction is catalytic so it is assumed that the source term, $S_{\mathrm{CH_3OH}}$, for methanol can be formulated in finite volumes strictly adjacent to the micro-reactor wall. The source of the decomposition of methanol ($S_{\mathrm{CH_3OH}}$) depends on the catalytic properties of the intermetallic phase of the Ni_3Al, and can be expressed via the following equation:

$$S_{\mathrm{CH_3OH}} = W_{\mathrm{CH_3OH}} \nu_{\mathrm{CH_3OH}} (k_{\mathrm{Ni_3Al}} [X_{\mathrm{CH_3OH}}] \frac{A_{\mathrm{Ni_3Al}}}{V_{cell}}) \tag{6}$$

where:, $\nu_{\mathrm{CH_3OH}\,(\alpha)}$ - the molar stoichiometric coefficient for $\mathrm{CH_3OH}$ and α -th non-reversible reaction and $W_{\mathrm{CH_3OH}}$ molecular mass of methanol, kg/kmol. To obtain the volumetric source term, $S_{\mathrm{CH_3OH}}$, the surface reaction rate should be divided by the height of the computational cell adjacent to the microreactor wall (A_{cat}/V_{cell}). It should be added that the source of methanol decomposition $S_{\mathrm{CH_3OH}}$ is then usually given as a function of methanol molar concentration $[X_{\mathrm{CH_3OH}}]$ and the forward catalytic rate constant k_{cat}. There exist no proper data for a full model validation. Partially, it can be done through the methanol conversion to reaction time characteristics diagram that is often available for a particular thermocatalytic micro-reactor. Kinetic mechanism was simplified by introducing one step reaction.

2.3. The diffusion of mixture components

For methanol and others component concentrations, it is necessary to consider the effect of diffusive fluxes, such as ordinary molecular diffusion, in gas channels as well as Knudsen diffusion. As was presented in the work [10,20], for the considered geometry of the thermocatalytic microreactor, Knudsen diffusion is neglected. In general, diffusion flux $\mathbf{J}_m$ depends on the diffusion velocity $\mathbf{V}_m$ in the following way:

$$\mathbf{J}_m = \rho Y_m \mathbf{V}_m. \tag{7}$$

The best proven closure on $\mathbf{V}_m$ is Dixon-Lewis' formula [6]:

$$\mathbf{V}_m = \frac{1}{X_m \overline{W}} \sum_{m' \neq m}^{NS} W_{m'} D_{mm'} \nabla X_{m'} - \frac{D_m^T}{\rho Y_m} \frac{\nabla T}{T}. \tag{8}$$

where: $D_{mm'}$ and D_m^T, are coefficients of respectively: multi-components and temperature diffusion, $W_{m'}$, molecular mass of component $m' \neq m$, the average molecular mass of components, $\overline{W}$, while T is the temperature in the reactor.

3. Analysis

3.1. Experiment and constant to calibration

In the examinations the thin Ni_3Al foils with thickness about 50μm and average grain size about 15μm were taken. The stability test for Ni_3Al catalyst in methanol decomposition to hydrogen has been carried out in the flow reactor, under atmospheric pressure, at 500 °C. In this case, Ni_3Al foil was placed in the reactor and vapour mixture containing 40 %vol. of methanol was introduced into reactor (W/F = 1.5 (g_{cat} s) / cm^3). The desired methanol concentration in the gas phase was obtained by bubbling helium through methanol saturator held at 43 °C. The reaction mixture was analysed *on-line* on TCD-GC (HP 5890 Series II) equipped with Porapak Q packed column. Helium was used as carrier gas in GC. All lines were heated above 100 °C in order to prevent water and methanol from condensation. The thermocatalytic decomposition rate of methanol $q_{\mathrm{CH_3OH}}$ presented in Table 1 has been employed for validation purposes. Others parameters have been treated as boundary conditions for CFD calculations.

Table 1. An experimental thermocatalytic decomposition rate of methanol $q_{\mathrm{CH_3OH}}$ over Ni_3Al foil estimated at $\dot{Q}_{mix}$ = 35.1 cm^3/min under the lowest possible temperature of the catalysis process which is sufficient to obtain a full conversion from the initial mole fraction i.e. $X_{\mathrm{CH_3OH}}$= 0.4.

Parameter	Unit	Value
Total mass of catalytic specimens	g	0.777
W/F (weight / flow)	(g_{cat} s) / cm^3	1.48
Conversion of methanol	%	100
Experimental reaction rate of methanol	g / (g_{cat} s)	1.39

3.2. Numerical boundary condition

The recalculation of the experimental reaction rate and diffusion flux were performed under the following assumptions: 1) constant temperature of the process; 2) monolithic structure of the reactor; 3) conversion parameters as in table 1; 4) geometric dimensions as were presented in Fig. 1 c). Therefore, temperature diffusion part is neglected and only $D_{mm'}$ is considered in numerical way.

Microchannels coupled at the inlet and outlet are shown in Fig. 1 b). The velocity was assumed at the level **v** =1 cm/s. It is also known that the lowest possible temperature of the catalysis process which is sufficient to obtain a complete conversion occurred at T =500 °C. For the steady state flow analysis a CFD commercial solver, namely Fluent, was employed. This finite volume based code permits one to solve the three-dimensional fluid and heat flow problems concerned with turbulent structures and chemical reactions [21-23]. However, it also allows for the addition of user defined subroutines programmed in C++ for problems that fall outside the capability of the standard version of code. As has been demonstrated in recent works [24-26], in modelling flows with a strong interaction of surface and liquid material, boundary conditions and appropriate closure in mathematical models are essential.

The grid used in the numerical calculations presented here consist of 1 600 000 finite volumes. This enables the model to maintain a high accuracy of the results, without consuming unnecessary computing power. The channels beginning at inlet into the package and terminating at the outlet from the package has been divided into some blocks that have been discretized by means of a structured numerical grid, steeply refined in the normal wall direction. Initial tests allowed to use the numerical grid to ensure that further refinement did not influence the computational results. In the microreactor, a k - ε turbulent approach was applied due to necessity of utilizing the Eddy-Dissipation model which was considered as the simplest one to work through using UDF. Regarding the fact that Reynold's numbers are very small we assumed that turbulence model does not affect results in significant way. Therefore, the k - ε model was chosen as the most wide applied one. It was also assumed that the surface structure of the micro-reactor can be treated as homogeneous.

The standard SIMPLE (semi-implicit method for pressure-linked equations) method has been used for pressure-velocity coupling. The second order upwind schemes have been employed for the solution of the convection term in the governing equations. The diffusion terms have been central-differenced with second order accuracy as well.

4. Results and discussion

Thermocatalytic decomposition of methanol in a single package micro-reactor is considered as a complicated phenomenon where the main products of the reaction are hydrogen and CO (see Eq. (2)). Considering obtained experimental results, with respect for start (TOS=0 min) and equilibrium state of process (TOS=270 min), it can be concluded high influence of diffusion flux (Fig.2.).

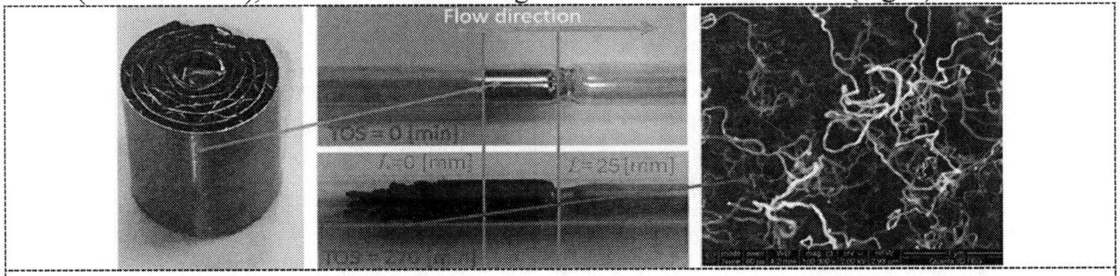

Figure 2. A honeycomb for start (TOS=0 min) and equilibrium state of process (TOS=270 min).

The distribution of the components due to the impact of different processes can be managed by: 1) diffusion effects; 2) mixing effects due to eddy turbulence; 3) thermocatalytic surface decomposition and 4) volumetric reactions. Hydrogen mole fraction distribution at the eight single microchannels with an inlet into the package and an outlet from the package has been presented in Fig. 3.

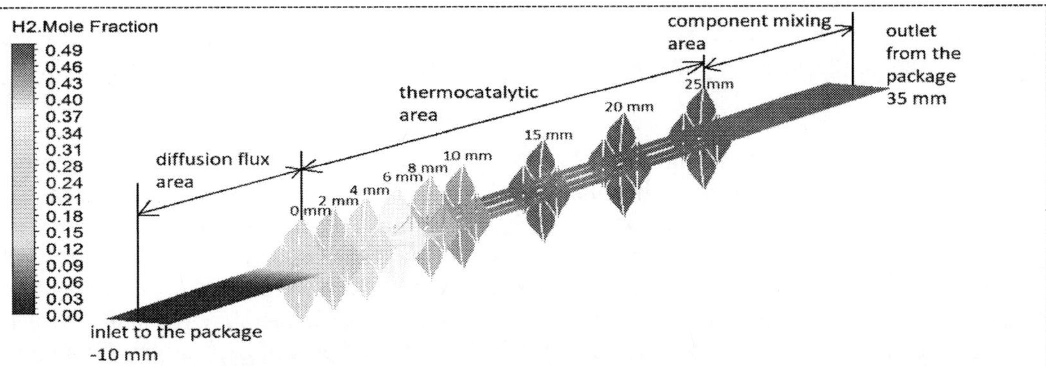

Figure 3. Hydrogen mole fraction X_{H_2} distribution at the eight single microchannels beginning at inlet into the package and terminating at the outlet from the package.

As is highlighted in Fig. 3 the three main areas can be distinguished for the whole package: Firstly, the diffusion flux area before microchannels (between $L = -10$ mm and $L = 0$ mm); secondly, the thermocatalytic reaction area (between $L = 0$ mm and $L = 25$ mm) and thirdly, the flow mixing area (between $L = 25$ mm and $L = 35$ mm). It is worth mentioning that average Reynolds number in domain is equal to 0.06 while maximum value is around 0.2. The maximum of the methanol mole fraction is located at the beginning of the inlet package and decreases between the centre of the diffusive flux area and the thermocatalytic area in the microchannels, while further diminishing due to the surface catalytic reaction on the Ni_3Al thin foil, as is presented in Fig.4. It should be understood that a decrease of methanol concentration in the diffusion flux area is connected with the diffusion fluxes of the other components, mainly hydrogen and carbon monoxide (see Eqs. (1) and (7)). A change in the average mole fraction of hydrogen, methanol, carbon monoxide, steam, methane and carbon dioxide as a function of the reactor length L is shown in Fig.4.

A decrease of CH_3OH concentration in the direction of the outlet from the package is related directly to the decomposition effects connected with the properties of the intermetallic phase of Ni_3Al, which is highlighted in Fig. 4. The view of the local changes in the field of the hydrogen mole fraction in both the axial-sectional and the cross-sectional directions is illustrated in Fig. 3, thus the effects of the diffusion and the chemical reaction are clearly demonstrated.

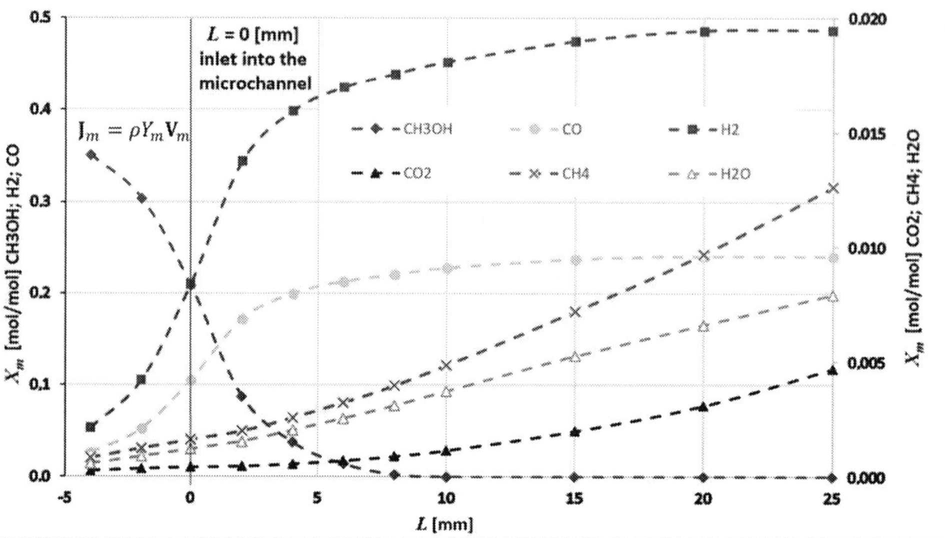

Figure 4. Change of the averaged cross-sectional mole fraction of hydrogen, methanol, carbon monoxide, steam, methane and carbon dioxide as a function of reactor length L.

5. Conclusions

This paper is concerned with of the modelling of hydrogen production by methanol decomposition in the thermocatalytic reactor based on the intermetallic phase of Ni_3Al. The numerical simulation has been performed via CFD procedure with extensions including the increased chemical reactions rate at the interphase between fluid and solid. Analysis presented should be considered as having demonstrated a promising concept in chemical processing possibilities. Results obtained through numerical calculations were calibrated and compared with experimental data to receive satisfactory agreement, however a more sophisticated approach will be considered which includes the effects of deposit growth on the reaction conditions and temperature field at the wall. Influence of the temperature will be added as soon as new experimental data will be available which will allow to enhance the code accordingly. Carbon formation could be modelled by introducing Lagrange formulation. Therefore, the boundary conditions including fluid-solid interactions should be extended with respect of carbon nanotubes formation.

The original methodology of the 3D numerical analysis for thermocatalytic micro-reactor used in the decomposition of methanol has been presented in this paper. The data extrapolated via the implemented numerical model have made it possible to assess the minimal length of the microreactor channels, which provide a full conversion of methanol at the system outlet.

Acknowledgments: The work results were obtained in studies co-financed by the National Research and Development Centre in the project PBS 3 ID 246201 titled .: "The development of innovative technology, thin foils of alloys based on intermetallic phase Ni_3Al with high activity thermocatalytic in the field of purification of air from harmful substances or controlled decomposition of hydrocarbons."

6. References

[1] Jóźwik P, Grabowski R, Bojar Z, 2010. Catalytic activity of Ni_3Al foils in methanol reforming *Mater. Sci. Forum* **636-637** p 895–900.

[2] Olafsen A, Daniel C, Schuurman Y, Raberg L B, Olsbye U, Mirodatos C 2006 Light alkanes CO_2 reforming to synthesis gas over Ni based catalysts *Catal. Today* **115** p 179-185

[3] Jóźwik P, Polkowski W, Bojar Z, 2015 *Materials* **8** p 2537-2568
[4] Moussa S O, El-Shall M S, 2007 *J. Alloys Compd.* **440** p 178-188
[5] Xu Y, Ma Y, Sakurai J, Teraoka Y, Yoshigoe A, Demura M, Hirano T 2014 Effect of water vapor and hydrogen treatments on the surface structure of Ni_3Al foil *Appl. Surf. Sci.* **315** p 475-480
[6] Badur J 2003 *Numerical modelling of sustainable combustion in gas turbines* IFFM Publlishers Gdańsk
[7] Kuo K K and Acharya R 2012 *Applications of turbulent and multiphase combustion* John Wiley & Sons, Inc. New Jersey
[8] Williams F A 1965 *Combustion theory* Addision Wesley Massachussets.
[9] Chen G-B, Chen C-P, Wu C-Y, Chao Y-C 2007 Effects of catalytic walls on hydrogen/air combustion inside a micro-tube *Appl. Catal. A* **332** p 89–97
[10] Jóźwik P, Badur J and Karcz M 2011 *Chemical and Process Engineering* vol 32 (3) p 215-227
[11] Aoki N, Yube K, Mae K, 2007 Fluid segment configuration for improving product yield and selectivity of catalytic surface reactions in microreactors *Chem. Eng. J.* **133** p 105-111
[12] Duran J E, Mohseni M, Taghipour F, 2010 Modeling of annular reactors with surface reaction using computational fluid dynamics (CFD) *Chem. Eng. Sci.* **65** p 1201-1211
[13] Mitani H, Xu Y, Hirano T, Demura M, Tamura R, 2017 Catalytic properties of Ni-Fe-Mg alloy nanoparticle catalysts for methanol decomposition *Catalysis Today* **281** p 669–676
[14] Michalska-Domańska M, Bystrzycki J, Jankiewicz B, Bojar Z 2017 Effect of the grain diameter of Ni-based catalysts on their catalytic properties in the thermocatalytic decomposition of methanol *Comptes Rendus Chimie* **20** p 156-163
[15] Badur J, Ziółkowski PJ and Ziółkowski P 2015 On the angular velocity slip in nano-flows *Microfluid Nanofluid* **19** p 191-8
[16] Badur J, Ziółkowski P, Sławiński D, Kornet S 2015 An approach for estimation of water wall degradation within pulverized-coal boilers *Energy* **92** p 142-152
[17] Modliński N, 2014 *Chemical and Process Engineering* **35** p 361-368
[18] Modliński N, Madejski P, Janda T, Szczepanek K, Kordylewski W 2015 A validation of computational fluid dynamics temperature distribution prediction in a pulverized coal boiler with acoustic temperature measurement *Energy* **92** p 77-86
[19] Weber R, Schaffel-Mancini N, Mancini M, Kupka T 2013 Fly ash deposition modelling: requirements for accurate predictions of particle impaction on tubes using RANS-based computational fluid dynamics *Fuel* **108** p 586-596
[20] Ziółkowski P, Stajnke M and Jóźwik P 2017 Modeling of a mixture flow of helium and methanol in thermocatalytic reactor and chemical reactions on the intermetallic phase of Ni_3Al *Transactions IFFM*, accepted to publication
[21] Flaszyński P, Doerffer P, Szwaba R, Kaczyński P, Piotrowicz M 2015. *Journal of Thermal Science* vol 24(6) p 510-515
[22] Badur J, Ziółkowski P, Zakrzewski W, Sławiński D, Kornet S, Kowalczyk T, Hernet J, Piotrowski R, Felicjancik J, Ziółkowski P J 2014 An advanced Thermal–FSI approach to flow heating/cooling *J. Phys.: Conf. Ser.* **530** 10.1088/1742–6596/530/1/012039.
[23] Pianko-Oprych P, Kasilova E, Jaworski Z, 2014 CFD analysis of heat transfer in a microtubular solid oxide fuel cell stack *Chemical and Process Engineering* **35** p 293-304
[24] Badur J, Karcz M, Lemański M, Nastałek L 2011 Enhancement transport phenomena in the Navier-Stokes shell-like slip layer *CMES: Computer Modeling in Engineering & Sciences* **73** p 299–310
[25] Ziółkowski P, Badur J 2018 A theoretical, numerical and experimental verification of the Reynolds thermal transpiration law. International Journal of Numerical Methods for Heat and Fluid Flow vol 28(1) p 64-80.
[26] Lewandowski T, Ochrymiuk T, Czerwińska J 2011 Modeling of heat transfer in microchannel gas flow *ASME Journal of Heat Transfer* **133** 022401-1

IOP Conf. Series: Journal of Physics: Conf. Series **1101** (2018) 012051 doi:10.1088/1742-6596/1101/1/012051

Applications of wave equations with logarithmic nonlinearity in fluid mechanics

Konstantin G. Zloshchastiev

Institute of Systems Science, Durban University of Technology, P.O. Box 1334, Durban 4000, South Africa

E-mail: k.g.zloschastiev@gmail.com

Abstract. We apply statistical mechanics and Madelung hydrodynamical presentation for an effective description of strongly-interacting many-body systems, such as Bose liquids or Korteweg-type fluids. The logarithmic nonlinearity is shown to appear in equations describing such fluids. The resulting equations describe the irrotational and isothermal flow of a two-phase barotropic compressible inviscid fluid with internal capillarity and surface tension. We demonstrate spontaneous symmetry breaking in this class of fluids, which leads to a number of wave-mechanical and topological effects. We show the relationship between the "logarithmic" fluids and those described by polynomially nonlinear wave equations, such as the Gross-Pitaevskii one.

1. Introduction

It has been recently proposed to use the barotropic Korteweg capillary fluids as a model of magma transport in a conduit [1, 2]. In such a class of models, often referred as diffuse interface models, the capillary interface is viewed as a diffusion transition domain of rapid smooth variation of density, while the surface tension is intrinsically incorporated [3, 4, 5, 6]. This allows to describe flows with a spontaneous nucleation, coalescence and breakdown of density inhomogeneities in the two-phase systems [7]. The Schrödinger-like equation with a logarithmic nonlinearity is shown to appear in equations describing such fluids [1, 2].

In what follows, we place statistical and thermodynamic foundations under this approach and deduce physical implications.

2. Wave equation

Suppose we have a fluid flowing in a cylindrical channel which is in thermal contact with a reservoir of infinitely large heat capacity as to maintain constant temperature; we assume also that its microscopic structure can be regarded as a many-body system of particles, atoms and molecules, which interaction energy is larger than their kinetic energy. Then the probability P of a microstate is given by a Boltzmann rule, in which the kinetic energy can be neglected compared to the potential one:

$$P \propto \exp\left(-\mathcal{E}/T\right) \approx \exp\left(-\mathrm{U}/T\right), \tag{1}$$

where T, $\mathcal{E}$ and U are the absolute temperature, energy and potential energy of a many-body system, respectively; we work in the units where the Boltzmann constant $k_B = 1$. At a microscopical level, this fluid is described by an extremely large set of particles' evolution equations, therefore certain simplifications must be made.

Namely, we employ the fluid approximation used in continuum mechanics: we must make a transition to a single equation for the fluid wavefunction written in a Madelung form [8, 1, 2]

$$\Psi = \sqrt{\rho}\exp(iS), \tag{2}$$

where $\rho = \rho(\mathbf{x}, t)$ is a fluid density, and $S = S(\mathbf{x}, t)$ is a phase which is related to the fluid velocity $\mathbf{u} = \nabla S$ (we assume the irrotational flow for simplicity).

This wavefunction should not be confused with particle wavefunctions in quantum mechanics, but rather it is a complex-valued function which stores macroscopic information about a fluid, such as the density and velocity of a fluid parcel, as discussed in Ref. [8]. Nevertheless, some mathematical similarities between these categories of functions do exist and will be used in what follows.

In particular, function Ψ obeys a normalization condition

$$\int_V |\Psi|^2 \mathrm{d}V = \int_V \rho\,\mathrm{d}V = M > 0, \tag{3}$$

where M and V are the total mass and volume of the fluid. This poses restrictions upon fluid wave functions which are similar to a quantum-mechanical case: the set of all normalizable fluid wave functions must constitute a Hilbert space, such as L^2.

As a first derivation step of an equation for Ψ, it is natural to relate the probability sample space of an underlying microscopic many-body system to the volume occupied by fluid, and thus relate the probability (1) to the fluid density:

$$P \sim \rho, \tag{4}$$

and take into account this statistical effect upon an energy of system's collective degrees of freedom.

Therefore, one obtains

$$|\Psi|^2 \sim \exp(-\mathrm{U}/T), \tag{5}$$

from which an expression follows for an operator of the potential U:

$$\hat{\mathrm{U}} \sim T\ln(A|\Psi|^2) = -K(T - T_0)\ln(\rho/\rho_0), \tag{6}$$

where $A = 1/\rho_0$ and K are some dimensionful and dimensionless proportionality constants, respectively, and ρ_0 and T_0 are critical values of, respectively, fluid density and temperature at which the statistical effect vanishes. From now on, we assume $K = 1$ for simplicity.

Furthermore, let us assume our flow to be Hamiltonian, and consider an operator form for the energy conservation, which can be written as the following identity:

$$\hat{\mathrm{H}}\Psi = (\hat{\mathrm{T}} + \hat{\mathrm{U}})\Psi, \tag{7}$$

where $\hat{\mathrm{H}} = i\eta\partial_t$ is a Liouville-type generator of time translations, η being a positive scaling constant, $\hat{\mathrm{T}} \propto \frac{1}{2}\hat{\mathbf{p}}\cdot\hat{\mathbf{p}}$ is a kinetic energy operator, $\hat{\mathbf{p}} \propto -i\nabla$ is a generator of spatial translations, and $\hat{\mathrm{U}}$ is a potential energy operator derived above. Altogether, this identity can be recast in the form of a logarithmic Schrödinger-like equation:

$$i\partial_t\Psi = \left[-\frac{\mathcal{D}}{2}\nabla^2 - b\ln\left(|\Psi|^2/\rho_0\right)\right]\Psi, \tag{8}$$

where $b = (T - T_0)/\eta$ and $\mathcal{D}$ are positive constants. This equation must be supplemented with the normalization condition, as discussed above.

Substituting Eq. (2) into Eq. (8), one can recover the hydrodynamic laws for mass and momentum conservation for a two-phase compressible inviscid fluid with internal capillarity which flow is irrotational and isothermal:

$$\partial_t \rho + \nabla \cdot (\rho \mathbf{u}) = 0, \tag{9}$$

$$\partial_t \mathbf{u} + \mathbf{u} \cdot \nabla \mathbf{u} - \frac{1}{\rho} \nabla \cdot T = 0, \tag{10}$$

with the stress tensor T of the Korteweg form with capillarity effects [4], which is used to model fluid mixtures with phase changes and diffuse interfaces [6, 7]:

$$T = -\frac{\mathcal{D}}{4\rho} \nabla \rho \otimes \nabla \rho - \tilde{p}\, I, \tag{11}$$

where I is the identity matrix,

$$\tilde{p} = p(\rho) - \frac{1}{4}\mathcal{D}\nabla^2 \rho = -b\rho - \frac{1}{4}\mathcal{D}\nabla^2 \rho$$

is a capillary pressure, and

$$p(\rho) = -b\rho$$

is a barotropic equation of state for the fluid pressure p [1, 2]. The stress tensor is also related to the chemical potential by the formula:

$$\nabla \mu = -\frac{1}{\rho} \nabla \cdot T = -\nabla \left(\frac{\mathcal{D}}{2} \frac{\nabla^2 \sqrt{\rho}}{\sqrt{\rho}} + b \ln(\rho/\rho_0) \right), \tag{12}$$

where a scale factor η is incorporated into μ for brevity, which means that the chemical potential can be derived from Eq. (8) for stationary functions $\Psi(\mathbf{x}, t) = \exp(-i\mu t)\Psi(\mathbf{x})$. Thus, Eq. (8) is a concise form of writing two hydrodynamic equations, which makes it useful for studies of the Korteweg-type fluids, especially considering a large information accumulated about properties of wave equations with logarithmic nonlinearity in different branches of physics [9, 10, 11, 12, 13, 14, 15], including a theory of logarithmic Bose-Einstein condensates and superfluids [16, 17, 18, 19, 20].

Besides, by using a standard averaging procedure with respect to the inner product in the Hilbert space of fluid wavefunctions, one can show that an averaged form of Eq. (8) can be written as a formula for a wave-mechanical internal energy of the fluid:

$$\mathcal{U} = \langle \hat{\mathrm{H}} \rangle = \mathcal{F} + TS, \tag{13}$$

where $\mathcal{F} = \langle \hat{\mathrm{T}} \rangle$ is a wave-mechanical free energy, and

$$S = -\int_V \rho \ln(\rho/\rho_0)\, \mathrm{d}V, \tag{14}$$

is an entropy function; according to aforesaid

$$S \sim -\int_V P \ln P \, \mathrm{d}V.$$

Thus, T and S must be thermodynamically conjugated:

$$T \sim b \sim \left(\frac{\partial \mathcal{U}}{\partial S} \right)_V, \tag{15}$$

if one treats $\mathcal{U}$ as a thermodynamical potential.

IOP Conf. Series: Journal of Physics: Conf. Series **1101** (2018) 012051 doi:10.1088/1742-6596/1101/1/012051

3. Symmetry breaking and solitary waves

Spontaneous symmetry breaking is a widely known phenomenon in physics, through which a system undergoes a transition from a more symmetric to less symmetric state. One must differentiate this from the dynamical symmetry breaking phenomenon, in which a state of a system, described as a solution of evolution equations, has a different (lesser) symmetry from what it is expected from the properties of evolution equations themselves. The latter are coming usually from Lagrangian or Hamiltonian formulations and written in a form of differential equations, with positions as variables.

Perhaps a simplest example of the spontaneous symmetry breaking phenomenon would be a long solid rod or bar made of elastic material, to which one applies a pressure force in the direction parallel to the main axis. Since a bar is rotationally invariant with respect to that axis, one could expect that it can not bend (since it cannot "decide" which direction to bend), only splat, i.e., expand in the direction perpendicular to the main axis. In reality however, most long bars are more likely to eventually bend at some point than to splat into a rotationally symmetric shape. The microscopic reasons for that are of course small anisotropies inside the molecular or crystal structure of real bars as many-body systems, but in the idealized model, *i.e.*, perfectly radial symmetric single body, the only explanation and description one can offer is the spontaneous symmetry breaking: a bar undergoes a transition from initially rotationally symmetric shapes to one of bent shapes which break this symmetry.

The most known mechanism by which the spontaneous symmetry breaking can be introduced or described is the Higgs mechanism. Originally proposed for purposes of physics of elementary particles, where a single-body idealized description and small fluctuations thereof are not only natural but also inevitable in most cases, it can be extended to condensed matter theory, mathematical ecology, financial mathematics, and, as we shall see below, fluid mechanics. Its basic idea is the following.

Suppose that our system's evolution equation can be somehow written in the form of Lagrange-Euler equations minimizing an action whose potential part has a shape with a few extrema. One of those extrema usually corresponds to a maximally symmetric configuration, while others do to configuration with less symmetry. A simplest example of such potential would be a Mexican-hat one where the very peak of the dome is placed in the origin (where we have a configuration which is symmetric with respect to Z_2 or $U(1)$ transformation of a variable), while an infinite number of possible minima at the perimeter does not obey this symmetry. Since the local minimum of the potential (hence energy) is realized on those non-symmetric solutions, it is more energetically favorable for a system to stay in a less-symmetric states even if it was originally in the symmetric state. The transition from the latter will readily occur, sooner or later, due to its instability against small fluctuations (in the elementary particles' case, quantum).

Thus, our task now to check whether our system belongs to this class of dynamical systems, which would allow us to reliably apply the above-mentioned formalism and make certain predictions.

Using a variational approach, Eq. (8) can be derived as an Euler-Lagrange equation for the Galilean-invariant Lagrangian density:

$$\mathcal{L} = \frac{i}{2}(\Psi\partial_t\Psi^* - \Psi^*\partial_t\Psi) + \frac{\mathcal{D}}{2}|\nabla\Psi|^2 + \mathcal{V}(|\Psi|^2), \tag{16}$$

with the potential density given by

$$\mathcal{V}(\rho) = -b\rho\left[\ln\left(\rho/\rho_0\right) - 1\right] + \mathcal{V}_0, \tag{17}$$

where $\mathcal{V}_0 = \mathcal{V}(0)$ being an arbitrary constant; for simplicity we can assume $\mathcal{V}_0 = 0$. For complex values of Ψ, this potential has an upside-down Mexican-hat shape, with local maxima at $|\Psi_e| = \sqrt{\rho_0}$, which is placed inside a square potential well; the latter occurs due to the normalization condition which bounds $|\Psi|$ from above.

Thus, the maximally symmetric state $|\Psi|$ is not the most energetically favorable one here, since there could exist one or more states which located between local maxima and one of walls of the above-mentioned square well. This indicates a possibility of spontaneous symmetry breaking (or restoration), therefore, Eq. (8) should have solutions of a solitary wave type, and indeed such solutions exist.

One solitary wave solution of Eq. (8) can be found exactly. It corresponds to the ground state (*i.e.*, the one with a lowest eigenvalue of the chemical potential μ) and can be written in the form

$$\Psi_{(g)}(\mathbf{x},t) = \exp\left(-i\omega t + i\mathbf{k}\cdot\mathbf{x}\right)\sqrt{\rho_g(\mathbf{x})}, \tag{18}$$

where $\mathbf{k}$ is a constant vector, $\omega = \mu/\eta$ is a wave frequency, and the density eigenfunction and frequency eigenvalue are, respectively:

$$\rho_g(\mathbf{x}) = \tilde{\rho}\exp\left[-\frac{(\mathbf{x}-\mathbf{x}_0)^2}{\ell^2}\right], \tag{19}$$

$$\omega_g = \mu_g/\eta = \frac{1}{2}\mathcal{D}\mathbf{k}^2 + b\left[d - \ln\left(\frac{\tilde{\rho}}{\rho_0}\right)\right], \tag{20}$$

where d is a number of spatial dimensions of the fluid, and $\mathbf{k} = 0$ if $b \neq 0$. The propagating solutions can be generated by means of a Galilean transformation $\mathbf{x} \to \mathbf{x} - \mathbf{u}t$.

The value of $\tilde{\rho}$ can be found from the normalization condition (3) and a formula for the d-dimensional Gaussian integrals $\int_V \exp\left(-\mathbf{x}^T\mathbf{A}\,\mathbf{x}\right)\mathrm{d}V = \pi^{d/2}/\sqrt{\det\mathbf{A}}$, $\mathbf{A}$ being a $d\times d$ positive-definite matrix. Thus

$$\tilde{\rho} = M/\tilde{V},\ \tilde{V} = \pi^{d/2}\ell^d,\ \ell = \sqrt{\mathcal{D}/(2b)}, \tag{21}$$

are the wave's peak density value, effective volume and Gaussian width, respectively.

It should be noted that, for the same set of boundary conditions, Eq. (8) allows also other normalized (eigen)solutions which correspond to excited states in the Hilbert space of the problem, with different eigenvalues for μ. This means that the Korteweg-type fluid can "choose" one of these solutions spontaneously. Though, the ground state is a preferable one as it corresponds to a minimum of the wave-mechanical energy.

Furthermore, the system's behavior changes drastically when the nonlinear coupling b changes its sign; this could happen if, in accordance with (6), it becomes proportional to $T - T_c$ where T_c is some transition temperature. In that case, the potential (17) acquire a conventional Mexican hat shape, so the states $|\Psi_e|$ become global minima and thus become energetically favorable. This leads to occurrence of multiple topological sectors as well as topological soliton solutions whose stability is ensured by a conserved topological charge [21].

Thus, the overall theory predicts two types of phase transitions in these fluids, at least: the transition related to the sign flip of the nonlinear coupling, and the transitions related to spontaneous symmetry breaking for a given value (sign) of this coupling.

4. Conclusion

In this paper, we have explored a mapping between nonlinear Schrödinger-like wave equations of a logarithmic type and flow equations of the Korteweg-type materials including fluids with internal surface tension and capillary effects. This mapping has allowed us to rewrite the flow equations in the form of a single Schrödinger-like equation which can be studied by means of the conventional wave mechanics which has been hitherto popular in describing mostly microscopical systems.

In particular, studying Korteweg fluids in terms of the Schrödinger-like wave equations reveals an existence of the Hilbert space of states represented by different solutions, for the same set of boundary conditions. This means that the fluid can randomly "choose" either of these states (with a preference to the lower-energy ones, of course), or it can even be in a superposition of these states (an analogue of an entangled pure state in quantum mechanics) or their statistical ensemble (an analogue of a mixed state in quantum mechanics).

Analogies of this kind were known in studies of propagation of classical electromagnetic waves in media [22] but not in a theory of Korteweg fluids. This open further perspectives for searching phenomena in classical fluids with internal surface tension and capillary effects, which are analogous to quantum-statistical phenomena in quantum dissipative systems where mixed states are inevitable due to interaction of a (sub)system with its environment.

Acknowledgments
This work is based on the research supported by the National Research Foundation of South Africa under Grants Nos. 95965 and 98892.

References
[1] De Martino S, Falanga M, Godano C and Lauro G 2003 *Europhys. Lett.* **63** 472
[2] Lauro G 2008 *Geophys. Astrophys. Fluid Dyn.* **102** 373
[3] Korteweg D 1901 *Arch. Neerl. Sci. Exactes Nat.* **6** 1
[4] Dunn J E and Serrin J B 1985 *Arch. Rat. Mech. Anal.* **88** 95
[5] Dell'Isola F and Kosiński W 1993 *Arch. Mech.* **45** 333
[6] Anderson D M, Mc Fadden G B and Wheeler A A 1998 *Annu. Rev. Fluid Mech.* **30** 139
[7] Antanovskii L K 1996 *Phys. Rev. E* **54** 6285
[8] Yu. A. Rylov, J. Math. Phys. **40**, 256 (1999).
[9] G. Rosen, J. Math. Phys. **9**, 996 (1968).
[10] I. Bialynicki-Birula and J. Mycielski, Ann. Phys. (N. Y.) **100**, 62 (1976).
[11] Zloshchastiev K G 2010 *Grav. Cosmol.* **16** 288 [arXiv:0906.4282].
[12] Dzhunushaliev V and Zloshchastiev K G 2013 *Central Eur. J. Phys.* **11** 325 [arXiv:1204.6380].
[13] Dzhunushaliev V, Makhmudov A and Zloshchastiev K G 2016 *Phys. Rev. D* **94** 096012
[14] T. C. Scott, X. Zhang, R. B. Mann, and G. J. Fee, Phys. Rev. D **93**, 084017 (2016).
[15] Znojil M, Růžička F and Zloshchastiev K G 2017 *Symmetry* **9** 165
[16] Zloshchastiev K G 2011 *Acta Phys. Polon.* **42** 261
[17] Avdeenkov A V and Zloshchastiev K G 2011 *J. Phys. B: At. Mol. Opt. Phys.* **44** 195303
[18] Zloshchastiev K G 2012 *Eur. Phys. J. B* **85** 273
[19] Bouharia B 2015 *Mod. Phys. Lett. B* **29** 1450260
[20] Zloshchastiev K G 2017 *Z. Naturforsch. A* **72** 677
[21] Zloshchastiev K G 2018 in preparation
[22] Zloshchastiev K G 2016 *Phys. Rev. B* **94** 115136

Institute of Physics
Dirac House, Temple Back
Bristol BS1 6BE UK

ISSN: 1742-6588
ISBN 978-1-5108-7373-5

9 781510 873735

CURRAN ASSOCIATES INC.
proceedings.com